P9-CLX-695
Kuala Terengganu
SOUTH CHINA SEA
AYSIA
Kuantan
JALA
MPUR
KEP. NATUNA UTARA
P. NATUNA BESAR
KEP. ANAMBAS
KEP. NATUNA SELATAN
Melaka
SINGAPORE
P. BATAM
P. BINTAN
Tanjung Pinang
KEP. RIAU
KEP. TAMBELAN
KEP. LINGGA
P. LINGGA
P. SINGKEP
JAMBI
S. Batang Hari
Bangko
Mentok
PANGKAL PINANG
Selat Bangka
Tanjung Pandan
Selat Gaspar
S. Musi
PALEMBANG
uk Linggau
urup
Perabumulih
P. BELITUNG
Lahat
Pagaralam
S. Mesuji
Baturaja
Kotabumi
Bukit Kemuning
Metro
Way Kambas
NGGANO
Krui
BANDAR LAMPUNG
Kalianda
Bakauheni
Bukit Barisan
Selat Sunda

The Ecology of Sumatra

The Ecology of Indonesia Series

Volume I

The Ecology of Indonesia Series

Volume I: The Ecology of Sumatra

Other titles in the series

Volume II: The Ecology of Java and Bali
Volume III: The Ecology of Kalimantan
Volume IV: The Ecology of Sulawesi
Volume V: The Ecology of Nusa Tenggara and Maluku
Volume VI: The Ecology of Irian Jaya
Volume VII: The Ecology of the Indonesian Seas - Part 1
Volume VIII: The Ecology of the Indonesian Seas - Part 2

Published by Periplus Editions (HK) Ltd.
First Periplus edition, 2000

ISBN 962-593-074-4

Publisher: Eric Oey
Typesetting and graphics: JWD Communications Ltd.
Copyediting: Sean Johannesen

Distributors:

North America
Tuttle Publishing
Distribution Center
Airport Industrial Park
364 Innovation Drive
North Clarendon, VT 05759-9436
Tel: (802) 773 8930
Fax: (802) 526 2778

Japan
Tuttle Publishing
RK Building 2nd Floor
2-13-10 Shimo-Meguro, Meguro-Ku
Tokyo 153-0064
Tel: (03) 5437-0171
Fax: (03) 5437-0755

Asia Pacific
Berkeley Books Pte. Ltd.
5 Little Road #08-01
Singapore 536983
Tel: (65) 280-3320
Fax: (65) 280-6290

Indonesia
PT Java Books Indonesia
Jl. Kelapa Gading Kirana
Blok A14 No.17
Jakarta 14240
Tel: (021) 451 5351
Fax: (021) 453 4987

Printed in Singapore

The Ecology of Sumatra

Tony Whitten
Sengli J. Damanik
Janzanul Anwar
Nazaruddin Hisyam

PERIPLUS

Table of Contents

Acknowledgements

This book was prepared as part of a Government of Indonesia/United Nations Development Programme (UNDP) Project INS/78/056 'Education and Training in Environment and Resources', executed by the World Bank with a subcontract to Dalhousie University. The project was intended to develop capabilities at two newly-formed environmental study centres, including the Centre for Resource and Environmental Studies (CRES) at the University of North Sumatra where *The Ecology of Sumatra* was written. This revised edition has received financial support from a project, Environmental Management Development in Indonesia (EMDI), which is implemented by the Indonesian Ministry of State for Population and Environment and the School for Resource and Environmental Studies, Dalhousie University, sponsored by the Canadian International Development Agency (CIDA). The preparation of a slightly revised edition of *The Ecology of Sumatra* in 1987 would not have been possible without the financial support from CIDA and the full agreement of UNDP and World Bank staff.

The writing of this book was conducted under the guidance and with the constant enthusiasm of Prof. Soeratno Partoatmodjo and Dr. Arthur Hanson who also masterminded the development of a concept into a reality. Others who gave exceptional support were Prof. Koesnadi Hardjasoemantri, Albert Howlett, Prof. A.P. Parlindungan, and Prof. Abu Dardak. Thanks are also due to Prof. Gabriel Horn and Dr. Ken Joysey of the Zoology Department, University of Cambridge, who provided A.J.W. with space and facilities for five months.

The whole of the book was conscientiously criticized by Kuswata Kartawinata, and a number of people, Ruth Chambers, Jim Davie, Kathy MacKinnon, Adrian Marshall, Edmund Tanner and Jane Whitten, who read one or more chapters.

Great thanks are expressed to H.J. Koesoemanto, Executive Director of Gadjah Mada University Press for his enthusiasm and encouragement.

Trips and expeditions made by teams from the Centre for Resource and Environmental Studies (CRES) at the University of North Sumatra were greatly assisted by the following: P.T. Semen Andalas Indonesia particularly Malcolm Llewellyn; P.T. Caltex Pacific Indonesia particularly Wisaksono Noeradi, L.L. Tobing, and R.B. Pulono Wahjukusumo; the present and former District Officers (Camat) of Muara Tembesi, A. Wahab Moehib and K.P.T. Ismail; P.T. Tambang Timah particularly Sumardekar and Iwa Wardiman; and staff of the Directorate-General of Nature Conservation in Medan, Pekanbaru, Jambi, and Sungaipenuh. The authors also wish to thank staff attached to CRES for their assistance in the field, namely Djuharman Arifin, Usman Rasyid and Hardy Guchi.

Many people have helped by sending material which was unpublished, not yet published or simply hard to obtain, by advising, by making helpful suggestions or by identifying specimens. They are: Hans Bänzinger, Eamonn Barrett, Roger Beaver, C.G.G. van Beek, Elizabeth Bennett, David Benzing, Philip Chapman, the Earl of Cranbrook, Glyn Davies, Geoffrey Davidson, John Dransfield, John Flenley, Ian Glover, Peter Grubb, Larry Hamilton, John Edwards Hill, Frank Howarth, Robert Inger, Daniel Janzen, Patricia Jenkins, Andy Johns, A.G. Kostermans, Yves Laumonier, Colin McCarthy, E. Edwards McKinnon, John MacKinnon, Bernard Maloney, Joe Marshall, Willem Meijer, John Miksic, Mohammad Amir, Robert Morley, Francis Ng, Harald Reidl, Yarrow Robertson, Carel van Schaik, John Seidensticker, George Sherman, Ian Spellerberg, C.G.G.J. van Steenis, Michael Tweedie, H.Th. Verstappen, Peter Waterman, David Wells, Tim Whitmore, W.J.J.O. de Wilde, Jane Wilson and the librarians of the Institute for Resource and Environmental Studies (Halifax), Rijksherbarium (Leiden), British Museum (Natural History) (London), Universiti Malaya (Kuala Lumpur), Environmental Studies Centre, National Biological Institute, Central Agriculture and Biology Library, Herbarium Bogoriense (all at Bogor), Scientific Periodicals Library, University Library, Balfour Library, Haddon Library, and Botany Library (all at Cambridge).

The typists for this book, Pam Ismail, Meribeth Schmidt, Jesaya Hutapea, Gladys Pudduck and Pat Reay, have worked with great diligence and care.

Finally and mainly, we thank Almighty God for His caring love shown throughout the preparation of this book without which it would never have been completed.

Yogyakarta, November 1987.

Anthony J. Whitten
Sengli J. Damanik
Jazanul Anwar
Nazaruddin Hisyan

Foreword to the Original Edition

Indonesia, a country which is developing rapidly, is actively engaged on a wide variety of development projects. The eventual aim of Indonesian National Development is to achieve harmony between man and his God, between mankind and his environment, between races and between human ideals in this world and joy in the world hereafter.

The increase in the pace of development brings the risk of pollution and environmental degradation such that the structure and function of ecosystems which support life can be irretrievably damaged. Wise development must be based on a knowledge of the environment as an entity so that we can achieve a guarantee for the well-being of this generation and for those generations to come.

Many ecology books have been written but until now there had been no book available specially written for a part of Southeast Asia. *The Ecology of Sumatra* is therefore extremely important for, armed with this, readers will be better able to understand the physical and biological characteristics of this island. If the environment is better understood then the exploitation of natural resources in the context of development can be managed in a more rational way.

The team which produced this book was comprised of the following authors: Dr. Jazanul Anwar, Ir. Sengly Janus Damanik, M.Sc and Ir. OK. Nazaruddin Hisyam, MS together with Dr. Anthony J. Whitten. The first three are Staff at the Centre for Environmental and Resource Studies, University of North Sumatra, and the last was consultant to the above centre under a United Nations Development Program/World Bank project (No. INS/78/056) for environmental education. The project was administered for the Government of the Republic of Indonesia by the State Ministry of Population and Environment.

We express great thanks to the authors for their many hours of work in the production of this book. We specially thank Dr. A.J. Whitten and his wife, Jane Whitten, M. Phil who have worked extremely hard in encouraging, guiding and helping the staff of the Centre for Environmental and Resource Studies and in producing the English edition of *The Ecology of Sumatra.* They receive our sincere congratulations.

Finally, we hope that this book will encourage the execution of detailed ecological investigations on Sumatra by a wide range of investigations.

April, 1984

Prof. Dr. A.P. Parlindungan, S.H.
Rector
University of North Sumatra
Medan

Preface to the Original Edition

It is with great pleasure that I write the preface to *The Ecology of Sumatra* because it has been one of the major activities of the Centre for Resource and Environmental Studies at the University of North Sumatra over the last two years.

At this time in Indonesia's history when environmental awareness has penetrated so far both into the minds of the common man and into government policy at the highest levels, we must remind ourselves that although ecology (unlike environmental science) does not always directly influence policy making, legislation, politics or planning, it is the *foundation* of environmental science. An understanding of the components of ecosystems and the manner in which they interact is central to studying the environment and to conducting resource management. The study of those interactions and their effects is the study of ecology.

The book has been written primarily for those concerned with conducting environmental impact analyses and related studies in Sumatra. Before now it has been more or less impossible for such people to find relevant information on Sumatran ecosystems to help them with their work and as such we hope that this book represents a significant step in the history of Indonesian environmental science. It is hoped that the book will also be enjoyed by those with a general interest in Sumatra and Southeast Asian natural history, such as lecturers, school teachers, students and travellers.

April, 1984

Prof. Dr. Abu Dardak
Director
Centre for Resource and Environmental Studies
University of North Sumatra
Medan

Notes

All the references used in the preparation of this book are, or soon will be, in the Documentation and Information Unit collection at the Centre for Resource and Environmental Studies, University of North Sumatra. It is hoped that this material will be consulted by readers to deepen their knowledge of particular aspects of Sumatran ecology, and to improve the standard of environmental impact analyses and similar studies. Readers are encouraged to consult or order this material: to assist them, the three-number classification of each book, paper or report is shown after the reference. Details of costs, etc., can be obtained by writing to the Centre's Director.

Reflections on Sumatra: 1983-1998

Context

It had been hoped by many, not least myself, that the re-publication of this book would allow a thorough revision and updating of the material and of the lessons. Unfortunately, I was unable to raise the funds to do this and the publishers eventually decided that, in order to meet the commitments to its partners to complete the *Ecology of Indonesia* series, the current book should be published. In reviewing the text it has been evident that there were not many mistakes, but it is obviously out-of-date. I hope the mistakes have been removed. The referencing system was cumbersome (though done for a good reason) and has been changed to a more conventional format.

It is very unfortunate that funds are not yet available for the republication of the Indonesian version which has also been out of print for many years. The Indonesian versions of the *Ecology of Indonesia* series have always been the primary reason for engaging in so much work, and it is hoped that a source of funds will emerge after the publication of this book. It is always most gratifying to see grubby, well-thumbed copies of the Indonesian versions in libraries. However, Indonesia's various problems at this time may make a translation an invariable proposition.

Good News

First, the good news. The awareness of, interest in and commitment to the wise use of biological resources on Sumatra have all grown tremendously. Non-governmental groups have become effective and increasingly-trusted partners of the government and international development agencies. The human capacity and environmental awareness in government has also increased substantially, though clearly not to the extent necessary. More Indonesian scientists are getting into the field in the major national parks and elsewhere, as well as publishing their results in good Indonesian journals such as *Tropical Biodiversity*. This is important since this scientific presence and publicity increases security for, and awareness of, these sensitive areas.

Some companies have shown foresight and responsibility by replanting and caring for mangrove and other forest trees after exploitation of the original resource, sometimes in cooperation with local communities, but these efforts are very much outweighed by other less thoughtful land practices. Other good news, such as fruitful cooperation between Indonesian and foreign scientists, the production of the RePPProT maps, a new national park, and the discovery of new species, are all described in the following annex on new publications.

Bad News – Land and Habitats

The bad, and desperately sad, news is that despite these improvements the exploitation of timber and every other biological and physical resource appears to have progressed without restraint, though not without complaint. Greed and haste have ruled supreme. If one could argue that the people of Sumatra had benefited, especially those who once used and lived near those resources, maybe the loss would be felt less acutely. But instead there are numerous reports of derisory or no compensation, intimidation, corruption, evasion and bending of regulations, and of opaque planning processes.

Gone are all but a very, very few of the grand and awe-inspiring dryland lowland forests. As Laumonier (1997) says in his new and authoritative book, *"The current situation of the forests of Sumatra is quite clear. There is no intact [dryland] lowland forest left. ... In a few decades, [even disturbed] lowland forests will have disappeared completely in Sumatra if logged-over forests are not maintained under sustainable management."* Hill forests have also been affected and are currently being decimated by the frenetic forest exploitation which precedes the impending disappearance of such a resource and by local agricultural transformation. The original, natural forests now remain simply as memories in the minds of those lucky enough to have seen them. Those who didn't experience them don't know what they have lost. Will it be the case in a few years, if a complete revision of *The Ecology of Sumatra* is produced, that it will have to be written of the lowland forests, as in the recent The Ecology of Java and Bali, that *"it is in some ways pointless to describe what is essentially an historical situation. There are remnant disturbed areas, however, and for the sake of encouraging some interest in and understanding of these the descriptions in this chapter are provided"* (Whitten et al. 1996)? One forest type which has now probably been erased totally from the landscape in its original and even modified form is the exceptional pure stands of ironwood forest. This very valuable resource has been squandered, its potential for regeneration bypassed, and the cultural links with Jambi lost.

I am writing this piece at a time when international attention is focused on Sumatra as never before because of the terrible forest fires and the thick, acrid smoke they produce. As this 'haze' sits over the land so transportation is disrupted, people's health suffers, and the grey cloak encourages those who might wish to add to the inferno to facilitate the clearance or deforestation of land before the rains come and the opportunity is lost. Never mind the wildlife. Debates rage, hosted by the world's media, as to whether the environmental disaster, during a severe El Niño-provoked drought period, is the fault of corporations and companies engaged in the large-scale conversion of land to tree-crops, timber plantations or transmigration settlements, or of farmers who are clearing land for crops. Supporters of the first reason are being branded as communists, and those supporting the latter are accused of being anti-people. Whatever the reason, it has been recommended for years that if farmers were given security of tenure on the land they farm they would be more likely to be concerned about its treatment. Since undisturbed tropical rain forest does not burn except under the most severe conditions, very few native Sumatra species of animals and plants are adapted to fire. As a result there is untold ecological damage and the extent of the burning means that forest regrowth will be very slow, even on those areas which are designated to be forest, such as protection forests and conservation areas, and even where social conditions permit it.

When *The Ecology of Sumatra* was originally written (1981-1983) industrial timber plantations were not in our vocabulary. The area of Sumatran forests cleared and replaced (or being replaced) with industrial timber estates now runs into the many hundred thousands of hectares. Some of these estates were established on degraded alang-alang lands where the conversion into productive lands is to be applauded. However, the majority of the estates have taken advantage of poorly-managed logging concessions, and managers and sponsors have even been accused of starting forest fires in order to lower the standing crop of the land below the level used as the criterion for conversion. These vast new areas are tied to pulpmills. Sumatra has the distinction of having the country's largest mill, which is currently being built by PT Tanjung Enim Lestari in South Sumatra at a cost of $1 billion. The factory alone required the clearance of over 1,000 ha of forested land. It will one day be supplied with 2 million m^3 of timber annually from a plantation, but until these trees are mature it will use trees from natural forests. Local communities, who have traditionally derived a good income from jungle rubber, have written letters of complaint to prominent decision makers in the civil and military administration not least because the project harms them economically. The people's 'lack of cooperation' has not been appreciated, and adverse reporting of the issue has closed the local newspaper. In addition to the infamous Indorayon mill on the shores of Lake Toba, another pulp mill where safeguards have not proved effective is Indah Kiat's mill in Perawang, Riau, currently the largest operating pulp mill in the world, producing over 1 million m^3 of pulp annually. Some sources believe that it is impossible to grow enough plantation timber on the land allocated to the mill. If this is so, it will have to take timber from the neighbouring peat forests which simply do not grow back in to the original forest, and many of which are on peat soils too deep for agriculture. Since second-hand mills have been installed, a cynic might suggest that when the natural forests have been cleared and pulped then the investors will have got the return on their investment and move elsewhere.

Even without fires and industrial timber estates, the lowland forests, selectively-logged and those under jungle-rubber use, would still be disappearing while areas under oil palm increase. Some of the growth in oil palm plantations is also due to the conversion of old rubber plantations. The sustainability of these enterprises may be questioned now that the gene for producing lauric acid (the product for which oil palm is primarily grown) has been successfully inserted in canola and has begun to be harvested. Given that this is a temperate crop which can be grown over large areas very efficiently it is not inconceivable that this will in due time displace the need for oil palm.

Large areas of mangroves all round Sumatra have been felled, partly for their fibre, partly as fuelwood, and partly to make way for shrimp ponds, in the hope of quick profit. It is ironic that only now that so much mangrove has been lost, have the long-term impacts been admitted by the Forestry Department's Director of Reforestation and Land Rehabilitation. Rehabilitation has been more difficult than was thought and protecting the areas from further attrition has also been problematic. After years of complacency it is now felt that mangroves should have protected status and licenses for aquaculture in mangrove areas should no longer be issued. Sadly the clock cannot be turned back. Apart from the problems of acid-sulphate soils, the devastating 'red virus' has knocked out 90% of the shrimp harvest in some areas and has discouraged consumers and traders. The investors will move on to some other project, leaving behind degraded land that will not quickly regenerate into the original, productive, diverse ecosystem.

National parks, the jewels in the crown of what should be one of the world's most significant protected area systems, are not secure. For example, the best parts (lowland, relatively undisturbed forest) of the original area of Leuser and Kerinci National Parks were given out as official and unofficial logging concessions. After many years of preparation and study, the government and the World Bank with the Global Environment Facility have agreed on a project to work with local governments and communities in and around Kerinci to safeguard its future and to help social and economic development. Barely had the project formally begun, however, than there were newspaper reports, later confirmed, that a local government road was cutting through the gazetted, mapped Park despite clear agreements that this would not be permitted. It had also been agreed that the logging companies around the present park would cooperate in the adoption of practices which would result in better forestry management, but even this component has been stalled. In addition, encroachment by cinnamon farmers appears to continue unhindered. It is important to remember in this regard that the people involved are not poverty-stricken farmers, but tenants acting for urban entrepreneurs.

Another internationally significant area is Siberut Island which is one of only four UNESCO-Man and Biosphere Program Biosphere Reserves in all of Indonesia. For over 15 years there have been persistent rumours and publicised plans for transmigration with plantations of oil palm or industrial timber. Vociferous international and national protests prior to an Asian Development Bank-funded conservation and development project on the island preceded the cancellation of logging permits, but by the time the companies actually left, most of the accessible and commercial timber had already been felled. The current developments are accompanied by tales of coercion and trickery in persuading local people to sign away their rights to land.

The exceptional Gunung Leuser National Park in northern Sumatra (van Schaik and Supriatna 1996) has also experienced decades of poor or no management despite considerable international NGO support, and as a result has suffered all manner of encroachments with no great protection from the conservation agency (PHPA), or others charged with enforcing policies and regulations (e.g., Whitten and Ranger 1986). It is to be hoped that things will change under its new, young, innovative management body, the International Leuser Foundation. This was established in place of, and with wider powers than, the government conservation agency. The Foundation has received a seven-year concession agreement to manage the 'Leuser ecosystem' which covers 1.8 million ha from which a new national park and surrounding buffer zone will be gazetted. The costs of management are shared between the Government's Reforestation Fund and the European Union. The involvement of the politicians, high-level provincial personalities, police, military, as well as Ministry of Forestry staff has resulted in perhaps the potential for the best management ever for the beleaguered park, and most other Indonesian parks. Their involvement has already prevented plans for roads and a transmigration scheme from proceeding, though this has led to pressure on the Foundation to develop alternatives. Some of this will be related to income-generating activities such as increased ecotourism and licensed exploitation of natural resources within the buffer zone. The Foundation and its program for integrated conservation and development are as yet young, and there are risks ahead, but their is great hope that it will be successful in its endeavours. Unfortunately it is impossible to envisage every conservation area receiving this type of attention.

BAD NEWS – WILDLIFE

With the loss of habitat, inevitably the populations of plants and wildlife have been reduced, separated and isolated. This, together with poaching, has taken some species to appallingly low numbers. For example, over the period 1990-1996 the numbers of Sumatran rhino in Kerinci National Park and adjacent forests fell from about 300 to just 30. The Sumatran rhino is the focus of a UNDP/Global Environment Facility project which is developing capacity to deal with rhino conservation in PHPA (the conservation agency) with the objective of arresting and reversing the decline due to poacher activity and habitat disturbance toward the national and global goal of recovery of viable populations of rhino species in Indonesia.

Disappearing only slightly less rapidly are tigers which have experienced a 95% decline throughout their range during this century. These 'protected species' are sought within Indonesia and beyond for their skins and body parts, and possibly just 500 remain of the Sumatran subspecies, many in areas now unable to sustain populations. While the trade in skins is largely domestic (highly skilled taxidermists can be found in major Sumatran towns), the illicit trade in parts focuses on China where, for example, tiger forelimbs sell for $1,000 per kg, and $320 buys a bowl of tiger penis soup for those with flagging libidos who are envious of the tiger's ability to copulate several times an hour (Plowden and Bowles 1997). If consumers realised that each copulation lasted a scant 15 seconds the bottom might fall out of the penis soup market.

Bones are believed to have healing powers, and whiskers and eyes also have uses, and are clearly easier to hide and trade than larger parts. Middlemen use skilled villagers, but have also turned to the most skilled of the forest peoples, the Kubu, to hunt for valuable species. The hunters are probably paid a derisory fee for their services despite the value of the product, but it compares well with the income derived from other activities. True to economic theory, the prices are rising as the beasts become rarer, and if it is true that even those in authority are guilty of complicity in the trade then the outlook is very bleak indeed. Certainly, the forces of the black market are so strong that all attempts from within and outside Sumatra have failed to stem the tide.

Sumatra is not the only Indonesian island where such stories can be related, and this is demonstrated in the recent volumes in the *Ecology of Indonesia* series (MacKinnon et al. 1996; Whitten et al. 1996; Monk et al. 1997; Tomascik et al. 1997). The nationwide progressive degradation and loss must be seen in its context: the government has supported the detailed work of RePPProT (see Annex following), the Indonesian Forestry Action Program, numerous bilateral and multi-lateral-supported projects related to rational land use, conservation and forests (Wells et al. 1998), the Biodiversity Action Plan (BAPPENAS 1993), the *Atlas of Biodiversity* (Anon. 1995), numerous wetland plans and strategies, out of all of which have come innumerable considered recommendations, proposals, arguments and pleas for sustainable development with restraint. A new conservation law has been passed and budgets to a larger number of national parks have increased.

And yet willful destruction continues. Those with little authority are easily scared off or won over when conflicts arise, the judiciary tends not to give high priority to such infringements, conservation budgets appear to be used up on prestigious activities rather than on the grunt work of patrolling and building relationships with those communities whose interest and cooperation are essential. It is easy to blame the private sector for many of the problems, but far more important is high-level political will. If this were truly engaged in commitment to a sustainable path for development, then the reduction of options through the loss of biodiversity, the direct and indirect degeneracy of land productivity and water quality, and the bypassing or unfriendly treatment of vulnerable people would not occur. What will it take to see meaningful change?

Acknowledgements

I wish to thank a small band of friends who agreed to read through this section and to give comments which have greatly improved it: Suraya Afiff, Sofia Bettencourt, Boeadi, Sengli Damanik, Wim Giesen, Andy Gillison, Kuswata Kartawinata, Ani Kartikasari, Margaret Kinnaird, Kathy MacKinnon, Kathyrn Monk, Øyvind Sandbukt, Carel van Schaik, Jito Sugarjito, David Wall, Tim Whitmore, and Jane Whitten (who did much else besides). Any mistakes and misinterpretations remain solely my responsibility.

March, 1998 Tony Whitten
Washington, D.C.
USA

Annex 1

Publications

A great deal has been written about Sumatra in the last 15 years and it is beyond the scope of this book to provide an exhaustive list of publications. However, a selection of books and papers appearing in peer-reviewed journals is given below to provide direction for those interested in the biology and fate of Sumatra.

Of singular importance and significance are the reports and sets of 1:250,000 maps produced by the RePPProT project (RePPProT 1988). This systematic baseline study mapped for the first time the biophysical environment, not only of Sumatra but later the entire country, and where and how it was being used. Areas appropriate and still available for different types of development or conservation were identified by comparing land qualities with development and conservation criteria and with existing land use and land status. One of its major contributions was in exposing gross discrepancies between existing forest function boundaries (as agreed by inter-Ministerial consensus in 1983) and what the boundaries should be rationally and ecologically. It also produced for the first time a comprehensive database and analysis of climatic data. In addition, the Sumatra study, together with seven other regional volumes were summarised in a National Overview with an atlas of thematic maps at 1:2,500,000 scale. Sadly and frustratingly, a major recommendation given by RePPProT, to remap the land use and land status of Indonesia every five years, has not been heeded and developments on Sumatra (as elsewhere in Indonesia), are taking place in ignorance of accurate and up-to-date land status and land use information.

Hundreds of papers are published each year on tropical ecology and many of them are relevant to the understanding of Sumatra. Two books on tropical forests by Whitmore (1984, 1990) are regarded as classics and should be consulted for solid information and interpretations. The long-awaited volume by Laumonier (1997) on the vegetation of Sumatra has recently appeared. It gives a critical appraisal of the methods used in vegetation surveys in the tropics. The small-scale eco-floristic physiographical classification of vegetation could form the basis of detailed monitoring of deforestation.

The coastal and marine environments around Sumatra are dealt with thoroughly by Tomascik et al. (1997), and the ecology, status and management of the peatlands and mangrove forests of eastern Sumatra have been a focus of interest by Asian Wetlands Bureau/Wetlands International in Bogor from which a list of the unpublished reports can be obtained. One widely-available publication is by Claridge (1994) concerning Berbak National Park.

In addition there are a number of coffee-table and travel books with exceptional photos and informative text (Griffiths 1989; Oey 1991; Stone 1994; Whitten and Whitten 1992, 1996a,b), the last two of which include some exceptional coverage of the Kubu people by Sandbukt (see also Sandbukt 1988). Ancient history and general human geography are dealt with in two new and very attractive encyclopaedias (Miksic 1996; Rigg 1996).

It is easier now to identify certain animal and plant species than it was in

1984. Larger trees can now be compared with the illustrations in Whitmore and Tantra (1986), and tree genera can in most cases be identified from the interactive key on the World Wide Web designed for use on Borneo (http://django.harvard.edu/users/jjarvie/borneo.htm) (Jarvie and Ermayanti 1996). Freshwater fish can be identified using the keys, descriptions and photographs in Kottelat et al. (1996 – this second printing also has an addendum of new species and name changes since the 1993 printing). The birds are now well served with a guide to the commoner species (Holmes and Nash 1990), and a guide to all species of the Sunda Region (MacKinnon and Phillipps 1993). There is also a bird watcher's guide to sites and species (Jepson and Ounstead 1997) and an annotated checklist (van Marle and Voous 1988). The mammals are well, if not perfectly, served by a guide to the species of Borneo (Payne et al. 1985). The snakes are now covered by an illustrated checklist (David and Vogel 1997). In addition, the first three guides from the Fauna Malesiana Project will soon be available and will be useful on Sumatra. These cover snails (Vermeulen and Whitten 1998), pest grasshoppers and allies (Willemse, in press), and flies (Osterbroek, in press).

Our knowledge of ecosystems, especially their dynamics, is generally rather weak. We are lucky that there are some areas in Sumatra where teams have worked and have had some significant impacts on the way the land is designated and managed.

Some globally significant studies are underway in central Sumatra by the Centre for International Forestry Research, based in Bogor, which is seeking indicators for the impact on land use on biodiversity by conducting baseline studies along disturbance and altitudinal gradients (Gillison 1996a,b: Gillison and Carpenter 1997). These are producing generic methods for rapid biodiversity assessment which potentially could provide a much needed information base for regional planners and managers.

A major team of natural and social scientists under NORINDRA (Norwegian-Indonesian Rain Forest and Resource Management Project) was stationed in Riau in 1991-92 and produced the most comprehensive analysis of people-forest interactions in Southeast Asia, major contributions to forest policy, understanding traditional and commercial use of forest resources, knowledge of biodiversity, and conservation. Indeed, the establishment of the Bukit Tigapuluh National Park was a direct result of the work and now conserves perhaps the only remaining major expanse (about 1,000 km^2) of lowland forest left on Sumatra. Three important books on forest and resource management have appeared (Sandbukt and Østergaard 1993; Angelsen 1994; Sandbukt and Wiriadinata 1994) the third of which contains 27 papers on a wide range of subjects. In addition, a number of interesting papers have been published (e.g., Danielsen and Schumacher 1997), and Norwegian and Indonesian interest in the area continues.

There has been a very fruitful cooperation between Andalas University and Japanese scientists, and many papers, mainly on plant and insect ecology, have resulted (e.g., Hotta 1986; Kohyama and Hotta 1986; Hotta 1987; Okada and Hotta 1987; Sakagami et al. 1989; Yoneda et al. 1990; Oi 1990, 1996; Itino et al. 1991; Kato et al. 1991, 1993; Kohyama 1991; Aimi and Bakar 1992; Mukhtar et al. 1992; Koike and Syahbuddin 1993; Sianturi et al. 1995).

Work in the Gunung Leuser National Park has continued to produce seminal papers on wildlife and ecology and its management (e.g., the many papers by van

Schaik and his colleagues, van Noordwijk 1985; Soegarjito 1986; Rachmatika and Wirjoatmodjo 1988; Griffiths 1989; te Boekhorst et al. 1990; Cant et al. 1990; Mukhtar 1994; Rijksen and Griffiths 1995).

The centennial anniversary of the massive eruption of Krakatau fell in 1983, and resulted in a rush of publications on a wide range of ecological topics related to the recolonisation of animal and plant species. The publications are too many to list but they are synthesised in a masterly book by Thornton (1996).

In southern Sumatra there has been some excellent work on traditional agroforestry and its ecological and social benefits (Torquebiau 1984, 1986; Michon and Bompard 1987a,b; Gouyon et al. 1993; Thiollay 1994, 1995; Levang et al. 1997).

Bukit Barisan Selatan National Park is now also a major long-term research site thanks to the work started by the national and international staff of Wildlife Conservation International (O'Brien and Kinnaird 1996).

Research in the field and in laboratories has resulted in the formal naming of many new species: insects, notably in the ongoing Heterocera Sumatrana series covering the moths (e.g., Holloway 1990; Holloway and Bender 1990), as well as in other groups such as land snails (Djajasasmita 1988; Dharma 1993), freshwater crabs (Ng 1993; Ng and Tan 1995), fish (see 1996 insert in the printing of Kottelat et al. 1993), and plants amongst which are numbered several new and apparently endemic genera (Hotta 1987; Stone 1988; Hyde 1989; Burtt 1990; Nagamasu 1990; Kostermans 1992).

Possibly the most dramatic new species will turn out to be the 'orang pendek' or 'short man' of the Kerinci area which has been generally dismissed in the past. Since 1995 small teams, supported by Fauna and Flora International, have been trying to collect solid support for this creature's existence. A very impressive cast of a large foot print has been obtained which baffles mammal specialists and field workers alike, and even those who started as sceptics have reported positive sightings in the montane. Photo traps have been set in the forest, but despite a fascinating range of exceptional photographs of rarely seen large forest animals, the orang pendek itself has remained elusive.

References

Abdulhadi, R., Mirmanto, E. and Kartawinata, K. (1987). Lowland dipterocarp forest in Sekunder, North Sumatra, Indonesia, five years after logging. In *Dipterocarps* (ed. A.J.G.H. Kostermans), pp. 255-273. MAB-UNESCO, Jakarta.

Aimi, M. and Bakar, A. (1992). Taxonomy and distribution of *Presbytis melalophos* group in Sumatra, Indonesia. *Primates* 33: 191-206.

Amato, G. et al. (1995). Assessment of conservation units for the Sumatran rhinoceros. *Zoo Biol.* 14: 395-402.

Andrew, P. (1992). *The Birds of Indonesia: A Checklist* (Peter's Sequence). Indonesian Ornithological Society, Jakarta.

Angelsen, A. (1994). *Shifting Cultivation and 'Deforestation': A Study from Sumatra.* Chr. Michelson Institute, Bergen.

Aumeeruddy, Y. (1995). Perceiving and managing natural resources in Kerinci, Sumatra. *Nature and Resources* 31: 28-37.

Bihari, M. and Lal, C.B. (1989). Species composition, density and basal cover of tropical rainforests on Central Sumatra, Indonesia. *Trop. Ecol.* 30: 118-137.

Bismark, M. (1991). Population analysis of longtail macaques *Macaca fascicularis* in various forest types. *Bul. Pen. Hutan* 532: 1-10.

Brady, M.A. (1989). A note on the Sumatra, Indonesia, peat swamp forest fires of 1987. *J. Trop. For. Sci.* 3: 295-296.

Brady, M.A. and Kosasih, A. (1991). Controlling off-site forest destruction during oil field development in Sumatra, Indonesia. Proceedings of the 20th Annual Convention, Indonesian Petroleum Association, Jakarta.

Burtt, B.L. (1990). Gesneriaceae of the Old World II. A new *Didymocarpus* from Sumatra. *Edinb. J. Bot.* 47: 235-238.

Cant, J.G.H. et al. (1990). Stress tests of lianas to determine safety factors in the habitat of orangutans *Pongo pygmaeus* in northern Sumatra, Indonesia. *Am. J. Phys. Anthropol.* 81: 203.

Claridge, G. (1994). Management of coastal ecosystems in eastern Sumatra: The case of Berbak Wildlife Reserve, Jambi Province. *Hydrobiologia* 285: 287-302.

Colfer, C.J.P., Gill, D.W. and Agus, F. (1988). An indigenous agricultural model from West Sumatra, Indonesia: A source of scientific insight. *Agric. Syst.* 26: 191-210.

Corbett, G.B. and Hill, J.E. (1992). *The Mammals of the Indomalayan Region: A Systematic Review.* Oxford University Press, Oxford.

Danielsen, F. and Schumacher, T. (1997). The importance of Tigapuluh Hills, southern Riau, Indonesia, to biodiversity conservation. *Trop Biodiv.* 4: 129-160.

Danielsen, F. et al. (1997). The Storm's stork *Ciconia stormi* in Indonesia: Breeding biology, population and conservation. *Ibis* 139: 67-76.

David, P. and Vogel, G. (1997). *The Snakes of Sumatra: An Annotated Checklist and Key with Natural History Notes.*

Dharma, B. (1993). Description of two new species of *Amphidromus* from Sumatra, Indonesia (Gastropoda: Pulmonata: Camaenidae). *Apex* 8: 139-143.

Djajasasmita, M. (1988). A new cyclophorid land snail from North Sumatra, Indonesia (Mollusca: Gastropoda, Cyclophoridae). *Treubia* 29: 271-274.

Dring, J., McCarthy, C. and Whitten, A.J. (1990). The herpetofauna of the Mentawai Islands, Indonesia. *Indo-Australian Zool.* 1: 210-215.

Faust, T., Tilson, R. and Seal, U.S. (1995). Using GIS to evaluate habitat risk to wild populations of Sumatran orangutans. In *The Neglected Ape; International Orangutan Conference and the Orangutan Population and Habitat Viability Analysis Workshop.* Fullerton, California, and Medan, Indonesia. Jan 18-20, 1993. Plenum Press, New York, London.

Fuentes, A. (1994). Social organization in the Mentawai langur. *Amer. J. Phys. Anthropol. Suppl.* 18: 90.

Fujisaka, S. et al. (1991). *Wild pigs, poor soils, and upland rice: a diagnostic survey of Sitiung, Sumatra, Indonesia.* IRRI Research Paper Series, 155: 1-9.

Giesen, W. and Sukotjo. (1991). *Lake Kerinci and the Wetlands of Kerinci Seblat National Park, Sumatra.* Asian Wetlands Bureau, Bogor.

Gillison, A.N., Liswanti, N. and Ismail Rachman (1996a). *Rapid Assessment, Kerinci Seblat National Park Buffer Zone: Preliminary report on plant ecology and overview on biodiversity assessment.* CIFOR Working paper No. 14.

Gillison, A.N., Liswanti, N. and Arief-Rachman, I. (1996b). Contributors. In:

Final Report, Rapid Ecological Assessment in HPH Pt Serestra II and HPH PT Bina Samaktha. Pre-implementation, Integrated Conservation and Development Project, Kerinci Seblat National Park, World Wildlife Fund for Nature, Indonesia Program. Bappenas, The World Bank. (Published 1997.)

Gillison, A.N., and Carpenter, G. (1997). A plant functional attribute set and grammar for dynamic vegetation description and analysis. *Functional Ecology* 11: 775-783.

Gouyon, A., de Foresta, H. and Levang, P. (1993). Does 'jungle rubber' deserve its name? An analysis of rubber agroforestry systems in southeast Sumatra. *Agrofor. Syst.* 22: 181-206.

Griffiths, M. (1989). *Indonesian Eden: Aceh's Rainforest.* Jakarta.

Hoggarth, D.D. and Utomo, A. (1994). Fisheries ecology of the Lubuk Lampam River floodplain in south Sumatra, Indonesia. *Fisheries Research* 20: 191-213.

Holloway, J.D. (1990). The Limacodidae of Sumatra. *Heterocera Sumatrana* 6: 9-77.

Holloway, J.D. and Bender, R. (1990). The Lasiocampidae of Sumatra. *Heterocera Sumatrana* 6: 137-204.

Holmes, D. (1996). Sumatra bird report. *Kukila* 8: 9-56.

Holmes, D. and Nash, S. (1990). *Birds of Sumatra and Kalimantan.* Oxford, Kuala Lumpur.

Hotta, M. (1986). *Diversity and Dynamics of Plant Life in Sumatra: Forest Ecosystem and Speciation in Wet Tropical Environments. Sumatra Nature Study* (Botany), Kyoto University.

Hotta, M. (1987). A new rheophytic aroid *Schismatoglottis okadae* new species from West Sumatra, Indonesia. *Contrib. Biol. Lab. Kyoto Univ.* 27: 151-152.

Hyde, K.D. (1989). Intertidal mangrove fungi from North Sumatra, Indonesia. *Can. J. Bot.* 67: 3078-3082.

Inoue, T. et al. (1993). Population dynamics of animals in unpredictably-changing tropical environments. *J. Bioscience* 18: 425-455.

Itino, T., Kato, M., and Hotta, M. (1991). Pollination ecology of the two wild bananas, *Musa acuminata* subsp. *halabanensis* and *M. salaccensis*: chiropterophily and ornithophily. *Biotropica* 23: 151-158.

Jepson, P. and Ounstead, R. (1997). *A Bird Watcher's Guide to the World's Largest Archipelago.* Periplus, Singapore.

Kato, M. et al. (1991). Pollination of four Sumatran *Impatiens* species by hawkmoths and bees. *Tropics* 1: 59-73.

Kato, M. et al. (1993). Inter- and intra-specific variation in prey assemblages and inhabitant communities in *Nepenthes* pitchers in Sumatra. *Trop. Zool.* 6: 11-25.

Kobayashi, M. et al. (1996). Indonesian medicinal plants. XVIII Kompasinol A, a new stilbeno-phenylpropanoid from the bark of *Koompassia malaccensis.* Chemical and Pharmaceutical Bulletin (Tokyo). 44(12):2249-2253.

Kohyama, T. (1991). Simulating stationary size distribution of trees in rain forests. *Ann. Bot.* 68:173-180.

Kohyama, T. and Hotta, M. (1986). Growth analysis of Sumatran *Monophyllaea,* possessing only one leaf throughout perennial life. *Pl. Sp. Biol.* 1: 117-125.

Koike, F. and Syahbuddin (1993). Canopy structure of a tropical rain forest and the nature of an unstratified upper layer. *Funct. Ecol.* 7: 230-235.

Kostermans, A.J.G.H. (1992). An important economical new *Durio* sp. from northern Sumatra. *Econ. Bot.* 46: 338-340.

Kottelat, M. (1994). Freshwater fish biodiversity in western Indonesia: State of the art, threats and need for pragmatic biodiversity approach of systematics. *Rev. Suisse Zool.* 101: 853.

Kottelat, M., Whitten, A.J., Wiryoatmodjo, S. and Kartikasari, A. (1993). *Freshwater Fishes of Western Indonesia and Sulawesi.* Periplus, Singapore. Reprinted in 1996 with a separate summary of new species, name changes, etc.

Kunkun, J.G. (1986). Ecology and behaviour of *Presbytis thomasi* in northern Sumatra, Indonesia. *Primates* 27: 151-172.

Laumonier, Y. (1990). Search of phytogeographic provinces in Sumatra, Indonesia. In *The Plant Diversity of Malesia* (ed. P. Baas, K. Kalkman and R. Geesink), pp. 193-212. Kluwer, Dordrecht.

Laumonier, Y. (1997). *The Vegetation and Physiography of Sumatra.* Kluwer, Dordrecht.

Levang, P., Michon, G. and Foresta, H.D. (1997). Forest agriculture or agroforestry? Indonesian examples of farmers' strategies adapted to environments with mediocre fertility. *Bois For. Tropiques* 251: 29-42.

Lim K.S. (1997). Further notes on the avifauna of the Riau archipelago. *Kukila* 9: 74-77.

MacKinnon, J. (1997). *Protected Areas Systems Review of the Indo-Malayan Realm.* World Conservation Monitoring Centre, Cambridge.

MacKinnon, J. and Phillipps, K. (1993). *A Field Guide to the Birds of Borneo, Sumatra, Java and Bali.* Oxford University Press, Oxford.

Meijer, W. (1997). Rafflesiaceae. *Flora Malesiana* I 13: 1-42.

Michon, G.M. and Bompard, J.-M. (1987a). Indonesian agroforestry practices: A traditional contribution to the conservation of rain forest resources. *Rev. Ecol. Terre Vie* 42: 3-38.

Michon, G.M. and Bompard, J.-M.(1987b). The damar gardens (*Shorea javanica*) in Sumatra. In *Dipertocarps* (ed. A.J.G.H. Kostermans), pp.3-7. MAB-UNESCO, Jakarta.

Miksic, J. (ed.) (1996). *Ancient History. Indonesian Heritage Encyclopaedia Vol. 1.* Archipelago Press, Singapore.

Mukhtar, A., Nazif, M. and Setiawati, T. (1990). Habitat and behaviour of red lutung (*Presbytis melalophos*) in Bukit Sebelah Protection Forest, West Sumatra. *Bul. Pen. Hutan* 528: 1-12.

Mukhtar, A.S. (1994). Some aspects of elephant ecology in Sekunder Forest Complex, Leuser Mountains National Park. *Bul. Pen. Hutan* 564: 1-23.

Mukhtar, E. et al. (1992). Regeneration process of a climax species *Calophyllum* cf. *soulattri* in tropical rain forest of West Sumatra. *Tropics* 2: 1-12.

Nagamasu, H. (1990). A new *Symplocos* (Symplocaceae) from Sumatra. *Blumea* 35: 229-232.

Ng, P.K.L. (1993). *Parathelphusa maindroni,* a peatswamp crab from Peninsular Malaysia and Sumatra (Crustacea: Decapoda, Bracyura: Parathelphusidae). *Malay. Nat. J.* 46: 189-200.

Ng, P.K.L. and Tan, C.G.S. (1995). *Geosarma notophorum* sp. nov. (Decapoda, Brachyura, Grapsidae, Sesarminae), a terrestrial crab from Sumatra with novel brooding behaviour. *Crustaceana* 68: 390-395.

O'Brien, T.G. and Kinnaird, M.F. (1996). Birds and mammals of the Bukit Barisan Selatan National Park, Sumatra, Indonesia. *Oryx* 30: 207-217.

Oey, E. (1991). *Sumatra.* Periplus, Singapore.

Ohsawa, M. et al. (1985). Altitudinal zonation of forest vegetation on Mt. Kerinci, Sumatra: with comparisons to zonation in the temperate region of East Asia. *J. Trop. Ecol.* 1: 193-216.

Oi, T. (1990). Population organization of wild pig-tailed macaques *Macaca nemestrina nemestrina* in West Sumatra, Indonesia. *Primates* 31: 15-32.

Oi, T. (1996). Sexual behaviour and mating system of the wild pig-tailed macaque in West Sumatra. In *Evolution and Ecology of Macaque Societies* (ed. J.E. Fa), pp.342-368. Cambridge University Press, Cambridge.

Okada, H. and Hotta, M. (1987). Species diversity of wet tropical environments. II. Speciation of *Schismatoglottis okadae,* an adaptation to the rheophytic habitat of mountain streams in Sumatra, Indonesia. *Contrib. Biol. Lab. Kyoto Univ.* 27: 153-170.

Osterbroek, P. (in press). *The Families of* Diptera *of NW Borneo, Indonesia, and East New Guinea.* Fauna Malesiana Foundation, Leiden.

Payne, J., Francis, C.M. and Phillipps, K. (1985). *A Field Guide to the Mammals of Borneo.* Sabah Society, Kota Kinabalu.

Plowden, C. and Bowles, D. (1997). The illegal market in tiger parts in north Sumatra. *Oryx* 59-66.

Rachmatika, I. And Wirjoatmodjo, S. (1988). Feeding ecology of *Glyptothorax major* (Bagaridae, Siluriformes) in the Alas River, South-east Aceh, Sumatra, Indonesia. *Berita Biol.* 3: 396-399.

Rampino, M.R. and Self, S. (1992). Volcanic winter and accelerated glaciation following the Toba super-eruption. *Nature (Lond.)* 359: 50-52.

Rao, M. and van Schaik, C.P.(1997). The behavioural ecology of Sumateran orangutans in logged and unlogged forest. *Trop. Biodiv.* 4: 173-186.

RePPProT (1988). *Sumatra.* Regional Physical Planning Programme for Transmigration, Natural Resources Institute and Direktorat Bina Program, Direktorat Jenderal Penyiapan Pemukiman, Departemen Transmigrasi, Jakarta.

Riggs, J. (ed.) (1996). *Human Environment. Indonesian Heritage Encyclopaedia Vol. 2.* Archipelago Press, Singapore.

Rijksen, H. and Griffiths, M. (1995). *Leuser Development Masterplan.* Institute for Forestry and Nature Research, Wageningen.

Ruedi, M. and Fumagalli, L. (1996). Genetic structure of gymnures (genus *Hylomys;* Erinaceidae) on continental islands of Southeast Asia: Historical effects of fragmentation. *J. Zool. Syst. Evol. Res.* 34: 153-

162.

Sakagami, S.F. et al. (1989). Nests of myrmecophilous stingless bee *Trigona moorei*: How do bees initiate their nest within an arboreal ant nest? *Biotropica* 21: 265-274.

Sandbukt, Ø. (1988). Resource constraints and relations of appropriation among tropical forest foragers: the case of the Sumatran Kubu. *Res. Econ. Anthrop.* 10: 117-156.

Sandbukt, Ø. and Østergaard, L. (1993). *Bukit Tigapuluh: Rain Forest and Resource Management*. WWF and Ministry of Forestry, Jakarta.

Sandbukt, Ø. and Wirtiadinata, H. (1994). Rain Forest and Resource Management: Proceedings of the NORINDRA Seminar. Indonesian Institute of Sciences, Jakarta.

Santiapillai, C. and Ashby, K.R. (1988). The clouded leopard in Sumatra. *Oryx* 22: 44-45.

Scheffrahn, W., de Ruiter, J.R. and van Hooff, J.A.R.A.M. (1996). In *Evolution and Ecology of Macaque Societies* (ed. J.E. Fa), pp.20-42. Cambridge University Press, Cambridge.

Shine, R. et al. (1995). Biology and commercial utilization of acrochordid snakes, special reference to karung (*Acrochordus javanicus*). *J. Herpetol.* 29: 352-360.

Shine, R., Harlow, P.S., Keogh, J.S. and Boeadi (1996). Commercial harvesting of giant lizards: the biology of water monitors *Varanus salvator* in southern Sumatra. *Biol. Cons.* 77: 125-134.

Sianturi, E.M.T. et al. (1995). Nest structure, colony composition and foraging activity of a tropical-montane bumblebee *Bombus senex* (Hymenoptera: Apidae), in West Sumatra. *Jap. J. Entomol.* 63: 657-667.

Siebert, S.F. (1989). The dilemma of a dwindling resource: rattan in Kerinci, Sumatra. *Principes* 33: 79-87.

Sourrouille, P. et al. (1995). Molecular systematics of *Mus crociduroides*, an endemic mouse of Sumatra. *Mammalia* 59: 91-102.

Stone, B.C. (1988). New and noteworthy Malesian Myrsinaceae II. *Emblemantha* new genus from Sumatra, Indonesia. *Proc. Acad. Nat Sci. Phila.* 140: 275-280.

Stone, B.C. (1990). Studies in Malesian Myrsinaceae V. Additional new species of *Ardisia*. *Proc. Acad. Nat. Sci. Phila.* 142: 21-58.

Stone, D. (1994). *Tanah Air: Indonesia's Biodiversity*. Archipelago, Singapore.

Sugarjito, J. (1986). *Ecological constraints on the behaviour of Sumatran orang-utans* (Pongo pygmaeus abelli) *in the Gunung Leuser National Park, Indonesia.* Pressa Trajectina, Utrecht.

Suharti, S. (1991). The profile of shifting cultivators in the nature reserve of Isau-Isau, Pasemah, South Sumatera. *Bul. Pen. Hutan* 544: 39-49.

Sukardjo, S. (1987). Natural regeneration status of commercial mangrove species *Rhizophora apiculata* and *Bruguiera gymnorhiza* in the mangrove forest of Tanjung Bungin, Banyumasin, South Sumatra, Indonesia. *For. Ecol. Manage.* 20:233-252.

Supriatna, J. et al. (1996). A preliminary survey of long-tailed and pig-tailed macaques in Lampung, Bengkulu, and Jambi provinces, Southern Sumatra, Indonesia. *Trop. Biodiv.* 3: 131-140.

Tan, S.-H. and Tan, H.-H. (1995). The freshwater fishes of Pulau Bintan, Riau Archipelago, Sumatera, Indonesia. *Trop. Biodiv.* 2: 351-368.

te Boekhorst, I.J.A., Schurmann, C.L. and Sugarjito, J. (1990). Residential status and seasonal movements of wild orangutans in the Gunung Leuser Reserve, Sumatra, Indonesia. *Anim. Behav.* 39: 1098-1109.

Thiollay, J.-M. (1994). Rain forest raptor community in Sumatra: The conservation value of traditional agroforests. *J. Raptor Res.* 28: 51.

Thiollay, J.-M. (1995). The role of traditional agroforests in the conservation of rain and forest bird diversity in Sumatra. *Conserv. Biol.* 9: 335-353.

Thornton, I. (1996). Krakatau: The Destruction and Reassembly of an Island Ecosystem. Harvard University Press, Cambridge, Mass.

Torquebiau, E.F. (1984). Man-made dipterocarp forest in Sumatra. *Agroforestry Syst.* 2: 103-127.

Torquebiau, E.F. (1986). Mosaic patterns in dipterocarp rain forest in Indonesia and their implications for practical forestry. *J.Trop. Ecol.* 2: 301-325.

Ungar, P.S. (1996). Feeding height and niche separation in sympatric Sumatran monkeys and apes. *Folia primatalol.* 67(3): 163-168.

Van Marle, J.G. and Voous, K.H. (1988). *The Birds of Sumatra: An Annotated Checklist. British Ornithologists' Union, Tring.*

van Noordwijk, M.A. (1985). *The socio-ecology of Sumatran long-tailed macaques* (Macaca fascicularis). *II. The behaviour of individuals.* Elinwijk, Utrecht.

van Schaik, C.P. (1985). *The socio-ecology of Sumatran long-tailed macaques* (Macaca fascicularis). *I. Costs and benefits of group living.* Elinwijk, Utrecht.

van Schaik, C.P. (1986). Phenological changes in a Sumatran rain forest. *J. Trop. Ecol.* 2: 327-348.

van Schaik, C.P. and Griffiths, M. (1996). Activity periods of Indonesian rain forest mammals. *Biotropica* 28: 105-112.

van Schaik, C.P. and Mirmanto, E. (1985). Spatial variation in the structure and litterfall of a Sumatran rain forest. *Biotropica* 17: 196-205.

van Schaik, C.P. and Noordwijk, M.A. (1985). Evolutionary effect of the absence of felids on the social organization of the macaque on the island of Simeulue. *Folia* primalol. 44: 138-147.

van Schaik, C.P. and Noordwijk, M.A. (1985). Interannual variability in fruit abundance and the reproductive seasonality in Sumatran long tailed macaques. *J. Zool.* 206: 533-549.

van Schaik, C.P. and Supriatna, J. (1996). *Leuser: A Sumatran Sanctuary. Yayasan Bina Sains Hayati, Depok.*

van Schaik, C.P. et al. (1995). Estimates of orangutan distribution and status in Sumatra. In *The Neglected Ape* (ed. R.D. Nadler), pp.109-116. Plenum, New York.

van Schaik, C.P., van Amerongen, A., and van Noordwijk, M.A. (1996). Riverine refuging by wild Sumatran long-tailed macaques (*Macaca fascicularis*). In *Evolution and Ecology of Macaque Societies* (ed. J.E. Fa), pp.160-181. Cambridge University Press, Cambridge.

Vermeulen, J. and Whitten, T. (1998). *Guide to the Land Snails of Bali.* Fauna Malesiana Foundation, Leiden.

Wallach, V. (1988). Status and redescription of the genus *Padangia* with comparative visceral data on *Collorhabdium* and other genera (Serpentes: Colubridae). *Amphib-Reptilia* 9: 61-76.

Wells, M., Khan, A. and Jepson, P. (1998). *Investing in Biodiversity: Integrated Conservation Development Projects in Indonesia.* The World Bank, Washington D.C.

Weaver, J.S. III (1989). Indonesian Lepidostomatidae (Trichoptera) collected by Dr. E.W. Diehl. Aquat. *Insects* 11: 47-63.

Whitmore, T.C. (1984). *Tropical Rain Forests of the Far East.* Clarendon, Oxford.

Whitmore, T.C. (1990). *Introduction to Tropical Rain Forests.* Oxford University Press, Oxford. (Second edition due 1998).

Whitmore, T.C. and I G.M. Tantra (1986). *Tree Flora of Indonesia.* Forest Research and Development Centre, Bogor.

Whitten, A.J. and Damanik, S.J. (1986). Mass defoliation of mangrove in Sumatra, Indonesia. *Biotropica* 18:176.

Whitten, A.J. and Ranger, J. (1986). Logging at Bohorok. *Oryx* 20: 246-248.

Whitten, J. and Whitten, A.J. (1987). Analysis of bark eating in a tropical squirrel. *Biotropica* 19: 107-115.

Whitten, T. and Whitten, J. 1992. *Wild Indonesia.* New Holland, London.

Whitten, T. and Whitten, J. 1996a. *Indonesian Heritage Encyclopaedia Vol. 4: Plants.* Archipelago, Singapore.

Whitten, T. and Whitten, J. 1996b. *Indonesian Heritage Encyclopaedia Vol. 5: Wildlife.* Archipelago, Singapore.

Whitten, T., Whitten, J., Mittermeier, C., Supriatna, J. and Mittermeier, R. (1998). Indonesia. In *Megadiversity: Earth's Biologically Wealthiest Nations* (ed. R. Mittermeier, P.R.Gl, and C. Mittermeier), pp. 74-97. Cemex, Prado Norte.

Whitten, T., Soeriaatmadja, R.E. and Afiff, S. (1996). *The Ecology of Java and Bali.* Periplus, Singapore.

Wibowo, A. (1990). The effect of fire on alang-alang grassland ecosystem in Benakat, South Sumatra, Indonesia. *Bul. Pen. Hutan* 521: 19-30.

Willemse, L.P.M. (in press). *Guide to the Pest Orthoptera of the Indo-Malayan Region.* Fauna Malesiana Foundation, Leiden.

Yoneda, T., Tamin, R. and Ogino, K. (1990). Dynamics of above ground big woody organs in a foothill dipterocarp forest, West Sumatra, Indonesia. *Ecol. Res.* 5: 111-130.

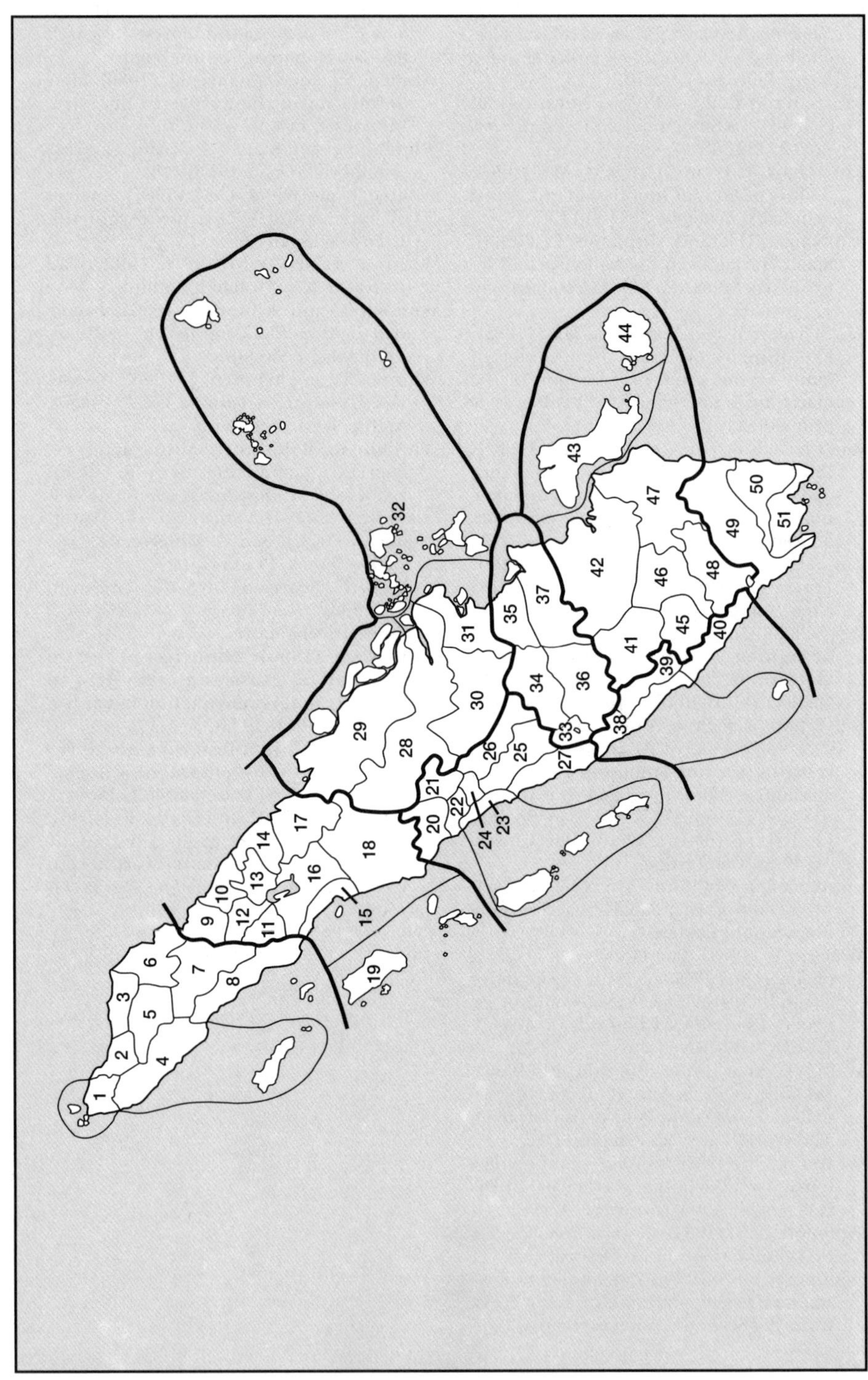
1
2
3
4
5
6
7
8
9
10
11
12
13
14
15
16
17
18
19
20
21
22
23
24
25
26
27
28
29
30
31
32
33
34
35
36
37
38
39
40
41
42
43
44
45
46
47
48
49
50
51

Counties of Sumatra by Province

Aceh

1. Greater Aceh
2. Pidie
3. North Aceh
4. West Aceh
5. Central Aceh
6. East Aceh
7. Southeast Aceh
8. South Aceh

North Sumatra

9. Langkat
10. Deli Serdang
11. Dairi
12. Tanah Karo
13. Simalungun
14. Asahan
15. Central Tapanuli
16. North Tapanuli
17. Labuhan Batu
18. South Tapanuli
19. Nias

West Sumatra

20. Pasaman
21. Limapuluh Kota
22. Agam
23. Padang Pariaman
24. Tanah Datar
25. Solok
26. Sawahlunto Sijunjung
27. Pesisir Selatan

Riau

28. Kampar
29. Bengkalis
30. Upper Indragiri
31. Lower Indragiri
32. Riau Archipelago

Jambi

33. Kerinci
34. Bungo Tebo
35. Tanjung Jabung
36. Bangko Sarolangun
37. Batanghari

Bengkulu

38. North Bengkulu
39. Rejang Lebong
40. South Bengkulu

South Sumatra

41. Musi Rawas
42. Musi Banyuasin
43. Bangka
44. Belitung
45. Lahat
46. Muara Enim
47. Lower Ogan Komering
48. Upper Ogan Komering

Lampung

49. North Lampung
50. Central Lampung
51. South Lampung

Part A

Introduction

The process of land formation, land movements, climatic change and evolution have been operating for thousands of millions of years and we see the result of them as natural ecosystems and as ecosystems disturbed or created by man. The chapter in this first section summarizes what is known about these processes in relation to Sumatra in order to provide a background against which later chapters can be viewed. As we learn about the past and its influence on the present we can better predict and understand the possibilities for the future.

Chapter One

Background

Geomorphological and Geological History

About 250 million years ago, at the beginning of the Mesozoic era (table 1.1), the earth's continents formed a single land called Pangea. During the Triassic, roughly 230 million years ago, Pangea broke into two, with Laurasia (North America, Europe and Asia) splitting off from Gondwanaland (India, Australia, Africa, South America and Antarctica). Gondwanaland began to split into its separate parts 200 million years ago during the Jurassic, and India floated towards Asia at about 10-18 cm per year (figs. 1.1 and 1.2). By the early Tertiary, about 70 million years ago, India and the plate on which it lay began to collide with and move under Asia. Its major thrust caused the uplift of the Himalayas, and one of the associated thrusts caused the uplift of the Barisan Mountains that run the length of Sumatra. The movement of the Indian Plate under Asia caused great earthquakes, and although the rate of movement is now less, almost all the earthquakes experienced in Sumatra (and Java, Burma, etc.) today are caused by those continuing movements. The main parts of Southeast Asia remained in more or less their present relative positions throughout this period although their absolute positions moved north and south of the equator.

As the Barisan Range buckled upwards, so a corresponding downward thrust formed the deep channel to the west of Sumatra, and the secondary upward thrusts that occurred at different times along its length formed the chain islands from Simeulue to Enggano. A very deep trough falling to the Indian Plate itself was formed to the west of these islands. A smaller trough was formed to the northeast of the Barisan Range and is represented today by low hills and plains (fig. 1.3). The alluvial eastern plains are narrower in the north of Sumatra either because sedimentation was slower or because subsequent subsidence of the land was faster (Verstappen 1973).

The southeast of Sumatra, including the islands of Bangka and Belitung, forms part of the stable Sunda Shelf (see fig. 1.9) and occasional ancient granite outcrops, particularly on Bangka, are indications that the geological foundation of the Sunda Region[1] is not far below the surface.

The rocks of the Barisan Range are largely sedimentary, laid down over about 100 million years between the late Palaeozoic and early Mesozoic eras

Table 1.1. Summary of geological timescale and events.

Era	Period	Epoch	Years from beginning of period/epoch to present	Geological & geomorphological events	Biological events
	Quaternary	Holocene	10,000	Volcanic activity based on mountains & usually associated with faults	Man appears
		Pleistocene	1 million	Some sedimentation & folding in E. Sumatra	Man's earliest ancestors appear
Cenozoic (age of mammals)	Tertiary	Pliocene	10 million	Final mountain-building faulting, rift valleys major upthrust,	
		Miocene	25 million	violent volcanic activity	Modern species of mammals evolve, extinction of large forms (e.g., mammoths); monkeys appear in Oligocene
		Oligocene	40 million	Sumatra subsides, formation of sedimentary rocks	
		Eocene	60 million	Barisan Range starts to be uplifted	Rise of birds and placental mammals
		Palaeocene	70 million		
Mesozoic (age of reptiles)	Cretaceous		145 million		Dominance of angiosperm plants, extinction of large reptiles
	Jurassic		215 million		Reptiles dominant; first birds and archaic mammals
	Triassic		250 million	Sedimentary rock of Barisan Range forming	First dinosaurs. Cycads & conifers dominant plant groups
	Permian		280 million 350 million		Increasing variation of reptiles displace amphibians as dominant group
Palaeozoic Carboniferous	Devonian		400 million		Ferns dominant plant group; sharks abundant; first reptiles
					Age of fishes (mostly freshwater), first trees, amphibia insects
					Invasion of land by plants & arthropods; primitive vertebrates appear

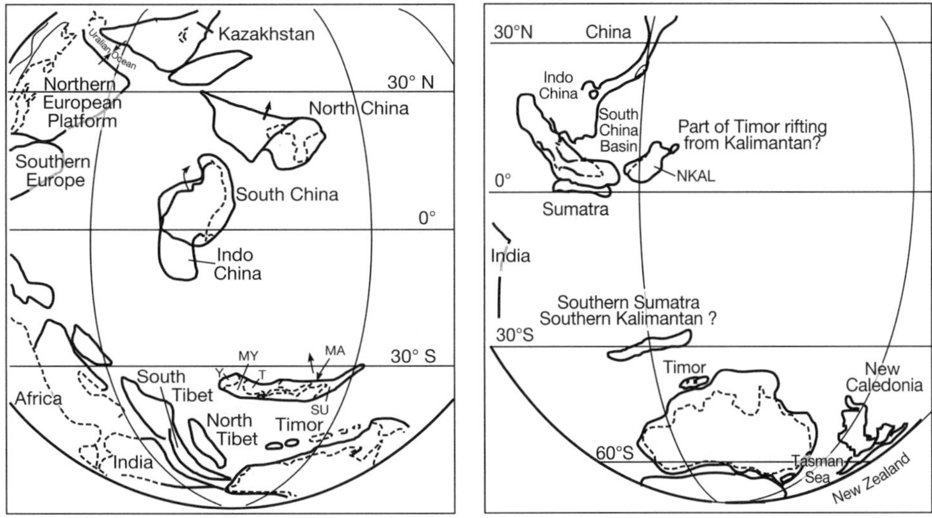

Figure 1.1. Geographical distributions of continents in the Jurassic period (±160-140 million years ago).

After Smith and Briden 1977

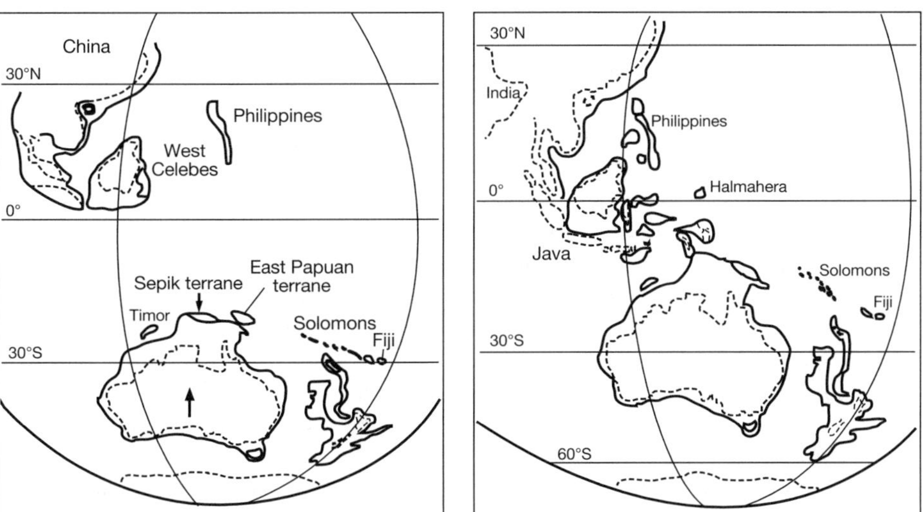

Figure 1.2. Geographical distributions of continents in the Palaeocene period (±70 million years ago).

After Smith and Briden 1977

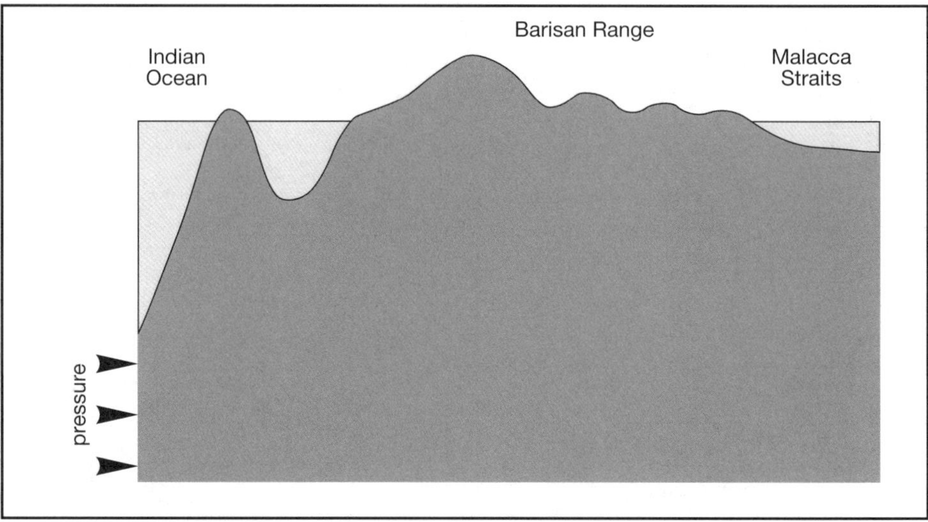

Figure 1.3. A generalised cross-section through Sumatra and the western islands to show the folding caused by the Indian Plate colliding with the Asian Plate.

up to 180 million years ago. The Barisan Range began to be lifted and formed in the early Palaeocene 60 million years ago, but between the Oligocene and Miocene epochs, 35 million years later, there was considerable subsidence. Sedimentary rocks laid down by the sea then covering much of Sumatra can be found today in west and east Sumatra and some places inland. At about that time andesitic vulcanism was important in south Sumatra. A major upthrust occurred 20 million years ago and this mountain building activity was accompanied by considerable faulting and violent volcanic activity. No significant faulting has occurred since then. A final period of mountain building activity occurred between the Pliocene and Pleistocene about 3 million years ago, and some faulting, block-faulting, creation of rift valleys and horizontal offsetting of land accompanied this. The Semangko Fault Zone, traceable from Semangko Bay in the south to Weh Island in the north, is the largest. At the same time there was sedimentation and folding in eastern Sumatra, generally parallel to the island's axis (Verstappen 1973).

The volcanoes of the Quaternary period (the last million years) are located in the mountainous areas and are usually associated with faults.

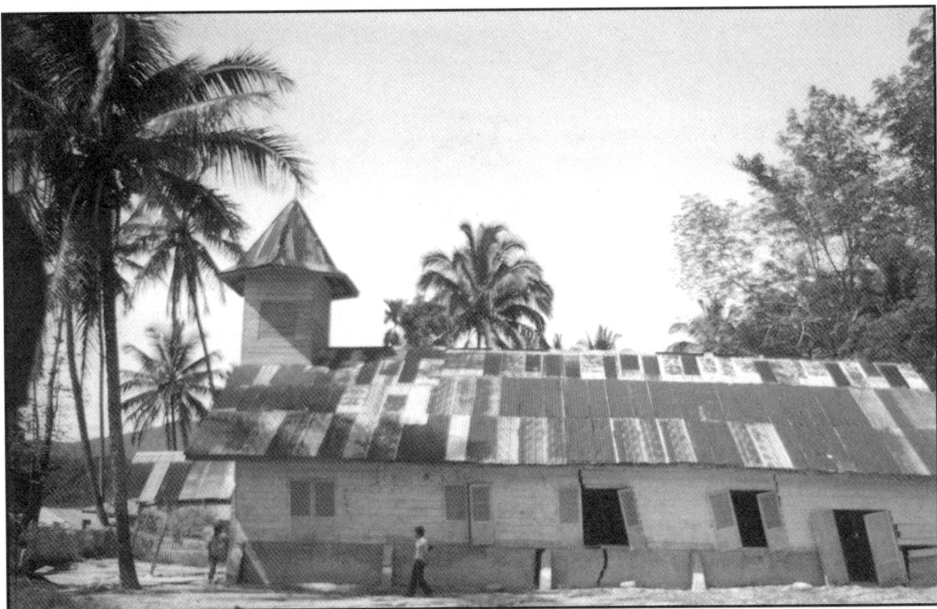

Figure 1.4. A church damaged by an earthquake near Lake Toba in 1983.
A.J. Whitten

Ash and other products from these volcanoes cover great areas of the higher Barisan area, particularly in the south. In general, the most recent volcanoes are in the northern half of Sumatra – Lembuh in Aceh; Sinabung, Sibayak and Sorikmerapi in North Sumatra; Merapi and Talong in West Sumatra – but volcanic activity continues at Kaba near Pasemah, hot springs occur in parts of Bengkulu, and the most famous of contemporary volcanic eruptions (1883) occurred at Krakatau (p. 343). In the same way that the andesitic products of the Krakatau eruption fertilised many of the previously infertile areas of Lampung, so the products of the (geologically) recent volcanic activity northwest of Lake Toba flowed both south and far to the east to give fertility to the Karo highlands and the plantation areas, respectively.

One of the greatest ever volcanic eruptions occurred 75,000 years ago when the formation of Lake Toba began. This is very recent in geological terms and stone-age axes have been found below the 'ash' that reached Peninsular Malaysia. The deep Semangko Fault allowed molten material to

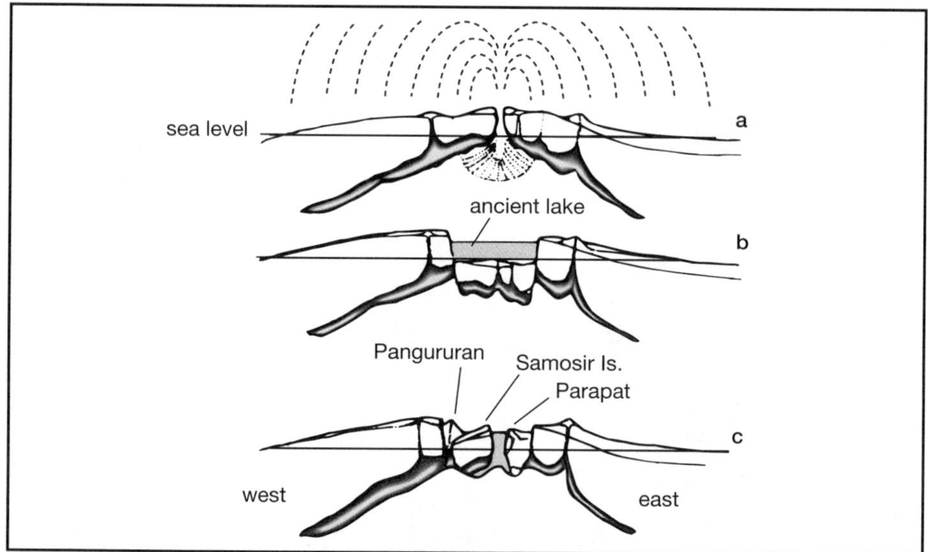

Figure 1.5. The formation of Lake Toba. (a) After pressure caused the earth's crust to bend upwards, a major vent opened and the major eruption occurred. (b) The loss of so much magma caused the volcanic cone to be forced up again, and a minor explosion occurred. (c) The second cone broke in two, forming Samosir Island and the Parapat Peninsula.

After Hehuwat 1982

flow up towards the earth's surface. This created enormous pressures and the land rose up as a result between where the Wampu and Baruman Rivers now flow and formed the Batak Tumour. Eventually the tumour exploded and the 'ash', or Toba tuff, ejected during the eruptions spread over 20,000-30,000 km^2, reaching Sri Lanka, the Bay of Bengal and the Andaman Islands. In some areas near Lake Toba the tuff is 600 m thick. About 1,500-2,000 km^3 of material was ejected, the greatest volume known from any explosion, and it has been estimated that over 1,000 km^3 of this ash was deposited in just nine days (Francis 1983). With so much material removed, the top of the volcano collapsed, forming a huge crater. A second minor series of eruptions built up a second volcano inside the crater, and when this too collapsed the volcano split into two halves – the western slope forming Samosir Island and the eastern slope forming the peninsula between the towns of Parapat and Porsea (fig. 1.5). The second explosion probably occurred as recently as 30,000 years ago (Francis 1983). The Toba crater and island are correctly called a 'resurgent caldera', and at 100 km long, Toba is easily the largest such in the world. For comparison, the extensively reported explosion of Mount St. Helen's in Washington State, U.S.A., ejected only 0.6 km^3 of material and left a crater only 2 km in diameter (Francis 1983). Many of the lakes in Sumatra such as Maninjau and Ranau occupy the tops of old volcanoes (Hehuwat 1982).

Soils

The first ever soil map produced for a tropical region was of part of Sumatra (*Groondsoortenkart van een gedeelte van Deli*), published in 1901 (Young 1976). Soil maps of various scales now exist for most of the island, and convenient maps with text have been published by BPPP/Soeprapto-hardjo et al. (1979) and Scholz (1983). A detailed description of the soil types will not be attempted here. Descriptions, uses and limitations of the various soils can be found in the above references.

The eastern fringe of Sumatra and the land on either side of the major rivers are dominated by hydromorphic soil, frequently with alluvial soils or grey hydromorphic soils (so important for the plantation industry) behind them. The swampy eastern portions of Riau, Jambi and South Sumatra are dominated by organosols, but these soils also occur in parts of southeast and south North Sumatra, West Aceh, northwest and south West Sumatra. The majority of lowland Sumatra is covered by yellow podzolic soils formed from a variety of parent materials. The soils of the mountainous areas have very complex distributions but, again, various forms of red-yellow podzolic soils associated with altosols or litosols predominate. Over the limestone areas brown podzolic and renzina soils occur, and andosols and brown-grey podzolic soil appears over the volcanic rocks. The western coast is fringed with sandy regosols for most of its length. Both the western and eastern islands are largely overlain by red-yellow podzolic soils, but the soils on the eastern islands are associated with podzols overlying Tertiary sandstone and with litosols over the intrusive granite.

Table 1.2 summarises the distributions of the different soil types.

Table 1.2. The area of different soil types in each province and in the whole of Sumatra.

Soil type	Aceh Ha	Aceh %	North Sumatra Ha	North Sumatra %	West Sumatra Ha	West Sumatra %	Riau Ha	Riau %	Jambi Ha	Jambi %	South Sumatra Ha	South Sumatra %	Bengkulu Ha	Bengkulu %	Lampung Ha	Lampung %	Sumatra Ha	Sumatra %
1. Organosol	436,508	7.9	402,359	5.7	435,542	8.7	4,827,972	51.0	866,986	19.3	1,966,035	19.0	4,246	0.1	—	—	8,985,678	19.0
2. Organosol and Gley humus association	—	—	—	—	—	—	145,542	1.5	32,580	1.5	32,340	0.3	—	—	—	—	210,463	0.5
3. Hydromorphic alluvium	66,689	1.2	216,905	3.1	—	—	498,857	5.2	141,179	3.1	1,363,846	13.2	—	—	163,444	4.9	2,450,920	5.2
4. Alluvium	277,869	5.0	178,947	2.5	158,930	3.2	—	—	319,464	7.1	588,807	13.2	161,826	7.6	52,300	1.6	1,598,224	3.4
5. Alluvium and Gley humus association	—	—	46,635	0.6	47,315	0.9	39,146	0.4	64,255	1.4	353,507	3.4	—	—	290,218	8.7	841,076	1.8
6. Grey Hydromorphic	145,503	2.6	412,120	5.8	—	—	—	—	—	—	107,056	1.0	—	—	79,627	2.4	744,306	1.6
7. Grey Hydromorphic and regosol association	—	—	80,255	1.1	—	—	—	—	—	—	—	—	—	—	—	—	80,255	0.1
8. Regosol	112,159	2.0	129,059	1.8	420,984	2.7	—	—	2,715	0.1	249,796	2.4	8,503	0.4	80,674	2.4	807,573	1.7
9. Andosol	205,118	3.7	244,019	3.5	420,984	8.5	—	—	293,219	6.5	162,814	12.4	262,339	12.4	209,544	6.3	1,798,037	3.8
10. Andosol and regosol association	—	—	54,226	0.8	—	—	—	—	—	—	—	—	—	—	—	—	54,226	0.1
11. Brown forest soil	—	—	81,339	1.1	—	—	—	—	—	—	10,036	0.1	3,015	0.2	—	—	94,390	0.2
12. Renzina	238,462	4.3	78,086	1.1	—	—	—	—	—	—	16,728	0.1	11,056	0.5	8,382	0.2	352,714	0.2
13. Brown podzolic	—	—	156,172	2.2	—	—	—	—	—	—	215,226	2.0	—	—	31,432	0.9	402,830	0.9
14. Grey-brown podzolic	—	—	339,457	4.8	—	—	—	—	—	—	—	—	—	—	—	—	339,457	0.7
15. Lateritic	—	—	—	—	—	—	—	—	—	—	—	—	—	—	8,382	0.2	8,382	0.1
16. Latosol	306,162	5.5	471,769	6.7	1,185,306	23.8	26,097	0.2	712,231	15.8	419,302	4.0	501,559	23.7	719,783	21.6	4,342,209	9.2
17. Latosol and regosol association	—	—	—	—	—	—	—	—	—	—	4,461	4.0	—	—	—	—	4,461	9.2
18. Latosol and andosol association	—	—	—	—	66,727	1.3	3,011	0.1	35,295	0.8	68,025	0.7	—	—	—	—	173,058	0.3
19. Latosol and red-yellow podzolic association	—	—	36,874	0.5	57,021	0.1	—	—	19,910	0.4	37,916	0.3	255,302	12.1	97,438	2.9	504,461	1.1
20. Red-yellow podzolic	1,458,056	26.3	1,980,343	28.0	1,023,949	20.5	3,162,773	33.4	1,397,314	31.0	2,968,568	28.6	309,579	14.6	1,522,336	45.7	13,522,918	29.2

Table 1.2. (Continued.) The area of different soil types in each province and in the whole of Sumatra.

Soil type	Aceh Ha	Aceh %	North Sumatra Ha	North Sumatra %	West Sumatra Ha	West Sumatra %	Riau Ha	Riau %	Jambi Ha	Jambi %	South Sumatra Ha	South Sumatra %	Bengkulu Ha	Bengkulu %	Lampung Ha	Lampung %	Sumatra Ha	Sumatra %
21. Red-yellow podzolic and grey hydromorphic	—	—	30,367	0.4	—	—	—	—	—	—	761,658	7.3	—	—	—	—	792,025	1.7
22. Red-yellow podzolic and regosol association	—	—	175,693	2.5	—	—	—	—	—	—	—	—	—	—	—	—	175,693	0.4
23. Red-yellow podzolic and litosol association	—	—	—	—	—	—	116,433	1.2	—	—	—	—	—	—	—	—	175,693	0.4
24. Red-yellow podzolic and litosol association	—	—	—	—	—	—	355,323	3.7	—	—	907,744	8.7	—	—	—	—	1,263,067	2.7
25. Red-yellow podzolic, latosol and litosol complex	—	—	156,171	2.2	—	—	40,149	0.4	184,619	4.1	53,528	0.5	22,113	1.0	67,054	2.0	523,634	1.1
26. Red-yellow podzolic, latosol and litosol complex	1,436,837	25.9	967,397	13.7	922,040	18.5	240,897	2.5	337,563	7.5	—	—	—	—	—	—	3,904,734	8.2
27. Red-yellow podzolic, brown and grey podzolic and litosol complex	12,125	0.2	10,845	0.1	12,132	0.2	—	—	—	—	28,994	0.2	141,273	6.7	—	—	205,819	0.4
28. Litosol, latosol and regosol complex	—	—	—	—	—	—	—	—	—	—	40,146	0.4	—	—	—	—	40,146	0.1
29. Brown podzolic, podzol and litosol complex	843,712	15.2	386,091	5.5	505,908	10.1	—	—	85,070	1.9	12,267	0.1	263,344	12.4	—	—	2,096,392	4.4
30. Other	—	—	443,571	6.3	7,279	0.1	—	—	—	—	—	—	38,195	1.8	—	—	489,045	1.0
Total	**5,539,200**		**7,078,700**		**4,977,800**		**9,456,200**		**4,492,400**		**10,368,800**		**2,116,800**		**3,330,700**		**47,360,600**	

Adapted from BPPP/Soepraptohardjo et al. 1979

CLIMATES

Palaeoclimate

During the last two million years there has been a worldwide alternation of colder and warmer climates. The colder periods, or glacials, were associated with lower sea levels (fig. 1.6) because cooling of areas that are now temperate locked up a great deal of water as ice, thus preventing its return to the sea (fig. 1.7). Minor changes in sea level have probably continued up to the present (fig. 1.8). The maximum sea level during the last interglacial was only 3 to 6 m higher than the present level (Jancey 1973; Haile 1975; Geyh et al. 1979) but Tjia (1980) reports sea levels up to 50 m higher than present at various times during the last two million years. The minimum sea level was about 180 m lower than at present, more or less outlining the Sunda Shelf and exposing three times more land in the Sunda Region than is visible now. The Sunda Shelf is an extension of the Asian continent (fig. 1.9) and lies between 40-200 m below present sea level. When exposed, the shelf was bisected by the Great Sunda River, which arose between Belitung Island and Borneo and flowed northeast between the North and South Natuna Islands (Tjia 1980). The extremes of sea level did not last very long and for most of the Quaternary the sea lay only some 30-80 m below present sea level.

There is no information about climates for any low-water (glacial) period other than the most recent, about 15,000 years ago. It has been estimated that the regional sea surface temperature was only about 2° to 3°C below today's temperature (Climap 1976) so that the lowlands, at least, remained quite warm. On the highest mountains, however, where ice very occasionally forms today, large glaciers formed (Walker 1982). Until recently, glaciers had been confirmed only on Mt. Kinabalu (Sabah) and on various peaks in New Guinea. Now, however, it has been shown that glaciers once existed on Mt. Leuser, Mt. Kemiri and Mt. Bandahara in the Mt. Leuser National Park (van Beek 1982). The largest of these, on Mt. Leuser, measured about 100 km^2 and it is calculated that the temperature at the snowline (then about 3,100 m) was 6°C colder than now. When the glaciers melted, peat began to form and the oldest peat at the bottom of a small lake near the summit of Mt. Kemiri has been given a radiocarbon date of 7,590 ± 40 years before present.

During low-water periods the greater land area and consequent deflection of ocean currents changed weather patterns and air circulation, with the result that rain fell more seasonally and total annual rainfall almost certainly decreased (Walker 1982).

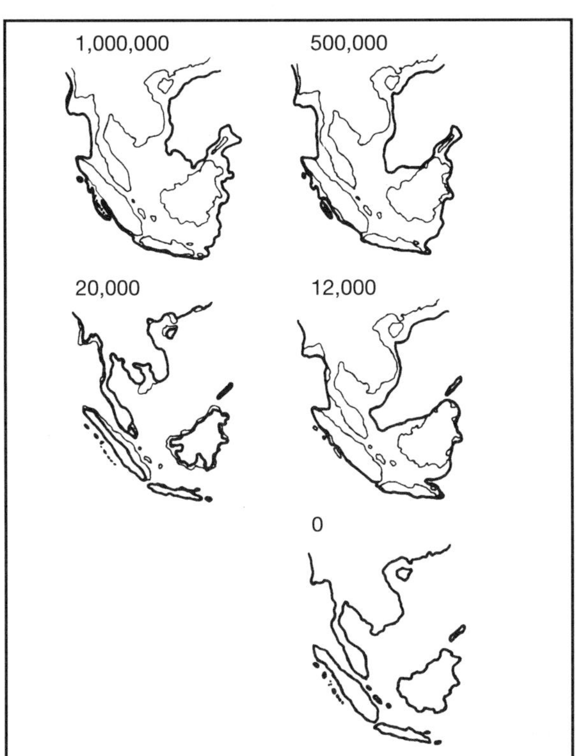

Figure 1.6. The location of past (thicker lines) and present (thinner lines) coastlines. Number represents years before present.

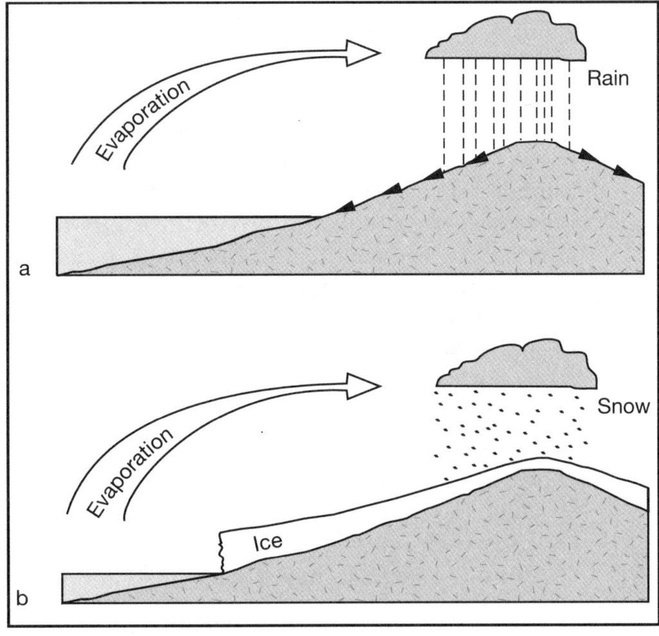

Figure 1.7. Hydrological cycles (a) in warm conditions, and (b) in cold conditions. Note the fall in sea level because water cannot flow to the sea.

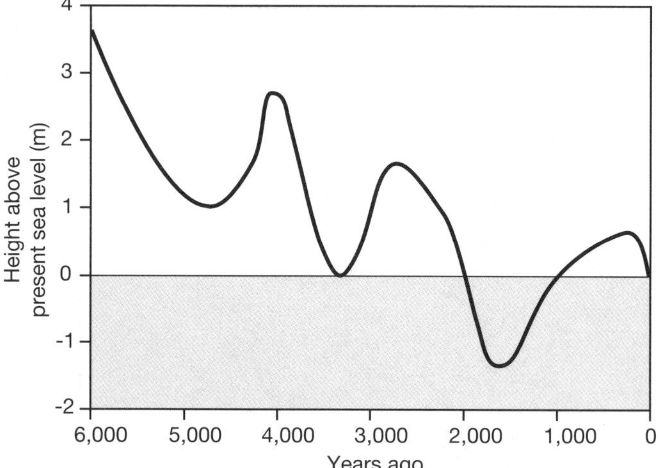

Figure 1.8. Changes in sea level during the last 6,000 years.

After Tjia 1980

Present Climate

The climate of Sumatra is characterised by abundant rainfall moderately well distributed through the year, with the wet and dry seasons much less clearly defined than in Java and eastern Indonesia.

Rainfall. Rainfall in Sumatra is very variable, dependent somewhat on the topography, and ranges from over 6,000 mm per year in areas to the west of the Barisan Range, to less than 1,500 mm per year in some areas of the east, which are blocked from humid winds by the Barisan Range and the Malay Peninsula. Out of 594 rainfall stations in Sumatra, 70% have annual rainfall of over 2,500 mm. Over much of the island the driest months are normally associated with the northeasterly monsoon between December and March, and the main rainy season usually falls during the transition period before the northeasterly monsoon and after the southwesterly monsoon which last from May to September. A secondary rainy period occurs in about April which is after the northeasterly and before the southwesterly monsoon. Southern Sumatra, however, has a single pronounced dry season around July, and the northern tip of Aceh has a pronounced dry season in February (Scholz 1983). Figure 1.10 shows the distribution of climatic zones, the definitions of which are as follows:

Zone A—more than the nine consecutive wet months, and two or less consecutive dry months.

Zone B—seven to nine consecutive wet months and three or less consecutive dry months.

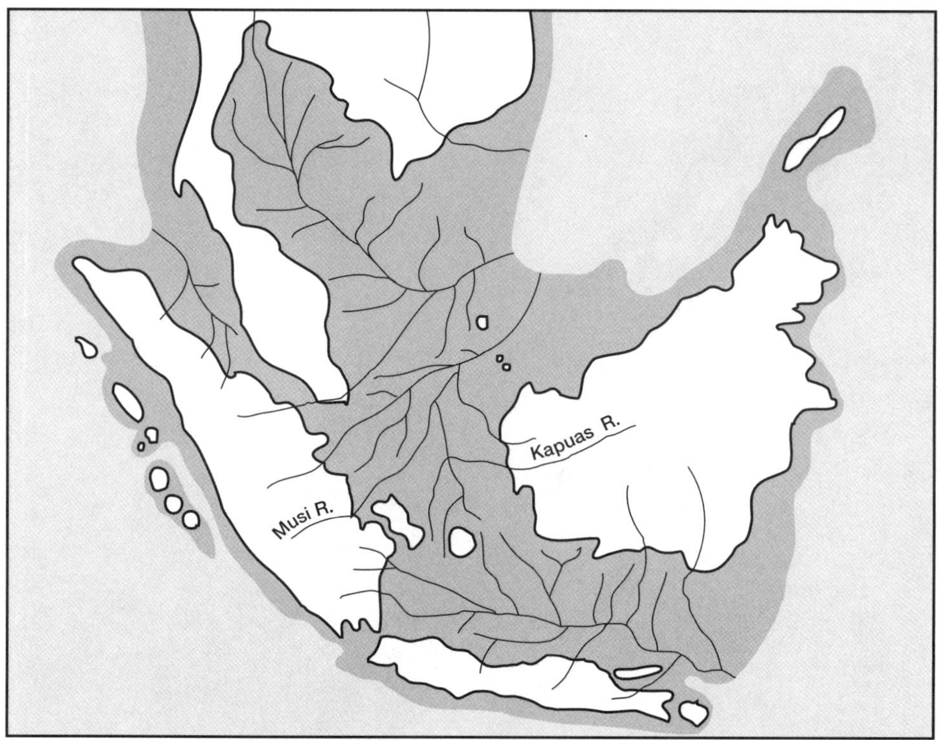

Figure 1.9. The Sunda Shelf showing greatest area of land exposed during the driest periods of the last two million years (shaded area), major rivers and present-day coastlines.
After Tjia 1980

Zone C—five to six consecutive wet months and three or less consecutive dry months.

Zone D—three to four consecutive wet months and two to six consecutive dry months.

Zone E—up to three consecutive wet months and up to six consecutive dry months.

"Wet" and "dry" are defined as more than 200 mm rainfall/month and less than 100 mm/month, respectively (Oldeman et al. 1979).

An important point to note is the considerable difference between the climates of Sumatra and Java (table 1.3). On Sumatra 71% of the land area has seven or more consecutive wet months and up to three consecutive dry months, whereas only 27% of Java experiences similar conditions (Oldeman et al. 1979).

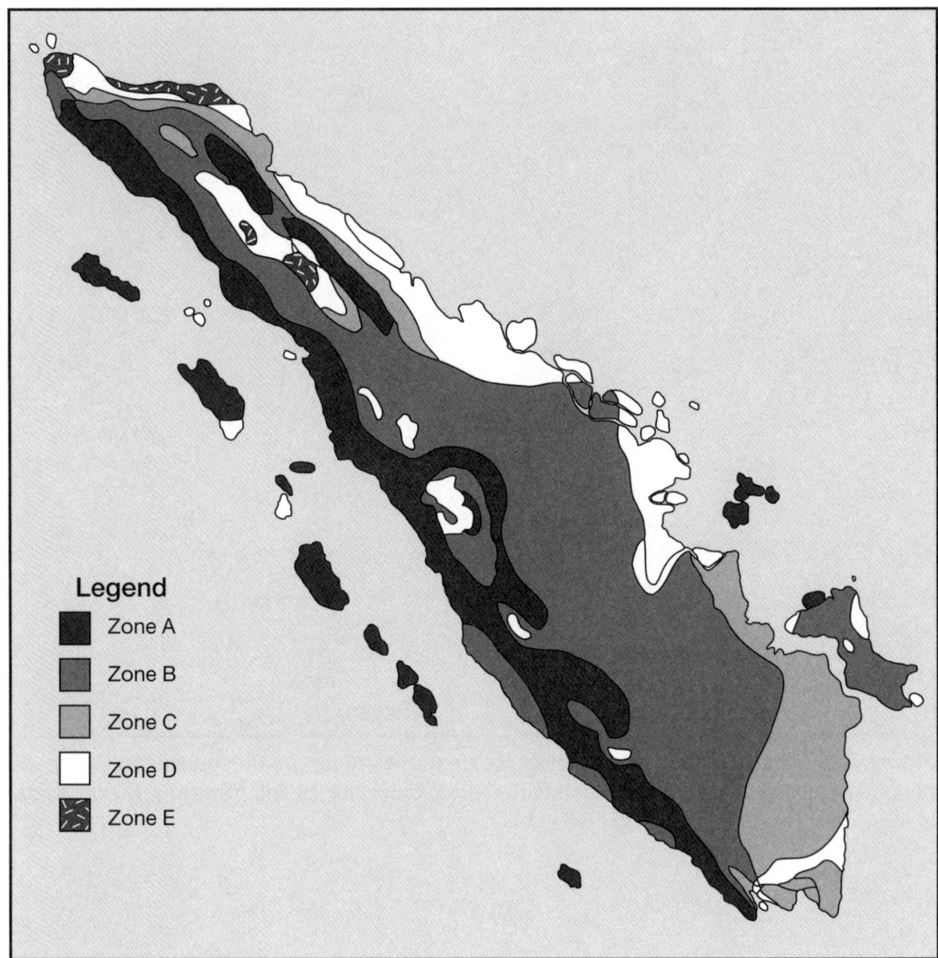

Figure 1.10. Climatic zones.

After Oldeman et al. 1979

Table 1.3. Percentage of land area on Sumatra and Java in five agroclimatic zones.

	Climatic zones				
	A	**B**	**C**	**D**	**E**
Sumatra	24	47	15	12	2
Java	4	23	39	25	9

After Oldman et al. 1979

Temperature. The annual fluctuations in temperature are very small for almost all locations in Sumatra. Daily variations are greatest in the drier months when the Sun is not obscured by clouds and when the heat escapes quickly at night because there are no clouds to insulate the earth. The differences in temperature between locations are caused mainly by altitude, and this is discussed further on page 277.

Wind. There is great variability in the winds but in general they blow from the north from December to March, and from the south from May to September. Where there are breaks in the Barisan Mountains, monsoon winds are channelled and wind speed may be greatly increased. Some westerly winds of Sumatra are particularly famous. For example:

Angin Depek—at Air Tawar near Takengon between April and October. This stormy wind is named after a fish that comes to the surface at this time and is easily caught.

Angin Bohorok—in areas of North Sumatra. It is not feared because of its strength but because it is very dry and can damage crops, particularly tobacco. Angin Bohorok is a föhn wind, that is, a depression (low atmospheric pressure) moving to the east of the Barisan Range, draws air into itself from the west of the range. The air ascends the western slopes, clouds form and there is heavy rain. When it reaches the eastern slopes, the air is still warm, but it has lost most of its moisture, so it blows down as a very warm, dry wind. It starts in the Karo highlands and is funnelled into upper Langkat and upper Deli Serdang through the Bohorok depression.

Angin Padang Lawas—east of Padang Sidempuan is a particularly serious föhn wind, drying the soil and making most forms of agriculture virtually impossible (Oldeman et al. 1979).

VEGETATION

Palaeovegetation

The oldest Sumatran forests for which we have evidence are those that formed the coal seams during the Carboniferous period, about 300 million years ago. Coal is the highly compressed remains of undecayed swamp vegetation, sandwiched between layers of marine sedimentary deposits. These primeval swamp forests covered much of the earth many millions of years before the first gymnosperm (e.g., conifer) or angiosperm (flowering) trees evolved. The earliest 'trees' were in fact primitive ferns, and some, such as the giant club-moss *Lepidodendron*, reached 30 m in height. By the

Tertiary many of the plants now found in forests had developed; for example, a leaf imprint of a *Dipterocarpus* leaf and a fossil dipterocarp fruit have been found in southern Sumatra dating from this era (Ashton 1982).

The tropical rain forests of today have been commonly regarded as relics from the Tertiary period, a kind of living museum, and phrases such as 'unchanged for thousands of years' are often written. Evidence from many tropical regions now show that this is wrong (Flenley 1980a,b). The climate has changed during at least the last 30,000 years (p. 12), both in temperature and rainfall, and this in turn affected the vegetation.

In the drier, more seasonal periods of the Pleistocene the areas with a wet, non-seasonal climate would have contracted as the area of seasonal climate extended, but the exact amount of change is unknown. Populations of plants restricted to the wet, non-seasonal forest would therefore have had their ranges constricted, but contact between populations could have been maintained to indeterminate degrees by occasional transfer between patches of suitable forest in the seasonal areas, as indeed occurs today in parts of Java (Whitmore 1982). It is possible that monsoon forest and perhaps even dry savannah forest (Whitmore 1984) would have occupied some of Sumatra and other parts of the Sunda Region. There is, however, no evidence from the few studies of preserved pollen from lowland sites in the region, of any replacement of non-seasonal forest by seasonal forest during or since the Tertiary (Maloney 1980a,b; Morley 1980, 1982; Whitmore 1982b), but this may yet be revealed by future research.

The palaeovegetation of Sumatra has been studied by investigating fossil pollen preserved in sediments of small lakes in the Kerinci and Toba regions. The sediment core taken from Lake Padang at 950 m south of Lake Kerinci revealed evidence of vegetation changes over the last 10,000 years. About 8,300 years ago an upper montane forest characterized by *Myrica* trees and the conifer *Podocarpus* gave way to a lower montane forest with abundant oaks (Fagaceae), suggesting that as the climate warmed up, so the upper montane species[2] retreated up the mountains (Morley 1980, 1982).

The sediment core taken from Pea Sim-Sim at 1,450 m on the Siborong-borong Plateau south of Lake Toba was found to cover the last 18,500 years. Up to 16,500 years ago a mosaic of subalpine and upper montane forest probably existed on the plateau (Maloney 1980a,b). Oak forest became established about 16,500 years ago and persisted until ± 12,000 years ago. 7,500 years ago montane vegetation becomes less important in the samples and *Eugenia* species, possibly from swamp forest, predominate. From these studies it appears that before the climate began to warm about 8,000-9,000 years ago, the vegetation zones (next section) had been lowered by 350-500 m. A review of the late Pleistocene vegetation in Sumatra has been published by Maloney (1983c).

Present Vegetation

Sumatra supports a wide range of vegetation types and, due to its recent connection to Asia (p. 12), these are very rich in species. For species diversity the Sumatran forests are comparable to the richest forests of Borneo and New Guinea and are richer than those found on Java, Sulawesi and other small islands (Meijer 1981). Sumatra has 17 endemic genera of plants (compared with 41 in Peninsular Malaysia/southern Thailand; 59 in Borneo; and 10 in West Java), and has some unique and spectacular species such as *Rafflesia arnoldii,* the largest flower in the world, and *Amorphophallus titanum,* the tallest flower in the world.

Sumatra supports a broad altitudinal range of vegetation types (chapter 9), and some other distinctive vegetation types characteristic of the soil or topography on which they occur (chapter 8). Dipterocarp trees such as *Shorea, Dipterocarpus* and *Dryobalanops* dominate much of the tall lowland forest, forming a continuous canopy, and are capable of producing over 100 cubic metres of high-quality logs per hectare (chapter 7). The peat swamps of the east coast show characteristic forest formations of great commercial importance and botanical interest (chapter 5). The sandy west coast supports various types of coastal forest (chapter 3), but the east coast is dominated by very extensive areas of mangrove forest (chapter 2). In the south, large peaty freshwater swamps are found (chapter 6). These vegetation types are summarised in table 1.4, and the distribution of the natural vegetation (i.e., not disturbed by man) of Sumatra is shown in figure 1.11.

The bibliographies by Tobing (1968) and Jacobs and de Boo (1983) are very useful as preliminary means of exploring the older literature on Sumatran vegetation and plants.

The distribution of vegetation types between provinces is uneven (table 1.5) – for instance, Riau has no montane forest but has the largest areas of peat swamp and mangroves, and West Sumatra has almost no mangroves and little swamp but considerable areas of montane forest.

The natural, remaining and protected areas of different vegetation types are shown in table 1.6 both for 1982 and for 1996. These figures have been calculated from a range of sources such as air photos, satellite images, vegetation maps, and consultative meetings (Laumonier 1997; MacKinnon 1997). The largest percentage drops are for heath forest and freshwater swamp, the first of which is unlikely to lead to any productive land use (see p. 253). The loss of mangroves is due to widespread conversion to coastal aquaculture. Also of major significance is the large loss of freshwater swamp probably to irrigated agriculture. A sense of these losses can be gained from the forest cover maps from 1932 and 1982 (fig. 1.12) and from 1996 (fig. 1.13). The extreme patchiness of lowland forests is shown in figure 1.14. Considering the fires of 1997-98 the current figures, if available, would show an even greater loss, particularly for peat swamp forest, freshwater swamp forest, and the other inland lowland forests.

Table 1.4. Major vegetation types found on Sumatra.

Soil water	Location	Soils		Vegetation type
Dryland	Inland	Zonal soils	Lowlands, up to 1,200 m	Lowland forest
			Mountains 1,200-2,100 m	Lower montane forest
			Mountains 2,100-3,000 m	Upper montane forest
			Mountains, 3,000+	Subalpine forest
		Podzolized sands	Mostly lowland	Heath forest
		Limestone	Mostly lowland	Forest over limestone
	Coastal			Beach vegetation
Water table high (at least periodically)	Salt water			Mangroves
	Brackish water			Brackishwater forest
	Fresh water	Oligotrophic peats		Peatswamp forest
		Eutrophic (muck and mineral soils)		Freshwater/Seasonal swamp forest

After van Steenis 1957, and Whitmore 1984. For a more detailed division see the chart by Kartawinata 1980

Table 1.5. Approximate areas in 1982 of natural vegetation types in each province (in 1,000 ha). 'Wet', 'Moist' and 'Dry' correspond to Zones A/B, C and D/E in figure 1.10.

Vegetation type	Aceh	North Sumatra	West Sumatra	Riau	Jambi	South Sumatra	Beng-kulu	Lam-pung
Moss forest	80	30	30	—	—	10	10	6
Montane forest on volcanic soils	950	250	230	—	—	—	—	20
Montane forest on limestone	310	120	560	—	350	570	220	170
Montane forest on other rocks/soils	330	1,300	320	—	160	110	220	—
'Wet' lowland forest on:								
volcanic soils	100	360	950	—	780	980	240	240
limestone	1,340	480	710	150	110	—	—	—
alluvium	30	120	80	100	200	690	260	20
other rocks/soils	450	2,130	1,160	3,100	1,020	1,740	1,120	260
'Moist' lowland forest on:								
volcanic soils	70	20	80	—	20	50	10	320
limestone	120	—	160	—	—	—	—	20
alluvium	230	330	—	—	40	140	—	180
other rocks/soils	450	620	320	320	170	440	—	1,640
'Dry' lowland forest on:								
volcanic soils	60	—	—	—	—	—	—	—
alluvium	260	50	—	—	—	—	—	—
other rocks/soils	110	—	20	—	—	—	—	30
Ironwood forest	—	—	—	—	500	820	—	—
Heath forest*	—	—	—	—	—	1,090	—	—
Peat swamp	20	200	—	4,500	840	1,720	—	—
Freshwater swamp on alluvium	—	460	20	380	110	1,640	10	250
Freshwater swamp on other soils	450	400	330	140	20	530	20	150
Mangrove forest	170	200	—	680	170	230	—	20
Beach vegetation	2	2	3	—	—	2	2	2
Totals	**5,532**	**7,072**	**6,973**	**9,450**	**4,490**	**10,362**	**2,112**	**3,328**

*The figure for heath forest may be a considerable overestimate (see p. 253).

After FAO/MacKinnon 1982a

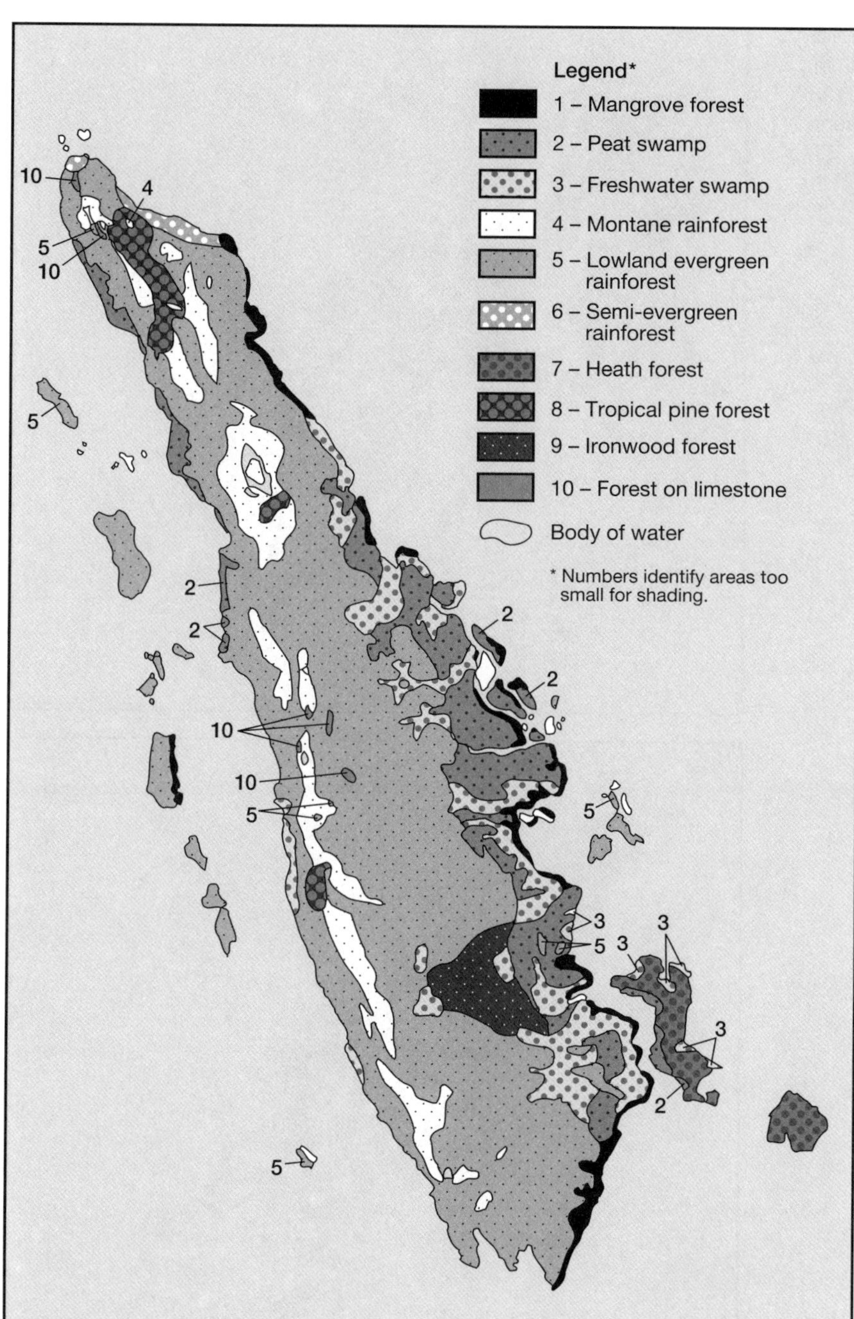

Figure 1.11. Natural vegetation.

After MacKinnon 1997

Figure 1.12.
A) Remaining forest in 1932;
B) remaining forest in 1982.

A) van Steenis 1935;
B) After FAO/MacKinnon 1982a

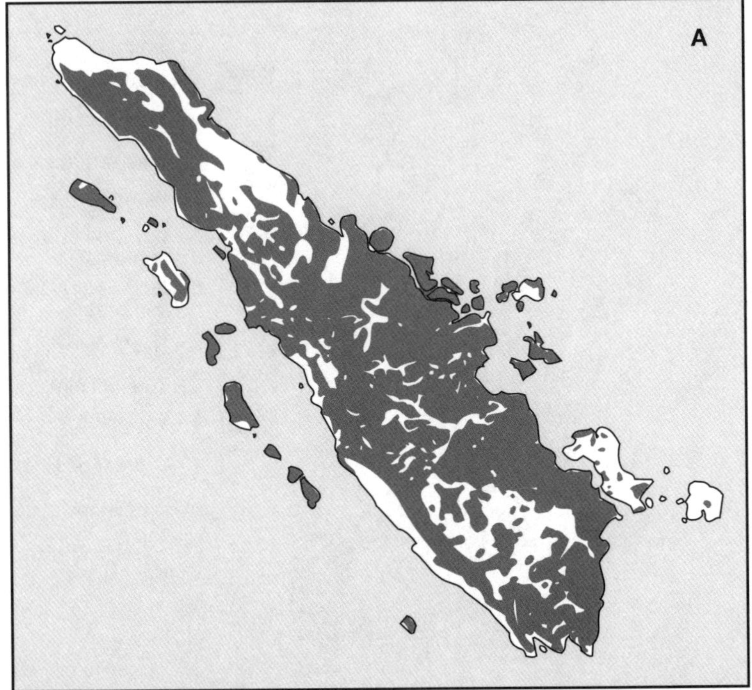

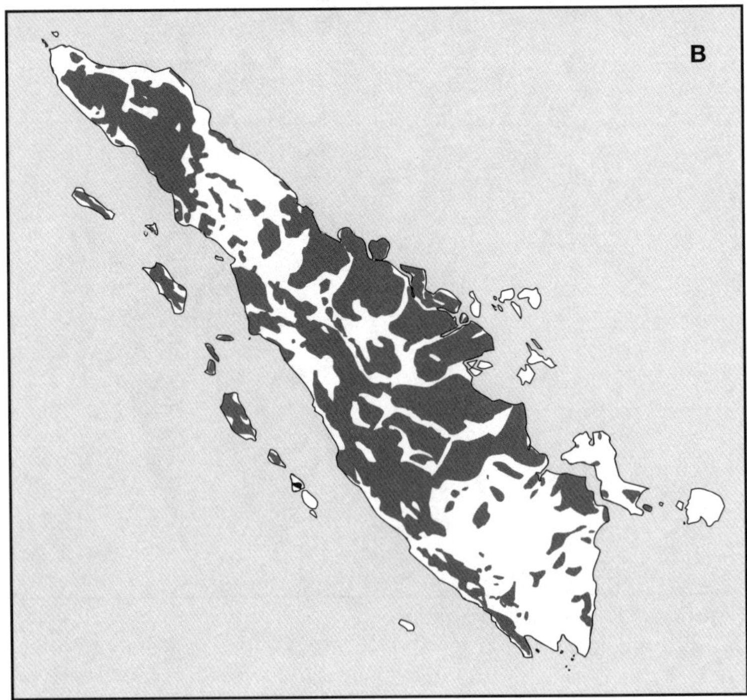

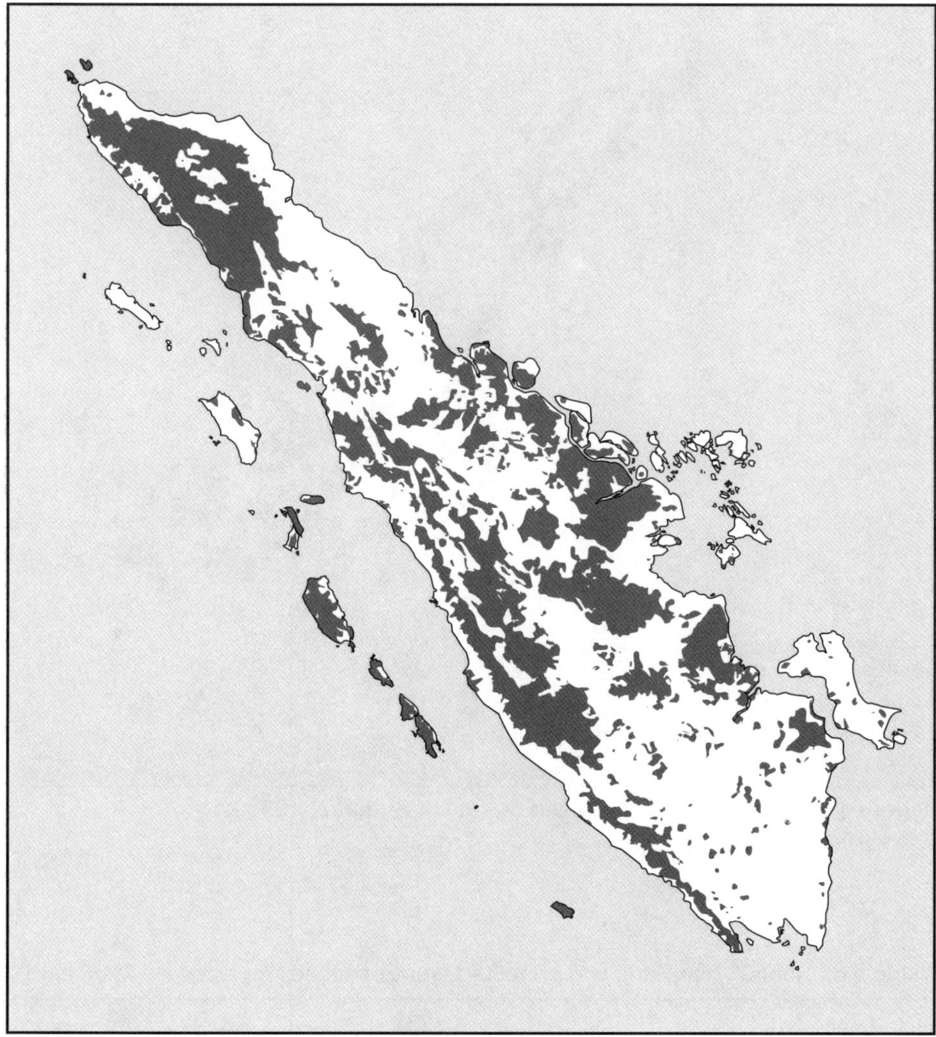

Figure 1.13. Remaining forest in 1996.
After Conservation Science Program—World Wildlife Fund–U.S.A.

The approximate 1996 remaining forest areas, existing reserves and proposed reserves in each of the provinces are shown in figures 1.15 to 1.22. It should be noted that it is very difficult to differentiate selectively logged/disturbed forest from primary forest by analysis of air photos or

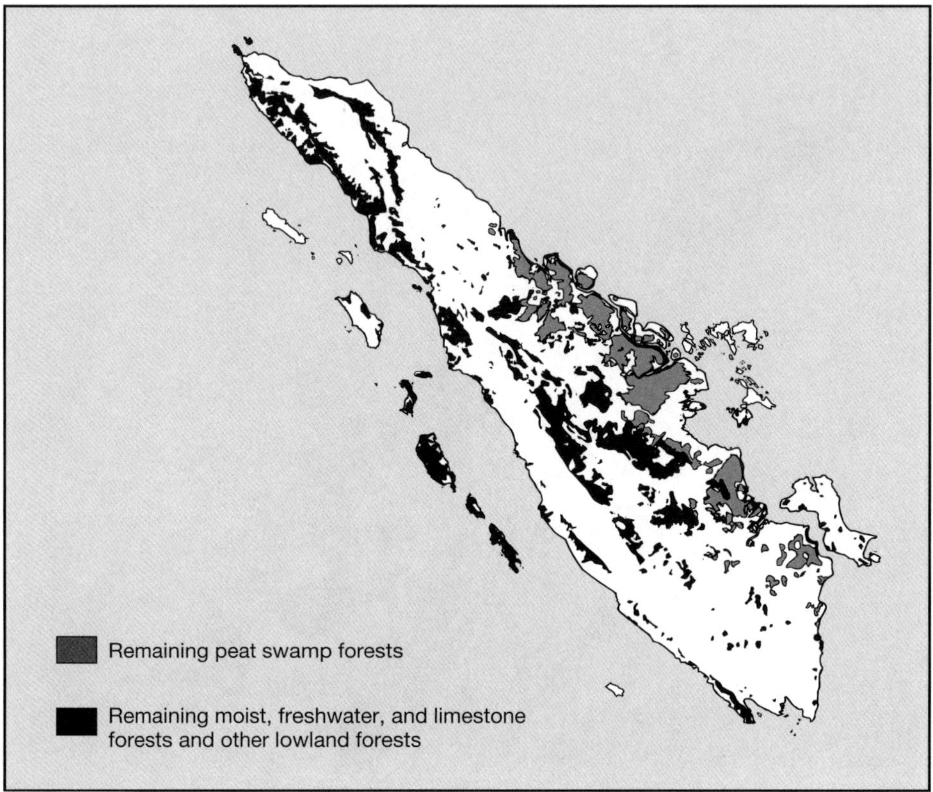

Figure 1.14. The remaining lowland forests of Sumatra, 1996.
After WWF-US

Table 1.6. Natural, remaining and protected areas of vegetation types (in 1,000 ha).

Vegetation type	Natural area	1982 Remaining area (and % of original area)		1996 Remaining area (and % of original area)	
Montane forest	5,680	3,951	(69)	3,426	(60)
Tropical evergreen lowland forest	25,154	8,716	(35)	7,961	(32)
Tropical semi-evergreen lowland forest	540	81	(15)	28	(5)
Ironwood forest	1,320	440	(33)	389	(29)
Heath forest*	1,090	200	(18)	48	(4)
Peat swamp	7,280	4,613	(63)	4,219	(58)
Freshwater swamp	4,910	1,090	(22)	385	(8)
Mangrove forest	1,340	385	(29)	268	(20)
Beach vegetation	13	7	(54)	6	(46)

*The estimate for heath forest may be too high - see p. 253.

After FAO/MacKinnon 1982; MacKinnon 1997

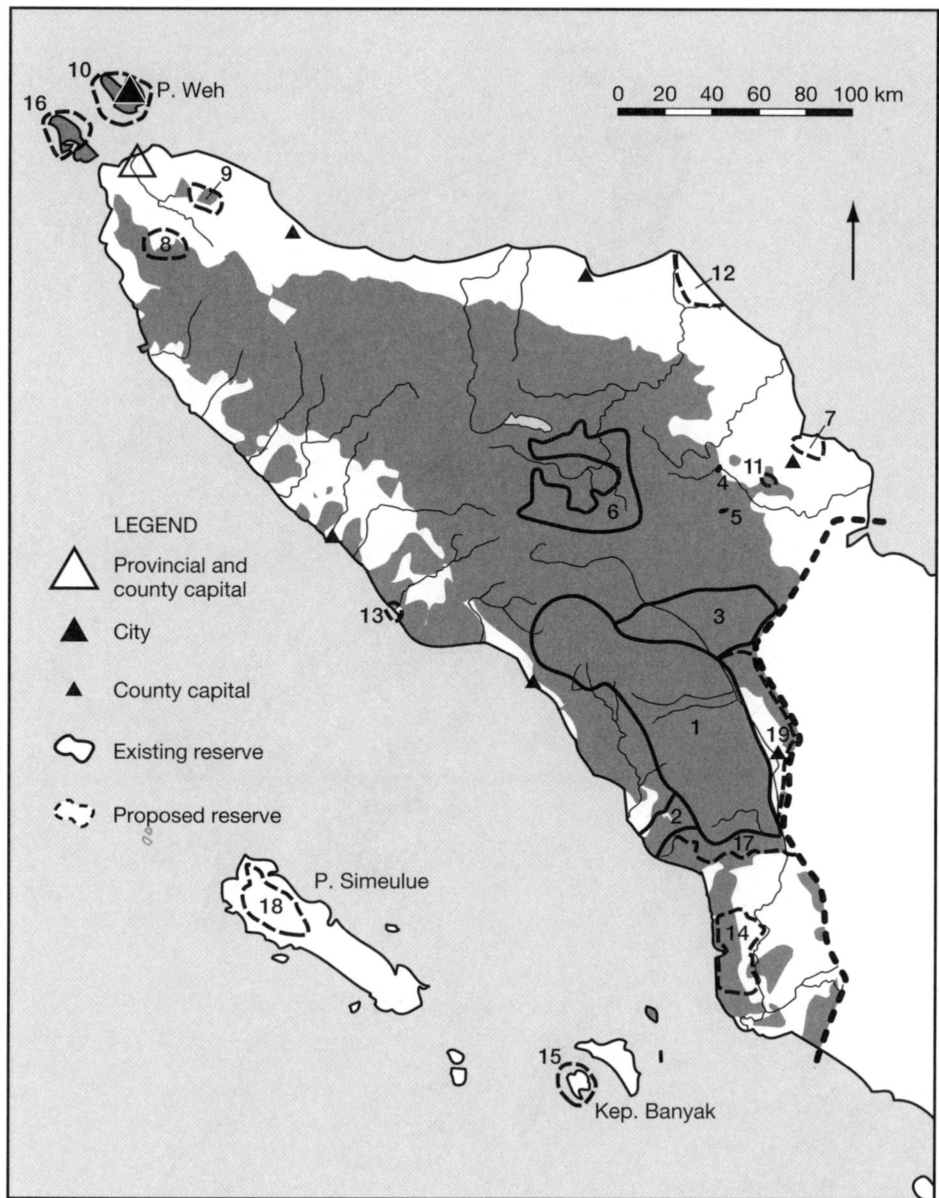

Figure 1.15. Remaining forest, reserves and proposed reserves in Aceh. 1, 2 and 3. Gunung Leuser; 4 and 5. Serbojadi; 6. Lingga Isaq; 7. Kuala Langsa; 8. Jantho; 9. Gunung Seulawah Agam; 10. Aneuk Laut; 11. Langsa Kemuning; 12. Kuala Jambu Air; 13. Rantau Pala Gadjah; 14. Singkit Barat; 15. Pulau Bangkaru; 16. Perairan P. Weh and P. Beras; 17. Perluasan Gunung Leuser Bengkong; 18. Pulau Simeulue; 19. Curah Serbolangit.

After FAO/MacKinnon 1982a

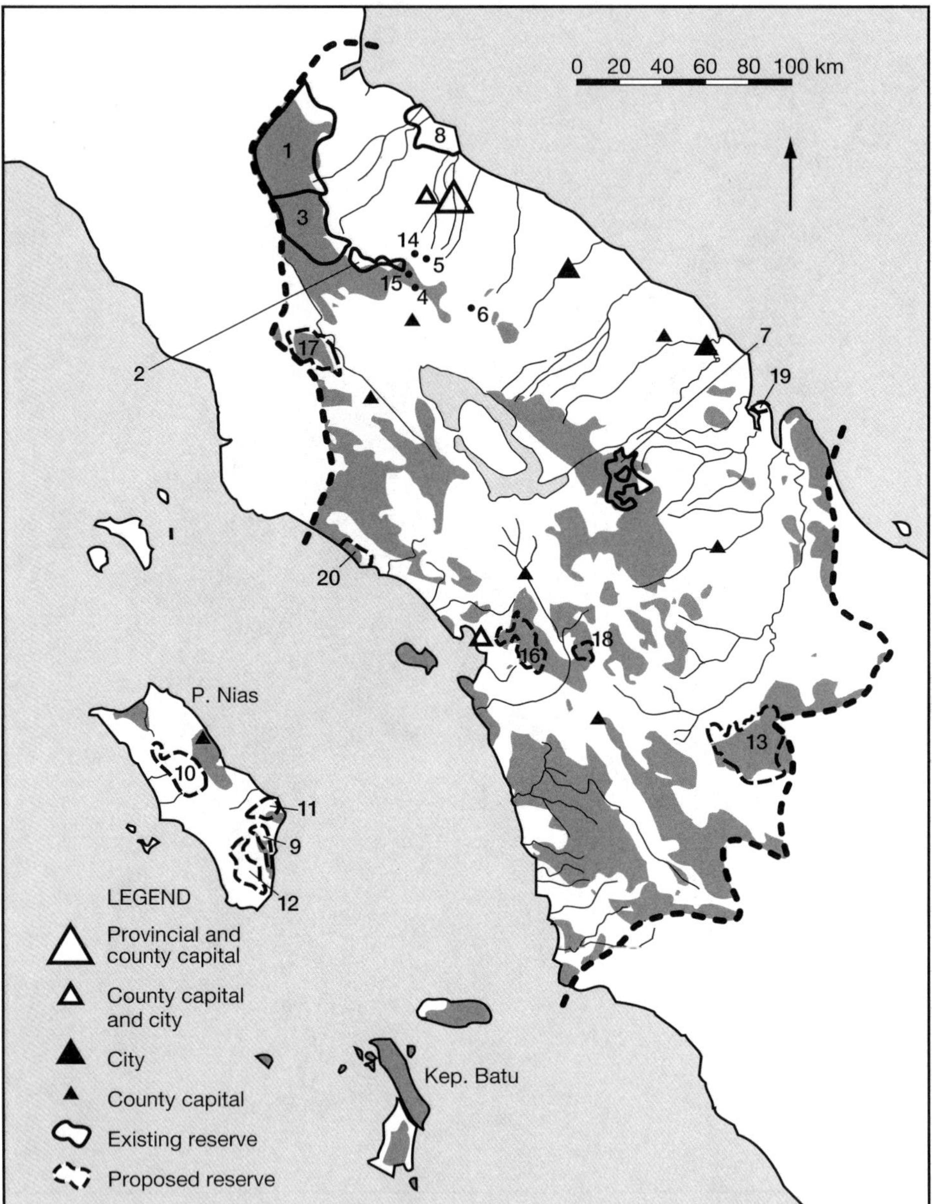

Figure 1.16. Remaining forest, reserves and proposed reserves in North Sumatra. 1. Sekundur; 2. Langkat Selatan; 3. Langkat Barat; 4. Lau Debuk-debuk; 5 and 14. Sibolangit; 8 Karang Gading/Langkat Timur Laut; 6. Dolok Tinggi Raja; 7. Dolok Surungan; 9, 10, 11 and 12. Nias; 13. Padang Lawas; 15. Bandar Baru; 16. Sibolga; 17. Dolok Sembilin; 18. Dolok Sepirok; 19. Sei Prapat; 20. Lau Tapus.

After MacKinnon 1997

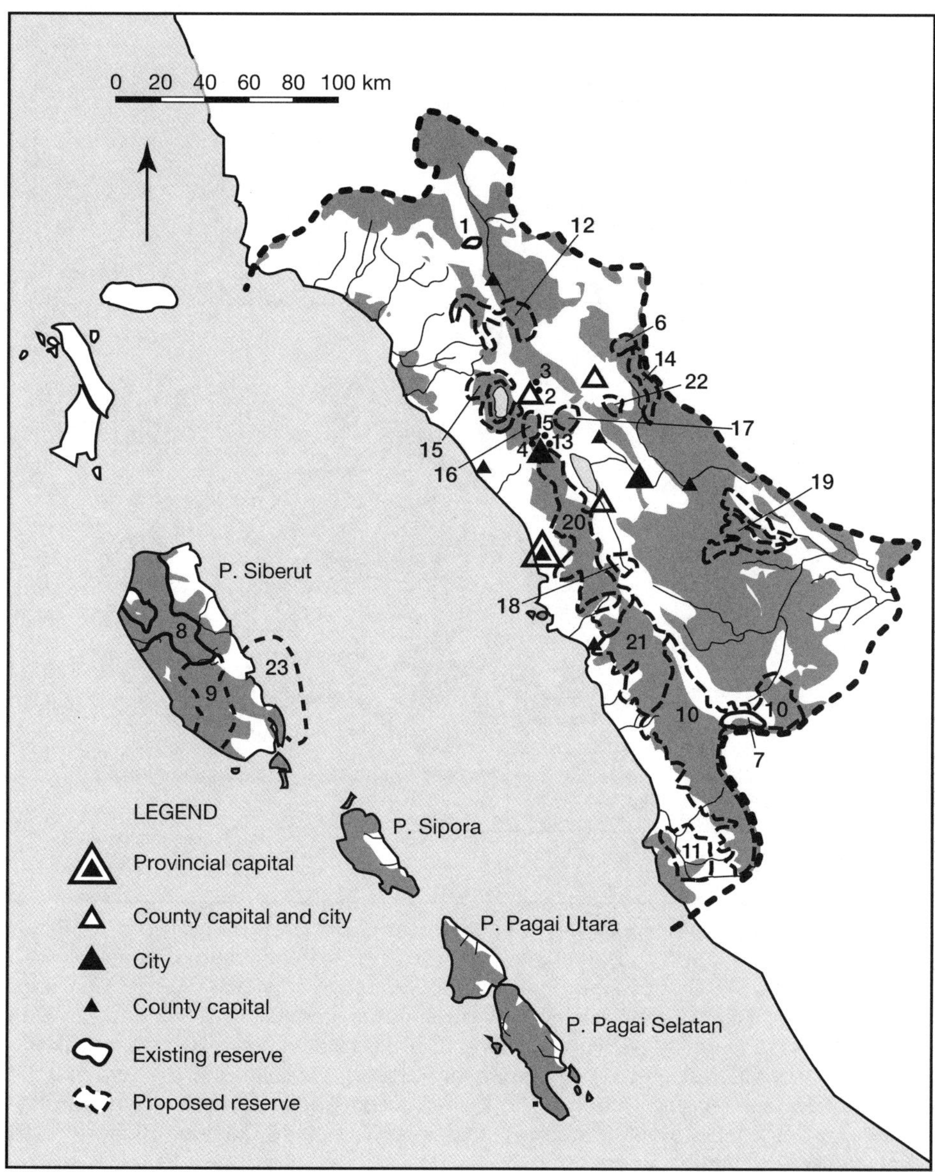

Figure 1.17. Remaining forest, reserves and proposed reserves in West Sumatra. 1. Rimbo Panti; 2. Batang Palupuh; 3. Beringin Sati; 4. Lembah Anai; 5. Megamendung; 6. Lembah Harau; 7 and 10. Indrapura (Gunung Kerinci); 8 and 9. Teitei Batti; 11. Lunang; 12. Malampah - Alahan Panjang; 13. Lembah Anai extension; 15. Danau Maninjau; 16. Gunung Singgalang; 17. Gunung Merapi; 18. Gunung Sulasihtalang; 19. Bukit Sebelah and Batang Pangeran; 20. Bajang Air Tarusan Utara; 21. Kambang/Lubuk Niur; 22. Gunung Sago/Malintang/Karas; 23. Muara Siberut.

After FAO/MacKinnon 1982a

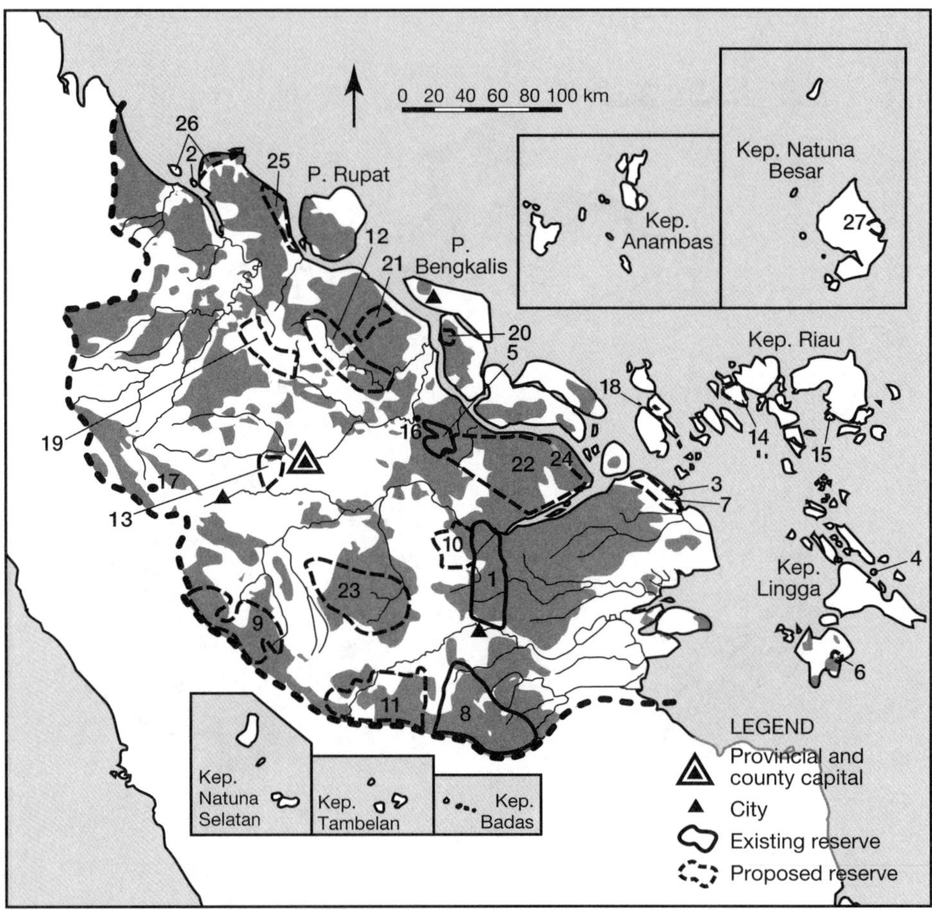

Figure 1.18. Remaining forest, reserves and proposed reserves in Riau. 1. Kerumutan Baru; 2. Pulau Berkeh; 3. Pulau Burung; 4. Pulau Laut; 5. Danau Bawah and Pulau Besar; 6. Bukit Muncung and Sel Gemuruh; 7. Muara Sungai Guntung; 8. Seberida; 9. Bukit Baling Baling; 10. Kerumutan Lama; 11. Peranap; 12. Siak Kecil; 13. Pantai Cermin; 14. Pulau Bulan; 15. Pulau Penyengat; 16. Istana Sulatan Siak; 17. Candi Muara Takus; 18. Pulau Pasar Panjang; 19. Gian Duri; 20. Buaya Bukit Batu; 21. Danau Tanjung Padang; 22. Danau Belat/Besar Sekok/Sarang Burung; 23. Air Sawan; 24. Bakau Muara; 25. Bakau Selat Dumai; 26. Tanjung Sinebu/Pulau Alang Besar; 27. Natuna Besar.

After MacKinnon 1997

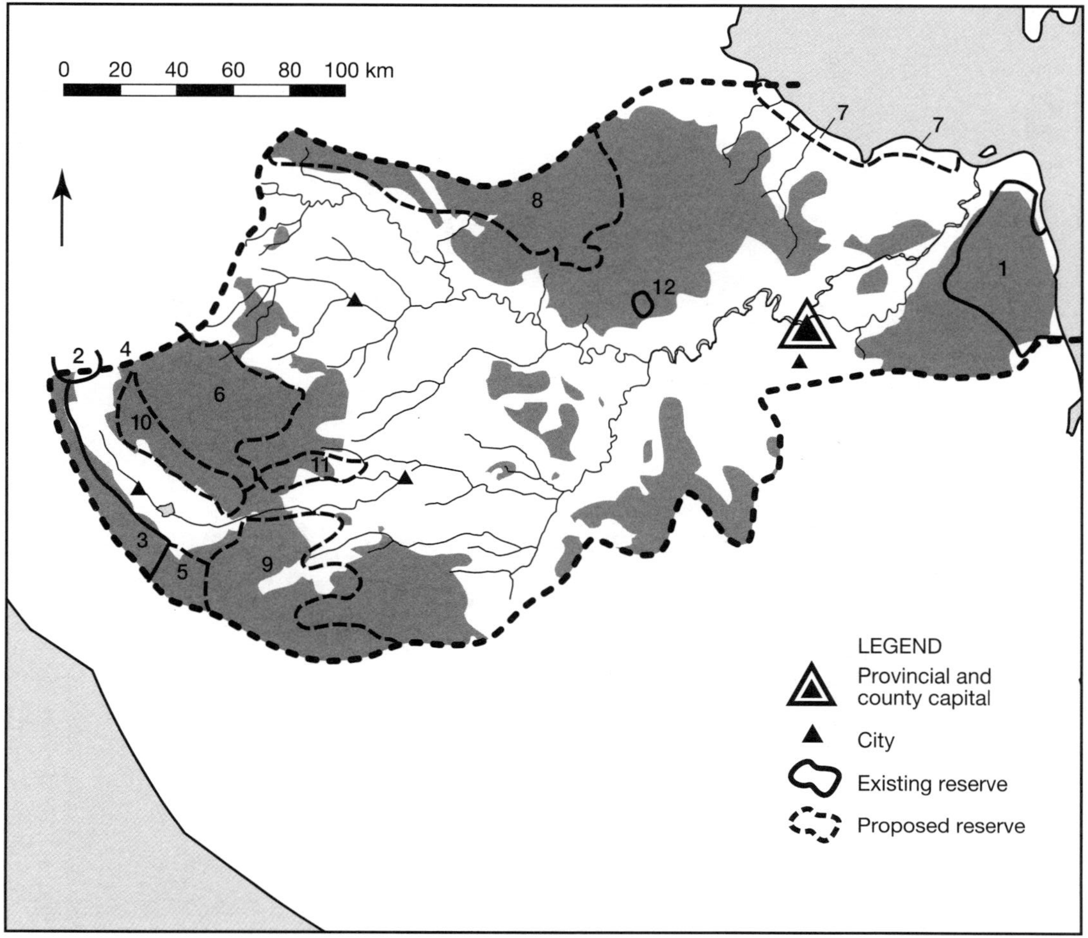

Figure 1.19. Remaining forest, reserves and proposed reserves in Jambi. 1. Berbak; 2. Gunung Indrapura; 3. Bukit Topan; 4. Gunung Tujuh; 5. Batang Marangin Barat/Menjuta Ulu; 6. Sangir Ulu/Batang Tahir; 7. Kelompok Hutan Bakau Pantai Timur; 8. Bukit Besar; 9. Gunung Sumbing/Masural; 10. Hutan Sinlah; 11. Batang Bungo; 12. Singkati Kehidupan.
After MacKinnon 1997

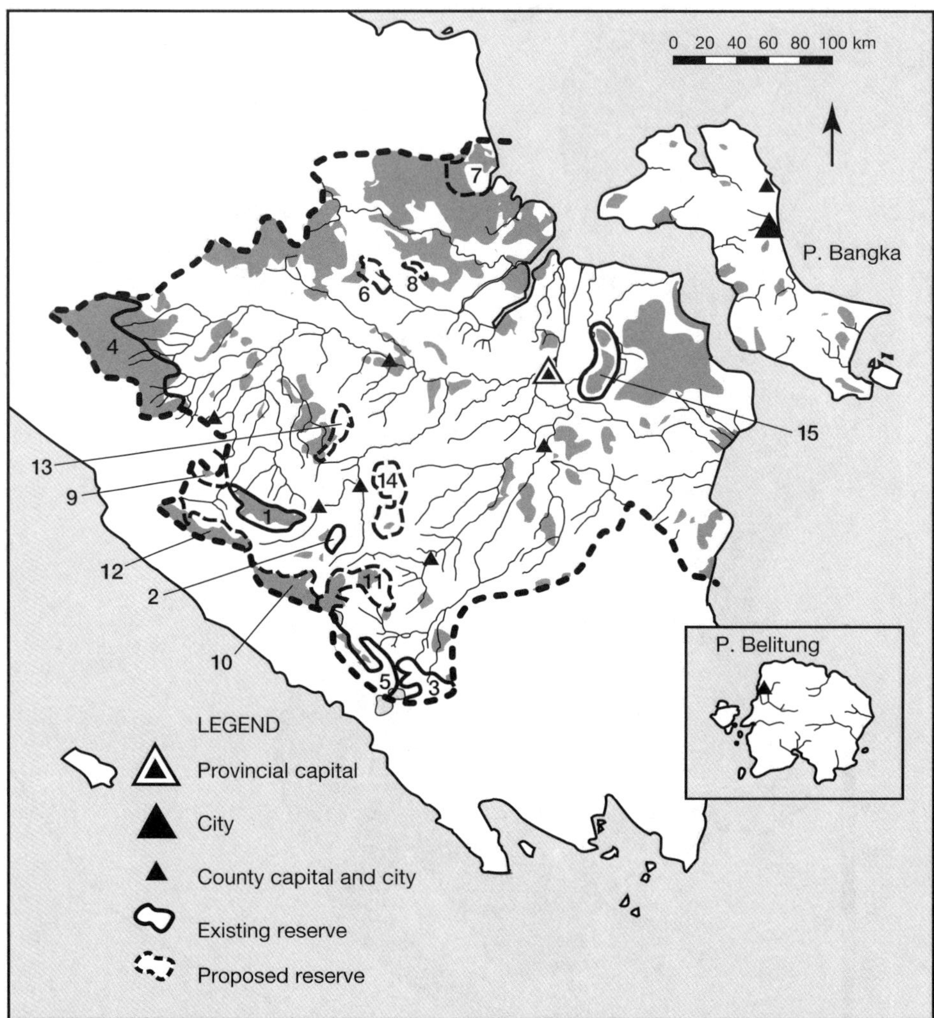

Figure 1.20. Remaining forest, reserves and proposed reserves in South Sumatra. 1. Gumai Pasemah; 2. Isau - Isau Pasemah; 3. Gunung Raya; 4. Rawas Hulu Lakitan; 5. Subanjeriji; 6. Dangkau; 7. Terusan Dalam; 8. Bentayan; 9. Paraduan Gistang; 10. Sungai Terusan Dalam; 11. Bukit Halal; 12. Gunung Patah; Bepagut, Muara Duakisan; 13. Bukit Natiagan Hulu/Nantikomering Hulu; 14. Bukit Dingin/Gunung Dempo; 15. Palembang Plabiu.

After MacKinnon 1997

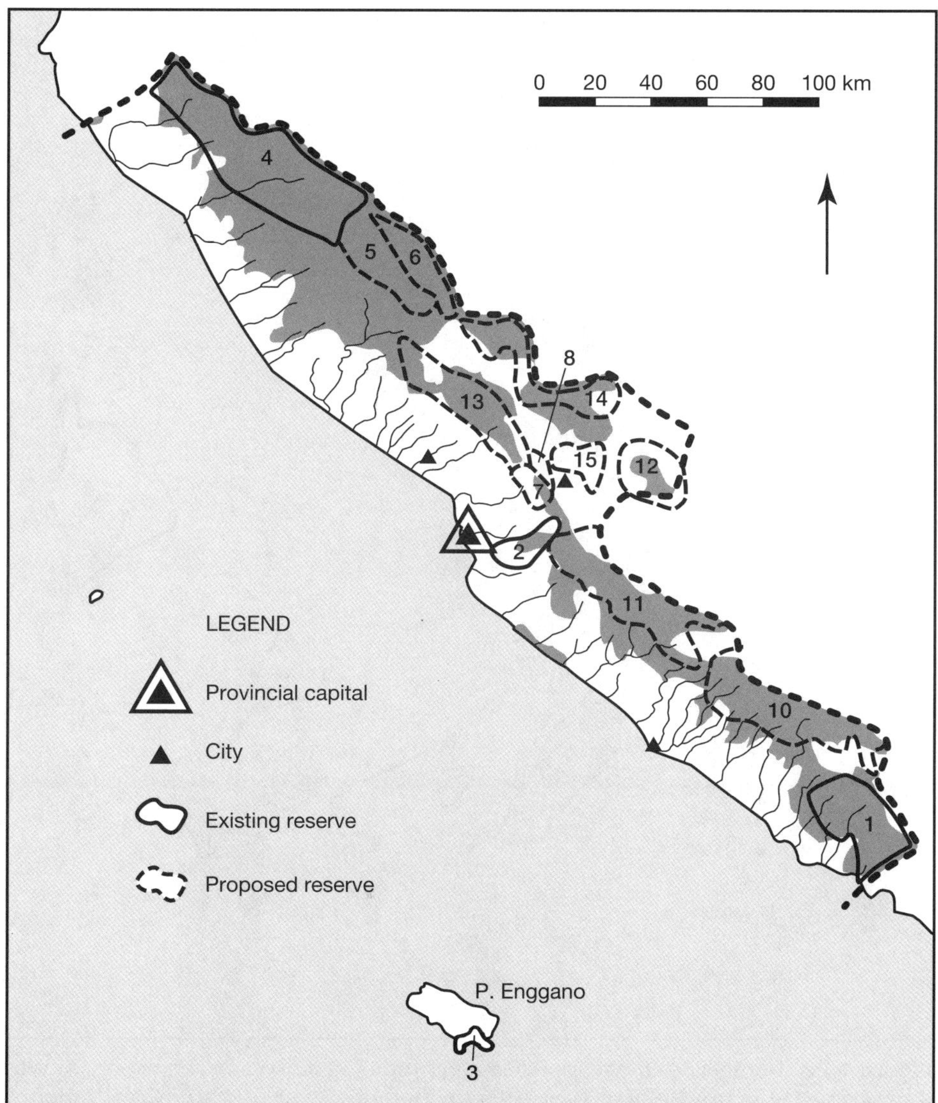

Figure 1.21. Remaining forest, reserves and proposed reserves in Bengkulu. 1. Sumatra Selatan I (part); 2. Semidang Bukit Kabu; 3. Nanuua; 4. Bukit Gedang Seblat; 5. Bukit Gedang Seblat (northern extension); 6. Bukit Gedang Seblat (southern extension); 7. Toba Penanjung; 8. Pungguk Bingin; 9. Dusun Besar; 10. Bukit Raja Mandara/Kaur Utara; 11. Bukit Hitam/Sanggul/Dingin; 12. Bukit Balairejang; 13. Hulu Bintuanan/Palik Lebang/Lair/Wais Hulu/Lekat/Bukit Daun; 14. Bukit Regas/Hulu Sulup; 15. Bukit Kaba.

After FAO/MacKinnon 1982a

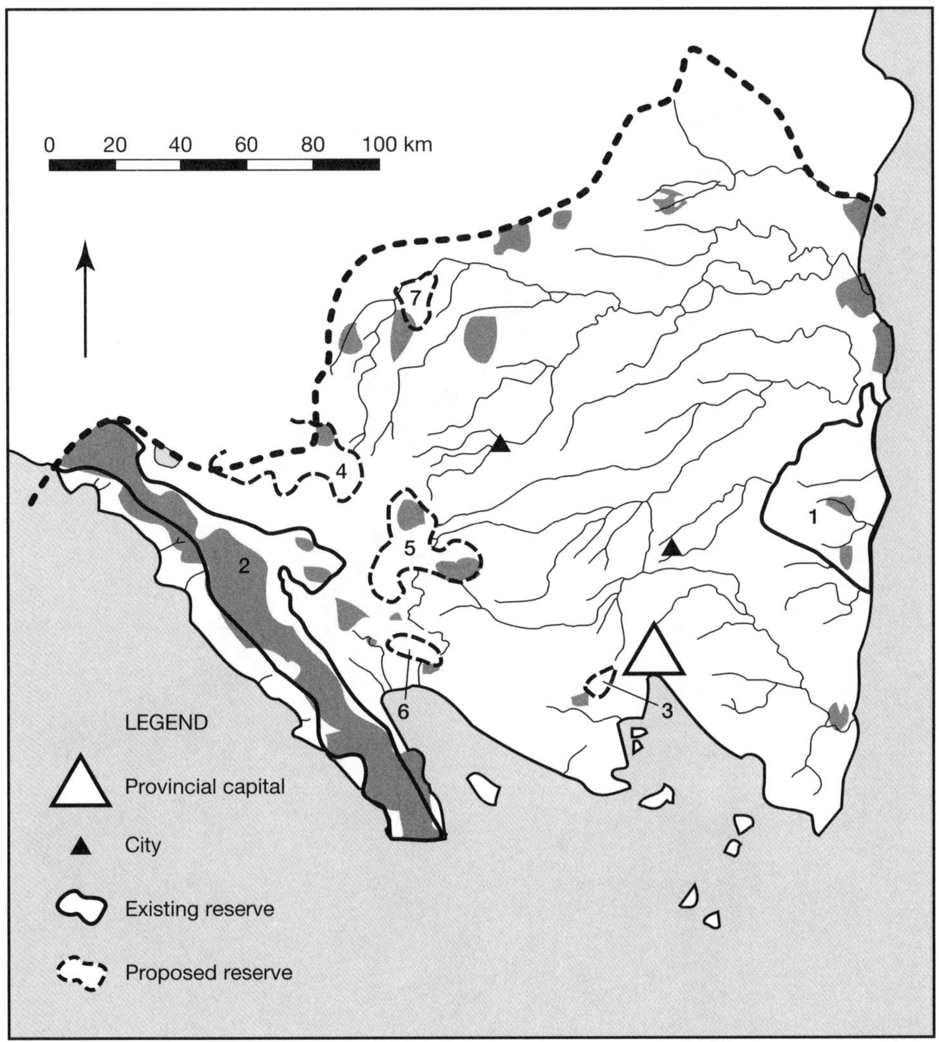

Figure 1.22. Remaining forest, reserves and proposed reserves in Lampung. 1. Way Kambas; 2. Sumatra Selatan I; 3. Gunung Betung; 4. Krui Utara/Bukit Punggul; 5. Tangkitebak/Kota Agung Utara/Wai Waja; 6. Tanggamus; 7. Rebang.

After FAO/MacKinnon 1982b

satellite images. The forest areas given are, therefore, where tree cover exists rather than where unspoilt natural ecosystems may be found. Most types of forest which have suffered limited disturbance are able to regenerate successfully if left alone (p. 339) and so the forest areas shown do have ecological meaning. Table 1.6 shows clearly that certain vegetation types have been reduced in area far more than others and that some vegetation types still have no protection. Montane forests have lost an average of one-third of their natural area, whereas between two-thirds and four-fifths of lowland forest have disappeared. This reflects the severity of land-use conflicts. Note that the lowland forests which have been reduced the most are those on the fertile alluvial and volcanic soils. Generally less than 10% of the natural vegetation types currently receive any protection, and those vegetation types with the least area in reserves are those that have been most reduced in area.

In terms of the loss of genetic diversity, the figures in table 1.7 show that in general the vegetation types under most pressure, such as lowland forest and heath forest, are also the richest or most diverse in terms of

Table 1.7. Average number of tree species greater than 15 cm trunk diameter found in 0.5 ha plots in different habitat types.

Vegetation type	Tree species
Upper montane forest	18
Fertile lower montane forest	66
Infertile lower montane forest	35
Fertile hill montane forest	75
Infertile hill forest	38
Fertile lowland forest	80
Infertile lowland forest	51
Heath forest	50
Beach forest	21
Volcanic scrub	20
Complex mangrove	14
Simple mangrove	6
Peat/freshwater swamp forest	46

After FAO/MacKinnon 1982b

plant species. Similarly, the vegetation types under most threat support the greatest diversity of plant species. For example, 245 Sumatran bird species live in lowland forest, 158 species in lower montane forest, 62 species in upper montane forest, and only two commonly live in the sub-alpine zones (p. 303) (FAO/MacKinnon and Wind 1979).

Fauna

Palaeofauna

Nothing is know for certain about the Sumatran Pleistocene fauna or the fauna of any earlier period. Dubois investigated some caves in the Padang highlands in 1890 and found subfossils of 15 mammal species from several thousand years ago (table 1.8) (Dubois 1891). The prehistoric remains of orangutan, gibbons, leaf monkeys, long- and pig-tailed macaques, tapir, banteng cattle and Javan rhinoceros show that all of these species are somewhat smaller in size today than at that time, and similar size differences have been found for animal remains in Niah caves, Sarawak (Hooijer

Table 1.8. Mammal species, remains of which were found in central Sumatra caves by Dubois (1891).

Leaf monkey	*Presbytis* sp.
Banded leaf monkey	*Presbytis melalophos*
Pig-tailed macaque	*Macaca nemestrina*
Long-tailed macaque	*Macaca fascicularis*
Siamang	*Hylobates syndactylus*
Gibbon	*Hylobates* sp.
Orangutan	*Pongo pygmaeus*
Common porcupine	*Histrix brachyura*
Sun bear	*Helarctos malayanus*
Leopard	*Panthera pardus*
Tiger	*Panthera tigris*
Tapir	*Tapirus indicus*
Javan rhinoceros	*Rhinoceros sondaicus*
Elephant	*Elephas maximus*
Wild cattle/banteng	*Bos javanicus*

1962). This reduction in size is probably a response to the warmer temperatures now prevailing (Edwards 1967). Larger animals have a lower area : volume ratio, thus reducing the rate of heat loss. As a result, larger animals are better able to survive at lower temperatures.

Orangutan are found only north of Lake Toba, leopard and banteng are now extinct in Sumatra and the Javan rhinoceros is almost certainly extinct in Sumatra. Inhabitants of Deli, near what is now Medan, are said to have known a 'buffalo rhinoceros' (peaceful) and a 'pangolin rhinoceros' (savage). The latter was probably the Javan rhinoceros and would doubtless have been hunted to ensure people's safety. Two Javan rhinoceros were shot at Tanjung Morawa (just outside the present limits of Medan) in 1883, and the last eight known from Sumatra were shot in South Sumatra between 1925 and 1982 (Hazewinkel 1933). An unconfirmed shooting of a Javan rhinoceros on the southern boundary of the Way Kambas National Park in 1961 has also been reported (FAO/Wind et al. 1979).

Some mammals of the Sunda Region used to be more widespread than they are today (Hooijer 1975) (table 1.9), perhaps because the flooding of land bridges prevented recolonisation by a species after it had become extinct on one island. It is important to remember that the low sea levels were accompanied by lower temperatures, thus allowing plants and animals of cooler regions, such as the mountain goat *Capricornis sumatraensis,* which is nowadays restricted to mountains, to expand their range and cross over to Sumatra from the Asian mainland (p. 291).

The Niah caves of Sarawak have revealed similar finds to the Padang caves, but in the lowest (= oldest = 30-40,000 years old) levels of cave sediment at Niah, remains of a giant pangolin *Manis palaeojavanica* have been found (Harrisson et al. 1961; Medway 1972c). This had previously been found in Java at the famous fossil sites of Trinil, Rali Gajah, Jebis, Ngandong and Sampung. These sites give the best impression in Southeast Asia of what the Pleistocene fauna was like.

Table 1.9. Extant (x) and extinct (o) mammals of the Sunda Region.

	Java	Sumatra	Borneo
Orangutan	o	x	x
Siamang	o	x	-
Tiger	x	x	?
Leopard	x	o	-
Sun bear	o	x	x
Elephant	o	x	?
Tapir	o	x	o
Javan rhinoceros	x	o	?
Sumatran rhinoceros	-	x	x
Wild cattle/banteng	x	o	x

After Hooijer 1975, and the Earl of Cranbrook, (pers. comm.)

Fossils only sample part of the fauna; for example, the larger animals are emphasised because they have more robust bones. In addition, fossils only form under certain conditions; most bones rot in tropical rain forest and so forest fauna are under-represented, but preservation can occur in open plains, along lake shores, along rivers and in dry caves. Considering the land bridges which have existed between Java and Sumatra, some if not most of the Javan fossil fauna probably once lived on Sumatra, and it is therefore worthwhile to consider the palaeofauna of Java.

Among the most interesting finds in the Javan fossil beds dating from about 700,000 years ago were various forms of man's immediate forbear, *Homo erectus*, who had probably arrived from southern China. *Homo erectus* was, as the name suggests, erect in stature, and ate mainly seeds and fruit, hunted opportunistically or scavenged, used simple stone tools and lived in groups. His demise was probably caused by an inability to compete with the more advanced *Homo sapiens*.

Remains of eight species of elephants, as many as three coexisting at one time, were found in the same fossil beds as the human remains, together with remains of tapir and three species of rhinoceros. There were usually three species of pig living in the same area at any one time, two species of antelope, a hippopotamus and three or four species of deer. These forms were preyed upon by hyenas, wild dogs, early forms of the tiger and leopard as well as sabre-toothed cats (Hooijer 1975; McNeely 1978; Medway 1972c).

In the absence of relevant information from Sumatra itself, one is left with many question about the changes – man-made, climatic, etc. – that have influenced the fauna which exists today. One such question is: why should the leopard have become extinct when it occurred on Sumatra until only a few thousand years ago, still exists on Java and in Peninsular Malaysia, is excellently adapted to hunting in mountains, forests or plains, and when there is an abundance of monkeys and other suitable prey?

Present Fauna

Sumatra is one of the richest islands in Indonesia for animals. It has the most mammals (201 species) and its bird list (580 species) is second only to New Guinea. Indeed, new mammal species are still being discovered or recognised (Bergmans and Hill 1980; Musser 1979). This great wealth is due to the large size of Sumatra, its diversity of habitats and also its past links with the Asian mainland (pp. 12 and 42). Nine species of mammal are endemic to mainland Sumatra and a further 14 species are endemic on the isolated group of Mentawai islands (table 1.10).

Sumatra has 15 other species confined only to the Indonesian region, including orangutan. The island also harbours 22 species of Asian mammals found nowhere else within Indonesia (table 1.11). In addition,

Table 1.10. Endemic mammals of Sumatra.

Pagai Islands horseshoe bat*	*Hipposideros breviceps*
Herman's mouse-eared bat	*Myotis hermani*
Mentawai macaque*	*Macaca pagensis*
Thomas' leaf monkey	*Presbytis thomasi*
Mentawai leaf monkey*	*Presbytis potenziani*
Snub-nosed monkey*	*Simias concolor*
Mentawai gibbon*	*Hylobates klossii*
Sumatran rabbit	*Nesolagus netscheri*
Aceh squirrel	*Callosciurus albescurus*
Loga squirrel*	*Callosciurus melanogaster*
Soksak squirrel*	*Lariscus obscurus*
Mentawai black-cheeked flying squirrel*	*Iomys sipora*
Mentawai orange-cheeked flying squirrel*	*Hylopetes sipora*
Mentawai civet*	*Paradoxurus lignicolor*
Giant Mentawai rat*	*Leopoldamys siporanus*
Mentawai forest rat*	*Maxomys pagensis*
Mentawai rat*	*Rattus lugens*
Hoogerwerf's rat	*Rattus hoogerwerfi*
Kerinci rat	*Maxomys hylomyoides*
Kerinci rat	*Maxomys inflatus*
Mentawai pencil-tailed tree mouse*	*Chiropodomys karlkoopmani*
Sumatran shrew-mouse	*Mus crociduroides*
Sumatran gymnure	*Hylomys parvus*

* Endemic to the Mentawai Islands.

Adapted from Corbett and Hill 1992; Ruedi and Fumagalli 1996

Table 1.11. Non-endemic Sumatran mammals not found elsewhere in Indonesia.

Dayak fruit bat	*Dyacopterus spadiceus*
Big-eared horseshoe bat	*Rhinolophus macrotis*
Glossy horseshoe bat	*Rhinolophus refulgens*
Arcuate horseshoe bat	*Rhinolophus arcuatus*
Sumatran pipistrelle	*Pipistrellus annectens*
Malaysian noctule	*Pipistrellus stenopterus*
Groove-toothed bat	*Phoniscus atrox*
Koka leaf monkey	*Presbytis femoralis*
White-handed gibbon	*Hylobates lar*
Siamang	*Hylobates syndactylus*
Bamboo rat	*Rhizomys sumatrensis*
Singapore rat	*Rattus annadalei*
Edward's giant rat	*Leopoldamys edwardsi*
Sumatran giant rat	*Sundamys infraluteus*
Kinabalu rat	*Rattus baluensis*
Mountain spiny rat	*Niviventer rapit*
Brush-tailed porcupine	*Atherurus macrourus*
Hog badger	*Arctonyx collaris*
Smooth-coated otter	*Lutra perspicillata*
Asian golden cat	*Felis temmincki*
Tapir	*Tapirus indicus*
Mountain goat	*Capricornis sumatraensis*

Figure 1.23. Young Mentawai gibbon, one of the four primate species endemic to the Mentawai Islands off West Sumatra.
A.J. Whitten

Sumatra has populations of several mammals which are virtually extinct in other parts of Indonesia (e.g., Sumatran rhinoceros, elephant, tiger and the *dhole* or forest dog *Cuon alpinus*) (FAO/van der Zon 1979).

Sumatra has an extremely rich bird fauna. Of its 580 species, 465 are resident and 21 are endemic (table 1.12). The Sumatran list includes 138 bird species confined to the Sunda Region, including 16 species found only on Java and Sumatra and 11 species occurring only on Borneo and Sumatra. Thirty-one Asian species found in Sumatra are not found on any other Indonesian islands (e.g., the great hornbill *Buceros bicornis*)

(FAO/MacKinnon and Wind 1979).

A phenomenon exhibited by birds is that of migration or making regional seasonal movements between two areas.[3] Sumatra has about 120 species of migrant birds (FAO/van der Zon 1979; King et al. 1975) but very little has been confirmed about their movements. Studies have been conducted in Peninsular Malaysia for several decades, however, and we can be sure that the general principles found for migrant birds there apply also to

Table 1.12. Endemic birds of Sumatra.

English name	Scientific name	Distribution	Habitat
Red-billed partridge	*Arborophila rubirostris*	northern and central	montane forest
Hoogerwerf's pheasant	*Lophura hoogerwerfi*	northern	montane forest
Salvadori's pheasant	*Lophura inornata*	West Sumatra and Aceh	montane forest
Bronze-tailed peacock pheasant	*Polyplectron chalcurum*	northern	montane forest
Sumatran ground cuckoo	*Carpococcyx viridis*	western and southern	hill and montane forest
Enggano Scops owl	*Otus egganensis*	Enggano	lowland forest
Mentawai Scops owl	*Otus mentawi*	Mentawai Islands	lowland forest
Kerinci Scops owl	*Otus stresemanni*	Mt. Kerinci	montane forest
Simeulue Scops owl	*Otus umbra*	Simeulue	lowland forest
Schneider's pitta	*Pitta schneideri*	northern	montane forest
Black-crowned pitta	*Pitta venusta*	western and southern	hill and lower montane forest
Striated bulbul	*Pycnonotus leucogrammicus*	northern	montane forest
Olive-crowned bulbul	*Pycnonotus tympanistrigus*	northern	montane forest
Sumatran drongo	*Dicrurus sumatransus*	widespread	lower montane forest
Blue-masked leafbird	*Chloropsis venusta*	eastern and southern	hill forest
Vanderbilt's jungle babbler	*Tricastoma vanderbilti*	northern	hill forest
Sumatran wren-babbler	*Napothera rufipectus*	northern	hill forest
Shiny whistling thrush	*Myophoneus melanurus*	widespread	montane forest
Sumatran cochoa	*Cochoa beccarii*	northern	lower montane forest
Rueck's blue flycatcher	*Cyornis ruckii*	northern	secondary forest
Eggano white-eye	*Zosterops salvadorii*	Enggano	lowland wooded areas

After Andrews 1992; van Balen (pers. comm.)

Sumatra. The majority of the information below comes from Medway (1974b) and Nisbet (1974).

About 60% of the migratory species that visit the area breed only in the Palaearctic (see fig. 1.24), 25% breed in the Palaearctic and tropical south and/or Southeast Asia, and 15% breed elsewhere in the Oriental Realm. It is estimated that perhaps 12-15 million birds leave the eastern Palaearctic for the south every year, although many species do not come further south than 10°N (e.g., India, the Philippines and Indochina). They fly south in order to avoid the winter of the northern regions when food is in short supply, and therefore they would be expected to be seen in Sumatra from September to February with a few present in March and August. It seems likely that many species merely pass through Peninsular Malaysia on migration and that Sumatra is their ultimate destination. It has been suggested by Nisbet (1974) that 11 species make Peninsular Malaysia and/or Sumatra their principal wintering areas (table 1.13).

Most birds migrate by night since they need the day to feed but some, such as birds of prey, which have to use hot air currents to gain height in the air, and barn swallows *Hirundo rustica,* which feed on small insects while flying, migrate by day. Huge congregations of swallows can be seen

Table 1.13. Palaearctic and temperate Oriental migrants for which Peninsular Malaysia and/or Sumatra provide principal wintering areas.

English name	Scientific name	Breeding area
Chinese egret	*Egretta eulophotes*	Southeast China and Korea
Crested honey buzzard	*Pernis apivorus*	Northeast Asia
Black-crested baza	*Aviceda leuphotes*	Northern Oriental tropics
Asian dowitcher	*Limnodromus semipalmatus*	South Siberia
Spotted greenshank	*Tringa guttifer*	Sakhalin
Eastern crowned leafwarbler	*Phylloscopus coronatus*	Northeast Asia
Korean flycatcher	*Ficedula zanthopygia*	North China, Korea, East Siberia
Blue-throated flycatcher	*Cyornis reculoides*	Himalayas to southern China
Migratory jungle flycatcher	*Rhinomyias brunneata*	Temperate southeast China
Javanese paradise flycatcher	*Terpsiphone atrocaudata*	Japan, Taiwan
Rufous-headed robin	*Erithacus ruficeps*	Central China
Thick-billed shrike	*Lanius tigrinus*	North China, East Siberia

After Nisbet 1974

roosting along telephone or electricity wires in many towns (p. 414) – for example, along Jl. Prof. M. Yamin in Jambi and the Jl. Sutomo area in Medan. Barn swallows have been studied in urban and rural Peninsular Malaysia by Medway (1973b) and comparative studies from Sumatra would be valuable.

The actual mechanism of migration has been studied a great deal and it seems that birds orientate themselves with reference to both star patterns and the earth's magnetic field (Wallraf 1978; Wallraf and Gelderloos 1978; Wiltschko and Wiltschko 1978). Part of the knowledge of migration is inherited but young birds frequently make errors of direction and distance. For these young birds in particular, migration is hazardous because they have no experience of the destination yet have to find a suitable location, find food in sufficient quantities to build up fat stores for the trip back north, and to leave on time. In the few species that have been studied in Peninsular Malaysia, the date of departure of a species varies by only three or four days from year to year. It seems that some environmental cues such as high or low insect abundance, leaf-fall of trees (see p. 219) or some climatic feature must be responsible. There is no shortage of possible environmental cues: what is needed is an investigation into their reliability, and observations of whether birds actually respond to them.

Reptiles, amphibians and fishes have been studied taxonomically in depth by authors such as de Beaufort, Bleeker, Fowler, van Kampen, Kottelat, Koumans, de Rooij, Tate Regan and Weber, but the amount of ecological knowledge on these animals is extremely small. There are enormous gaps in our knowledge of invertebrates, not just in knowing something about their ecology, but also in knowing simply what exists. The bibliographies by Tobing (1968) and Jacobs and de Boo (1983) are extremely useful as preliminary means of exploring the earlier literature on the Sumatran fauna.

Biogeography

Differences between Realms/Zones

For the purposes of zoogeography the world is divided into six realms (fig. 1.24) and Sumatra falls within the Oriental Realm. The majority of animal species in the Oriental Realm are not found elsewhere and in some cases whole subfamilies, families, and even orders are confined within its boundaries, such as leaf monkeys (subfamily Cercopithinae), gibbons (family Hylobatidae), and flying lemurs (order Dermoptera). Only one family of birds (which are obviously more mobile than mammals)

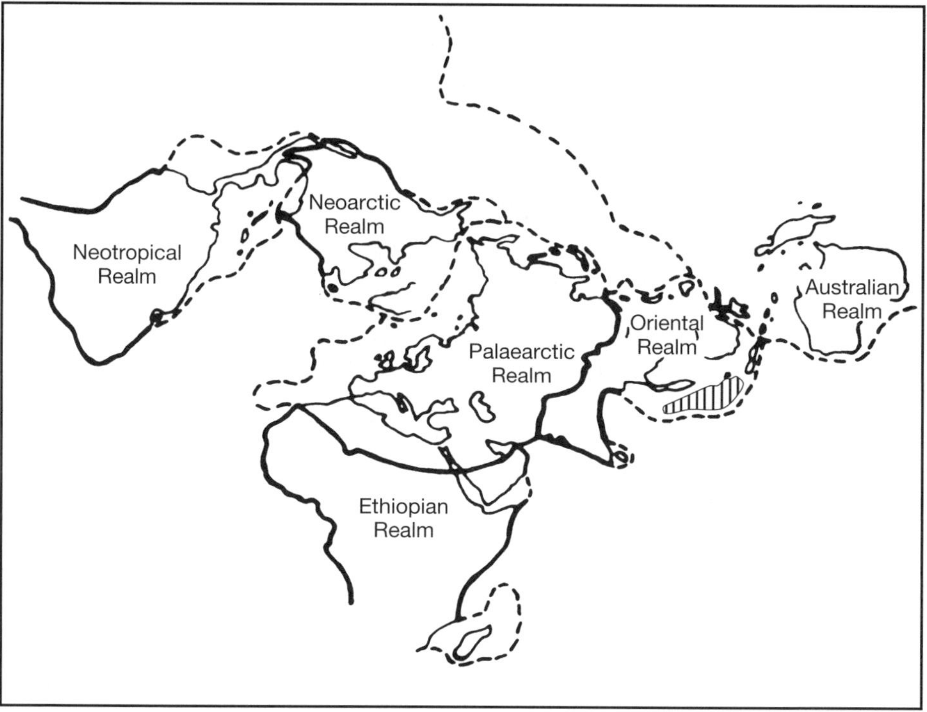

Figure 1.24. Zoogeographical realms of the world.

is endemic to the Oriental Realm and that is the Irenidae (fairy blue-birds *Irena,* leafbirds *Chloropsis* and ioras *Aegithina*) (King et al. 1975). This family seems to have its centre of evolution in Sumatra (Dunn 1974).

Within the Oriental Realm there are distinct differences between regions; the Isthmus of Kra in southern Thailand, which is usually taken as the boundary between the Sunda Region (which includes Sumatra) and mainland Asia, is one of the more obvious regional boundaries. It coincides with a change in vegetation (dry to the north and wet to the south) which is reflected in the fauna (fig. 1.25).

A peculiar zoogeographic distribution has been described by Hoffman (1979). The millipede *Siphoniulus albus* used to be the only species of a monotypic genus, family and order, and until recently was known from only one specimen found near Lake Maninjau in 1890. A second member of the genus has now been found in Guatemala and Hoffman argues that human intervention in this distribution is extremely unlikely. How one

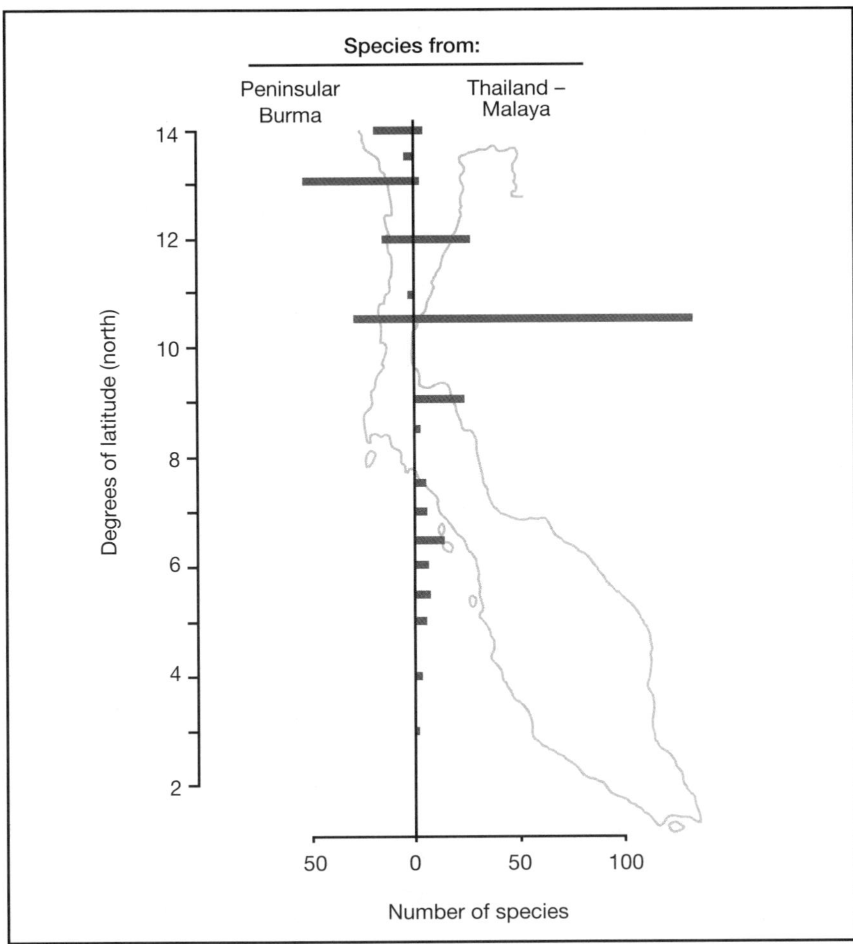

Figure 1.25. Range limits of lowland forest birds in the Malay Peninsula. The length of the horizontal bars is proportional to the number of species/subspecies reaching their limit at the different latitudes.

After Wells 1974

explains this widely separated distribution has not yet been established.

For the purposes of phytogeography, Sumatra falls within the floral region of Malesia (which includes all the Indonesian islands, northern Borneo, Papua New Guinea, southern Thailand and the Philippines) and its flora is distinct from that of Asia and Australia. The disjunction at the

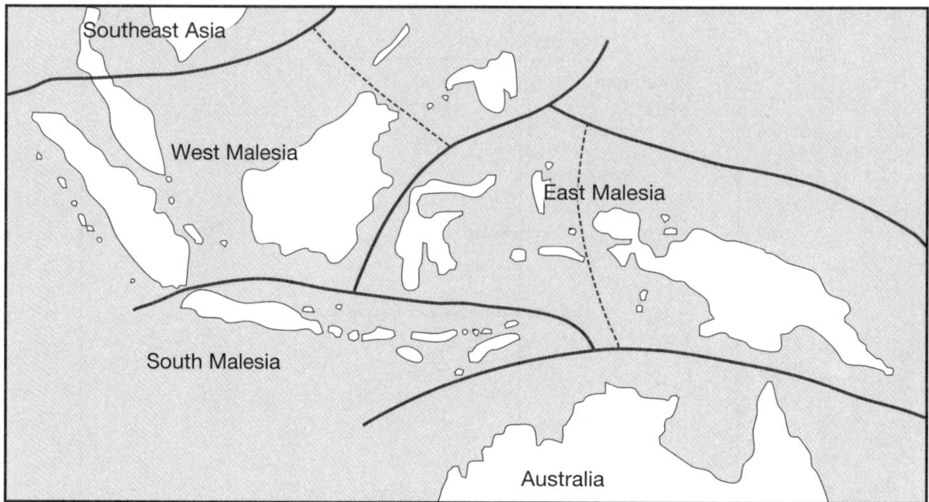

Figure 1.26. Floral region of Southeast Asia.
After van Steenis 1950

Isthmus of Kra is very marked, with 375 genera of plants reaching their northern limit there and 200 genera their southern limit. Within the Malesian region, Sumatra is classified as being part of West Malesia together with the Philippines, Borneo, part of southern Thailand, and Peninsular Malaysia (van Steenis 1950) (fig. 1.26). There are some similarities with the western part of Java but that is classified as a sub-region of South Malesia.

Differences within the Sunda Region

It has already been shown that the flora and fauna of the Sunda Region are similar largely because of the land bridges that once connected the various parts (pp. 12, 17, 34). When the sea level last rose, Sumatra was cut off first from Java, then Borneo, and lastly Peninsular Malaysia, and this is reflected in the degree of similarities between the biota (figs. 1.27, 1.28, 1.29). For both animals and plants, Java is more different from Sumatra than either Borneo or Peninsular Malaysia.

Not all species arrived in, or moved through, the Sunda Region by the same routes or at the same time, and this is reflected in their distributions (table 1.14). With our present knowledge it is not possible to state precisely why the tiger never managed to get to Borneo or the lesser

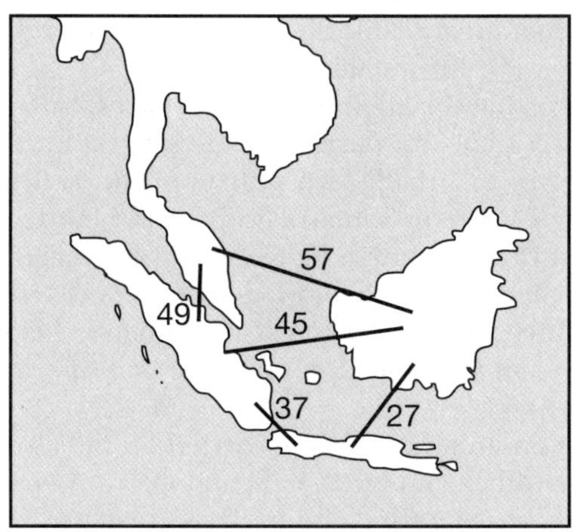

Figure 1.27. Percentage of combined totals of plant species shared between the major parts of the Sunda Region.

After FAO/MacKinnon 1982b

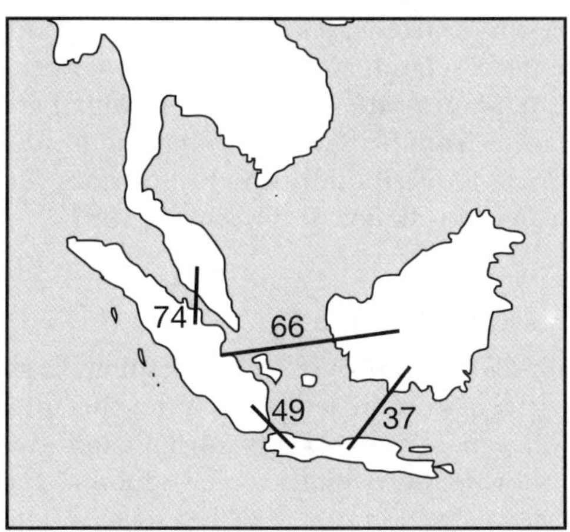

Figure 1.28. Percentage of combined totals of bird species shared between the major parts of the Sunda Region.

After FAO/MacKinnon 1982b

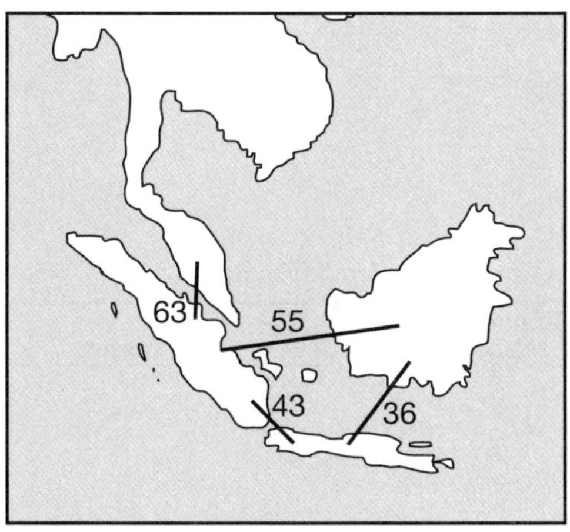

Figure 1.29. Percentage of combined totals of mammal species shared between the major parts of the Sunda Region.

After FAO/MacKinnon 1982b

mouse deer to Java but the information available about climate, sea levels and vegetation allows hypotheses to be formulated.

Care must be taken when discussing the apparent absence of a species. For instance, it was only recently that one particular biogeographical anomaly was resolved. *Isoetes* is a type of aquatic fern with a worldwide distribution but it had never been collected on Sumatra until it was found at the small lakes of Danau Sati and Danau Landah Panjang near Kerinci in 1972 (Flenley and Morley 1978). Similarly, the bent-winged bat *Miniopterus pusillus* was known from most of the Oriental Realm except Sumatra (Lekagul and McNeely 1977) until a team from CRES caught one in a cave in North Sumatra in 1982 (Hill 1983).

Many animals are dependent on walking or being carried on rafts for their dispersal from one area to another, but birds, bats, and many insects can fly and can be blown to new areas by strong and perhaps freak winds. Some spiders can also be dispersed by wind. For example, three small nets were held 18 m above the sea surface from a boat sailing in part of the East China Sea, 400 km from the nearest land, and after about two weeks 105 young spiders had been caught. Many of the species caught which are known from Sumatra are also known from Sri Lanka, China, the Philippines and even Australia and Africa, a distribution which illustrates the effectiveness of their means of dispersal (Okuma and Kisimoto 1981).

Differences within Sumatra

It is well known that remote, small islands support fewer species than large islands close to the mainland. After a certain length of time the total number of species on an island will remain more or less constant, and this total number of species is an equilibrium between the colonisation of the island by immigrant species and the extinction of its existing species. The rate of colonisation is clearly higher when an island is near the mainland

Table 1.14. Examples of mammal distributions in the Sunda Region.

	Sumatra	Peninsular Malaysia	Java	Borneo
Tiger	+	+	+	-
Greater mouse deer	+	+	-	+
Lesser mouse deer	+	+	-	+
Mountain goat	+	+	-	-
Orangutan	+	-	-	+

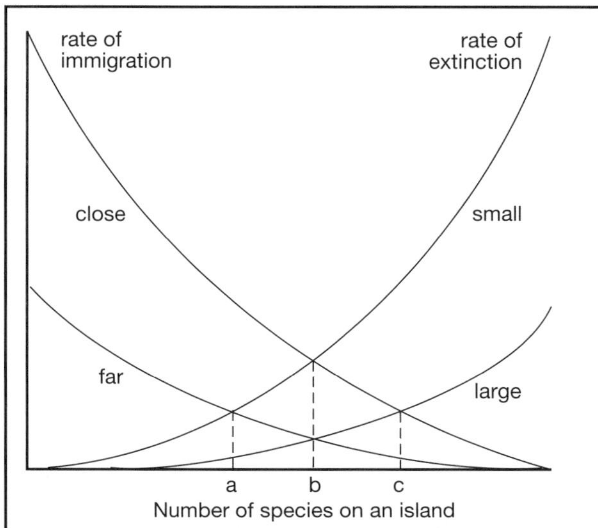

Figure 1.30. The relative number of species on small, distant islands (a) and large, close islands (c). The number of species on large, distant islands or small, close islands, (b) is intermediate.

because more species stand a chance of crossing the gap. Also, the rate of extinction is clearly greater when an island is smaller because the population of any species will be smaller and the chance of disease and other events reducing the population to zero or a non-viable number will be greater. These effects are illustrated in figure 1.30 and represent the foundation of the Theory of Island Biogeography.

The relationship between island size and number of species is relatively constant for a given group of animals and plants. An example for birds is shown in figure 1.31 and an analysis for mammals has recently been published (Heaney 1984). In general, however, reducing the island area by 10, halves the number of species. The depauperisation of species is not random and the large animals (which have relatively large range requirements and low population densities) are usually the first to be lost. Thus there are no tigers, clouded leopard, elephant or Sumatran rhinoceros on any of the islands west of Sumatra. It often happens on islands, however, that a few species are more abundant and fill a wider niche than they do on the nearby mainland where they have more competitors for the same resources (Whitten 1982a; Whitten 1980).

Island size also affects the body size of a particular species. In Prevost's squirrel *Callosciurus prevosti,* for example, which is found in Sumatra, Peninsular Malaysia and Borneo, the smallest individuals are found on the smallest islands. Body size increases with island size and the largest individuals are found on Penyelir Island (280 km^2) and Rupat Island (1,360 km^2) in Riau. On progressively larger islands body size decreases, but the Prevost's squirrels found on mainland Sumatra and Borneo are not

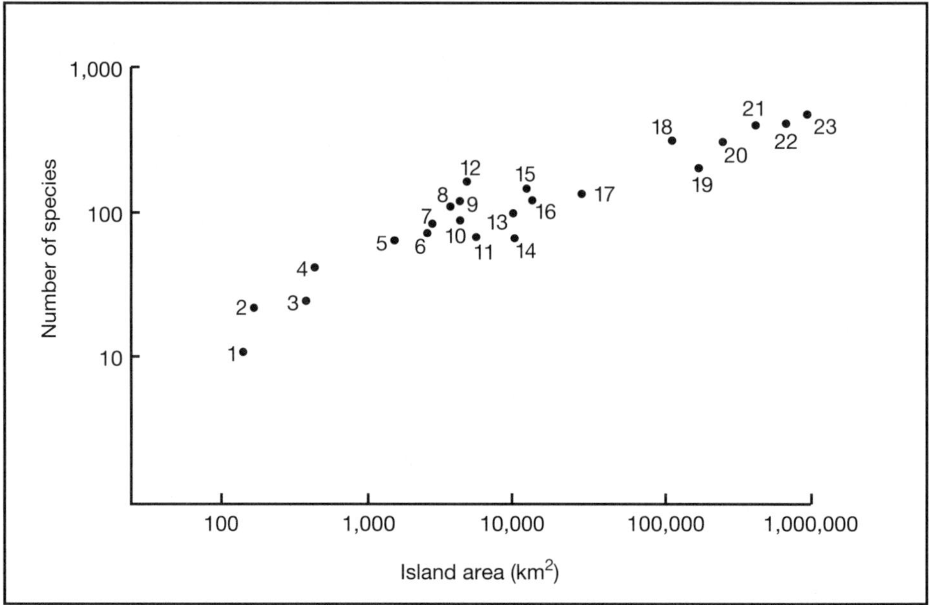

Figure 1.31. The number of land and freshwater birds on various islands and archipelagos in Indonesia, together with the Philippines and the whole of Irian. 1. Christmas; 2. Bawean; 3. Enggano; 4. Sawa; 5. Simeulue; 6. Alor; 7. Wetar; 8. Nias; 9. Lombok; 10. Belitung; 11. Mentawai; 12. Bali; 13. Sumba; 14. Bangka; 15. Flores; 16. Sumbawa; 17. Timor; 18. Java; 19. Sulawesi; 20. Philippines; 21. Sumatra; 22. Borneo; 23. New Guinea.

as small as those found on very small islands. The reasons for this bell-shaped curve are the differences found between islands in predation pressure, food limitations, competition between species, and selection for physiological efficiency (Heaney 1978).

It was clear from table 1.10 on page 37 that the Mentawai Islands have a large number of unique mammals found nowhere else in the world. In fact, 85% of the non-flying mammals are endemic at some level (Anon. 1980a). The reason for this can be seen in figure 1.6: it is more than half a million years since the Mentawai Islands had a land connection via the Batu Islands to the mainland. Long geographical isolation of the islands has allowed the evolution of endemic species and the survival of relicts of an early Indo-Malayan fauna and these 'primitive' forms are extremely significant in studies of the evolution of present Sumatra and Asian biota (Anon. 1980a).

The islands to the east of Sumatra have recent connections with the larger landmasses (Dammerman 1926). They may even have acted as stepping stones to other islands for animals drifting across the sea on rafts of rotting vegetation carried down rivers during floods. From island biogeographic theory we would expect these islands to have relatively few species and not a great number of endemic forms. This appears to be the case: the Riau/Lingga archipelago and Anambas/Natuna archipelago have no endemic mammal species (although they have some endemic subspecies), and their mammals are far more similar to those on Sumatra and Borneo than are those of Mentawai (table 1.15). It also appears that the mammals of Bangka and Belitung Islands and the Anambas/Natuna archipelago have a greater affinity with Borneo than with Sumatra.

Two Sumatran islands, Simeulue and Enggano, have probably never had land connections with the mainland and have extremely impoverished faunas–for instance, there are no squirrels on either island. Simeulue has three endemic species of snake, one endemic bird (and 20 endemic subspecies–one of a species not found on the Sumatran mainland), and a type of macaque and a type of forest pig that may be distinct species (Mitchell 1981). Enggano has 17 species of mammal (a similar number to Simeulue) of which three are endemic (De Jong 1938; Sody 1940). Two of the 29 bird species are endemic as is one of the snakes (Lieftinck 1984).

The contrast between the biota of Sumatra's islands and that of Sumatra itself is by no means the only variation that exists. Within the mainland of Sumatra there are barriers caused by rivers that are too wide, or mountains that are too high, for animals to cross. One of the best illustrations of this is the distribution of the many species and subspecies of leaf monkeys (fig. 1.32). The major boundaries limit the distributions of whole species or separate ecologically equivalent species. The minor boundaries limit gene exchange and separate subspecies.

Perhaps the most surprising zoogeographic boundary, because it is not obviously caused by a river or strait, is that which runs SW/NE through Lake Toba. Seventeen bird species are found only to the north of Lake Toba and 10 only to the south (J. MacKinnon, pers. comm.). Table 1.16 shows some examples of split distributions among the mammals. In the cases of the tapir and tarsier there is nothing that entirely replaces them in their ecological

Table 1.15. Number (and percentage of the combined totals) of mammals on four groups of Sumatra islands shared with the Sumatran mainland and Borneo.

	Sumatra	Borneo
Riau/Lingga archipelago	48 (26)	39 (25)
Bangka and Belitung Islands	39 (26)	36 (32)
Anambas/Natuna archipelago	50 (21)	48 (24)
Mentawai Islands	26 (13)	21 (13)

Data from FAO/van der Zon 1979

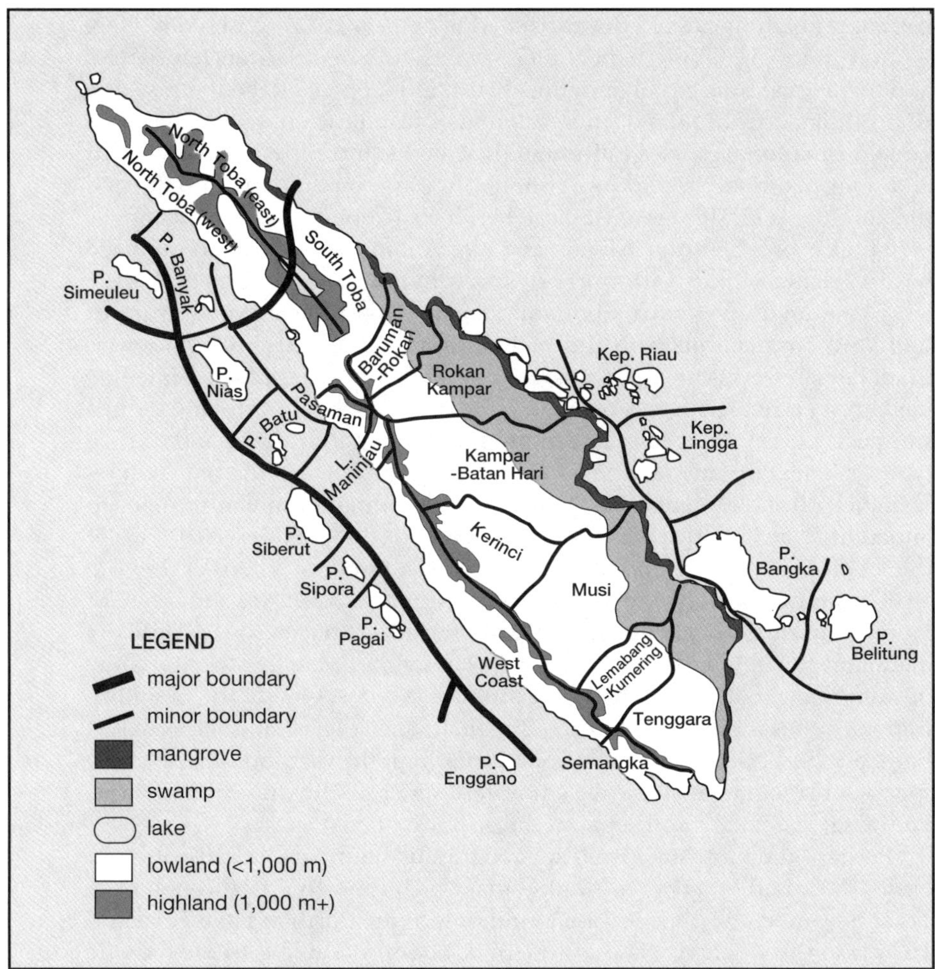

Figure 1.32. Zoogeographic boundaries and units. Thicker lines indicate major boundaries, thinner lines indicate minor boundaries.

After MacKinnon, unpubl.

Table 1.16. Some examples of split distributions of mammals north and south of the zoogeographic boundary through Lake Toba.

North only		South only	
Thomas' leaf monkey	*Presbytis thomasi*	Tarsier	*Tarsius bancanus*
		Banded leaf monkey	*Presbytis melalophos*
White-handed gibbon	*Hylobates lar*	Sumatran rabbit	*Nesolagus netscheri*
Orangutan	*Pongo pygmaeus*	Tapir	*Tapirus indicus*

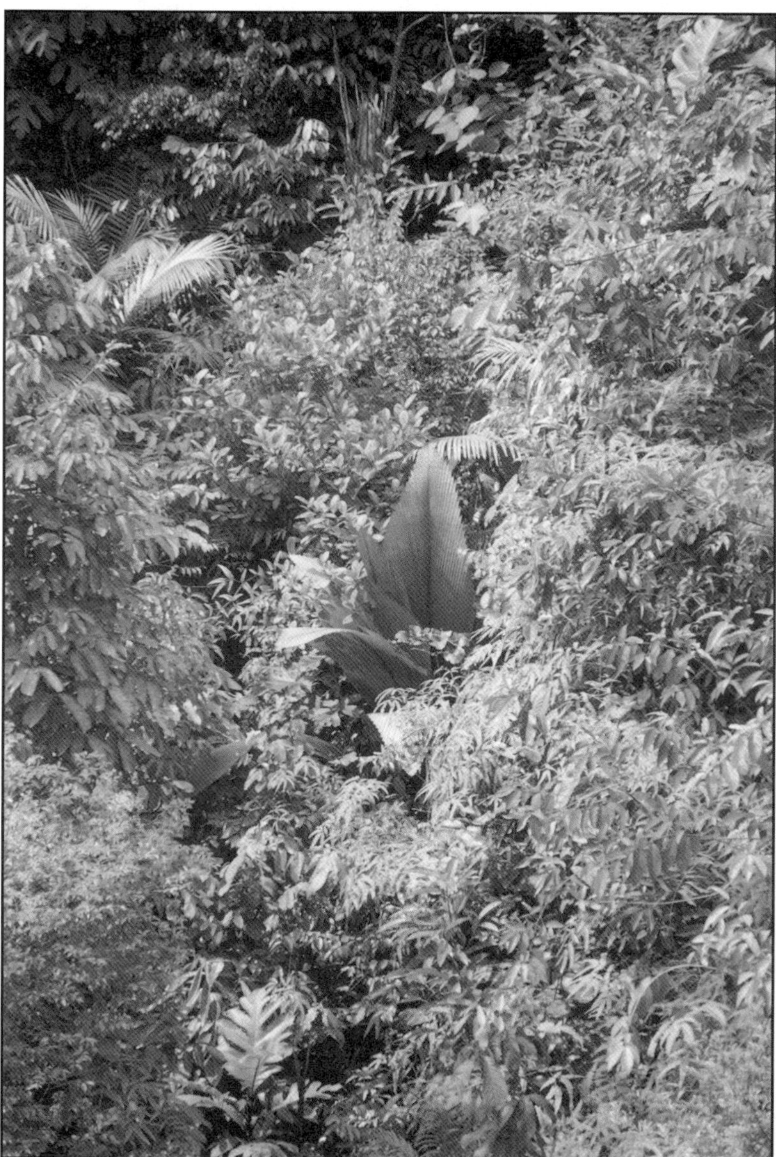

Figure 1.33. The distinctive rhomboidal leaves of the rare palm *Johannesteijsmannia altifrons* in logged-over forest in Sekunder, North Sumatra.

A.J. Whitten

Figure 1.34. The dried leaves of the rare lowland forest palm *Johannesteijsmannia altifrons* used as roofing material near Sekunder, North Sumatra.
A.J. Whitten

niche in the north, but Thomas' leaf monkey is more or less the ecological equivalent of the banded leaf monkey, as the white-handed gibbon is to the dark-handed gibbon (Gittins 1978). Clearly there must have been a cessation of gene flow along the islands that led to such isolated and differentiated faunas. The division between the northern and southern faunas approximately coincides with the position of Lake Toba, and is possibly caused by the wide barrier of bare, volcanic material resulting from the huge eruption. Alternatively, the area affected by drying föhn winds (p. 17) near Padang Lawas may once have been covered with species-poor heath forest (p. 253) and may have represented a significant barrier to animal movement, particularly during the drier periods of the Pleistocene. The Sumatran range of the tapir (found also in Peninsular Malaysia) and of the tarsier (found also in Borneo), is restricted to the lowland forests south of Lake Toba, and presumably these never reached further north.

Many species of plants and animals are restricted to specific habitats within a zoogeographic region, such as dry lowland, mountaintops or mangrove, either because they are unable to compete successfully against other species in other habitats, or because they are tolerant of only a certain range of conditions. One plant with such subtle restrictions to its distribution is the handsome palm *Johannesteijsmannia altifrons.* This palm has been collected from various localities in the northern half of Sumatra but it is most accessible in the Sekundur portion of the Mount Leuser National Park, around which local people use the leaves for roofing their houses (fig. 1.34). It is found only on hillsides, not ridges or valley bottoms, and may be present in one valley but absent in the valleys on either side, despite seemingly identical geology, vegetation and topography (Dransfield 1972).

Prehistory and History

Prehistory of Man

The prehistory of Sumatra is poorly known, even by comparison with that of neighbouring landmasses. Unlike Java, Sumatra has as yet produced no fossil remains of early hominids (human ancestors). Unlike Java, Borneo, the Malay Peninsula, Sulawesi, Flores and Luzon, it contains few if any non-marine fossil deposits of pre-Holocene/post-Mesozoic date.

It is not known when stone-age man arrived in Sumatra, but what may be rough stone tools found in Peninsular Malaysia beneath ash from the Toba eruption some 30,000 years ago might indicate the presence of stone-age man in the Sunda Region from at least that time. What effect the fall of dust had on those people is hard to imagine.

Early man throughout the world inhabited caves. Caves provided shelter, warmth during the periods of lower temperatures, a certain degree of security, and were the focal points of activity (Harrisson 1958). In the Sunda Region the most extensive archaeological excavations of caves have been at Niah, Sarawak, but a similar body of knowledge could result from work at Tiangko Panjang cave in western Jambi. This just one of about 10 sites in southern Sumatra discovered at the start of this century that has revealed stone and wood tools (fig. 1.35), but surveys showed it was the site most likely to contain information spanning a long period (Bronson and Asmar 1976).

Tiangko Panjang cave is about 13 m above the valley of the Tiangko River, a tributary of the Mesumai River. The rock face above the cave entrance is steep, and although the descent to the valley floor is less steep, it is sufficiently difficult to make the location secure from unwanted intrusions. The cave is a tunnel 24 m long and 5-8 m high and wide. With almost 200 m^2 of floor space, clear headroom, good ventilation, and a well-lighted and sheltered courtyard as well as a secure location, the cave

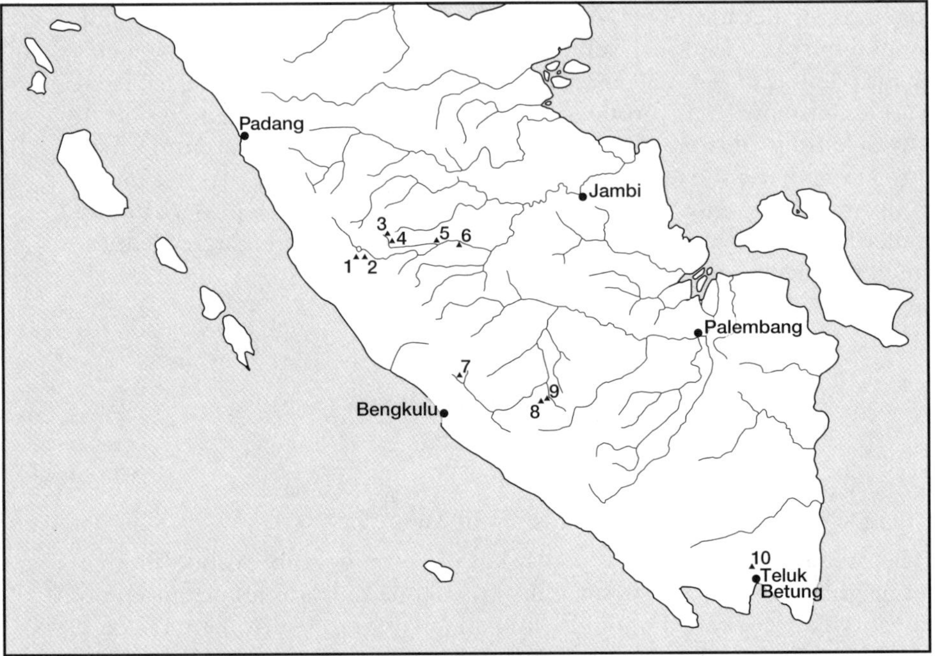

Figure 1.35. Sites in southern Sumatra where prehistoric stone tools have been found. Caves in the Payakumbuh Region were probably inhabited by prehistoric man; a CRES team visiting caves at Lho'Nga, North Aceh, found sea shells in cave deposits which were probably taken there by prehistoric man.

After Bronson and Asmar 1976

would seem to be a desirable dwelling place for early men.

A few small pits have been dug in the cave floor, and radiocarbon dating of charcoal samples from different depths showed that man was inhabiting the cave at least 10,000 years ago when the cave floor was 135-160 cm below its present level.

The floor deposits contained hundreds of small stone flakes, many of which had been used as tools. This indicates a culture rather more advanced than that which produced large rough flakes (Palaeolithic), but not as advanced as that which produced, for example, smooth axe heads (Neolithic). Remains of vertebrate animals included a single human tooth, a few deer teeth, a few bone fragments (usually burned) from deer-sized animals, numerous bat bones, many turtle shell fragments, moderate numbers of vertebrae and long bones from small-chicken sized birds, a mod-

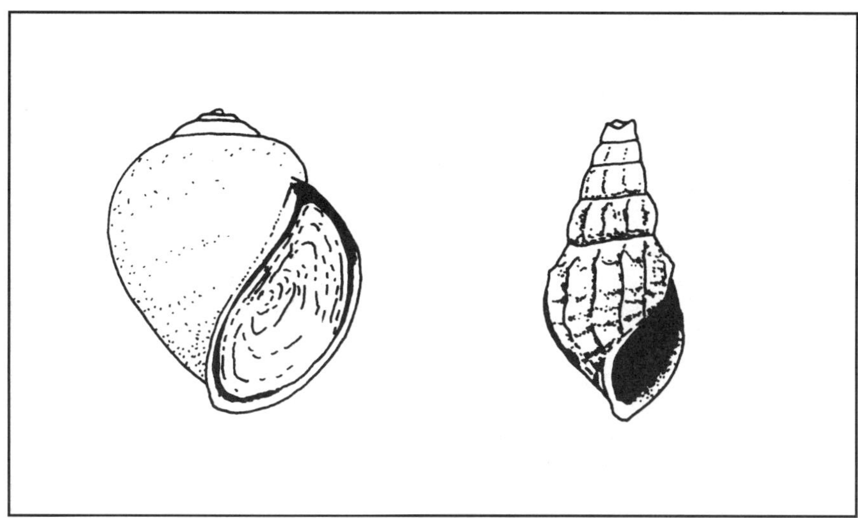

Figure 1.36. Shells of *Brotia* and *Pila,* both of which were eaten by prehistoric man. Each is about 1 cm long.

erate number of fish vertebrae, and teeth of rat-sized rodents. The most common invertebrate remains were snails, the majority of them from two edible freshwater species, *Brotia costula* and *Pila* sp. (fig. 1.36). *Pila* is now uncommon near the Tiangko River but *Brotia* is often eaten by the present inhabitants of the area. Remains of freshwater snails have also been found in Niah cave (Harrisson 1958) and in caves in Peninsular Malaysia (Evans 1918; Medway 1966a).

There is no sign of agriculture or domestication of animals in any of the remains. We do not know if these early men had begun to experiment with agriculture or animal husbandry as there are no archaeological remains to offer proof. Indications of rice in cave deposits are generally clear enough and have been found in caves in Sulawesi dating from 6,000 years ago (Glover 1979a). It seem that the inhabitants of Tiangko Panjang were successful enough at exploiting wild forest produce not to need true agriculture until long after other peoples of the region had fully entered the Neolithic culture.

Exploration and analysis of data from Tiangko Panjang cave is not complete but it has many similarities to Niah cave, where considerable data have been collected. Because of this it is felt justifiable to consider findings from Niah. The stages of man's development there have been described as Early Stone Age (30,000-60,000 years ago) with chopping tools; Neolithic

(± 4,000 years ago) with axes, pottery, mats, and nets; Intermediate II (±2,000 years ago) with elaborate pots and the first metal; and Iron Age (±1,300 years ago) with iron, glass beads, and imported ceramics (Harrisson 1958).

At Niah, remains of 58 species of mammals have been found among the archaeological remains (Medway 1979). It is assumed that no large animals other than flying or scansorial (climbing) types, or natural cave-dwellers, could freely enter the Niah cave (which is more difficult to enter than Tiangko Panjang cave) and hence that all remains of such large animals owe their presence in the cave to the intervention of man. In addition, the skeletal remains are dispersed and fragmented and many of them are charred – all suggesting that man was the cause of their presence. Most of the orangutans were juveniles and this can also be interpreted as proof that they were killed by man since, while it is possible that old, dying animals might climb into the cave, it is difficult to imagine why young ones would enter and die there (Medway 1979).

It has been suggested by palaeontologists working in different parts of the world that man has played a significant role in the late Pleistocene or early Holocene extinction of certain animal species – particularly of the gigantic types (either absolutely large, or large by comparison with modern relatives), collectively called the Pleistocene 'mega-fauna'. Opinion is not unanimous but man is now often accepted as a central contributive factor in the progressive decline of many species that occurred in late- and post-glacial Europe and northern Asia, and in the abrupt elimination of the North American mega-fauna about 11,000 years ago and of giant birds in New Zealand much more recently.

What impact did early man in the Sunda Region have on the fauna? Palaeolithic people would probably have set traps and snares to catch medium- and large-sized mammals, and bearded pig *Sus barbatus* was consistently the most commonly caught. Other prey were rhinoceros, tapir, mouse deer, and sambar deer. As with any trapping effort, as distinct from deliberate hunting, success depended on an animal coming into the trap and the impact on even local populations of a species was probably negligible.

With the advent of Mesolithic technology, with perhaps a greater use of bone as tips for arrows and spears, there was an increase in the numbers of arboreal mammals (such as monkeys and squirrels) caught. Even so, pigs remained the major source of meat. With the exception of the pigs (which are anyway very fecund), the pre-Neolithic man did not concentrate on any one species and his prey species were taken in an opportunistic rather than deterministic fashion. It seems likely that his impact on the wild populations was trivial (Medway 1979).

By the time the Neolithic culture had begun (about 4,000 years ago at Niah) (Harrisson 1958), man was beginning to make use of a few crop

Figure 1.37. On Siberut, the largest of the Mentawai Islands, monkeys are traditionally hunted with bow and poisoned arrows. The monkey skulls are cleaned, decorated, and hung in houses and it is hoped that the monkey souls will remain in them happily and help the hunters hunt more monkeys. The gibbon (below) is rarely hunted.

A.J. Whitten

plants such as sago, and to have domestic animals such as pigs *Sus scrofa* (Medway 1973a) and dogs *Canis familiaris* (Clutton-Brock 1959; Medway 1977). He then needed to clear land for his agriculture but as the human population was still very low, this probably had little impact on the forest. It was, however, the beginning of the end.

As the population grew, so a greater area of land had to be cleared. Loss of forest is the single greatest threat to the majority of forest biota and should not be underestimated (Olson and James 1982). In Java, for example, less than half the mammal fauna survived the mid-Pleistocene and the arrival of modern man *Homo sapiens* (McNeely 1978). For some species, however, there would have been benefits from limited clearance. Herbivores such as deer, tapir, Javan rhinoceros and elephant are relatively common in 'edge habitats', but their presence in or near crops would have encouraged a certain amount of hunting.

There is evidence of forest clearance 11,000 years ago on Taiwan, and of agriculture between 14,000 and 8,000 years ago in Thailand. Archaeological evidence for agriculture dating from about 9,000 years ago has been found in New Guinea and rice was probably cultivated in Sulawesi 6,000 years ago (Glover 1979a). There is little data available on the early activities of man in Sumatra but the studies of pollen samples from Sumatran lakes described on page 18 provide some information. Morley (1980, 1982) found some evidence of forest disturbance near Kerinci from about 7,500 years ago but firmer evidence was found from about 4,000 years ago. Maloney (1980), studying in the Toba highlands, found strong evidence of forest clearance about 6,200 year ago but changes in the pollen record 8,000 and 9,200 years ago may have been a result of man's activities. Rather less convincing evidence from one of the small lakes in the Toba highlands suggests forest clearance began about 17,800 years ago, but this may have been due to fire rather than felling. However, the possibility that man was using fire to clear forest in order to make hunting game easier should not be discounted (p. 282).

More information about prehistoric man comes from a number of sites in East Aceh and eastern North Sumatra where enormous middens (rubbish piles) consisting largely of cockle shells have been found (Schurmann 1931). These are the remains of the distinctive 'Hoabinhian' culture which is known from sites in southern China, through Indochina to Peninsular Malaysia and Sumatra (Glover 1977). Amongst the millions of cockle shells are found stone tools, bones of both animals and man, and other artifacts such as pottery in the upper (newer) layers. The range of dates for the Sumatran sites are not certain but 8,000-3,000 years before present seems likely. Very little is known about these people but their culture seems to have spanned the stages of hunting-gathering (hunting a wide range of animals and harvesting useful wild or semi-domesticated plants) in the early period, to being the prime movers for the development of agriculture in Southeast Asia. The information from the sites is not unequivocal but the people do seem to have brought many species of plants such as candlenut *Aleurites,* betel nut *Areca,* kenari *Canarium,* cucumber *Cucumis,* gourd *Lagenaria,* betel pepper *Piper,* and mentalun *Terminalia* into cultivation (Glover 1979a). Their exploitation of cockles probably resulted in changes in the cockle populations (Swadling 1976) and this may have caused the settlements to move in search of unexploited areas. Most of these middens have now been more or less destroyed for the manufacture of slaked lime for whitewash (McKinnon 1974).

In later prehistoric periods, the story of Sumatra is no more complete or clear; many finds do not have recorded locations and the total are relatively few. Bronze-Iron Age bronze kettledrums have been found near Lake Kerinci, Bengkulu and Lake Ranau, other bronze vessels from the Kerinci area, and statues from Bangkinang in West Riau (van Heekeren 1958).

Megaliths – large stones either set standing, hollowed out or carved into a bas relief showing scenes of animals and men – have been found in Lampung (e.g., at Kenali and Bojong), Riau and West Sumatra (on the banks of the River Kamparkanan), and in Jambi (near Kerinci), but the most famous megaliths are those on the Pasemah Plateau (Bengkulu/South Sumatra), and on Samosir and Nias Islands. It used to be thought that the stones at Pasemah dated from historic times, but it is now known that they were made much earlier (perhaps 100 A.D.) during the Bronze-Iron Age (Glover 1979b; van Heekeren 1958; Heine-Gelde 1972). The megalithic culture of Samosir continued until quite recently and on Nias certain vestiges remain today. Virtually nothing is known of how people in the Bronze-Iron Age culture lived in Sumatra nor how they interacted with their environment, other than that they were agriculturists.

The last remaining living elements of the megalithic culture on Nias, of hunter/gatherer culture of the Kubu people of Jambi/Riau/West Sumatra (Löeb 1972), and of Neolithic/Bronze Age culture on Siberut Island (Schefold 1980; Whitten 1982b) are still available for environmental study now. Although there is a new awareness of the information that can be learnt from people who live close to, and rely upon, the natural environment, it is already too late to learn anything much from the orang Sakai and Akit of Siak, the orang Lubu and Ulu of South Tapanuli, the orang Benua or Mapur of the Riau/Lingga archipelago and Bangka Island or from the orang Enggano (Löeb 1972; Jaspan 1973).

History – Its Effects on Natural Ecosystems

The transition between prehistory (with no written records) and history (written records) will clearly depend upon the people under consideration; for some, such as those on Sumatra's east coast, history began perhaps a thousand years earlier than for some of the inland tribes. In comparison with Java, there is as yet relatively little known about the early historical period of Sumatra, except about the economics and politics of the Srivijaya Empire (Wolters 1970). A review of Sumatra's early history is given in Miksic (1979).

Sumatra became known to Indian traders as the Island of Gold and settlements may have been established on the Sumatran coast by 200 A.D. The earliest Hindu-Buddhist artefact found in Sumatra (in Palembang) is a statue of Buddha, which dates from the late fifth century. The earliest known written inscriptions (in fact the oldest in Indonesia) date from 683-686 A.D. (i.e., the early period or Sailendra Dynasty of the Srivijaya Empire) and were found near Palembang and at Kota Kapur on Bangka Island (Heine-Gelde 1972).

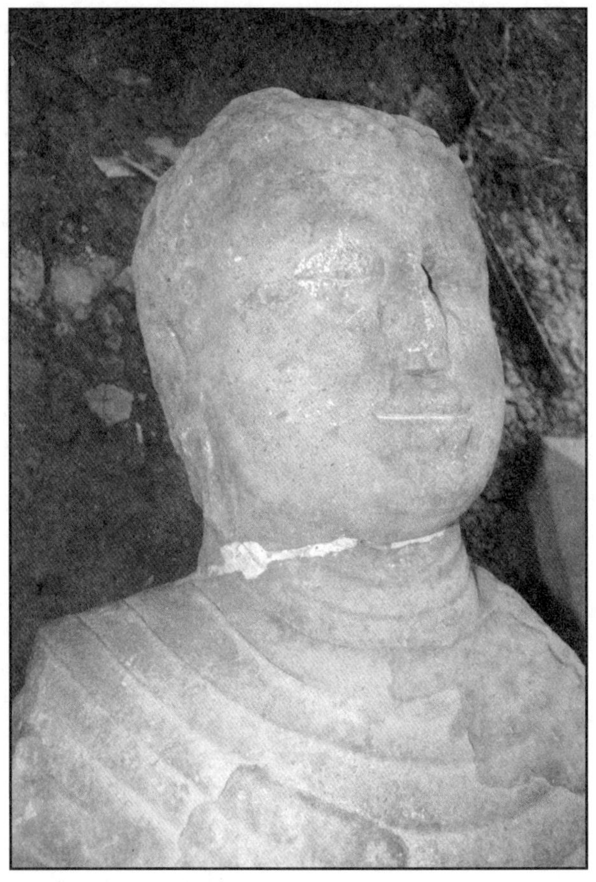

Figure 1.38. An early Buddha from near Palembang.
A.J. Whitten

A few statues of Buddha and other remains from before 1000 A.D. have been found in Jambi, Simangambat (South Tapanuli) and by the Lematang River near Palembang, but most Hindu-Buddhist remains of temples or inscribed stones on Sumatra date from the eleventh to thirteenth centuries. Two sites with remains of temples or biara are worthy of particular mention: Muara Takus on the Kampar Hulu River near Bangkinang, Riau, and Padang Lawas in South Tapanuli (Schnitger 1938). Sumatra also had close trading ties with South India, and a Tamil inscription written in 1088 A.D. found near Barus, now a small port in Central Tapanuli, describes one such commercial relationship (Heine-Gelde 1972). In about 1281 the first Islamic kingdoms were established on Sumatra's coast (although Islam probably entered Sumatra through Barus) but it was a few centuries before their influence affected the rest of Sumatra (Löeb 1972).

The environmental impact of those and other inland kingdoms during the first half of the second millennium is not certain. The temple sites at Muara Takus and Padang Lawas were not centres of population and probably had only a ceremonial function although they may have been on trading routes (Miksic 1979). Hindus from India introduced wet rice culture, the plough, cotton and the spinning wheel to Sumatra. We may therefore envisage small, more or less permanent, rural communities surrounded by a limited area of sawah. A gold mine may have been nearby. Around these there would have been some areas in which people practised genuine shifting agriculture (i.e., leaving fields fallow for 25-30 years before using them again). This type of land use creates a floristically diverse forest edge and favours many species of animals as well as maintaining and conserving soil structure and nutrients. Towns probably only existed as trading centres, primarily on the east coast.

The earliest substantial settlement found in the whole of Indonesia is Kota Cina, halfway between Medan and Belawan in North Sumatra (Manning et al. 1980). This may date back to 1000 A.D. At that time, seagoing vessels would have been able to anchor in the immediate vicinity of Kota Cina itself, but coastal accretion, a common phenomenon at many places on Sumatra's east coast, has resulted in the site being progressively isolated from navigable waters. Chinese documents written in 1282 mention a locality that is possibly Kota Cina. It was a dependency of Srivijaya (although by then Srivijaya had lost most of its power), and was known as a source of fragrant resins (damar) (Milner et al. 1978). Kota Cina appears to have been a very cosmopolitan town, for remains of Chinese and Persian ceramics and other pottery, Chinese coins, Indian Buddha statues, and coins from Sri Langka have been found. Kota Cina was abandoned by the late fourteenth century at the same time that Islam was beginning to become established, Malacca was becoming the region's major trading centre, and relations between Sumatra and Java were disturbed because of the increasing power of the Majapahit Empire.

It is not known precisely what was exported from Kota Cina but damar resin, camphor and beeswax from the highlands were likely products, and gold, lead ore, ivory and fine timber probably contributed to the trade. There is no evidence of Sumatran agricultural produce being exported, and the impact of Kota Cina and similar settlements on natural ecosystems was probably minimal.

A settlement from the same period as Kota Cina may have existed at Muara Kumpeh, 72 km downstream along the Batanghari from the city of Jambi (p. 85). This settlement may have been similar in its function then as Sungsang, at the mouth of the Musi River in South Sumatra, is now (McKinnon 1982).

In northern Southeast Asia, archaeological remains have been found for many cities, their satellite towns and hinterlands. In southern Southeast Asia, however, with the exception of Java, no such early cities have been found. Yet Srivijaya, the hypothetical capital of the Sumatra-based empire of the same name, should rival Central Java and southern Thailand for producing the first capital city in the Sunda Region. It was the administrative and intellectual centre of an empire which is known from Chinese historical records to have ruled the seas of Southeast Asia and to have lasted for over five hundred years from before 700 A.D.

Palembang has always been identified as the location of Srivijaya. It used to be a port nearer the sea, it has a strategic position for local and international trade, Chinese sailing directions indicate Srivijaya was in the Palembang area, and some ancient inscriptions had been found there that mention Srivijaya. In 1974, thorough archaeological excavations were conducted and not a single piece of evidence could be found for a settlement or related hinterland dating from before the fourteenth century – when the empire had already begun to crumble (Bronson and Wisseman 1978).

This has some important environmental/historical implications. Palembang was supposed to be the only early (pre-1000 A.D.) urban site south of the Thailand/Malaysia border. Now that Palembang, and probably the whole area drained by the Musi River, has been ruled out, there remains no example, even supposed, of an early urban culture in Sumatra.

Srivijaya did exist, but it must have been a form of habitation that would have been described as a city by a visitor, yet not leave obvious remains. One of the four requirements suggested by Bronson and Wisseman for such a city is that it exist in comparative isolation from its hinterland (Bronson and Wisseman 1978). This seems unlikely at first but at the peak of the history of Malacca in the fifteenth and sixteenth centuries, this important port was surrounded by uninhabited forest, and its population of about 20,000 people depended on supplies from Java and Thailand for its subsistence. Thus, even though the Srivijaya Empire was being coordinated from Sumatra, its emphasis was on controlling and profiting from trade rather than on controlling the interior, and so the impact on the natural environment may have been very slight. Some trade in Sumatran goods may have been conducted from the town of Srivijaya (as at Kota Cina later) but it could have served as an entrepot, similar to Singapore at its founding.

The ineffectiveness of Srivijaya as a cultural missionary, except at a few places along the coast, is illustrated by the survival in nearby areas of forest people with primitive cultures: the orang Mapur in the interior of Bangka Island, the Kubu of the Jambi/Riau forest, the Akit of the Siak and Bengkalis Region, and the nomads of Batam Island (Sopher 1977).

The history of the islands to the east of Sumatra was also not greatly affected by the Srivijaya Empire: as stated before, Srivijaya was more concerned with trade than with territory. Instead, the Orang Laut, specifically the Celates, were far more important. These were groups of nomadic people, not necessarily pirates, whose culture and economy were based on the coasts and seas. They tended to build settlements away from established centres of population (Sopher 1977) and, because they were concerned only with marine products, they never strayed inland from the coast. For example, the hilly interior of Belitung Island was completely deserted until tin mining began in 1851, although there were numerous settlements of Orang Laut around the coast (Sopher 1977).

So, it was with the arrival of Portuguese, Dutch and British in the sixteenth, seventeenth, eighteenth and nineteenth centuries that most of the natural ecosystems of Sumatra received their first major shocks. Before then the majority of the landscape probably consisted of low-density, small-scale agriculture among a scarcely-disturbed forest and with virtually no urbanisation (Marsden 1811). To supply the spice trade, land had to be cleared and early plantations of tobacco and other crops were also established. However, not all large-scale land clearance was directed towards the plantation industry or even conducted in response to Christian missionary influence. Burton and Ward (1827) were the first Europeans to enter the Toba Region and the description of their journey from Sibolga in 1824 makes instructive reading because they depict a largely deforested land, much of it planted with rice. For example:

> The woodland had already given place to grassy plains; and the mountain on which we stood had been cleared on every side for cultivation, merely retaining its original forest in a tuft at the top... The soil now became sandy and grey and the hills were entirely free from wood, and planted with the sweet potato in many instances to the very tops... The principal object of the picture was an uneven plain, ten or twelve miles long and three broad, forming a vast unbroken field of rice... The plain was surrounded by hills from five hundred to one thousand feet high, in a state of cultivation; and the whole surrounding country was perfectly free from wood, except the summits of two or three mountains said to be the abode of monstrous serpents and evil spirits.

Descriptions, land-use maps and photographs of the Toba, Alas and Karo areas at the turn of the century are surprisingly similar to today (von Hügel 1896; Volz 1909, 1912).

Part B

Natural Ecosystems

Sumatra, area 476,000 km^2, probably has over 10,000 species of higher (seed) plants, most of which are found in lowland forest. For comparison, the British Isles is 65% of the area but has only about 13% of the number of species. The number of tree species per unit area in the lowland forest (p. 189) of Malesia and particularly Western Malesia (which includes Sumatra) is probably greater than in similar forest in west Africa or South America. Indeed, these Sunda Region forests are probably the most species-rich plant communities in the world (Whitmore 1975).

Part of the reason for the very large number of plant species is that in the near-perfect growing conditions of warmth and adequate moisture, the tall trees provide a large framework for a wide and environmentally diverse range of structural niches. These are filled by smaller trees, shrubs, herbs, scramblers, climbers, epiphytes, parasites, etc. The underlying causes of high plant diversity are discussed on pages 189-192.

The vast number of plant species present in Sumatra and elsewhere in the Sunda Region represent a magnificently rich source of natural resources, many of which have proven or potential economic value (Williams 1975). About one-third of the 7,500 species of plants known from Peninsular Malaysia are listed by Burkill (1966) as having some economic value, and a similar proportion would be expected from Sumatra's plants. The dipterocarp timber trees are the most obvious example and these constitute about one-quarter of all the hardwood timber on international markets. Unfortunately, this reflects the productivity of timber companies rather than the productivity of the trees themselves, and accessible supplies of dipterocarp timber will be exhausted by the end of this century or before (Ashton 1980). In addition to the timber species already exploited, there are many tree species whose potential for timber, fibre or cellulose remains unexploited. In addition to these are hundreds of species in the forests whose economic potential for food or chemicals remains unassessed (Whitmore 1980). People of rural areas generally have a deeper understanding of the value of wild or semi-domesticated forest plants but, as Sastrapradja and Kartawinata (1975) have pointed out, knowledge of these can be lost within the timespan of a generation. Useful summaries of species with economic potential in Southeast Asian forest and of the need for genetic resource conservation are given by Sastrapradja et al. (1980), Whitmore (1980) and Jacobs (1980).

The economic value of forest vertebrates is less well studied but the vital role that macaque species played in the early stages of the mass production of polio vaccine should not be forgotten (Medway 1980). That role was largely unanticipated and it would be a brave person who would claim now that some other wild vertebrate could not be of similar value to human welfare. In addition, cave bats are vital to the production of durians (p. 329). The potential uses and values of invertebrates – which are more numerous, more diverse, less well known – are clearly even greater. The economic value of the goods and services provided by coastal swamps and natural ecosystems, particularly lowland forest, is described elsewhere (Burbridge 1982; Farnworth et al. 1981; Krutilla and Fisher 1975).

Considering the aesthetic value, economic importance, and diversity of Sumatra's forest and other natural ecosystems, it is surprising that studies of their components and the way these components function and interact has not been particularly intense or widespread. From discussions with staff at the University of North Sumatra there seem to be three major reasons for this situation – tigers, snakes and leeches – all of which have reputations unrelated to the facts.

a) **Tigers.** The tigers reported in the newspapers as killing stock animals or even humans are rogue individuals and are not typical of the species. Adventurous young or ailing old animals may take to seeking food around villages, at least in part because their forest habitat is being reduced in area. Field workers who have spent over two years conducting research in the forest of the Mount Leuser National Park, where tigers are regarded as common, have counted themselves 'lucky' to have had a fleeting glimpse of a fleeing tiger. It is worth reporting that during the CRES ascent of Mt. Kemiri, one or two tigers visited the first camp site (at 1,500 m) on two consecutive nights – passing within 30 cm and 50 cm of the sleeping party – and the only loss was a pair of sandals, later found thoroughly chewed at the side of the path.

b) **Snakes.** It surprises many people to learn that less than 18% of the world's 2,300-2,700 snake species are poisonous, and only a fraction of these actually endanger human life (Duellman 1979; C. McCarthy, pers. comm.). Sumatra has about 150 snake species (de Hass 1950). Of this total 6% are poisonous sea snakes, 67% are harmless land snakes, and 14% are poisonous land snakes. Of this last group, however, most are too small or weak to present any hazard to humans. However, the reticulated python, *Python reticulatus*, which constricts its prey, must be regarded as dangerous to man. In fact, the percentage and density of snakes that pose any sort of threat to people is probably considerably higher in towns and villages than in forest. The information above should not lead to carelessness with snakes in forests; however, they are only rarely seen.

c) **Leeches.** Most leeches, whose closest relatives are earthworms, are found in fresh water but some species have adapted to life on land or in the sea. Some of the larger freshwater (swamp) species are known to transmit protozoan blood parasites to their hosts (usually mussels, fish and frogs), or cause serious blood loss to cattle or buffalo which drink from infested water. The leeches seen in lowland forest are small (2-4 cm) and are not known to transmit any disease to humans or other animals (Sharma and Fernando 1961). Of the two commonest species, the dull-coloured *Haemadipsa zeylandica* gives a painless bite, but the brighter-coloured *H. picta* has a fiery bite. Bleeding from a land leech bite looks spectacular but the amount of blood lost is very small. Proper dress and perhaps a repellent (based on dimethyphthalate) provides protection, but if a leech manages to bite, a dab of antiseptic to prevent secondary infection is all that is required.

The concern for and interest in natural ecosystems expressed by some people (often foreigners) baffles others. "Why bother?" they might ask. Van Steenis (1971) has replied to such a rhetorical question by answering, "Come with an observant eye and the question will answer itself." In the next nine chapters it is hoped that the reader will be able to 'observe' a little from his chair and then be sufficiently enthused to venture out and find out more himself about the forest and natural ecosystems of Sumatra.

Chapter Two

Mangrove Forests

INTRODUCTION

Mangrove forests[1] are sometimes regarded as a very distinct ecosystem virtually requiring a separate discipline of science. In fact, they are just one of the many types of forest in Sumatra having a closed, even canopy made up of tree species which are predominantly evergreen. The environmental conditions in which they grow are extreme, however, because they are subject to soil-water salinity and waterlogging. Mangrove forests have important environmental roles concerned with land, wildlife and fisheries management (table 2.1) and have been exploited by man for numerous natural products (table 2.2). This exploitation by man has had important effects on the ecosystem.

On Sumatra, mangrove forests have been studied more than almost any other natural ecosystem. Every Sumatra province has mangrove fringing some of its shores, if only in a few sheltered bays or river entrances or around offshore islands, but of the 1,470,000 ha of mangrove forests in 1982, over 60% were in Riau and South Sumatra (p. 20) and this is where the studies have concentrated. Major studies have been conducted by Sabar et al. (1979), Soeriaatmadja (1979), Sukardjo (1979), and Sukardjo and Kartawinata (1979).

Mangrove forests form a protective and productive margin to much of Sumatra's coastline. It is important, from many points of view, that the ecology of mangrove areas should be understood as fully as possible and as soon as possible, but particularly because so many development projects are in the coastal region and create serious conflicts of land use.

An excellent review of mangrove ecosystems in Indonesia is given by Kartawinata et al. (1979). They conclude:

> Almost all species of plant which make up Indonesian mangrove swamps are now known. However, many fields such as variations in species composition, forest structure, seed dispersal, phenology, biology, flowering and fruiting, composition of the fauna, mineral cycling, productivity and ecosystem dynamics are barely understood. Because of this, studies of the above topics need to be encouraged. Exploration, inventory and mapping must be given priority too.

As with other chapters, results of studies from Peninsular Malaysia and elsewhere are used below in order to increase the depth of the information.

Although there is no generally accepted definition of a mangrove ecosystem, Saenger et al. (1981) proposed that it should consist of the following:

1) one or more of the exclusive mangrove tree species (see p. 79 and table 2.4);
2) any non-exclusive plant species growing with (1);
3) the associated biota – the terrestrial and marine animals, the lichens, fungi, algae, bacteria, etc., whether temporary, permanent, casual, incidental or exclusive in the area of (1);
4) the processes essential for its maintenance whether or not within the area of (1).

To this should be added:

5) and the foreshore below the margin of vegetation where colonization by pioneer mangrove trees will occur and where the fauna has a terrestrial component.

An extremely comprehensive bibliography of references to mangroves totalling 5,608 entries has been published by UNESCO (Rollet 1981).

Table 2.1. Roles of mangrove ecosystems.

Roles of mangrove ecosystems
Physical functions
Stabiliser of coastlines
Accelerator of land extension
Buffer against waves and storms
Protector of beaches and riverbanks
Assimilator of waste material
Biological functions
Nursery ground for fish, prawns and shellfish of the open sea
Sanctuary for large nesting birds
Natural habitat for many forms of wildlife
Potential commercial functions
Aquaculture
Salt ponds
Recreation
Timber

From Saenger et al. 1981

Table 2.2. Products of the mangrove ecosystem. See also table 27 in Soegiarto and Polunin (1980) for a species-by-species list.

MANGROVE FOREST PRODUCTS	
FUEL	**TEXTILES, LEATHER**
Firewood (cooking, heating)	Synthetic fibers (e.g., rayon)
Charcoal	Dye for cloth
Alcohol	Tannins for leather preservation
CONSTRUCTION	**FOOD, DRUGS & BEVERAGES**
Timber, scaffolds	Sugar
Heavy construction (e.g., bridges)	Alcohol
Railroad ties	Cooking oil, vinegar
Mining pit props	Tea substitute
Boat building	Fermented drinks
Dock pilings	Dessert topping
Beams and poles for buildings	Condiments from bark
Flooring, panelling	Sweetmeats from propagules
Thatch or matting	Vegetables from propagules, fruit or leaves
Fence posts, water pipes, chipboards, glues	Cigar substitute
FISHING	**HOUSEHOLD ITEMS**
Poles for fish traps	Furniture
Fishing floats	Glue
Wood for smoking fish	Hairdressing oil
Fish poison	Tool handles
Tannins for net and line preservation	Rice mortar
Fish-attracting shelters	Toys
AGRICULTURE	Matchsticks, incense
Fodder, green manure	**PAPER PRODUCTS**
OTHER PRODUCTS	Paper of various kinds
Packing boxes	Wood for smoking sheet rubber
Wood for burning bricks	Medicines from bark, leaves and fruits

OTHER NATURAL PRODUCTS	
Fish	Birds
Crustaceans	Mammals
Shellfish	Reptiles and reptile skins
Honey	Other fauna (amphibians, insects)
Wax	

From Saenger et al. 1981

Important Physical Features

Tides

The most important feature of the physical environment for the mangrove biota is the tide. Tides usually occur twice each day and are caused by the gravitational pull of the moon and to a lesser extent the sun. In the open ocean the amplitude (height) of the tides is not more than about 0.5 m but in shallow seas around Sumatra it is commonly up to 3 m. The tides of greatest amplitude, the spring tides, occur when the earth, sun and moon are in a straight line (i.e., about the times of the new moon and the full moon). Neap tides, those of smallest amplitude, occur when the sun is opposed (is at right angles) to the gravitational pull of the moon (i.e., about the times of the first and last quarters of the moon).

As the moon orbits the earth in the same direction as the earth's rotation, a period of rather more than one day (in fact, 24 hours 50 minutes) elapses between successive occasions when the moon is vertically above the same meridian. Thus, the interval between successive high or low tides is generally half this period (i.e., about 12 ½ hours). This general pattern is modified when the tidal waters move through complex archipelagos or irregularities in the coastline, and the fauna and flora of different locations will respond to these different regimes of inundation.

The tidal regimes for the waters around Sumatra are shown in figure 2.1, and it can be seen that Sumatra experiences four types of tide which are described below and illustrated in figure 2.2.

- Semi-diurnal tide – two high waters and two low waters daily with similar amplitudes;
- Mixed tide, mainly semi-diurnal – two high waters and two low waters daily with different amplitudes;
- Mixed tide, mainly diurnal – sometimes only one high water and one low water daily, but sometimes also two high waters which differ greatly in height;
- Diurnal tide – only one high water and one low water daily.

Some of the mangrove fauna do not avoid the alternate wetting and exposure caused by tides. Others climb into the trees during high tides and then descend to feed during low tides. One such animal is the snail *Cerithidea decollata* whose pattern of activity follows the tidal rhythm and is not associated with light or dark (Cockcroft and Forbes 1981). The snails descend during low tide periods and ascend before the next high tide. Note that the rising tide does not force them up, but that the tidal rhythm is part of their own internal rhythm.

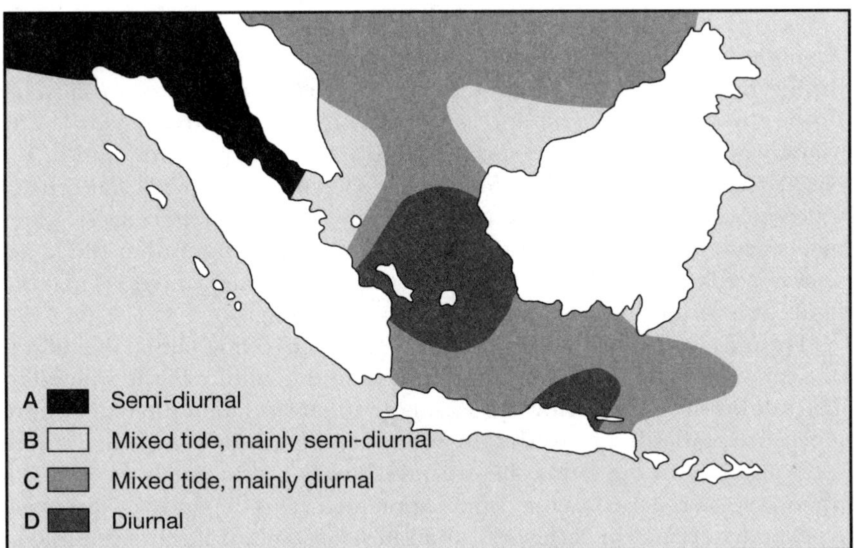

Figure 2.1. Distribution of tidal types around Sumatra. Legend as in figure 2.2.
After Wyrtki 1961

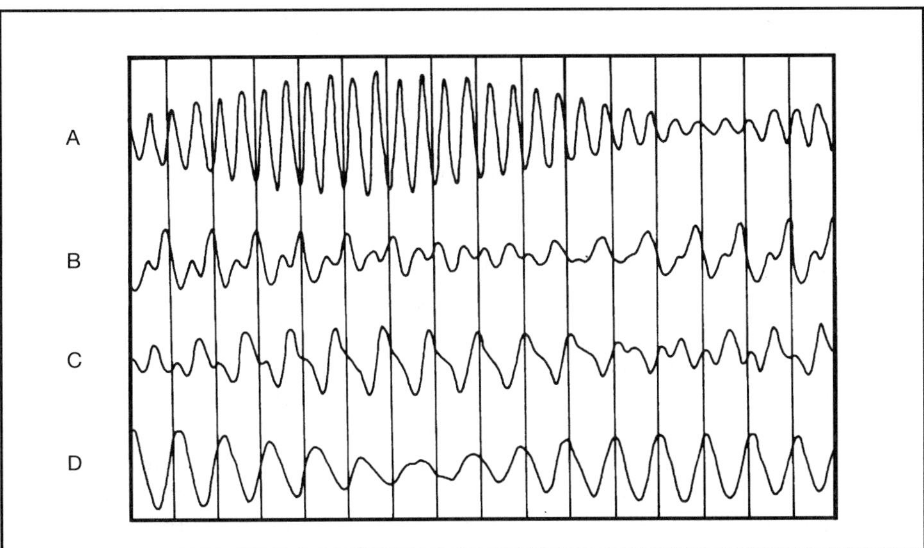

Figure 2.2. Fluctuations in tide level through a cycle of 16 days for four different tidal regimes. A - semi-diurnal tide; B - mixed tide, mainly semi-diurnal; C - mixed tide, mainly diurnal; D - diurnal tide. Refer to text for further explanation.
After Wyrtki 1961

Salinity

As might be expected, the salinity of seawater around the coast decreases during the rainy season because of the increased volumes of fresh water flowing out of the rivers, and is greatest during the dry seasons. The average annual variation in surface salinity (in parts per thousand or ‰ NaCl) for the waters around Sumatra is shown in figure 2.3. Patterns of water movement in estuaries are extremely complex. Where fresh water meets seawater, mixing does not automatically occur but rather the heavy seawater sinks below the fresh water to form a 'saltwater wedge'. This configuration varies with tides and also with river flow.

For most of the mangrove forest trees and burrowing fauna, the salinity of tidal water is probably less important than the salinity of the soil water. The salinity of this interflow (p. 133) is generally less than the water above it because of dilution by fresh rain water seeping through the soil. For tree roots and burrowing fauna the crucial factor is not so much the concentration of NaCl alone but the osmotic potential. This depends partly on soil type, being greater in clayey soils than in sandy soils. If it is not feasible to measure osmotic potential with the equipment available, then salinity and conductivity measurements represent a good second best.

Temperature

In tropical coastal waters the surface temperature is generally between 27° and 29°C. In shallow water, however, temperatures can reach at least 30°C, and the mud flats may become much hotter than this (Stebbins and Kalk 1961). Inside the actual mangrove forest the temperature is lower (dela Cruz and Banaag 1967) and the variation is more or less the same as for other shaded coastal areas. The forest areas therefore represent a rather more equitable environment for the mangrove biota than the mud flats.

Surface Currents

In general terms, seawater currents change with the rainy seasons but figure 2.4 shows that the Straits of Malacca experience a southeast current throughout the year and the west coast of Sumatra has a north/northwest current. However, the coast of southeast Sumatra has a north/northwest current in February and a south/southeast current in August. The western coast of Aceh and some of North Sumatra experience the opposite change in current. It can be seen that most of the water from rivers on the east coast will flow to the northwest, and water from rivers on the west coast to the southeast.

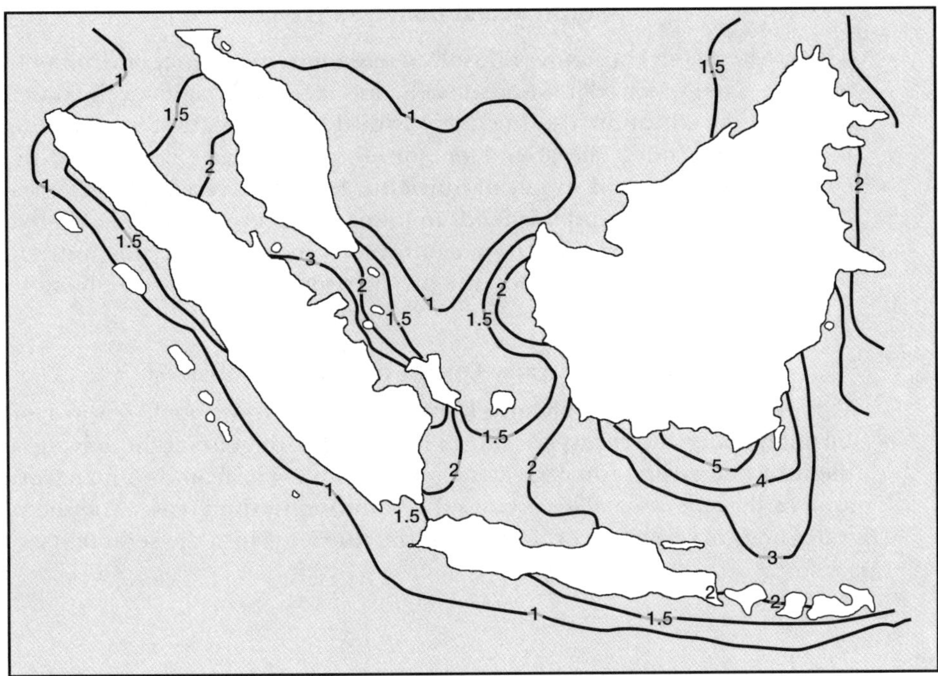

Figure 2.3. Average annual variation of surface salinity (‰)
After Wyrtki 1961

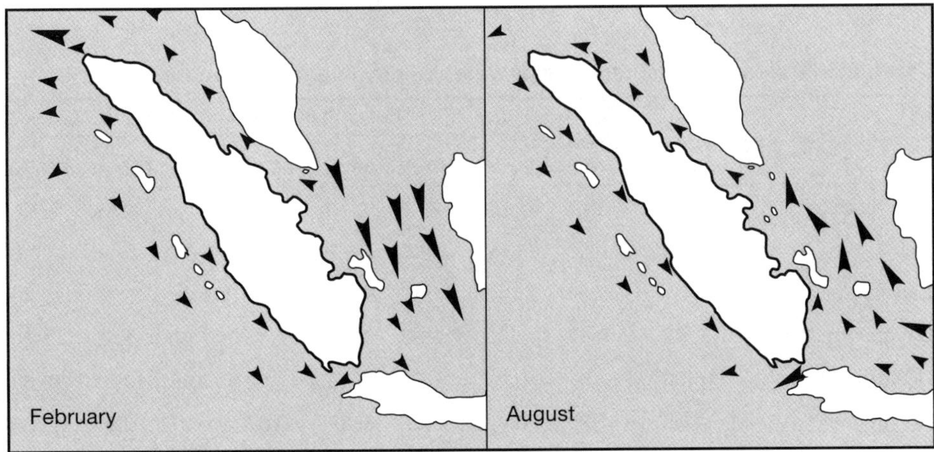

Figure 2.4. Surface currents around Sumatra in February and August.
After Wyrtki 1961

Nutrients and Dissolved Oxygen

The greatest concentration of dissolved oxygen in the coastal environment is at the open water edge where wave action repeatedly agitates the water. Oxygen concentration also fluctuates with tides, the highest values being found at high tide. The abundance of life in mangrove forests and the constantly replenished supply of nutrients, leads to a very high biological demand for oxygen, and this tends to lower the levels of available oxygen. Nevertheless, neither oxygen concentrations in free water nor nutrient levels impose any limitations on the productivity of the mangrove biota.

Summary of Water Quality in a Mangrove Forest

Variations in water quality at different locations from open sea to land through mangrove forest are shown in table 2.3. In general there is a gradient of decreasing nutrient concentrations moving from the mangrove area to the open sea. This is caused by dilution in the greater volume of water and increased incorporation of the nutrients into the sediments.

Table 2.3. Average measurements of water quality in a mangrove forest.

			Metres from forest margin						
	Open area	Forest margin	20 m	40 m	60 m	80 m	100 m	120 m	140 m
Temperature	29.5	29.1	28.7	28.2	27.9	27.8	27.6	27.5	27.6
pH	7.2	7.0	6.9	6.8	6.7	6.7	6.7	6.6	6.6
Salinity dissolved	22.21	22.04	21.46	21.18	20.96	20.67	19.87	18.62	19.49
Oxygen (mL/L)	4.40	3.37	2.99	2.68	2.36	2.15	1.99	1.55	1.70
COD (mg/L)	16.00	17.91	18.88	17.73	16.48	16.82	19.88	19.82	20.62
PO_4 (mg P/L)	1.18	1.29	0.97	0.97	0.82	0.84	0.87	0.83	0.86
NO_3 (mg N/L)	2.03	2.22	2.39	2.53	2.42	2.36	2.46	2.41	2.59

After Gomez 1980

Energy Flow and the Importance of Mangrove Vegetation to Fisheries

Mangrove forests are highly productive ecosystems but only about 7% of their leaves are eaten by herbivores (Johnstone 1981). Most of the mangrove forest production enters the energy system as detritus or dead organic matter (fig. 2.5). This detritus plays an extremely important role in the productivity of the mangrove ecosystems as a whole (Lugo and Snedaker 1974; Ong et al. 1980a,b). Its importance to offshore ecosystems is not clear (Nixon et al. 1980). Leaves and other small litter (twigs, flowers, and fruit) fall throughout the year and are broken down by macroorganisms (mainly crabs) and microorganisms (mainly fungi) into smaller particles that form the detritus. The detritus becomes rich in nitrogen and phosphorus because of the fungi, bacteria and algae growing on and within it and is therefore an important food source for many 'detritivore' animals such as zooplankton, other small invertebrates, prawns, crabs, and fish. These detritivores are eaten in turn by carnivores which are dependent to varying degrees on these organisms. Those carnivores not dependent on detritivores would be directly or indirectly dependent on planktonic benthic algae (p. 134). It is known that in the turbid coastal waters characteristic of at least the northern half of the Malacca Straits, the productivity of phytoplankton is low. It is probable, therefore, that most of the micro- and macrofauna in the mangroves and surrounding coastal areas are dependent on the productivity of litter from mangrove forests (Ong et al. 1980a,b).

The high productivity of mangroves and the physical structure and shading they provide form a valuable habitat for many organisms, some of which are of commercial importance. The most valuable mangrove-related species in Indonesia are the penaeid prawns, which support an export market worth over $150 million annually (Anon. 1979a). The juvenile stages of several of these prawn species live in mangrove and adjacent vegetation, while the adults breed offshore (Soegiarto and Polunin 1980).

The influence of mangroves extends far beyond the prawn fisheries themselves. In Sumatra, for example, the profits from the export of prawns subsidises the sale of fish which are caught in trawlers' nets (Turner 1977). The average coastal fishpond (tambak) produces 287 kg fish/ha/yr, which is more than the offshore shrimp yield, but the loss of one hectare of mangrove to tambak actually leads to an approximate net loss of 480 kg offshore fish and shrimp per year. Thus, widespread loss of mangrove to tambak is likely to cause the loss of jobs (Turner 1977). An assessment of the impact of a mangrove reclamation project on the south Java coast conservatively estimated that the development would lead eventually to loss of employment for 2,400 fishermen in the area and the loss of $5.6 million in annual income (Turner 1975).

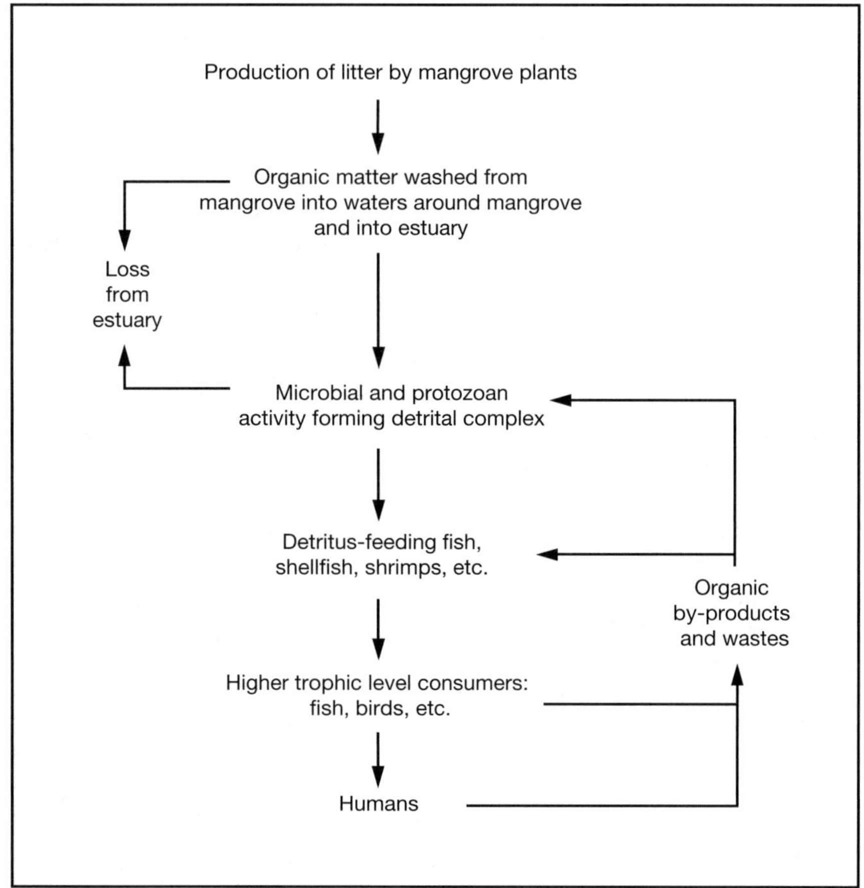

Figure 2.5. Major pathways of energy flow in a mangrove-fringed estuary.
After Saenger et al. 1981

Apart from the prawns, many other species of economic importance are associated with mangroves. These include the edible crab *Scylla serrata*, the shrimp *Acetes*, from which 'belacan' paste is made, and fishes such as *Chanos*, *Mugil* and *Lates* (MacNae 1968; Soegiarto and Polunin 1980).

The Vegetation

Tree Species

Mangrove forests are found in almost all tropical areas and it is likely that the centre of evolution was the Indo-Malesian region. This region (India, Burma, Thailand, Malaysia, Indonesia and northern Australia) has the most species-rich mangrove forests (Chapman 1977a,b) but even so contains many fewer species than other natural forest types. There are only 17 tree species (belonging to seven genera and only four families) which would be generally encountered in Sumatran mangroves (table 2.4), but over 20 different plant communities, many dominated by a single tree species, have been identified in Southeast Asian mangroves (Chapman 1977b). Details of the species found at different locations along the east coast of Sumatra and west coast of Peninsular Malaysia can be found elsewhere (van Bodegom 1929; Bunning 1944; DKPD I Riau 1979; Endert 1920; Jonker 1933; Lugo and Snedaker 1974; Luythes 1923; Ong et al. 1980a; Samingan 1980; Soeriaatmadja 1979; Steup 1941; Sukardjo 1979; Sukardjo and Kartawinata 1979; Tee 1982a; Versteegh 1951). Further Dutch references can be found in Kartawinata et al. (1979). A practical key to mangrove and estuarine trees is available (Wyatt-Smith 1979).

Table 2.4. Principle species of mangrove trees in Sumatra. For fuller lists, see Kartawinata et al. (1979). Old names which are sometimes still used are shown in brackets.

Rhizophoraceae	Avicenniaceae
Rhizophora apiculata (*conjugata*)	*Avicennia alba*
R. mucronata	*A. marina* (*intermedia*)
Bruguiera cylindrica (*caryophylloides*)	*A. officinalis*
B. gymnorrhiza	Meliaceae
B. parviflora	*Xylocarpus* (*Carapa*) *granatum* (*obovata*)
B. sexangula (*eriopetala*)	*X. moluccensis*
Ceriops tagal (*candolleana*)	
Kandelia candel	
Sonneratiaceae	
Sonneratia acida	
S. alba	
S. caseolaris	
S. griffithii	

Mangrove trees are 'halophytes'. That is, they are able to withstand the saline soil in which they are rooted, and repeated inundation by sea-water. This does not imply that they can grow only under such conditions; three species thrive next to a freshwater pond at the Botanic Gardens in Bogor (Ding Hou 1958). The occurrence of mangrove forest only along coasts indicates that these species exist there because of a lack of successful competition from other plant species which have not adapted to the salty environment.

Clearly the mangrove species are in turn unable to compete successfully against freshwater vegetation in other locations. This illustrates well the difference between fundamental niche and realised niche which is discussed on page 159. Some mangrove trees can be found growing on riverbanks over 100 km from the sea. The water they grow in may appear to be fresh, but a wedge of salt water (heavier than fresh water) may extend a great distance inland. Thus the roots of a plant may be in saline water although it grows through fresh water. A review of the response of mangrove plants to salt water is given by Walsh (1974).

A number of mangrove forest trees have peculiar rooting systems (fig. 2.6). *Rhizophora* spp. have stilt roots which help to support the trees and possibly prevent other seedlings from becoming established too close. The spike pneumatophores (roots that stick out of the ground) of *Sonneratia* and *Avicennia,* and the bent pneumatophores of *Bruguiera,* allow oxygen to enter the root system. The mangrove soil is generally waterlogged and unless there is extensive burrowing by crabs and other animals, roots a few centimetres below the soil surface are in essentially anoxic (no oxygen) conditions.

The roots of *Sonneratia* and *Avicennia* are composed of four distinct parts. The cable root runs beneath the soil surface and is held in place by anchor roots which grow downwards. The pneumatophores grow up from the cable root and small nutritive roots grow horizontally from these (fig. 2.6). In addition to the function of pneumatophores in gas exchange, it is thought that they also allow nutritive roots to grow quickly into new sediments should the soil surface suddenly rise, causing the nutritive roots to be buried within the anoxic layer (MacNae 1968; Troll and Dragendorf 1931).

The pollination ecology of mangrove trees is extremely diverse with wind, bats, birds, butterflies, moths and other insects all being pollinators. The pollination ecology of Rhizophoraceae (*Rhizophora, Bruguiera,* etc.) is described by Tomlinson et al. (1979).

A few mangrove tree species have evolved an unusual, though not unique, form of reproduction. Generally speaking, a fruit develops on a plant and when it is ripe or fully developed, the fruit or the seed(s) inside it is then dispersed and the seed germinates when or if it comes to rest in suitable conditions. In most of the Rhizophoraceae such as *Rhizophora* and

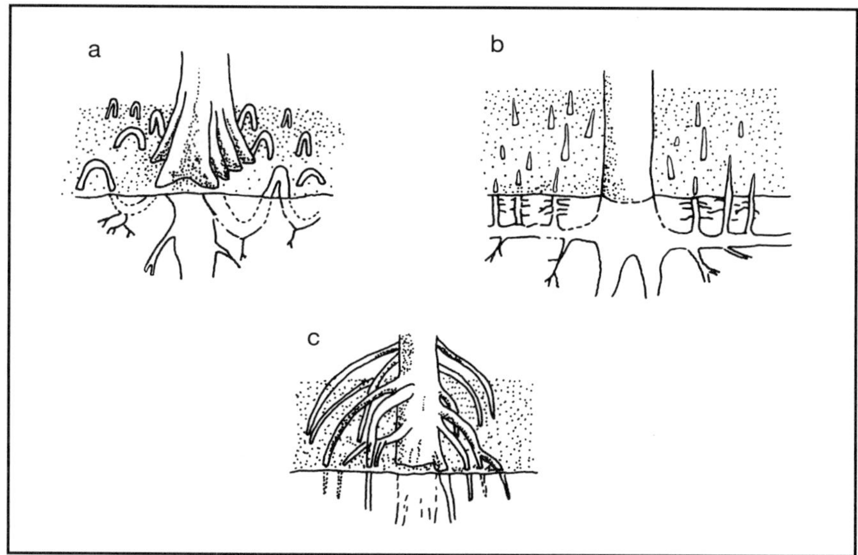

Figure 2.6. Different types of roots in mangrove trees: a- knee roots as found in *Bruguiera* spp.; b- spike roots as found in *Sonneratia* spp., *Avicennia* spp. (and sometimes *Xylocarpus moluccensis*); and c- still roots as found in *Rhizophora* spp.

Bruguiera, however, the fruits ripen and then, before leaving the parent tree, the seeds germinate inside the fruit, possibly absorbing food from the tree. The embryonic root or hypocotyl of the seedling pierces the wall of the fruit and then grows downwards. The cotyledons (first leaves) remain inside the fruit. Eventually, in *Rhizophora mucronata,* for example, the root may reach a length of 45 cm. When the seedling is fully grown it drops off by separating itself from the cotyledon tube, the scar of which forms a ring around the top of the fallen seedling, and the small leaf-bud can be seen above this scar (fig. 2.7). *Bruguiera* behaves similarly, but the break occurs at the stalk of the fruit. These types of fruit are described in more detail by MacNae (1968).

Zonation

Watson recognised five major divisions of mangrove forest in Peninsular Malaysia based on tidal regimes, such that the first division (nearest the sea) was inundated by all tides, and the fifth division was inundated only by abnormal tides (Watson 1928). These tidal zones supported different

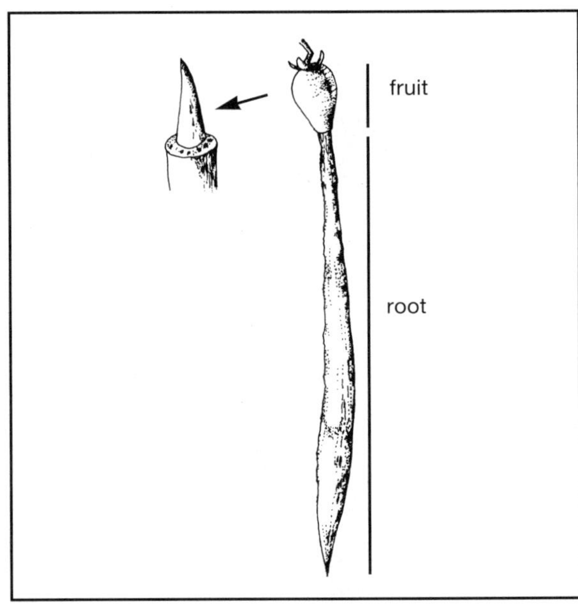

Figure 2.7. The propagule of *Rhizophora mucronata* showing the root (often mistaken for part of the fruit) and the top of the part that detaches itself from the parent plant.

types of vegetation, the species composition of which varied with the position relative to rivers and the sea (fig. 2.8). The major divisions were:

1. Forest nearest the sea dominated by *Avicennia* and *Sonneratia,* the latter growing on deep mud rich in organic matter. *A. marina* grows on comparatively firm clayey substrate which is easy to walk on, whereas *A. alba* grows on softer mud (Ding Hou 1958; Sukardjo and Kartawinata 1979; Troll and Dragendorf 1931).
2. Forest on slightly higher ground is often dominated by *Bruguiera cylindrica* and can form virtually pure stands behind *Avicennia* forest. It grows on firm stiff clays out of the reach of most tides.
3. Forest further inland dominated by *Rhizophora mucronata* and *R. apiculata,* the former preferring slightly wetter conditions and deeper mud. These trees can be 35-40 m tall (Ding Hou 1958). Also present are *Bruguiera parviflora* and *Xylocarpus granatum.* Mounds of mud built up by the mud lobster are often colonised by the large fern *Acrostichum aureum.*
4. Forest dominated by *Bruguiera parviflora* can occur in pure stands although it often invades *Rhizophora* forest after it has been clear-felled.
5. The final mangrove forest is that dominated by *Bruguiera gymnorrhiza.* The seedlings and saplings of this tree are tolerant of shade in conditions where *Rhizophora* cannot perpetuate itself. Like the she-oak

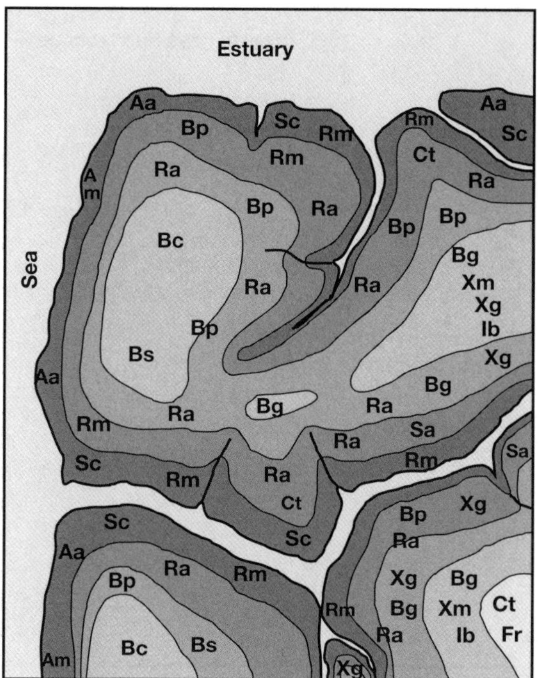

Figure 2.8. Typical but not invariable distribution of mangrove tree species near the mouth of a large river. Aa- *Avicennia alba*, Am- *A. marina*, Bc- *Bruguiera cylindrica*, Bg- *B. gymnorrhiza*, Bp- *B. parviflora*, Bs- *B. sexangula*, Ct- *Ceriops tagal*, Fr- *Ficus retusa*, Ir- *Intsia bijuga*, Ot- *Oncosperma tigillaria*, Ra- *Rhizophora apiculata*, Rm- *R. mucronata*, Sa- *Sonneratia alba*, Sg- *S. griffithii*, Xg- *Xylocarpus granatum*, Xm- *X. moluccensis*. Note that some of the names used by Watson have been modernized for this diagram. See table 2.4.

After Watson 1928

Casuarina (p. 117), however, seedlings of *B. gymnorrhiza* will not develop under the canopy of their parents. The transition between this type of forest and inland forest is marked by the occurrence of *Lumnitzera racemosa*, *Xylocarpus moluccensis*, *Intsia bijuga*, *Ficus retusa*, rattans, pandans and, at the inland margin, the tall palm *Oncosperma tigillaria*. This succession is not always visible, particularly where man has disturbed the forest; in such conditions the fern *Acrostichum aureum* is very common (p. 347).

The above descriptions refer mainly to gently sloping, accreting shores, but where there are creeks, bays or lagoons, *Rhizophora* is usually the pioneer tree. This is common in areas near coral reefs (p. 124).

The succession generally followed by mangrove vegetation under different conditions is shown in figure 2.9.

The tendency of the mangrove forest to occur in distinct floral zones has been interpreted by various authors in many different ways. These have been summarised by Snedaker in the following four categories: plant succession, geomorphology, physiological ecology, and population dynamics (Snedaker 1982b).

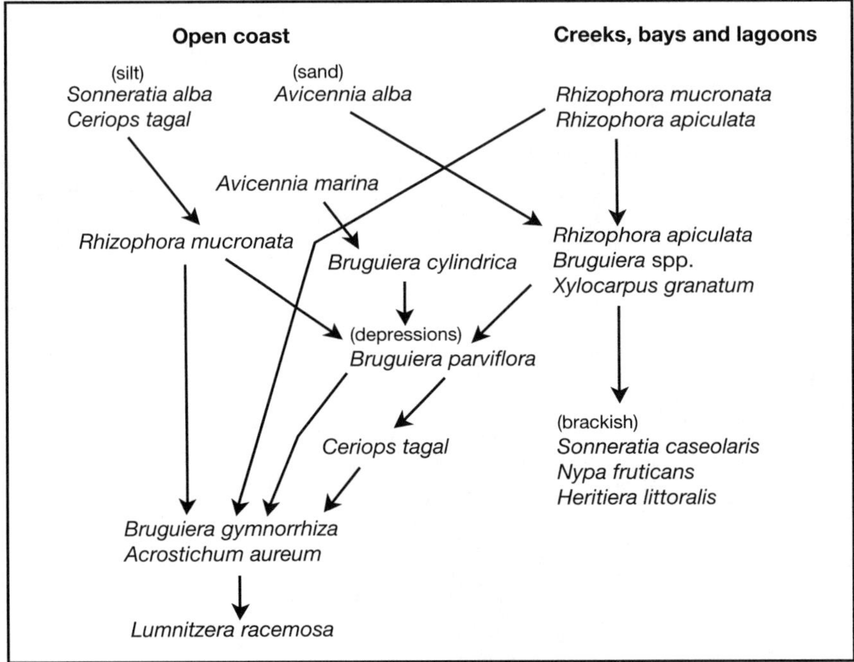

Figure 2.9. Succession in mangrove forests.
After Chapman 1970 in Walsh 1974

Plant Succession

Plant succession is a classic ecological concept and is defined as being the progressive replacement of one plant community with another (e.g., bare ground–alang-alang–scrub–secondary forest–primary forest [see p. 340]). Much of the early work on mangrove forests focused on its supposed land-building role. It seemed clear from this that one species colonised an exposed bank of mud, and as conditions changed (such as the increase in the organic debris of the mud) so other species took over. Thus at any one spot, a succession of species or communities of species would be observed over a period of time. It is clear however that in most places, including Sumatra, the stages of succession are not always consistent and different local environmental conditions and man's impact on those have an influence.

Budowski (1963) compiled a list of 20 characteristics of secondary succession in tropical forests which distinguish between early and late stages. If this is compared with the stages of succession for mangroves, only seven of the characters apply, nine do not, and four are inconclusive. This suggests that attributing the apparent zonation to succession is not the whole story (Snedaker 1982b).

Geomorphological Change

Early workers on mangrove felt that it was mangroves that 'built land'. However, it is clear from observations that in some areas such as the huge deltas of the Ganges, Indus and Irrawaddy, it is the process of sediment decomposition that builds land. It appears to most workers now that mangroves do not have any influence on the initial development of the land forms. Mangroves may accelerate land extension but they do not seem to cause it (p. 88) Van Steenis tells of engineers working at Belawan (the port north of Medan) who planted mangrove seedlings on a new port extension to stabilise it. The attempt failed, thereby demonstrating that it is not possible to force accumulation of silt (Ding Hou 1958).

From a geomorphological perspective, it is the shape, topography and history of the coastal zone that determine the types and distributions of mangrove trees in the resulting habitats. The position of species relative to tidal levels (and thus soil type) is obviously important, and Watson (1928) considered the pattern of tidal inundation and drainage to be the major factor in mangrove zonation. This idea was developed by Lugo and Snedaker (1974) to include variation in the salinity of the tidal water and the direction of its flow into and out of the forest. Although good correlations exist between salinity and zonation, they are not proofs of direct cause and effect.

Physiological Response to Soil-Water Salinity

If salinity is an actual cause of zonation in mangrove forests, it needs to be shown that the plants actually respond through their physiology to salinity gradients and not to factors such as oxygen levels, which under certain conditions will fall with decreasing salinity (Ding Hou 1958). It has already been mentioned that mangroves are able to live in fresh water (i.e., they are facultative not obligate halophytes) but each species probably has a definable optimum range of salinity for its growth. Indeed, Lugo et al. (1975) found that within each zone the characteristic species had apparently maximised its physiological efficiency and therefore had a higher metabolic rate than any invading species. These invading species would be at a competitive disadvantage due to their lower metabolic efficiency in that habitat. Similarly, it has been found that each species of the mangrove forest grows best under slightly different conditions, such as the amount of water in the mud, the salinity, and the ability of the plant to tolerate shade (Lind and Morrison 1974). This means that the various species are not mingled together in a haphazard way but occur in a fairly distinct zonation.

As was shown on pages 74 and 76, salinity can vary considerably between high and low tides and between seasons, and thereby presents a confusing picture to a scientist conducting a short-term study. Thus, to identify the salinity levels to which the different species are optimally adapted requires long-term and detailed measurements to determine long-term averages and ranges of salinity.

Differential Dispersal of Propagules

The suggestion that differential dispersal of propagules (fruit, etc.) influences zonation of mangroves rests on the idea that the principal propagule characteristics (e.g., size, weight, shape, buoyancy, viability, numbers, and location of source areas) result in differential tidal sorting and therefore deposition. There are as yet few data to support this hypothesis (Rabinowitz 1978) and interested readers should refer to Snedaker (1982b) for a thorough discussion.

Geomorphology and environmental physiology thus appear to be the most relevant of the topics discussed above to further the understanding of zonation and plant succession in the mangrove forest. However the impact of human activities, so ubiquitous in coastal regions, plays a major role in modifying species composition, and physical conditions, and so should also be considered in any study of zonation.

The term 'zone' itself needs to be clarified. Bunt and Williams (1981) suggest that recognisable zones may arise through at least two causes: situations where neighbouring vegetation associations have little or no floristic affinity although growing in the same environmental conditions (continuous variation); and situations where environmental gradients exist at a scale to permit sudden changes between related vegetation associations (discontinuous variation). Thus vegetational changes can be continuous, discontinuous or a combination of both. This, they argue, is why it is crucial to consider scale before attempting an analysis of mangrove forest and why, in the absence of such consideration, data from isolated transects, even from the same area, are so hard to interpret. Comparisons of transect data from different sites are useful in compiling inventories and noting similarities but are not the basis for a discussion of zonation.

Biomass and Productivity

Biomass is a term for the weight of living material, usually expressed as a dry weight, in all or part of an organism, population or community (Ricklefs 1979). It is commonly expressed as the biomass density or biomass per unit area. Plant biomass[2] is the total dry weight of all living plant parts and for convenience is sometimes divided into above-ground plant biomass (leaves, branches, boughs, trunk) and below-ground plant biomass (roots).

It appears that no study of mangrove biomass has been conducted in Sumatra, but several studies have been conducted on the other side of the Straits of Malacca in Peninsular Malaysia. Thus the biomass of trees in the Sungai Merbok Forest Reserve was between 122 and 245 t/ha (Ong et al. 1980a). In the Matang Forest Reserve, however, which has been exploited for timber on a sustained basis for 80 years, and which receives silvicultural management, the biomass of trees was 300 t/ha (Ong et al. 1980b). As explained below, the higher biomass in managed forest is not unexpected.

Biomass is a useful measurement but it gives no indication of the dynamics of an ecosystem. Ecologists are interested in productivity because if the dry weight of a community can be determined at a given moment and the rate change in dry weight can be measured, these data can be converted into the rate energy flow through an ecosystem. Using this information, different ecosystems can be compared, and their relative efficiencies of converting solar radiation into organic matter can be calculated (Brock 1981).

Plant biomass increases because plants secure carbon dioxide from the atmosphere and convert this into organic matter through the process of photosynthesis. Thus, unlike animals, plants make their own food. The rate at which biomass increases is the 'gross primary productivity'. This depends on the leaf area exposed, amount of solar radiation, temperature, and upon the characteristics of individual plant species (Whitmore 1984). During the day and during the night, plants, like all other living organisms, respire and use up a proportion of the production formed through photosynthesis. What remains is called 'net primary productivity' and over a period of time this is termed 'net primary production'. Net primary productivity is obviously greatest in a young forest which is growing, and it should be remembered that a dense, tall forest with a high biomass does not necessarily have a high net primary productivity. Large trees may have virtually stopped growing. Indeed in an old 'overmature' forest, the death of parts of the trees and attacks by animals and fungi may even act to reduce the plant biomass while net primary productivity remains more or less constant. The major aim of silvicultural management in forests or timber plantations is to maximise productivity, and the trees are usually harvested while they are still growing fast and before the net primary productivity begins to decrease.

One means of assessing net primary production is to measure the rate at which litter is produced. The production of litter at Merbok Forest Reserve was found to be 8.1 t/ha/yr for leaf litter and 12.0 t/ha/yr for total small litter (mainly leaves but with twigs, flowers and fruit) (Ong et al. 1980a). At Matang, Ong et al. obtained figures for production of 8.2 and 10.4 t/ha/yr (Ong et al. 1980b). These figures are similar to those obtained in lowland forest (p. 205).

The approximate annual accumulation of litter on the mangrove forest floor at Merbok was calculated to be 0.33 t/ha for leaf litter and 1.13 t/ha for total litter. An experiment at the same site revealed that 40%-90% of fallen leaves were lost after 20 days on the forest floor. The major agents in the disappearance were probably crabs which either bury them (eventually to form peat) or eat them, later to be excreted as detritus. At Matang, the time required for total leaf decomposition in areas with virtually no crabs (i.e., with only microorganisms involved in decomposing) was 4-6 months (Ong et al. 1980a).

In calculations of total productivity the contribution made by soil algae, algae growing on roots, and phytoplankton should not be overlooked, as their combined contribution to the total can be quite significant (Walsh 1974).

Tannins

It used to be thought that the leaves of mangrove trees were exceptionally rich in tannins, a view originating perhaps from the commercial use of mangrove bark for producing tannins for the leather industry (Walsh 1974, 1977). Recent analyses, however, suggest that tannin content of mangrove tree leaves exhibits a wide spectrum of concentrations, with *Bruguiera* having the highest levels. It may well be that mangroves have a higher-than-average concentration of tannins, but not significantly more than in other forests growing under somewhat adverse conditions (P.G. Waterman, pers. comm.) (pp. 175 and 256). Bark of trunks usually contains more tannins than bark of twigs, which has in turn more tannins than the leaves. Mangrove barks often contain 20%-30% dry weight of tannins (Walsh 1977), and although high, this it is not outside the range found in lowland forest barks (Whitten, unpubl.).

Tannins may play an extremely important role in the interactions between herbivores and leaves (pp. 110, 177, 230 and 258).

The Coastline of Eastern Sumatra and the Role of Mangroves in Land Extension

It used to be thought that in historical times the southern Sumatran coast comprised a series of large bays. Obedeyn (1941, 1942) studied old maps dating from 1030 to 1600 A.D. and proposed that both Jambi and Palembang had been ports at the end of promentories. Thus the dashed line in figure 2.10 would correspond to the coast in about 1000 A.D., and the dot-dash line to the coast in about 1600 A.D. The supposed infilling of the bays by silt since those times required the shoreline to have advanced at a rate of over 100 m per year, but recent studies in the Musi-Banyuasin area had

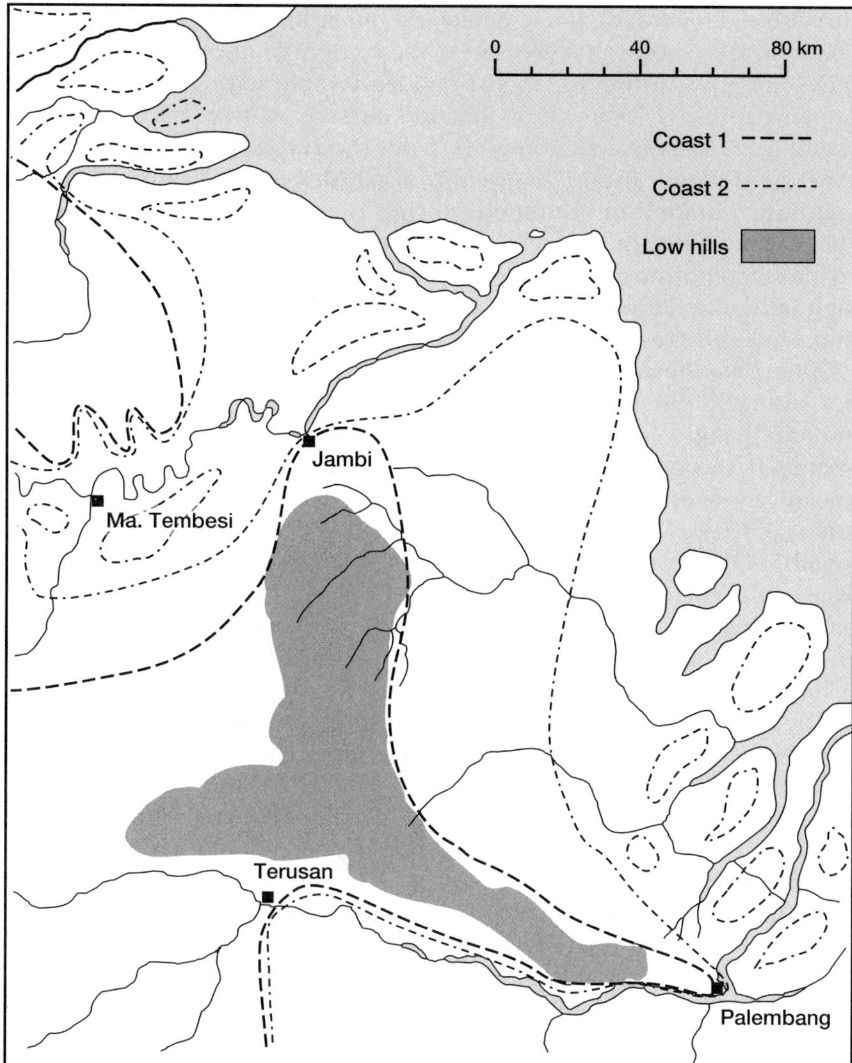

Figure 2.10. Changes in the coastline of part of eastern Sumatra.
After Obedeyn 1941, 1942

shown that present rates of accretion were over five times less, at only 20 m per year (Chambers and Sobur 1977). This discrepancy was not easy to explain, but Chambers (1980) suggested debris from volcanoes active during earlier periods could have caused the higher accretion rate.

It now seems, however, that the coastlines proposed by Obedyn certainly did once exist, but in geological rather than historical time. Investigations have shown that areas near the coast of Jambi had habitation at times when, according to Obedyn, no land would have existed. Cultural remains found 1.75 m below the soil surface at Muara Kumpeh, for example, have revealed stoneware from the twelfth to fourteenth centuries, and a coin from the seventh or eighth century. Chinese sources mention 'Kompel' or 'Kumpeh' during the sixth century (McKinnon 1982). These and other archaeological finds suggest that early rates of coastal accretion were not the same as present rates. Indeed, the current high rates of soil erosion caused by extensive bad land use have doubtless dramatically increased the rates of coastal accretion.

In the southern part of Sumatra, the areas of maximum accretion are not, strangely, directly related to the major drainage systems. For example, opposite Bangka Island lies one of the broadest stretches of 'tidal swamp' in Sumatra, extending over 100 km inland. This appears to be because of the longshore drift carrying sediment eastwards from the Musi/Banyuasin drainage system, and such eastward currents occur in the region for nine months of the year (Wyrtki 1961). Chambers and Sobur (1977) suggest that over half of the deposition of sediment from Musi/Banyuasin may occur outside the river delta itself, because the relatively exposed delta experiences considerable erosion. Thus, no conventional delta can be formed (Chambers 1980).

As described on page 85, mangrove forests do not 'build' land but they do have a vital role in colonising emerging mud banks, consolidating sediments and thereby accelerating the process of land extension (Bird 1982; Bird and Barson 1977; Walsh 1974). The pioneer colonisation by species of *Sonneratia* produces a network of pneumatophores which have three indirect functions:

- they help to protect the young trees from wave damage,
- they entangle floating vegetation, and
- they provide a habitat for burrowing crabs which help to aerate the soil (Chambers 1980).

If the mud banks were not colonised, it is likely that shifting currents, seasonally increased flows, or strong wave action could wash them away. Thus, at the mouth of rivers, deposition which typifies coastal extension is in the form of gradually enlarging offshore islands which ultimately coalesce (although occasionally separated from the mainland by small tidal channels [Chambers 1980]). This phenomenon can be seen near the mouths of the rivers Besitang, Bile/Barumun, Rokan, Siak, Kampar, and Inderagiri, as well as in the Musi/Banyuasin area.

Along the shore line between the major rivers, the sediment deposition is more or less parallel to the established shoreline, particularly in sheltered bays where wave action is minimal. Species of *Sonneratia* in these areas are again important as pioneer colonisers (Chambers 1980).

FAUNA

The fauna of mangrove forests can be divided conveniently into two groups: the essentially aquatic component of crabs, snails, worms, bivalves and others which depend directly upon the sea in various ways; and the terrestrial, often 'visiting', component including insects, spiders, snakes, lizards, rats, monkeys and birds (see pp. 109-111) which do not depend directly upon the sea. The majority of what follows, based on Berry (1972), deals with the aquatic component.

The animals mentioned below represent only a fraction of the total mangrove fauna, but the important points to be understood are that there are many different habitats within the mangrove forest, each of which supports a distinct community of animals, and that a proportion of the fauna in mangrove forests is more or less restricted to mangroves. Knowledge of the coastal fauna of the Sunda Region is not yet sufficiently advanced to decide whether any of the mangrove fauna is absolutely restricted to mangroves (Chapman 1976, 1977a; Hutchings and Reicher 1982; MacNae 1968).

The Challenges of Living in Mangroves

In comparison with other shore fauna, the mangrove forest fauna is heavily dominated by many species of crabs and snails. There are few types of worms, and very few bivalve molluscs, coelenterates (sea anenomes), echinoderms (starfish, sea urchins), etc. (fig. 2.11). The dominance of crabs and snails may be largely ascribed to the peculiar conditions found in mangrove forests to which these two groups of animals have evolved successful adaptations.

Animals living on the soil of the intertidal area are subjected to long periods during which they are not covered by the sea. At the mean high water level of spring tides (MHWS), for example, the soil is left exposed for about 270 days per year, and for up to 25 consecutive days; at the lower terrestrial margin, soil is left exposed for about 320 days per year, and for up to 30 consecutive days. Most marine animals cannot stand this because:

- they quickly dry out,
- most cannot respire in air,
- many can only feed on waterborne food, and
- many must release spermatozoa, eggs and larvae into seawater.

Crabs and snails, however, have:

- impervious exoskeletons/shells to restrict water loss,
- many can breathe air,
- many feed on microorganisms or organic materials,
- many climb into the trees to find these foods, and

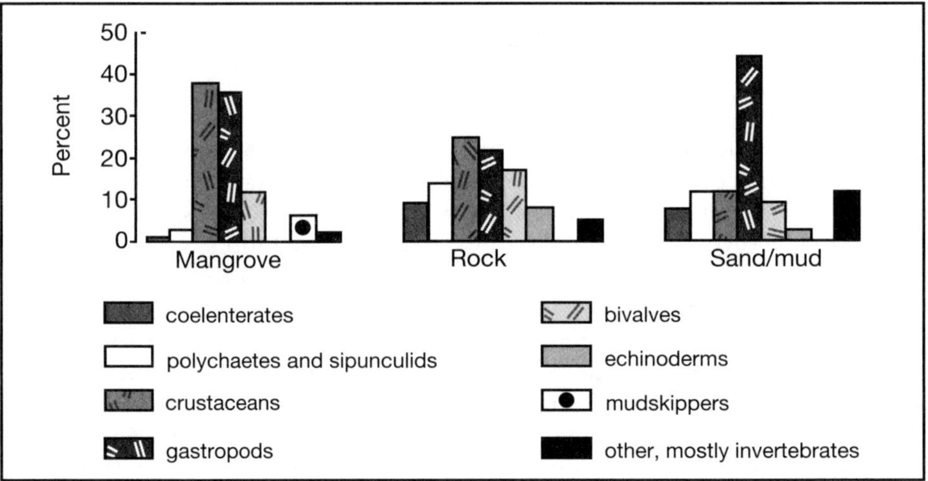

Figure 2.11. Percentages of total aquatic animal taxa recorded at three types of shore (omitting microscopic forms). Note that crustaceans and gastropod molluscs account for 74% of the total on the mangrove shore. Note also that these data are not based on complete lists or on equally exhaustive surveys, but they serve to illustrate the major differences.

Data from Berry 1972

- they have internal fertilisation and protect their eggs and early developing young in capsules or brood pouches.

The soil in mangrove forests is subjected to salinities ranging from nearly zero to 50‰ (Sasekumar 1974). During spring tides, salinities approximate those of tidal seawater, that is, about 27‰-32‰. During neap tides, however, when the landward mangrove soil is not covered by the sea for days at a time, rainfall may reduce salinity to less than 15‰. Conversely, evaporation without rainfall can increase salinity to more than 32‰.

Most marine animals are only able to withstand very minor variations in salinity such as are experienced in the open sea (see again fig. 2.3) because they cannot regulate the salt/water balance of their body fluids except within quite narrow limits. Many crabs are able to cope with salinity variation by precise osmoregulation, whereas many snails can allow the salt concentration of their body tissues to vary without ill effect. These adaptations thus allow crabs and snails to predominate in the mangrove fauna.

Zonation and Characteristics of the Aquatic Fauna

Faunal zonations within mangrove forest are effectively three-dimensional. The vertical component is shown, for example, by the worms and crabs bur-

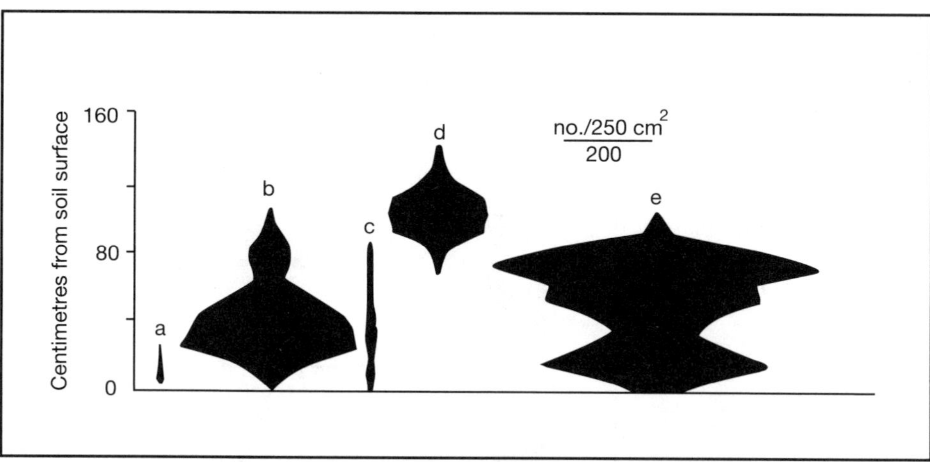

Figure 2.12. Relative density of several species of tree-dwelling fauna distributed vertically on trees in an *Avicennia* forest, 10 m from the sea. a-snail *Thais tissoti;* b-barnacle *Balanus amphitrite;* c-bivalve mollusc *Crassostrea cucullata*; d-barnacle *Chthamalus withersii*; e-mussel *Brachyodontes variabilis.*

After Tee 1982a

rowing into the mud, some snails living on the mud surface, and other snails and barnacles living in the trees (fig. 2.12). One horizontal component is shown by the different species observed on the foreshore, the seaward mangrove forest and at the limit of tidal influence (fig. 2.13). The other horizontal component is represented by the different features such as small streams, eroding banks, and mud mounds that occur irregularly through the mangrove forest. These forms of zonation set mangrove forest apart from other shore line ecosystems such as sandy shores or rocky cliffs, which lack the large-scale dimension of distance from the sea which can be over 50 km for mangroves on the east coast of Sumatra. The most complete studies of zonation of mangrove fauna in or near the Straits of Malacca are those of Berry (1972), Sasekumar (1974) and Frith et al. (1976).

The distribution of the macrofauna (animals visible to the naked eye) is influenced by the nature of the soil (particle size, consolidation, and content of organic matter and moisture), the vegetation and tidal factors. The true mangrove forest (Zone 4 below) has the highest animal abundance and species diversity because of its shade, complexity and abundance of organic detritus (Frith et al. 1976).

Ecologists have found that dividing ecosystems into recognisable subdivisions or zones is fraught with difficulties because vegetation types often intergrade with each other. Nevertheless, as long as the subdivisions are not

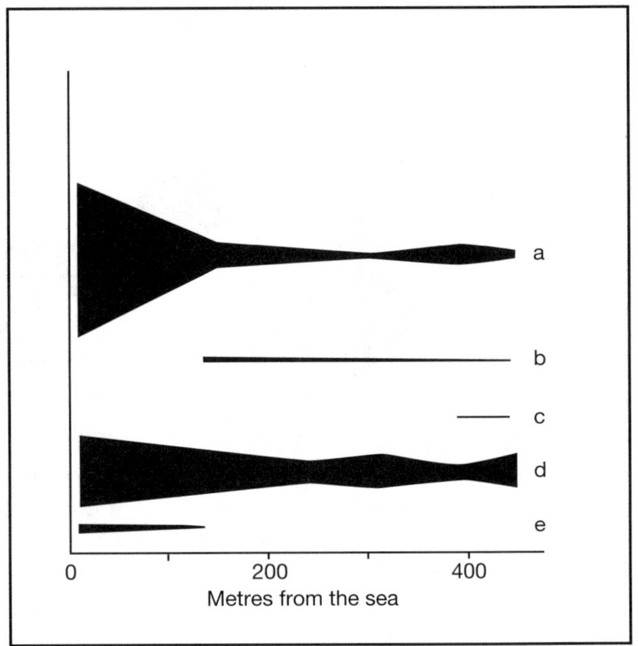

Figure 2.13. Relative abundance of several species of tree-dwelling fauna along a transect through mangrove forest.
a-barnacle *Balanus amphitrite;*
b-snail *Nerita birmanica;*
c-snail *Littorina scabra*;
d-barnacle *Chthamalus withersii*;
e-polychaete worm *Nereis* spp.
After Tee 1982a

regarded as rigid, and it is clear on what criteria they are based, such an approach provides a convenient means of describing the different areas (fig. 2.14). The mangrove zones described by Berry (1972) are as follows:

Zone 1: Seaward foreshore. As those field scientists who have tried to work in mangroves know to their discomfort, the seashore below the mangrove tree edge is usually extremely soft mud. This soil commonly comprises about 75% very fine sand (particles of 20-210μ), most of the remainder being even finer silt (Sasekumar 1974). It extends down from the mean low water level of neap tides (MLWN) and so is covered by every tide of the year and never left exposed for many hours. The fauna is essentially marine with certain crab species, a few snails such as *T. telescopium* (fig. 2.15) and two or three species of polychaete worms (fig. 2.16) predominating. A variety of mudskippers – an unusual group of fish capable of living out of water for short periods – occur commonly along the water's edge and in burrows in the mud. An excellent review of general mudskipper biology can be found in MacNae (1968).

Underwater, mudskippers breathe just like other fishes but in air they obtain oxygen by holding water and air in their gill chambers. This is supplemented by gas exchange through their skin and fins (Stebbins and

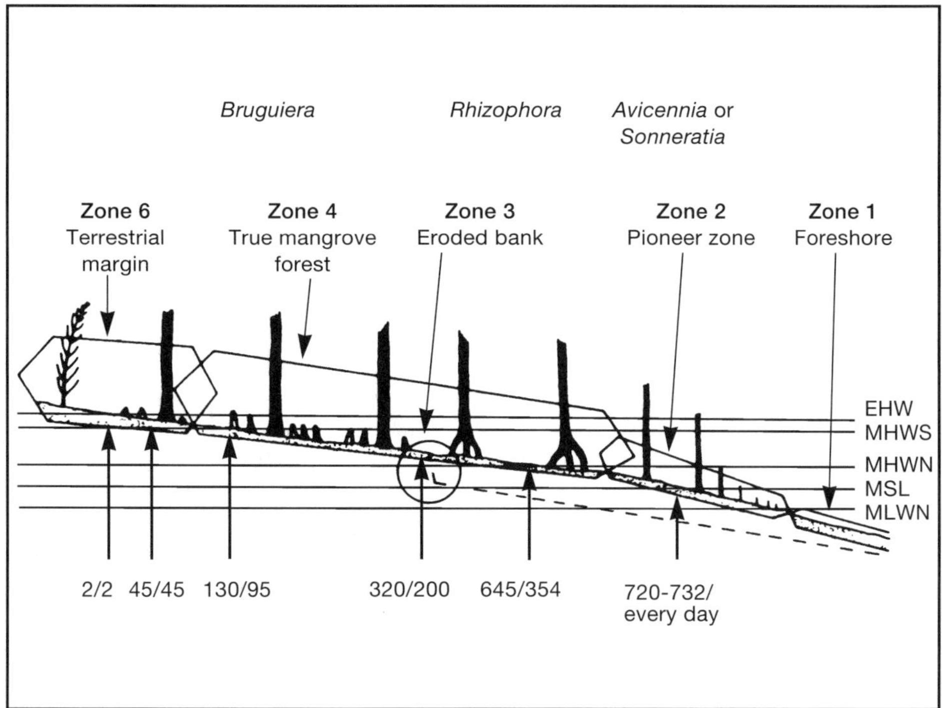

Figure 2.14. Major zones in a mangrove forest. Horizontal distances extremely reduced for clarity. Streams (Zone 5 in the text) are omitted. Figures represent tide coverings per year and days per year with tidal wetting. An eroded bank (Zone 3 in the text) is shown superimposed in dashed lines. Common tree species are indicated above. EHW = extreme high water, MHWS = mean high water of spring tides, MHWN = mean high water of neap tides, MSL = mean sea level, MLWN = mean low water of neap tides.

After Berry 1972

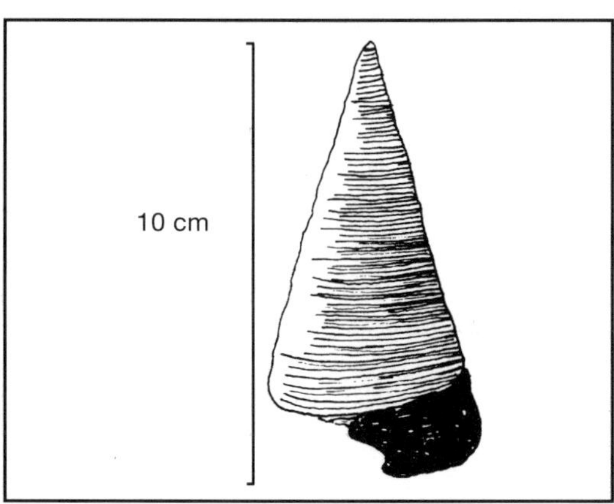

Figure 2.15. *Telescopium telescopium.*

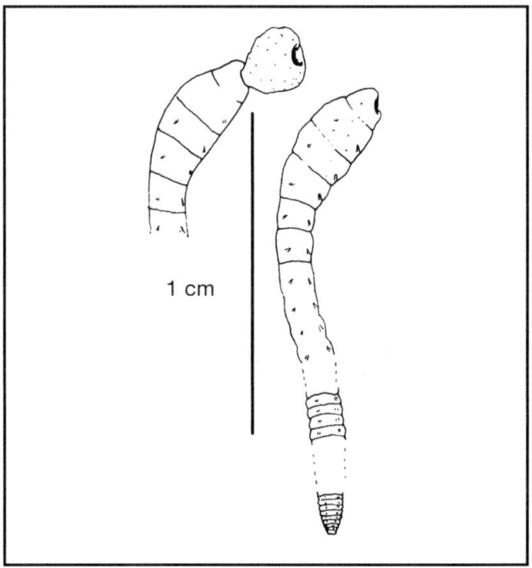

Figure 2.16. A polychaete worm common in mangrove soil – *Leiochrides australia*; this species commonly reaches 14 cm long.

Kalk 1961). On land mudskippers move in a variety of ways: they lever themselves forward on their pectoral fins, which move in synchrony with each other (i.e., not true walking), leaving characteristic tracks in the mud, they skip over the mud by flicking their tails, and they climb on vegetation using their pectoral and pelvic fins.

Although they look very similar, the different species of mudskipper have very different diets; some take mud into their mouths, retain algal material, and blow out the rest; some are omnivorous, eating small crustaceans as well as some plant material; and some are voracious carnivores, feeding on crabs, insects, snails and even other mudskippers (Burhanuddin and Martosewojo 1979; MacIntosh 1979).

It can be safely assumed that the modern mudskippers evolved from the same single ancestral mudskipper. It is not possible to determine what its feeding habits were, but it is likely to have been a generalised detritus-feeder. Genetic variations in individuals of this mudskipper made it possible for them to specialise in exploiting one particular food resource or group of food resources with more efficiency and thus greater competitive advantage. Although a taxonomic/ecological study of Sumatran mudskippers has not been undertaken, a study in Queensland has shown that at least five species of predatory mudskippers can live in the same area by successfully partitioning their habitat and its resources between them (Nursall 1981) (see also pp. 155, 240 and 326).

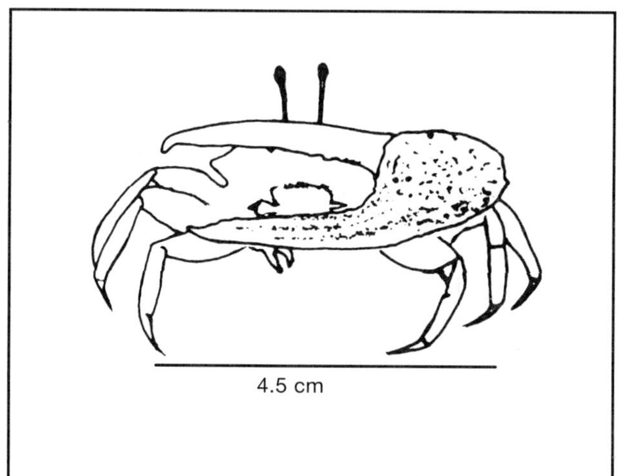

Figure 2.17. Fiddler crab *Uca dussmieri.*
After Berry 1972

If an observer stands still on the foreshore, animals which have been scared into their burrows by his approach will start to reappear. Hundreds of fiddler crabs *Uca* emerge, in densities as high as 50 per m^2. The larger mudskippers are their main predators (others include snakes, spiders, birds and monkeys) (MacIntosh 1979) and it would seem reasonable to ask why the mudskippers 'allowed' such high populations of potential food to exist.

Rates of prey exploitation depend upon the adaptations of both predator and prey. If predators evolved prudence so that they deliberately avoided overexploiting their prey, it would both oppose the pre-eminent goal of evolution which is to maximise the fitness or advantage of the individual, and also require that predator populations were regulated altruistically to maintain an optimum density. Natural selection should favour maximum exploitation of prey by each individual mudskipper at the expense of optimum prey management.

The level of exploitation of a prey population is determined by the ability of the predator to capture prey, balanced by the ability of the prey to avoid being captured. In this case, the speed of attack by mudskippers is balanced by the awareness and speed of retreat into burrows by the crabs. Both of these skill are evolved and tend to result in a more or less stable equilibrium between the numbers of predators and prey.

In an area of mangrove foreshore 10 m wide and 20 m from the sea edge to the first mangrove seedlings, it would not be unusual to find an average of about 10 *Uca dussumieri* fiddler crabs (fig. 2.17) per square metre – therefore 2,000 over the entire area – and a total of four large mudskippers

Periophthalmodon schlosseri. This gives a predator-prey ratio of 1:500. Ricklefs (1979) lists various predator-prey ratios taken from the literature, and their range is 1:30-1:1,263.

Most mudskippers occupy deep burrows, the tops of which are turreted in some species. Their inter- and intraspecific relationships are complex but burrows are usually evenly spaced on the mud, thereby reducing the likelihood of conflicts (Stebbins and Kalk 1961). Males seem to dig the burrows and make displays to attract passing females by jumping, and by erecting their dorsal fins. After pairing, eggs are laid on the sides of the burrow which is then defended by the male. Such territorial defence is rare or absent outside the breeding season (Nursall 1981).

Zone 2: Marine pioneer zone. Actively accreting pioneer zones rise gently through the region of neap low tide and mid-tide levels as bare mud gives way to seedlings and saplings, often of *Avicennia* and *Sonneratia.* The soil differs little from that of Zone 1. Most of this zone except the highest parts is covered by all the high tides of the year. This zone includes fewer common marine animals and begins the more typical mangrove fauna. Snails such as *Syncera, Salinator,* and *Fairbankia* may occur on the wet soil (fig. 2.18). Mudskippers are common, and the large fiddler crab *Uca dussumieri* and the smaller *U. mani* can be very numerous.

Horseshoe crabs (fig. 2.19) are periodically seen in this zone. They are distantly related to spiders but are the sole representatives of a group of animals which dominated the sea during the Palaeozoic period over 250 million years ago. Fossil horseshoe crabs from the Silurian period 400 million years ago are virtually indistinguishable from the five species surviving today. The reason for the lack of change in these animals is hard to explain but it may be that they simply have not experienced effective competition or predation from other species to select for refinements or other adaptations.

Horseshoe crabs live in shallow water along sandy or muddy shores. They can swim through the water by flapping appendages on their abdomens. They feed on soft-shelled molluscs and worms, digging for them using shovelling movements of their shell. When the moon is full and the tide is high, horseshoe crabs come up to the beach to breed. The females, which are larger than the males (van der Meer Mohr 1941b), dig a shallow nest for their thousand or so eggs. The male clings to the back of the female's shell, ready to release his sperm when the eggs are laid. One month later, again at high tide, the eggs hatch and the small larvae join the other marine zooplankton. Sumatran horseshoe crabs are of two species: one with a tail triangular in cross-section, *Tachypleus gigas,* and one with a tail circular in cross-section, *Carcinoscorpius rotundicauda.*

An extract of horseshoe crab tissue assists medical science in two unlikely ways. It is used in an extremely sensitive test for checking whether

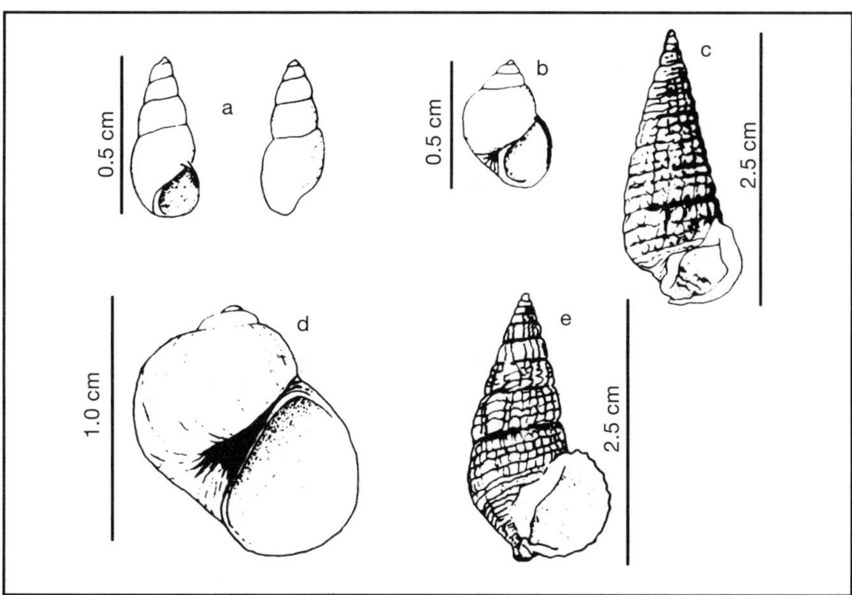

Figure 2.18. Some snails from mangrove soil. a-*Fairbankia* sp.; b-*Syncera brevicula*; c-*Cerithidea cingulata*; d-*Terebralia sulcata*; e-*Salinator burmana* (a lung-bearing pulmonate snail).

From Berry 1872

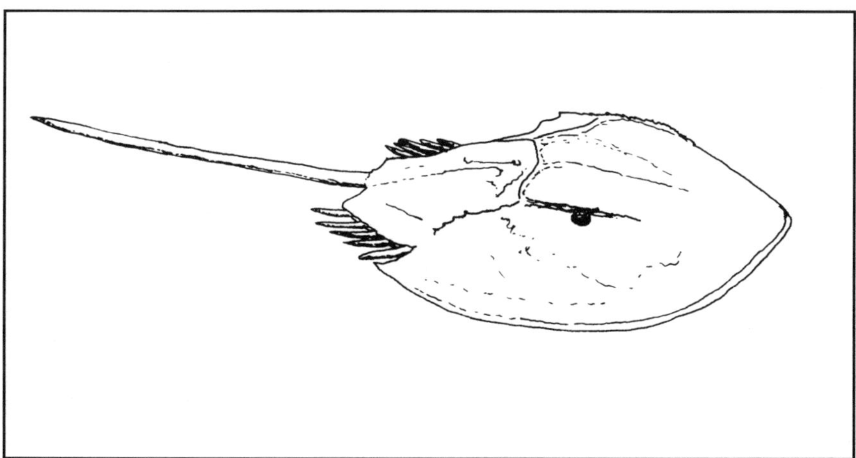

Figure 2.19. Horseshoe crab.

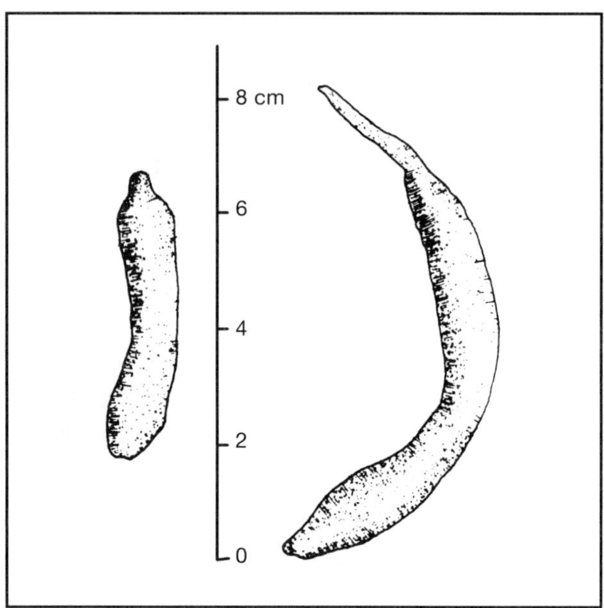

Figure 2.20. Peanut worm *Phascolosoma.*
After Berry 1972

vaccines and intravenous fluids are contaminated by bacterial toxins and in a test for gonorrhoea.

Various species of polychaete worms, occasional bivalve molluscs and the peculiar sipunculid peanut-worm (fig. 2.20) live permanently in the soil of this zone. Peanut worms rest vertically in the soil and sweep their long proboscis over the soil surface, picking up particles of organic matter.

The abundance and type of fauna living on the vegetation depends largely on the age of the tree – older ones have denser populations of more species. *Littorina* snails (fig. 2.21) occur on almost all the vegetation (Lim 1963), sometimes up to 2 m above the soil surface. Populations of sedentary animals encrust the lower stems of trees as they grow. These animals typically include: barnacles, with the larger ones *Balanus amphitrite* below and the smaller *Chthamalus withersii* extending higher; oysters, commonly *Crassostrea cucullata;* and the small black mussels *Brachyodontes* sp. which are attached to the tree by 'byssus' threads. A total of 15,401 animals, 9,199 of which were mussels of the genus *Brachyodontes,* were found on a single *Avicennia* tree (Tee 1982b). These attached fauna may be eaten off the lower stems by the carnivorous snails *Thais* and *Murex* (fig. 2.22). Barnacles and mussels sometimes suffer 50% mortality in the first 10 m above the soil surface, but the number of dead animals decreases upward, indicating

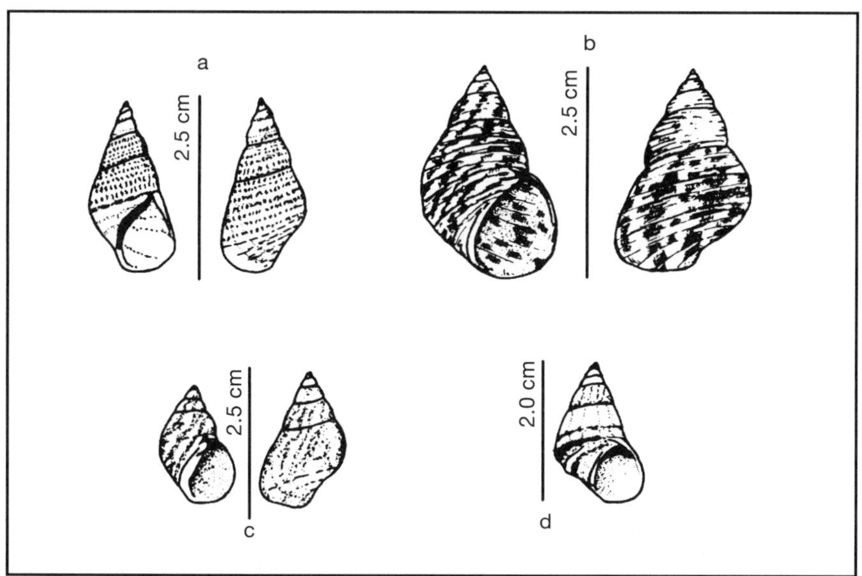

Figure 2.21. Four species of *Littorina* commonly found in mangrove forest near the seaward edge. a - *L. melanostoma*; b - *L. scabra*; c - *L. nudulata*; d - *L. carinifera.*

From Berry 1972

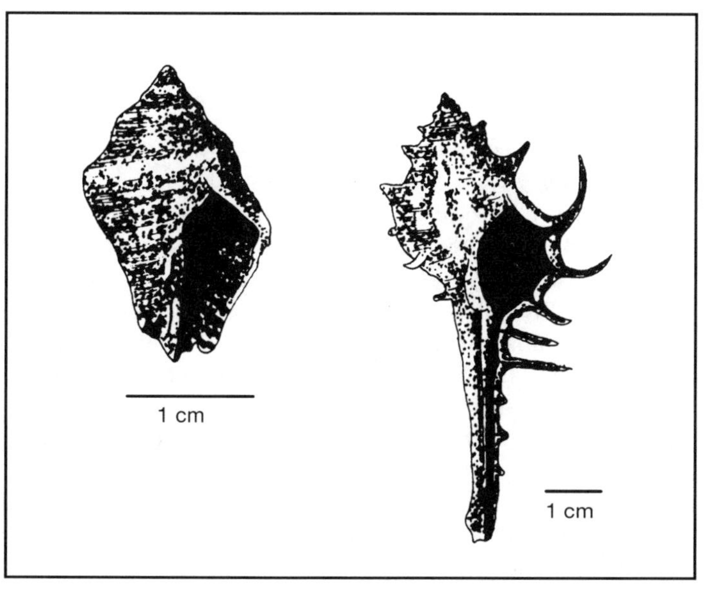

Figure 2.22. Two predatory snails *Thais* sp. (left) and *Murex* sp. (right).

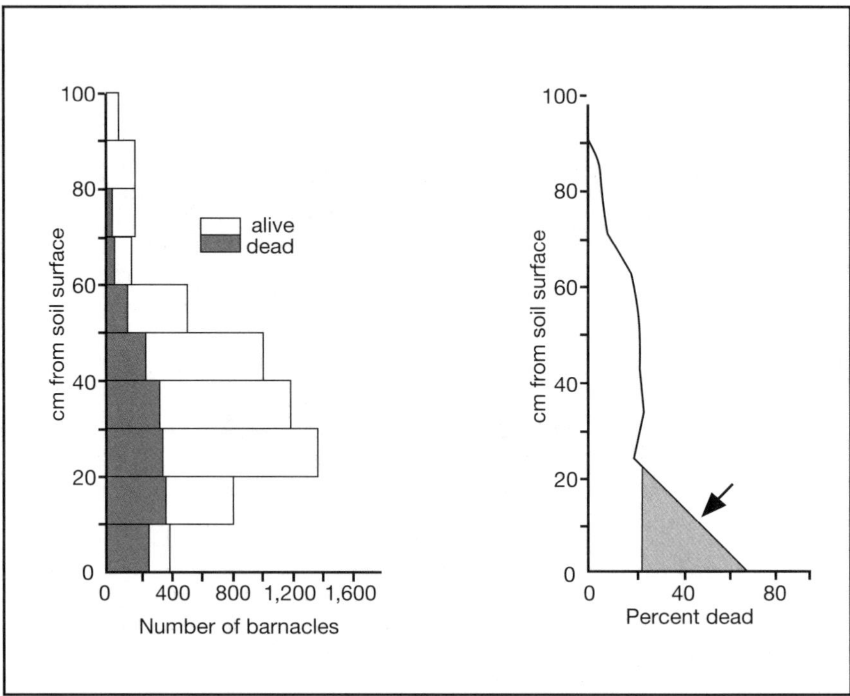

Figure 2.23. The numbers of the barnacle *Balanus amphritrite* found dead and alive at different heights up a mangrove tree stem, with the percentage of dead animals. The shaded portion indicated by the arrow represents those animals probably predated upon by the predatory snail *Thais tissoti.*

After Tee 1982b

that predation higher up the tree is less (Tee 1982b). Figure 2.23 shows the effect of predation by *Thais tissoti* on the barnacle *Balanus amphritrite.*

Two or three species of hermit crab are found in this zone. These crabs have lost the hard protective carapace over the rear part of their bodies and depend on finding empty shells for protection. The combination of security and mobility of the adopted shells is clearly advantageous and hundreds of hermit crabs can sometimes be seen on a beach. Suitable shells are, unfortunately for the crabs, a limited resource and this affects growth and reproduction. It has been found that crabs with roomy shells put their energy into growth and do not reproduce, whereas crabs with tight shells, stop growing and put their energy into reproduction (Bertness 1981a). The scarcity of shells leads to active competition between species of hermit crabs and this is described by Bertness (1981b). A review of hermit crab ecology is given by Hazlett (1981).

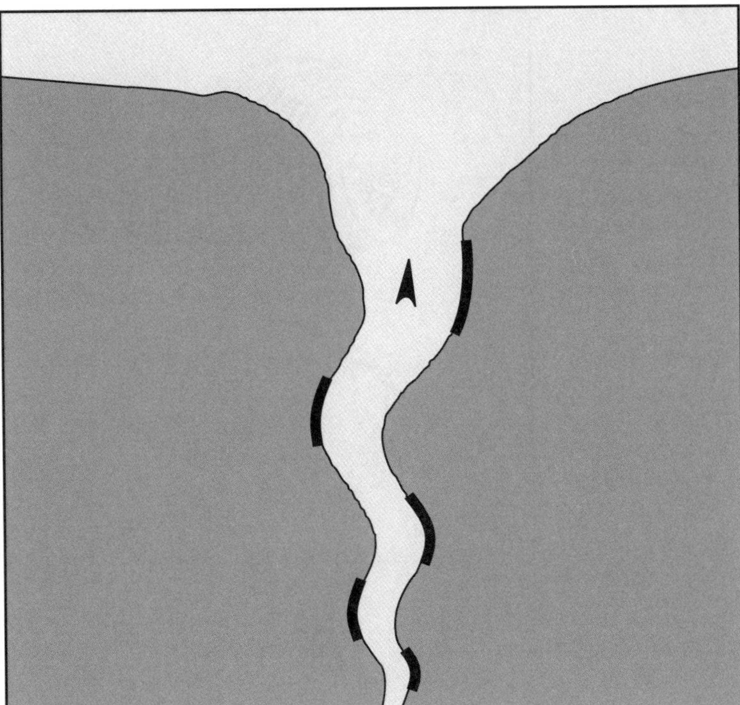

Figure 2.24. Schematic diagram of a meandering river near the sea to illustrate the position of eroded banks (heavy lines).

Zone 3: Eroded banks. The seaward edge of mangrove forest is often marked by a nearly vertical bank 1-1½ m high instead of a sloping pioneer zone. This is caused by current sweeping away the silt and is most obvious on the outer bend of rivers near the sea, where the current flows faster than on the inner bend (fig. 2.24). The bank may be broken in places and mangrove trees at its edge often fall into the sea as the soil is slowly eroded away. The top of this bank is usually at or slightly above the mean high water of neap tides, and may sometimes be left 9-10 days without tidal cover. The soil of an eroded bank resembles that of the mangrove forest floor behind it, with less fine sand (commonly about 65%) than in Zones 1 and 2.

The bank is burrowed into by various crab species (Berry 1963).

Zone 4: True mangrove forest. The mangrove fauna of this zone differs in several respects from the preceding zones. Most of this zone is very flat and the soil surface is exposed to the air for an average of 27 days per month. However, since the trees provide heavy shade, the humidity is very high, so

Figure 2.25. *Nerita birmanica.*

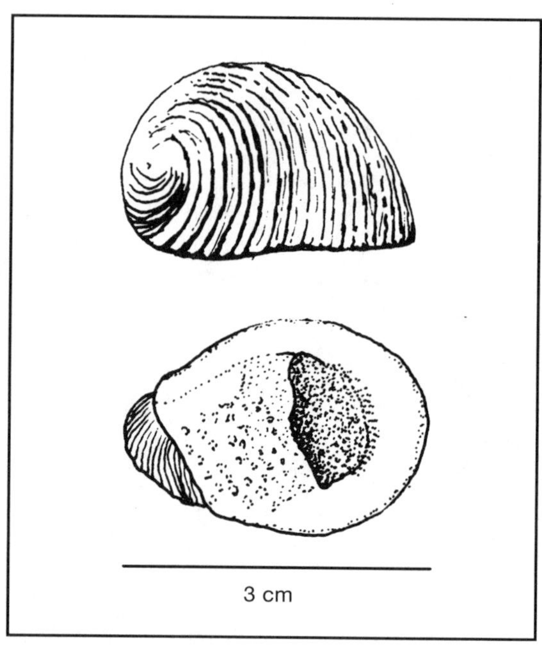

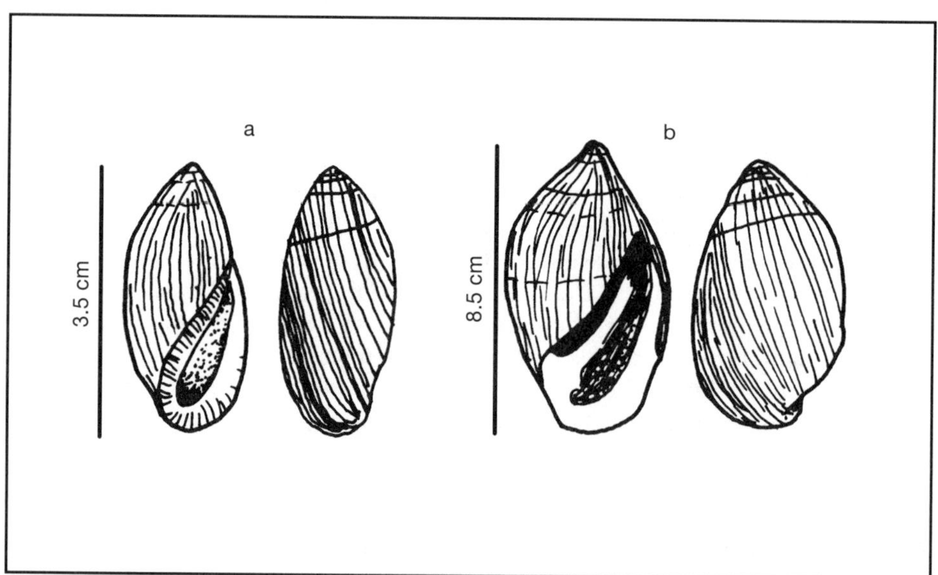

Figure 2.26. Two ellobid snails. a - *Ellobium aurisjudae*; b - *E. aurismidae*.
From Berry 1972

the soil rarely dries out. There is also abundant leaf litter and other organic matter so that detritus-feeders abound. The soil contains even less fine sand and more of the finer silt and clay particles. There is also more organic matter than in soils near the water's edge.

About 75% of the fauna of this zone is not found in the other zones (Frith et al. 1976), and is best divided into three groups: tree fauna, soil surface fauna and burrowing fauna.

Tree fauna. Perhaps the most striking change is the rapid decrease of encrusting animals – which rely on frequent inundation – on the lower stems and trunks of the trees. The remaining (mobile) tree fauna is largely composed of snails such as species of *Littorina* and *Nerita birmanica* (fig. 2.25), which are also found in Zone 2. Further back from the sea *Cerithidea obtusa* and *Cassidula* are found, and even further inland, species of *Ellobium* occur. (fig. 2.26) (Lim 1963). Most of these feed on algae and move up trees when tides wet the soil, but *Littorina* very rarely leaves the tree trunks. All these snails are able to breathe efficiently in air and some, the pulmonates such as *Ellobium,* have lungs.

Soil surface fauna. These animals comprise mostly crabs (see below) and snails, although the medium-sized mudskippers *Periophthalmus vulgaris* can be common. Among the snails, actual distance from the sea seems to matter less than details of ground conditions. In wetter areas such as where water drains into small gullies, *Syncera brevicula* is more common than anywhere else in the mangrove forest.

Molluscs at the seaward edge (Zone 1 and 2) of the mangrove forest comprise a mixed sample of gastropods and bivalves. Further back in the mangrove forest, however, carnivores and filter-feeding molluscs disappear. The vast majority of the molluscs in the true mangrove forest feed by grazing on algae or microorganisms on the soil surface. Little is known about molluscs' reproductive behaviour in mangrove forests but most have internal fertilisation of eggs and, unlike many aquatic snails, have eggs that develop directly into small snails rather than waterborne larvae.

Burrowing fauna. Nearly all the mangrove crustaceans and worms make burrows which reach down to the water table. Many different types of tunnels are constructed (fig. 2.27), and the elliptical tunnels of the edible crab may slope down from the bank of a river for as far as 5 m.

In general, these burrows serve as:

- a refuge from predators at the surface,
- a reservoir of water,
- a source of organic food,
- home for pairing and mating, to be defended, and
- a place for brooding eggs and young,

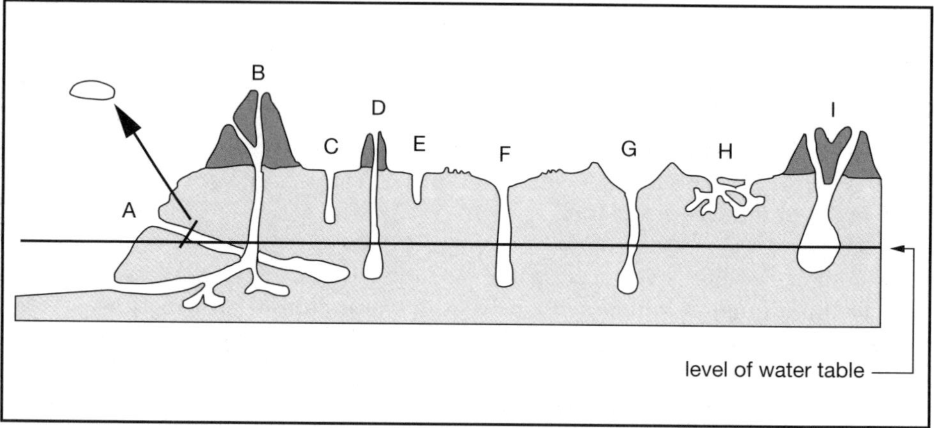

Figure 2.27. Different types of animal burrows in mangrove soil. A- crab *Scylla serrata* (with transverse section of burrow); B- mud lobster *Thalassina anomala* (not all burrows open below ground as shown here); C- crab *Uca* spp. (burrows may reach water table nearer the low-tide level); D- crab *Sesarma* spp.; E- peanut worm *Phascolosoma;* F- large mudskipper *Periophthalmodon schlosseri*; G- smaller mudskipper *Boleophthalmus boaerti*; H- pistol prawns *Alphaeus* spp.; I- smaller mudskipper *Periophthalmus vulgaris*.

From Berry 1972

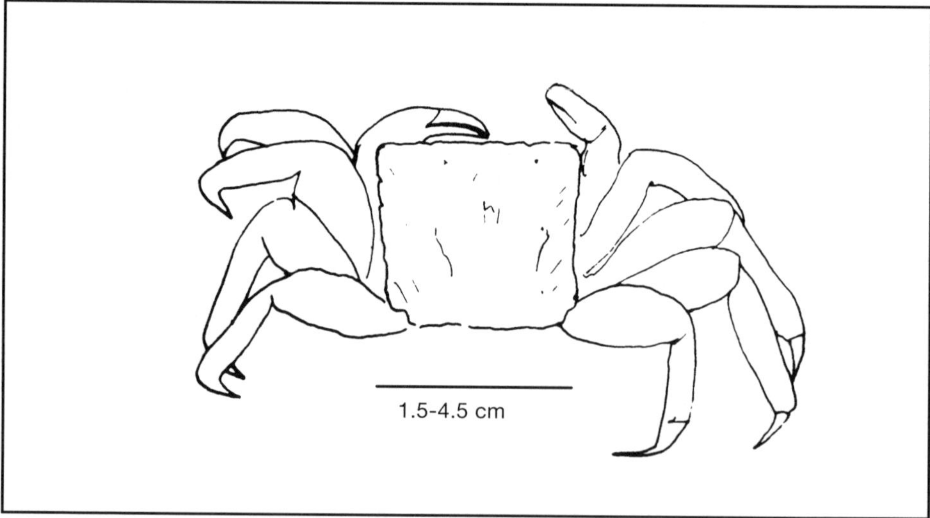

Figure 2.28. *Sesarma* sp.

although no one species uses a burrow for all these purposes.

As mentioned on page 91, most mangrove crabs are able to tolerate long periods out of water, and their respiratory systems are obviously specialised. Typical crabs inhale water under the sides of the carapace, between the legs and into the gill chambers. Deoxygenated water is expelled on either side of the mouth. Out of water, sesarmid crabs (small crabs with a characteristic square carapace) allow air into the moist gill chambers. *Sesarma* (fig. 2.28) can even recirculate the water around the carapace where it becomes reoxygenated (Malley 1977).

The most abundant, though largely invisible, source of food on the forest floor is a mixture of organic deposits, much of it originating from decomposed leaves and other vegetable material, mixed with a flora of diatoms, bacteria and other microorganisms growing on the deposits. Many crabs and other mangrove animals feed upon either the organic matter, the microorganisms, or both, picking up this 'mud' with their pincers or chelipeds. Male fiddler crabs have one huge pincer which is useless for feeding and the males are forced to eat half as fast as the females, which have two normal-sized pincers; presumably the males have to eat for twice as long. This seemingly unnecessary encumbrance is used to attract females for mating and for warning away males. It would serve a mutant male fiddler crab little if its larger pincer was reduced in size and could be used for feeding, if the animal could not then attract females for mating.

In the landward areas of true mangrove forest, the first signs are seen of an animal that is itself rarely seen, the mud lobster. This animal builds volcano-like mounds of mud which can reach over a metre high (see again fig. 2.27), and feeds on mud, digesting algae, protozoa and organic particles (Johnson 1961). The burrow below the mound is 1-3 m long, extending down to the water table. The entrance leading to the main burrow is generally plugged with layers of earth. The habit of burrowing deeply in anoxic mud, closing itself off in poorly oxygenated air and water, suggests that it may have evolved means of anaerobic respiration (Malley 1977). Research into this secretive creature would not simply be for academic interest, because its mounds are often regarded as a nuisance. Managing the numbers of distribution of mud lobsters requires a knowledge of their biology and ecology.

The bivalve mollusc *Geloina* (fig. 2.29) lives buried in mud and can occasionally be found in this zone, but is more common in mangrove forests on islands in or near river deltas. *Geloina* is remarkable in its ability to feed, respire and breed so far from open water and at levels where it is sometimes not covered by tidal seawater for several weeks at a time.

Zone 5: Rivers, streams and gullies. The banks and beds of water courses in the mangrove forest have a fauna which is generally distinct from that on the general forest floor. For example, many of the forest floor species of

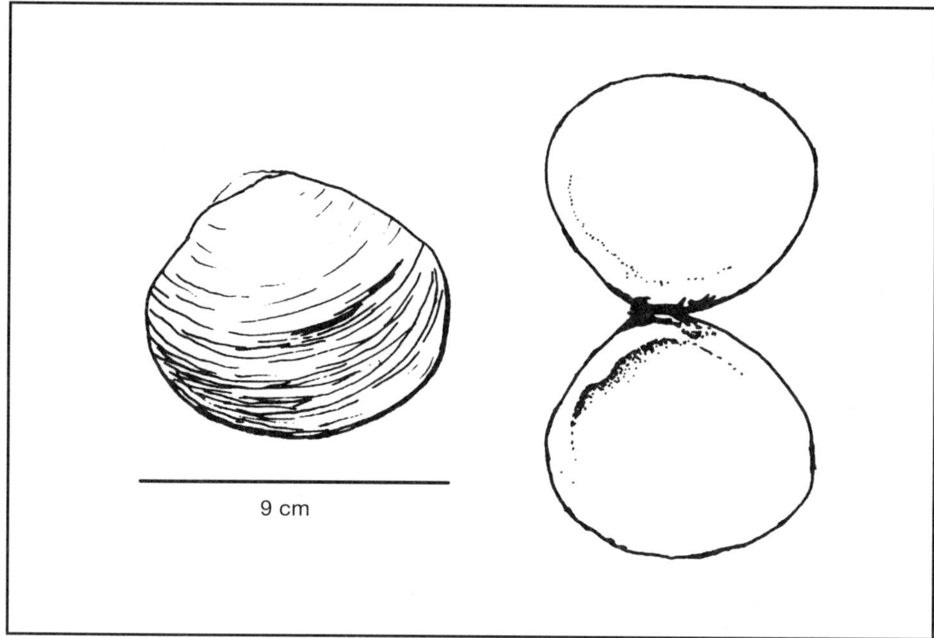

Figure 2.29. The bivalve mollusc *Geloina ceylonica*.
From Berry 1972

polychaete worms and crabs are missing, whereas juvenile fiddler crabs and some snail species are more common. The edible crab with paddle-like rear legs occurs in the larger streams and rivers in the mangrove forest where it is caught in traps.

Zone 6: The terrestrial margin. Far back from the sea, mangrove soil is covered by fewer and fewer tides and suffers longer and longer intervals of exposure. Unlike other types of shore, there is virtually no wave action in mangrove swamps to carry seawater higher than the true tidal level, because the wave energy is absorbed by the abundant vegetation. Thus, animals in this zone live on a salt-impregnated soil but are covered by seawater only at irregular and infrequent intervals. Insects, snakes, lizards and other typically terrestrial animals are much more common here than further seaward. *Sesarma* and some large crabs occur here and in the *Nypa* palm swamps behind. Their basic requirement is that their burrows must reach down to the water table. The mud lobster has one of the deepest burrows and is therefore seen some distance from the mangrove forest, for instance in coconut plantations. Where there is moving water some snails, such as *Fairbankia* and *Syncera,* and the large mudskipper *Periophthalmodon schlosseri* may be found.

Biomass of Aquatic Fauna

Only one estimate of biomass of aquatic mangrove fauna seems to have been determined in the Sunda Region. Tee (1982b) calculated the dry-tissue biomass of tree-dwelling aquatic fauna in four mangrove forest zones in Selangor, Peninsular Malaysia (table 2.5). The results show a general reduction of biomass with increasing distance from the sea. It is supposed that the turnover rate (time for one generation to replace another) must be quite fast because most of the tree-dwelling aquatic fauna are in the lower trophic group (i.e., filter-feeders), none of which have long life spans.

Terrestrial Fauna

Insects, birds and mammals live chiefly in the canopy of mangrove forest. Ground-living animals such as wild pig, mouse deer, wild cats, rats and lizards only venture into the landward edge of mangrove for brief periods. The two mangrove mammals seen most often are the silvery leaf monkey *Trachypithecus cristata* and the long-tailed macaque *Macaca fascicularis.* The leaf monkeys are largely arboreal and eat leaves, shoots and fruit, whereas the macaques are mainly ground-dwelling and eat crabs, peanut worms and small vertebrates as well as some leaves, shoots and fruit. Faecal analyses have also shown that both species eat the nectar-rich flowers of *Sonneratia* (Lim and Sasekumar 1979).

Leaves are a poor source of food for most mammals because the cellulose molecules are very long and are therefore difficult to break down without special digestive adaptations. Leaf monkeys have evolved a highly sacculated stomach, like a cow's, in which special bacteria ferment and break down cellulose so that it becomes digestible.

Table 2.5. Estimated biomass of tree-dwelling aquatic fauna in four types of mangrove forest.

Forest type	*Avicennia* forest	*Sonneratia* forest	*Bruguiera* forest	*Rhizophora* forest
Distance from sea (m)	10	235	310	445
Estimated biomass (kg)	1,014.0	272.4	29.7	37.9

From Tee 1982b

Many tree leaves are defended against being eaten by herbivores by defence compounds, the most common of which are fibre (lignin) and phenols, particularly tannins (p. 88). In a normal acidic digestive tract (the content of a human stomach are generally at pH 3), tannins will bind strongly onto any protein, both food and enzymes, thereby hindering the digestive process[3]. In the leaf monkey's complex stomach, however, the pH is probably maintained between 5.0 and 6.7 (Bauchop and Marrucci 1968), and in such conditions not only are the fermentation bacteria able to thrive, but the attachment of tannins onto proteins is weaker. Both silvery leaf monkeys and long-tailed macaques have been observed to eat bark from the twigs of *Bruguiera* (Lim and Sasekumar 1979), which have considerable quantities of tannins – perhaps 30% dry weight. For the macaques, which have a simple stomach, this may have been eaten as a stomach purgative (p. 232).

Fruit bats, particularly the flying fox *Pteropus vampyrus,* roost in the mangrove forest canopy and other bats such as the cave fruit bat *Eonycteris spelaea* and the long-tongued fruit bat *Macroglossus sobrinus* are very important in the pollination of the flowers of *Sonneratia* (Start and Marshall 1975) (p. 329).

The mangrove frog *Rana cancrivora* is exceptional among amphibians in being able to live and breed in weakly saline water. The tadpoles are more resistant to salt than the adults and metamorphosis into adults will only occur after considerable dilution of the salty water (MacNae 1968).

Mangroves are inhabited by a variety of reptiles such as the monitor lizard *Varanus salvator,* the common skink *Mabuya multifasciata,* mangrove pit viper *Trimeresurus pupureomaculatus* and the common catsnake *Boiga dendrophila.* Most snakes seen in or near mangroves are not in fact sea snakes (Hydrophidae), which live primarily in open water (Voris et al. 1978; Voris and Jayne 1979; Voris and Moffett 1981). Potentially the largest animal of the mangrove swamps is the estuarine crocodile *Crocodilus porosus.* Persecution for centuries has reduced its numbers to a very low level and it is possible that large specimens (they can exceed 9 m) no longer exist around Sumatra (p. 185).

The most conspicuous of the insects are mosquitoes, but only one of these, *Anopheles sundaicus,* carries malaria. The larvae of this species can live in water with a salinity of 13‰ (Chapman 1977b). Species of *Aedes* mosquitoes have been seen feeding on mudskippers, but they are also attracted to human skin. Another potential insect hazard is the leaf-weaving ant *Oecophylla smaragdina,* the workers of which, if their nest is disturbed, can inflict a very painful bite. The making of the nest is extraordinary because the adult ants have no means of producing silk to join the leaves of the nest together. The larvae do have silk glands, however, originally intended so they could weave cocoons to pupate within, but they are not now used for this. Instead, when a leaf is to be added to the nest or a tear repaired, some of the worker ants seize the edges to be joined and hold them in the required position. Other workers enter the nest and collect a larva. The larva is held in the worker's jaws and while it produces the sticky silk, it is moved back and forth between the leaf edges, like a tube of quick-drying glue (Tweedie and Harrison 1970).

One insect, the caterpillar of the moth *Olethreutes leveri,* has managed to extend its niche and avoid competition by spinning together leaves of *Sonneratia alba.* It is able to survive periodic immersion in seawater because air is trapped within the web. Experiments showed it could withstand 8 1/2 hours immersion within the web, but died quickly if immersed in seawater without the web (Lever 1955).

BIRDS

Mangrove forests on the west of Peninsular Malaysia are used by at least 121 species of birds which are probably almost identical to those found in Sumatran mangroves. The most important trees for nesting are probably *Sonneratia,* which are both tall (±30 m) and have occasional holes in their trunks which are used by woodpeckers. The landward side of a mangrove forest generally has more birds of more species than the seaward side, but birds are found throughout.

Nisbet (1968) divided the list of bird species observed in mangrove forests on the west coast of Peninsular Malaysia into 10 groups (table 2.6) according to the manner in which they use the ecosystem. Mangroves are thus important to birds in at least four ways:

- they provide nesting sites for a number of large species of herons and storks (fig. 2.30), and also for at least three birds of prey and two owls. Many of these species nest nowhere else, and presumably would be subjected to much greater predation (certainly by young humans) if they attempted to do so.
- they provide roots for many migrant species (and a few residents) which feed on or over the tidal mud flats. These roosts are particularly important for the migrant waders.
- they are visited seasonally by a number of pigeons and parrots, both from inland and outlying islands. One of these, the pied imperial pigeon *Ducula bicolor* (fig. 2.31), is thought to be the main source of seeds reaching Jarak Island in the Straits of Malacca. Of the plants growing on the island, 90% have fruit dispersed by birds (Wyatt-Smith 1951).
- they support a varied fauna of resident and migrant land-birds. Eight species (groups A and B in table 2.6) occur in no other habitat and 17 others (group E) occur largely within mangrove forest. This

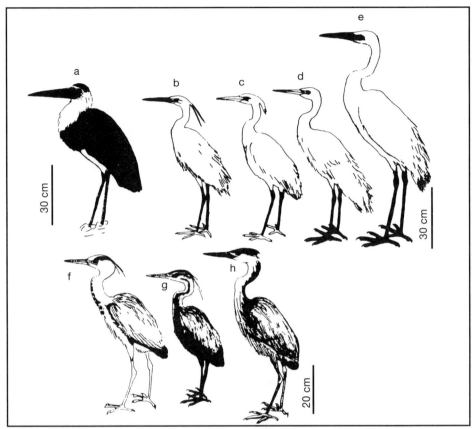

Figure 2.30. Storks, egrets and herons commonly seen near mangrove forests. a-woolly-necked stork *Ciconia episcopus*; b-little egret *Egretta garzetta*; c-Chinese egret *Egretta eulophotes*; d-plumed egret *Egretta intermedia;* e-great egret *Egretta alba*; f-grey heron *Ardea cinerea*; g-purple heron *Ardea purpurea;* h-great-billed heron *Ardea sumatrana.*

Figure 2.31. Pied Imperial Pigeon *Ducula bicolor.*

assemblage of birds is distinctive in that it is largely made up of predatory kingfishers, insectivorous woodpeckers and warblers and nectar-drinking sunbirds. Some of the dominant frugivores of adjacent forest (pheasants, hornbills, barbets, leaf-birds, bulbuls and babblers) are more or less absent. Interestingly, most of the species in groups A and B occur outside mangrove forests in other parts of their range.

A significant feature of the resident land-bird fauna is that most of the common species (in group E and I) are those which, in other parts of Malaysia, and indeed Sumatra, are characteristic of semi-open habitats – villages, towns, gardens, riversides and clearings. This is perhaps most surprising because mangrove forest is dense and much more similar in structure to secondary forest. Since semi-open habitats are largely man-made, it is likely that many of these species were, until recently, confined to a narrow coastal strip of mangroves and scrub and to clearings and river-banks in the forest. It should be said, however, that the species which have benefited most from the clearance of inland forest are not very numerous in mangrove, while most of the dominant mangroves species are still primarily coastal in distribution.

Some of the species which are most specialised for life in mangrove forest are replaced in inland areas by an obvious potential competitor. For example, the ruddy kingfisher *Halcyon coromanda* breeds in forest in the

Table 2.6. The number and percentage of birds observed in mangrove forest in Peninsular Malaysia, grouped according to the manner in which they use the habitat.

Group A.	Species which are confined exclusively to mangroves	4.0% (5)
Group B.	Residents confined to mangrove but migrants found elsewhere	2.5% (3)
Group C.	Nesting more or less exclusively in mangroves but feeding outside	7.5% (9)
Group D.	Migrants which depend largely on mangroves for roosting	7.5% (9)
Group E.	Coastal species for which mangrove is a major habitat	14.0% (17)
Group F.	Species which make seasonal movements to and from mangroves	7.5% (9)
Group G.	Migrants which use mangroves as well as lowland areas	14.0% (17)
Group H.	Aerial species which feed over mangroves as well as land	4.0% (5)
Group I.	Species of semi-open country which are also found in mangroves	28.0% (34)
Group J.	Species of forest which are also found in mangroves	11.0% (13)
Total		121 species

From Nisbet 1968

northern part of its range (India, E. China, Japan, Taiwan) but is confined to mangroves as a breeding species in Peninsular Malaysia and Sumatra where the forest is inhabited by the rufous-collared kingfisher *Halcyon concreta*. Thus it seems that, as inland forest was opened up by man, the unspecialised coastal species (group I) were able to spread inland, whereas the species which were more adapted to life in mangrove forest remained confined to the mangroves and nearby coastal scrub.

Effects of the Fauna on the Vegetation

Very little is known about the effects of the activities of animals on the mangrove forest. The holes constructed by crabs and mudskippers may affect rates of accretion or erosion and increase aeration of the soil, the mounds of the mud lobster and the activities of polychaete worms may significantly influence the rate of nutrient/energy flow in the ecosystem, inshore fish that graze on young seedlings may cause slower colonisation, etc., but there have been no studies in the Sunda Region on these or related topics.

Some of the known effects are rather surprising. Root tips of *Rhizophora* trees which have not yet reached the ground are sometimes attacked by small isopod crustaceans. They burrow into the soft root tip, form a chamber and breed. Such destruction of growing points, as with most plants, results in the formation of multiple side roots. It had been suggested that the activity of these isopods may benefit *Rhizophora* by increasing the number of roots and therefore the number of channels by which nutrients can reach the tree. A higher production of roots per tree would either result in a faster expansion of adjacent areas of *Rhizophora* or in a higher number of roots per unit area. In Florida, the number of roots per unit area was indeed higher where these trees were parasitised by isopods than at sites where they were absent (Ribi 1981, 1982). The effect of the increased root density on silt settlement and thence on land formation has not been measured.

Chapter Three

Other Coastal Ecosystems

Introduction

In addition to mangrove forests, five other types of ecosystems are found around the coasts of Sumatra. These are: beach vegetation on accreting coasts (pes-caprae formation), beach vegetation on abrading coasts *(Barringtonia* formation), brackishwater forests, rocky shores and coral reefs. The last of these is not strictly within the scope of this book as it is a marine ecosystem, but it is included as background material for the reader who may wish to pursue studies on coral reefs.

None of the coastal ecosystems is known particularly well, except for the economically important coral reefs, and so the following sections serve only as introductions upon which further investigations can be based.

Beach Vegetation

Pes-caprae Formation

Along accreting coasts, that is, where new sand is being deposited, the sandy beach is colonised by a form of vegetation known as the 'pes-caprae' formation. Such vegetation is found throughout the Indonesian Archipelago and the western Pacific (Richards 1952). The name refers to *Ipomoea pes-caprae*, a purple-flowered creeper related to convolvulus, which is just one of a number of low, sand-binding herbs, grasses and sedges which advance over the sand with long, deep-rooting stolons or stems. Other species expected in this formation would include the legume *Canavalia*, sedges *Cyperus pedunculatus* and *C. stoloniferus*, and the grasses *Thuarea involuta* (fig. 3.2) and the prickly *Spinifex littoreus*. Full lists are given by Whitmore (1984), Wong (1978) and Soegiarto and Polunin (1980).

These plants are well adapted, being tolerant of periodic drought, salt spray (although dependent on non-saline ground water), wind, low levels of soil nutrients and high soil temperatures. They also produce small seeds

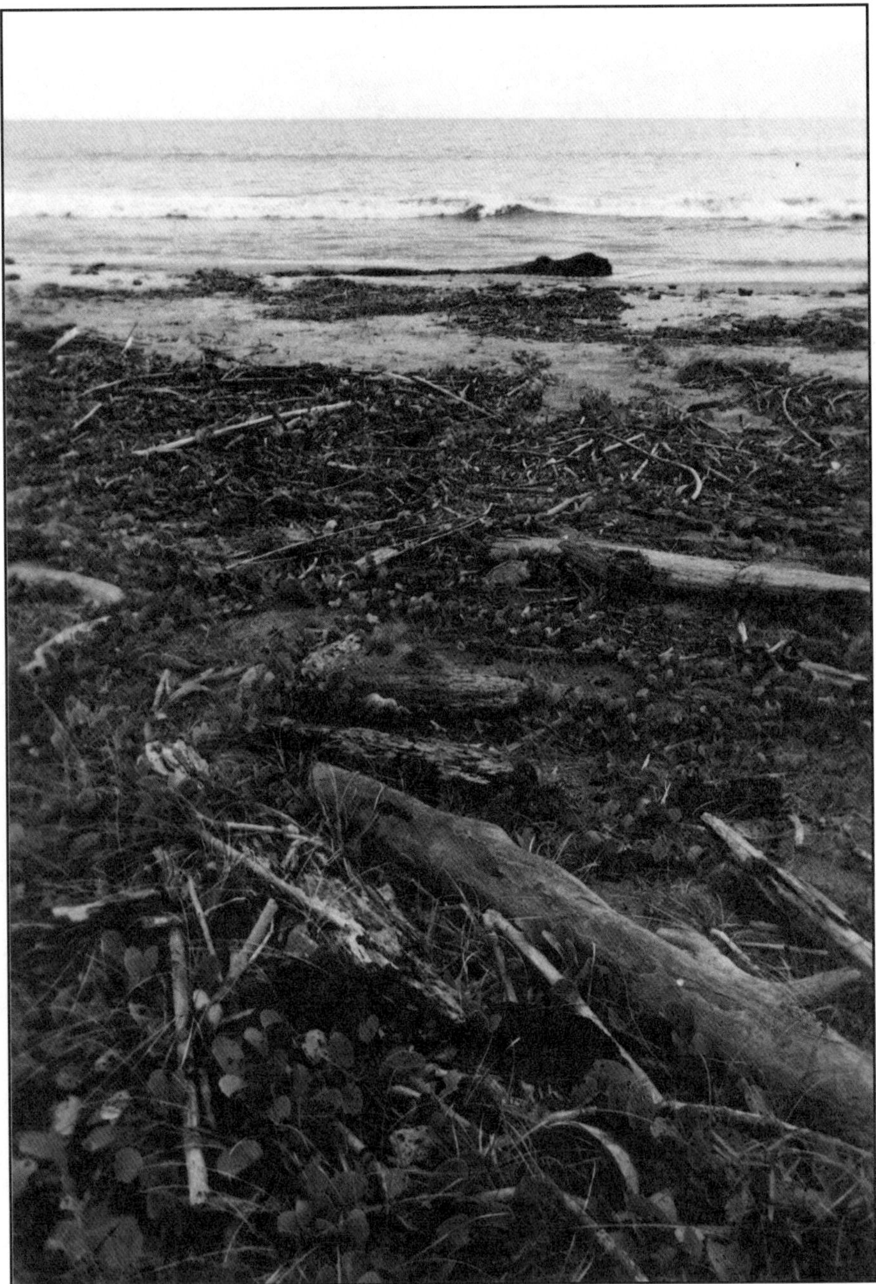

Figure 3.1. Pes-caprae association on a western beach in southern Aceh.
A.J. Whitten

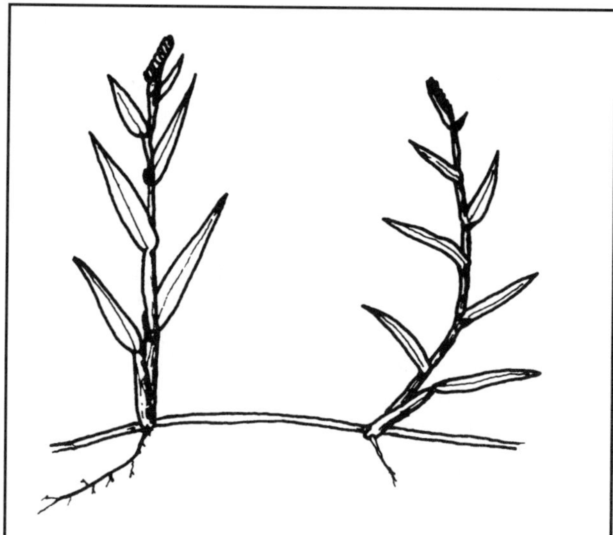

Figure 3.2. *Thuarea involuta,* a creeping species of grass found in the pes-caprae association.
From Holttum 1977

which are dispersed by floating on water; for example, the small seed of *Thuarea* is surrounded by an air-filled chamber which floats easily and is dispersed by tidal currents (Holttum 1977). The mat of low vegetation of the parent plants traps dead leaves and other organic material blown by the wind or tossed there by the tides, and provides a little shelter for animals. Thus nutrients increase and later stages of the plant succession (p. 80) can occur, and the landward fringe is colonised anew. She-oak *Casuarina equisetifolia* is found at the inner edge of the pes-caprae formation where it represents a late stage in the plant succession. She-oak can form pure stands but is unable to regenerate in closed forest or even in open forest on the carpet of litter formed by its own dead twigs. Thus, unless the coastline advances and provides fresh habitat, the belt of she-oak forest will be supplanted by other species (Corner 1952). The 30 m wide mature she-oak forest examined by a CRES team near Singkil, South Aceh, had a basal area of 32.7 m^2/ha for trees of 15 cm diameter and over.

The environment of exposed sandy beaches is particularly hostile to animals because of the instability of the sand, and the wide variations in temperature, salinity and humidity. A great many small animals live permanently within the sand where they have some protection, but they have to cope with the difficulties of finding food, reproductive partners and the low oxygen concentrations in the spaces of air between the sand grains (Brafield 1978). Near the high water level, burrows about 18 cm across can be found with small piles of sand around them. Their occupants are adult, beige-coloured sand crabs *Ocypode ceratophthalma*, which are rarely seen during the day. It is difficult to dig one of these crabs out of its burrow or to catch it as it runs across the sand. Young *Ocypode* are very numerous and can be seen on the sand surface both by day and night. *Ocypode* feed mainly on organic material from the sand but are sometimes predatory (Tweedie and Harrison 1970).

On wider beaches the small ghost crabs *Dotilla mictyroides* may occur in thousands with densities of over 100/m^2 (Hails and Yaziz 1982; McIntyre 1968). Although some of the larger individuals are coloured light blue with pinkish legs, the majority are sand-coloured. As the tide rises and covers the beach, each ghost crab builds a shelter of wet sand pellets over its back. The crab traps air beneath itself and as it burrows down, so the air pocket is carried down with it. When the crabs emerge as the tide falls, they are followed by a stream of small bubbles (Tweedie and Harrison 1970). Isopod crustaceans can be found by careful examination of the sand and the organic material washed ashore onto the upper sandy reaches (Jones 1979), and wading birds can sometimes be seen feeding on these and other small animals.

Lower down the beach a variety of molluscs occur but are rarely seen because they burrow beneath the surface. Examples of the bivalves are the white and pinkish *Tellina*, the large *Pinna*, and the economically important edible cockle *Arca granosa*.

The sandy beaches are used as nesting sites by sea turtles. Sumatra is visited by at least three species which nest on its sandy beaches:

- the most abundant is the green turtle which weighs up to 100 kg and has a carapace up to 1 m long. Its eggs are commonly seen on sale in western towns such as Padang;
- the hawksbill turtle, also common, is exploited for its shell, with the main trading centre at Sibolga. This species is slightly smaller than the green turtle, weighing up to 80 kg and with a carapace length of 90 cm;
- the rarer, massive leatherback turtle which can weigh up to 1 ton.

The distribution of a fourth species, the loggerhead turtle, may include Sumatra (fig. 3.3). There is hardly any information on nesting sites of turtles on Sumatra but the more important areas are almost certainly on the offshore islands and the west coast of the mainland (Soegiarto and Polunin 1980; van der Meer Mohr 1928; Rappart 1936). However, in an article on the shore life of Pantai Cermin in Deli Serdang, van der Meer Mohr (1941a) shows a picture of green turtle tracks on the beach, indicating that some turtles bred there in 1940. Turtles used to lay eggs on Berhala, a small island off the coast of North Sumatra in the Straits of Malacca (A. Jazanul, pers. comm.; van der Meer Mohr 1928), but there is no recent information. Turtles also nest at a small beach near Kuala Kambas, eastern Lampung, but since all the eggs are taken for local consumption, the turtles may not

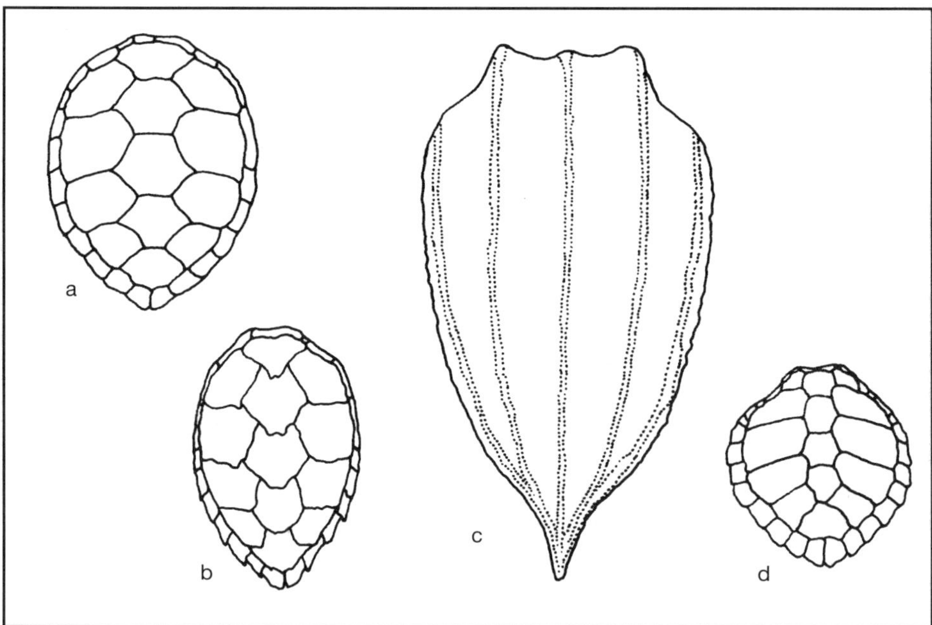

Figure 3.3. Shells of four species of sea turtle found around Sumatra. a-green turtle, *Chelonia mydas;* b-hawksbill turtle, *Eretochelys imbricata*; c-leatherback turtle, *Dermochelys coriacea*; d-loggerhead turtle, *Caretta caretta.*

From Anon. 1979b

return for much longer (FAO/Wind et al. 1979) (p. 349). Any information on nesting turtles in Sumatra is valuable so that their distribution and conservation status can be more accurately determined.

Barringtonia Formation

The *Barringtonia* formation is found behind the pes-caprae formation on sandy soils. It is also found on abrading coasts where sand is either being removed by unhindered ocean swells or where sand has at least ceased to accumulate; in such areas a beach wall about 0.5-1 m tall is formed. On this wall and inland, the second type of vegetation, the *Barringtonia* formation, is found. This formation is also tolerant of salt spray, nutrient-deficient soil and seasonal drought. This belt of vegetation is not very wide, usually between 25 and 50 m where the lie of the land allows it, but narrower where the coast is steep and rocky. This type of forest merges with lowland

Figure 3.4. The fruit of *Barringtonia asiatica.*

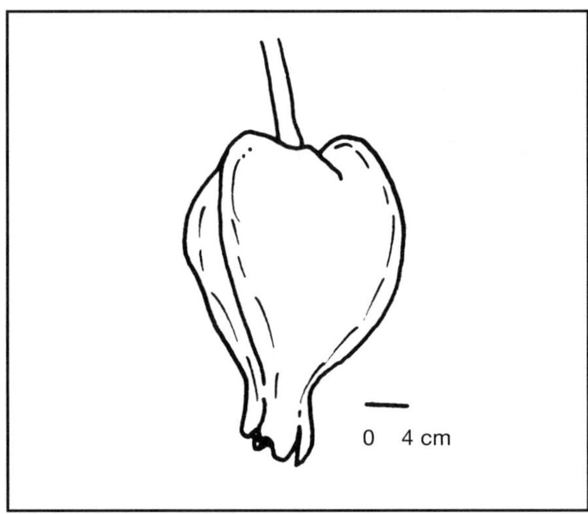

Figure 3.5. The fruit of *Heritiera littoralis.*

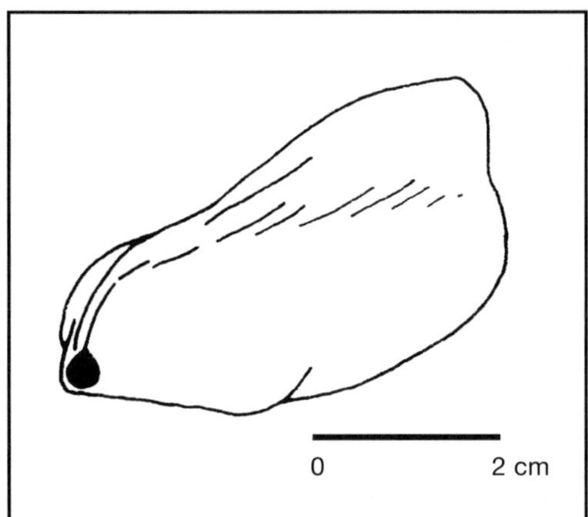

rainforest inland. Large trees sometimes sprawl across the upper parts of the beach, and as the beach wall is eroded away so these eventually fall over and die. The larger trees of the *Barringtonia* association are of two main species: *Barringtonia asiatica* (Wallwork 1982), which has huge 15 cm wide feathery flowers and unusual fruit (fig. 3.4), and *Calophyllum inophyllum,* which has transparent yellow sap and 3 cm round fruit, the seeds of which

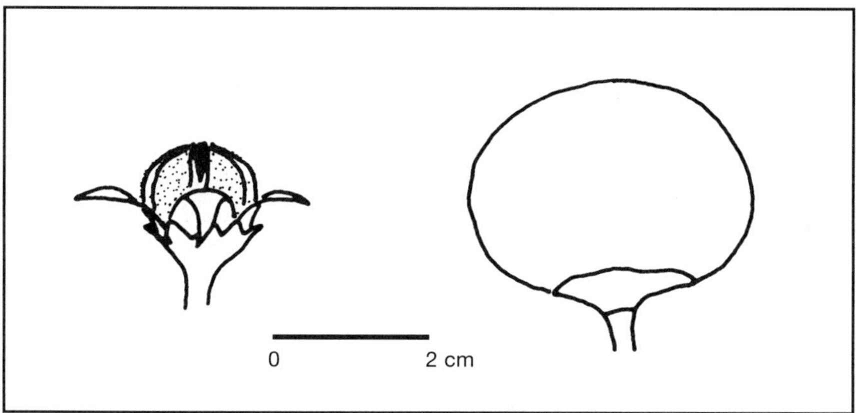

Figure 3.6. The fruit of *Hibiscus tiliaceus* (left), *Thespesia populnea* (right).

are dispersed by bats. It should be noted that *Barringtonia* itself is not invariably present in the formation which bears its name (van Steenis 1959; Steup 1941). The plants found in this (and also the pes-caprae) type of beach vegetation are found in similar locations throughout the Indo-Pacific region, and some are typical of sandy shores throughout the tropics. Many of the species are not found outside these formations.

In addition to the trees mentioned above, other typical species include the large bush *Ardisia elliptica* with its pink young twigs and leaves, *Heritiera littoralis* with its peculiar boat-shaped floating fruit (fig. 3.5), and other trees such as *Excoecaria agallocha*, pandans (Stone 1983), particularly *Pandanus tectorius, Scaevola taccada, Terminalia catappa,* and two types of hibiscus *Hibiscus tiliaceus* and *Thespesia populnea.* Both hibiscus have large yellow flowers with purple bases but the former species has slightly hairy lower leaf surfaces, heart-shaped leaves as long as they are broad, flowers which fall off as soon as they have dried and smaller fruit. The latter has smooth leaves longer than they are broad with a sharper tip, flowers which remain on the plant for some days after they have died and larger fruit (fig. 3.6). *Hibiscus tiliaceus* is commonly planted in towns and villages. The cycad *Cycas rumphii* is also sometimes found in the *Barringtonia* formation. Despite their appearance, cycads are not palms, neither are they ferns, but they are related to the now-extinct seed-ferns that flourished roughly between 280 and 180 million years ago (Corner 1964). Cycads are commonly planted in gardens. In addition to the species above, certain species from the pes-caprae association can also be found, particularly on the beach wall.

Barringtonia formation forests have been cleared in many areas to make way for coconut groves but some excellent examples still exist on the west coast of Siberut (Anon. 1980a; Whitten 1982b) and in remote parts of the western coast of northern Sumatra.

Almost nothing has been written about the animals of *Barringtonia* formation forests but Simakobu leaf monkeys *Simias concolor* have been observed in them on Siberut Island (Whitten 1982b), and the CRES team investigating the rocky shores near Painan, West Sumatra, observed silvered leaf monkeys *Trachpithecus cristata* in the *Barringtonia* trees. The fauna of the sand in front of the *Barringtonia* formation is more or less the same as that described for the sand in and below the pes-caprae formation.

Productivity

Little is known about the productivity of beach vegetation but there are obviously great differences between the low-lying pes-caprae and the trees and shrubs of the *Barringtonia* formations. On the beach below the vegetation zone, the majority of energy inputs originate from the sea, and flotsam such as wood, sea-grass leaves and algal fronds may be an important source of food at the upper limit of tides (Soegiarto and Polunin 1980). Details of plant succession on the shores of Krakatau are given on pages 343-347.

Brackishwater Forests

Brackishwater forest can be found at the inner boundary of mangrove forest, and at the upper tidal limit of rivers. It may also occur where the seaward beach is merely a sand barrier thrown up by the waves and formed by the currents. Behind such barriers the land is usually flat and low and streams often flood, thus creating marshy lagoons suitable for brackishwater forest. Examples of these are shown in figure 3.7. Brackishwater forest is characterized by nipa *Nypa fructicans* which forms pure, often extensive stands, but most of the other elements of the vegetation are commonly found in either mangrove forests or the *Barringtonia* formation (Whitmore 1984; Wyatt-Smith 1963).

Only one plant, the mangrove date palm *Phoenix paludosa*, appears to be restricted to this type of forest. It is easily recognised because it is the only feather-palm in Sumatra whose leaflets, when viewed from above, form a trough rather than a ridge at the junction with the midrib (Jochems 1927). The mangrove date palm is the most easterly relative of the date palm of north Africa and Arabia. Although brackish water and deserts are very different habitats, neither has fresh water freely available to plant roots (Whitmore 1977).

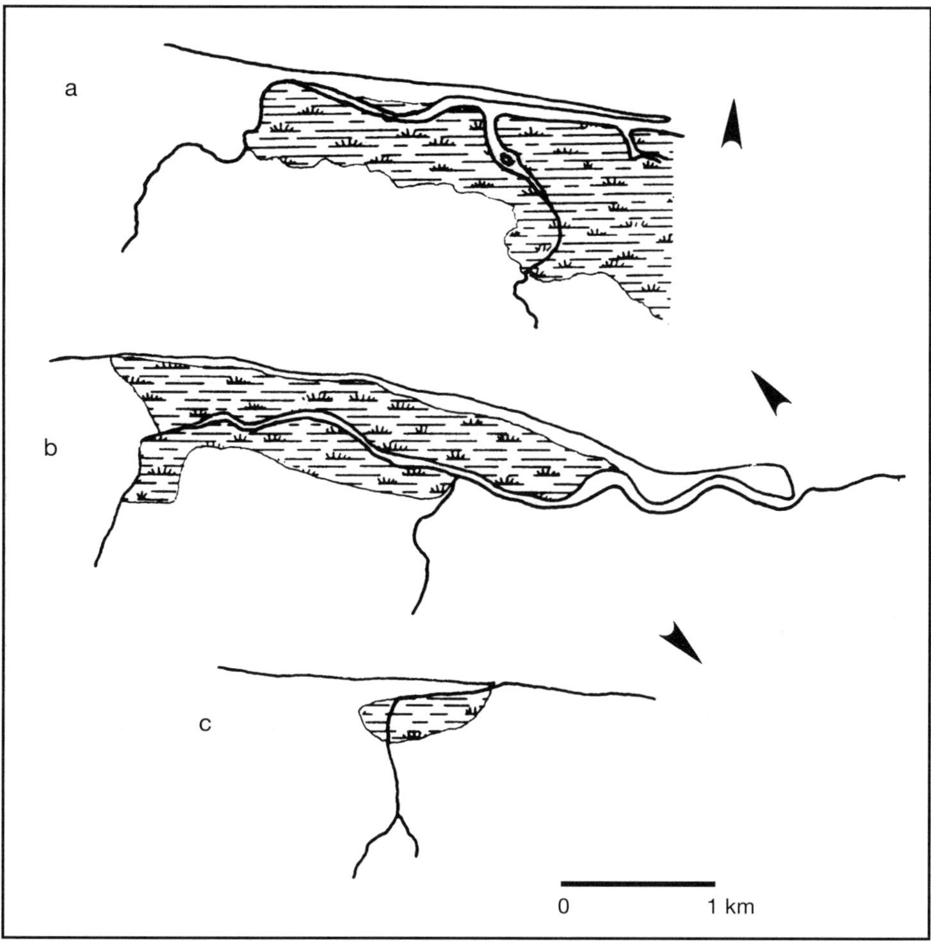

Figure 3.7. Locations of three brackishwater forests behind sand bars: a) River Rangau, Bangka Island, b) River Perbaungan, Deli Serdang, c) River Lebekiu, Siberut Island. The arrows point north.

Rocky Shores

Rocky shores occur where hard and resistant rock faces the sea in such a manner that the results of weathering are swept out to sea rather than deposited to form a wide beach. Such shores are usually steep, with the scarp face continuing below the sea surface. There is commonly, however, a narrow shelf or beach with shingle rather than sand in the upper tidal regions. Such steep coasts and cliffs are usually formed of old limestone (e.g., west of Banda Aceh), volcanic rock (e.g., south of Padang), granites and Tertiary sandstone (e.g., Bangka, Belitung and Bintan Islands and the Riau/Lingga, Anambas, Natuna and Tambelan archipelagos) (Sopher 1977).

There is no one type of vegetation peculiar to rocky shores, but a few specimens of trees normally found in the *Barringtonia* formation, such as *Barringtonia, Casuarina,* and *Calophyllum,* and pandans may be seen clinging to the rocks. Where the slope lessens and the salt spray is less, some form of lowland forest would be expected to grow.

The fauna of rocky shores is rather poor but has representatives from a very wide range of phyla. The animals are adapted to withstand the force of the waves, periodic desiccation, high temperatures and variable salinity. They all have efficient means of retaining a grip on the rocks; some are permanently fixed but others can move around to forage or graze. Barnacles are common and, among the molluscs, small oysters, cap-shaped limpets and *Nerita* snails are usually found (Tweedie and Harrison 1970). Medium-sized, long-legged rocky crabs (probably *Grapsus grapsus*) will be seen near the water's edge, and at the rocky shore examined by a CRES team near Painan, small blenny fish jumped from rock to rock in the spray zone. The zonation of animals up a rocky shore has been studied in Singapore (Purchon and Enoch 1954), and the principles of zonation are discussed by Brehaut (1982).

Most of the limestone or granite cliffs that have clefts or small caves in them will support colonies of swiflets (p. 317). One species, *Aerodramus fuciphaga,* builds its cup-like nest purely from hardened saliva but others mix in bits of vegetable material (Medway 1968).

Coral Reefs

The conditions necessary for coral to grow are:

- warm water (above 22°C)
- clear water
- water with near to normal seawater salinity, and
- light.

All the seas around Sumatra are warm enough for coral but the second and third conditions are not met where large rivers flow into the sea. These rivers reduce the salinity and their sediment loads cause increased turbidity which in turn reduces light penetration. Thus the coasts of eastern Sumatra, and the coasts along the plains in western Sumatra, have no coral (fig. 3.8).

Most of reefs around Sumatra are fringing reefs (i.e., growing out from land), but atolls (islands formed by waves throwing up piles of coral

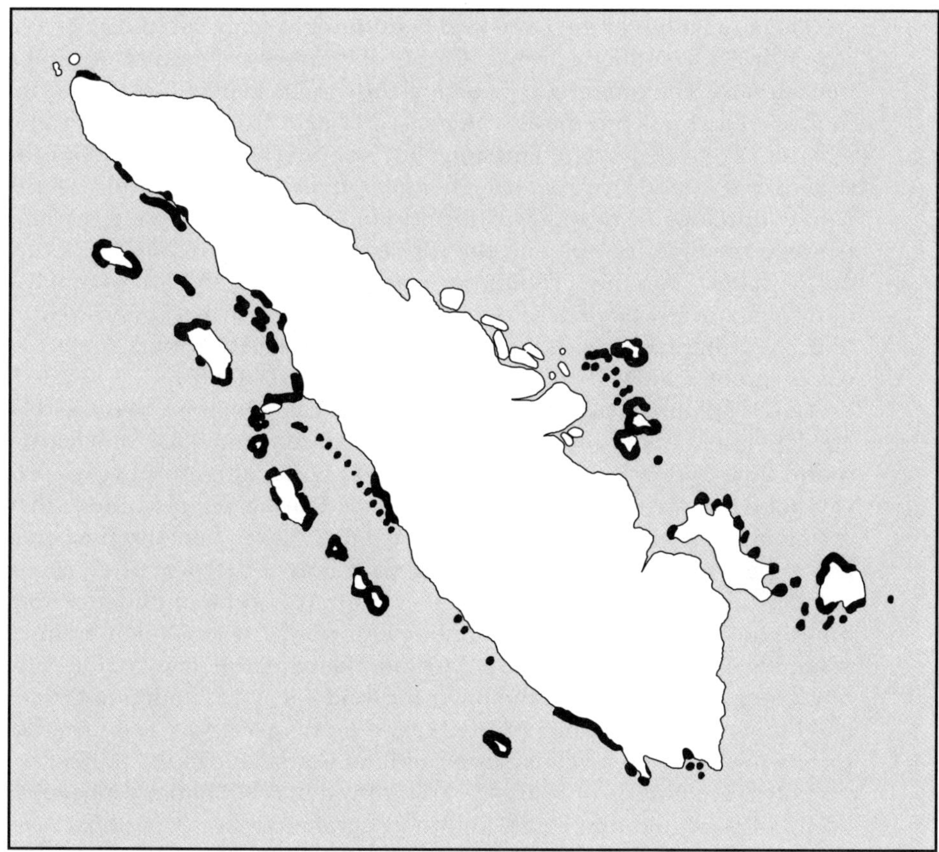

Figure 3.8. Distribution of coral reefs in Sumatra.

fragments) and barrier reefs (away from land where the seabed is near enough to the surface for coral to form) are found between the west coast of Sumatra and the islands of Simeulue, Nias and the Mentawai Islands. The fringing reefs found around the islands of the Riau/Lingga, Natuna, Anambas and Tambelan archipelagos are shelves exposed briefly at low tide. The surface is flat and slopes gently towards the open sea. Loosened coral blocks are tossed by the breakers and are piled up to form a low ridge at the edge of the coral shelf.

The animal life of the coral reef is stunningly beautiful in colour and shape, and astoundingly diverse. Corals themselves are examples of colonial animals. The outer layer of a block of coral is made up of numerous 'polyps'. Each polyp removes calcium carbonate from the seawater and secretes its own skeleton of limestone but is connected to its neighbours by strands that extend laterally through minute holes in their skeletons. As the colony develops so new polyps form, often on the connecting sections between the older polyps, and the new skeletons grow over and stifle the polyps below. Thus the reef-building corals grow. Each species of coral has its own characteristic pattern of budding and so constructs its own particular, and characteristic, shape. Many of the species around Sumatra's coasts can be readily identified (Henry 1980; Searle 1956).

Corals are unable to live deeper than light can penetrate, because they are dependent upon single-celled algae, called zooxanthellae, which grow within their bodies. The algae need light to photosynthesise and thus provide food for themselves and for the polyp. During photosynthesis they release oxygen which helps the coral to respire. During the day the corals may appear to be nothing but dead lumps of limestone, some of which are coloured, but most of which are rather dull. At night and under certain other conditions, however, the millions of polyps extend their brightly coloured, feathery tentacles into the water, blurring the shape of the coral block they form as they sift the water for food particles. Optimum growth conditions for reef-forming coral occur when vigorous wave action causes turbulence and a continuous supply of food is available to the polyps.

Polyps are not the only organisms involved in reef-building; a peculiar group of pink-coloured algae, known as coralline algae (Corallinaceae), also contribute. These plants are red algae which deposit calcium carbonate in their cell walls. Coralline algae exist as crusts, fronds, segmented branches, epiphytes, parasites on other coralline algae, or as gravelly nodules, and are the world's hardest plants; it is impossible even to scratch them with a fingernail. Nearly 22% of the surface of a coral reef off Hawaii was found to be covered by several species of coralline algae, each of which had its own distinct habitat preference (Johansen 1981).

There are innumerable types of organisms associated with a coral reef. Some of these are attached, such as sedentary worms and sea anemones, while some, including certain fish and molluscs, hide in crevices. Inshore fish and other wide-ranging organisms live close to the reef and exploit its high biological productivity (De Silva et al. 1980). Of 132 fish species listed as being of 'economic value' in Indonesia, 32 are associated with coral reefs (Soegiarto and Polunin 1980).

Green plants can also be found on and around coral reefs. Some are green seaweeds, such as *Ulva, Halimeda* and *Padina,* but more common are seed-plants known collectively as seagrasses. Most of the seagrasses to be found off the coast of Sumatra, particularly in the Riau Archipelago, belong to the family Hydrocharitaceae, which includes the cosmopolitan Canadian pond weed *Elodea canadensis.* Unlike *Elodea,* however, most seagrasses look remarkably grass-like, with long, narrow leaves (one important exception to this is *Halophila,* which has spoon-shaped leaves). Seagrasses form mixed-species meadows, particularly of *Thalassia hemprichii* and *Enhalus acoroides* (den Hartog 1958), or pure stands, depending on the substrate. Distinct zonation of species can also be observed (McComb et al. 1981).

Seagrass leaves may carry very heavy loads of epiphytes, so many in fact that half the above-ground biomass of a seagrass meadow can be accounted for by epiphytes such as bacteria, algae and various sedentary marine animals. These epiphytes tend to thrive in nutrient-enriched water caused by human activity, but the enhanced growth of epiphytes reduces the amount of light reaching the seagrass, sometimes to its ultimate demise.

The flowers of seagrasses open underwater and the pollen is dispersed either by water currents or, if floating, by being blown across the water surface. Some seagrasses do, however, exhibit various forms of self-pollination (den Hartog 1958).

Seagrass meadows have a high standing crop, high productivity (about one-third of a lowland forest), and they are able to concentrate available nutrients. They would therefore be expected to be important in food chains and this is indeed the case, but primarily through their detritus rather than through being grazed. The detritus is eaten by crabs and molluscs which are in turn eaten by fishes. Seagrass is grazed by green turtles and dugong *Dugong dugon,* but they take only a small proportion of the productivity (McComb et al. 1981). Dugong are now extremely rare and it is doubtful whether a viable population exists anywhere off Sumatra.

Productivity

The net primary productivity of coral reef ecosystems can be as much as 2,000 g carbon per m^2 per year. This is higher than almost any other terrestrial or aquatic ecosystem anywhere in the world. For example, the equivalent average figure for tropical forest and estuaries is 1,800 g, for open ocean 125 g and for savanna 700 g (Whittaker and Likens 1973). In reef lagoons (i.e., where there is little turbulence and thus little mixing of water), the net primary productivity is much lower. For example, Nontji and Setiapermana (1980) recorded values of 39-96 g carbon per m^2 per year for lagoons in the Seribu Archipelago off Jakarta.

Chapter Four

Rivers and Lakes

INTRODUCTION

Sumatran lakes (together with those in Java and Bali) have the distinction of being the first tropical lakes to be studied in great detail. In contrast to many other aspect of Sumatran biology, an enormous amount has been written about a number of Sumatra's lakes. Between 1928 and 1929 a German expedition spent 10 months conducting thorough studies of lakes (and to a lesser extent rivers) in Java, Sumatra and Bali. The results, written by over 100 experts, total an amazing 7,920 text pages with 3,055 tables and figures and were published in *Archiv für Hydrobiologie* (Supplement) between 1931 and 1958. Most of the papers are taxonomic (over 1,100 new species were described), and this gives a wonderful base for freshwater ecology studies. Very little work has been conducted on Sumatran rivers and lakes since that expedition, except for the work by the National Biological Institute on the River Alas (LBN 1980) and by the Bogor Agricultural Institute on the largely brackish Upang-Banyuasin delta (IPB 1975).

One of the major values of the German work for Sumatran environmental science today is in studying changes due to human settlement, irrigation schemes, etc. Studies of lakes in Java and Bali 45 years after the German surveys showed surprisingly little change, but the impact of settlements and recent introductions of foreign fish species was evident (Green et al. 1976, 1978).

Little recent work has been conducted on the ecology of Sumatran freshwaters (with the obvious exception of that conducted by the Faculty of Fisheries, University of Riau [see their journal *Terubuk*] so it is necessary to seek information from other parts of the Sunda Region. The detailed book on the limnology of the Gombak River in Peninsular Malaysia by Bishop (1973) is especially useful.

Identification keys and information in English are available on various aquatic plant and animal groups of the Sunda Region: Euglenidae (Prowse 1958a), Flagellata (Prowse 1962a), Desmida (Prowse 1957), bluegreen algae (Johnson 1970), Rotifera (Karunakaran and Johnson 1978), snails and bivalve molluscs (Berry 1963), bivalve molluscs (Berry 1963, 1974), leeches (Sharma and Fernando 1961), Cladocera (water fleas) (Johnson 1956, 1962), Gerridae (water skaters) (Cheng 1965), fishes (Kottelat et al. 1993), and freshwater turtles (Iskandar 1978).

SUMATRAN RIVERS AND LAKES

Major lakes occur in every province in Sumatra and these are listed in table 4.1 with their approximate surface areas and depths. It is important to remember that no two lakes and no two rivers are the same since their biological and physical characteristics are strongly affected by geological, topographical, and climate factors. The Indonesian saying, "*Lain lubuk, lain ikannya*" ("Different ponds have different fish"), is absolutely true and should be remembered by those conducting studies of aquatic ecosystems.

The 'lebak' lakes of South Sumatra, along the Ogan and Komering Rivers near Padamaran, Tanjungraja and Kayuagung, are particularly interesting. True lebaks are situated close to and connected with a river, and thus receive an annual flow of silty water during the rainy season when the adjacent river level rises and water flows over the banks (Vaas et al. 1953). The use that man has made of these special lakes for rice growing and farming is also worthy of further study.

When Sumatra and the other parts of the Sunda Region were connected by dry land the present rivers in western Peninsular Malaysia and eastern Sumatra were tributaries of the same, much larger river system that flowed towards the east (see fig. 1.9). This is now reflected in the similarity between

Table 4.1. The major lakes of Sumatra.

Province	Lake	Area (km^2)	Depth (m)
Aceh	Tawar	55	—
	Realoib	3	—
North Sumatra	Toba	1,146	450+
	Hulu Batumundam	2	—
West Sumatra	Maninjau	98	169
	Singkarak	110	269
	Di Bawah	15	—
	Di Atas	12	44
Riau	Lakes along the Siak Kecil River	—	—
Jambi	Kerinci	42	—
Bengkulu	Tes	2	—
	Dusun Besar	2	—
South Sumatra	Ranau	80	229
	Jemawan	8	—
Lampung	Ranau	45	229

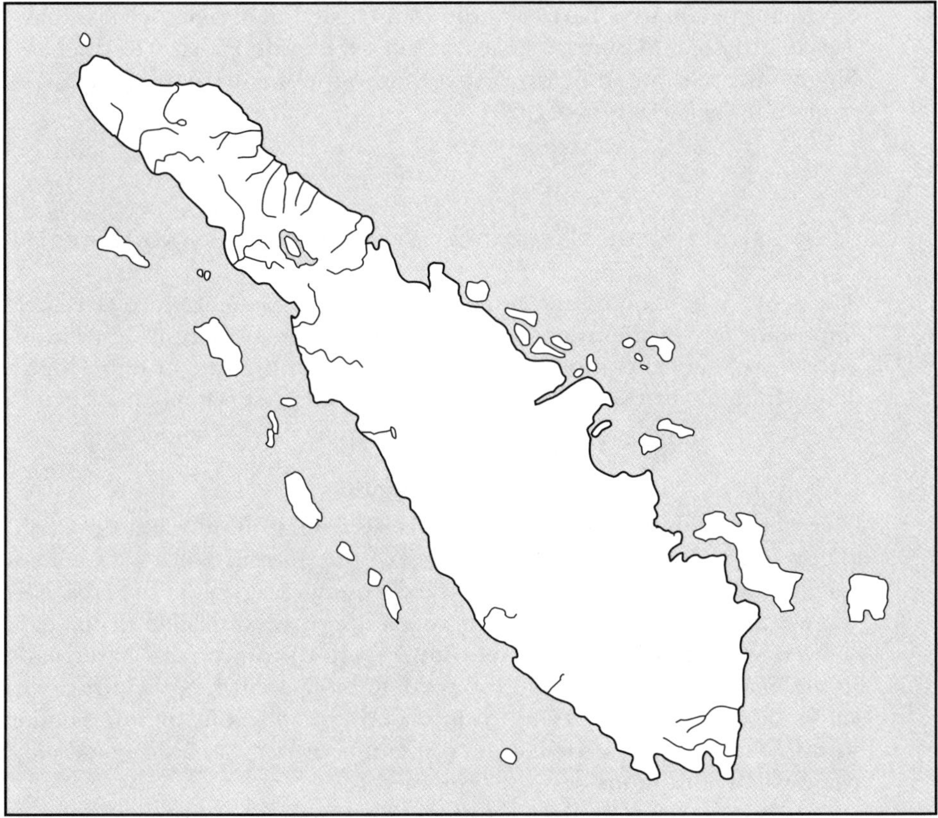

Figure 4.1. Rivers in which freshwater eels are recorded.
After Delsman 1929

the aquatic faunas of the regions. For example, Taniuchi (1979) found the same species of freshwater stingray *Dasyatis* sp. in the Indragiri River (eastern Sumatra) and the Perak River (western Peninsular Malaysia), although it is intolerant of seawater (Mukhtar and Sonoda 1979). When the sea level was lower, the Indragiri and Perak Rivers were tributaries of the same northwest-flowing river, and the present two populations of stingray would have been one. Another interesting example of fish distribution is provided by certain species of freshwater eels which breed in the Indian Ocean in water several thousand metres deep. The young larvae swim towards land and when they reach water about 200 m deep, they metamorphose into 'elvers'. These then enter rivers and grow into adults before swimming back to the ocean to breed. It appears that only rivers immediately adjacent or near to the ocean are entered by eels (fig. 4.1). Lake Toba contains no such eels because the

Siguragura Falls are a barrier to the young fish, but eels do occur in Lake Tawar and Lake Maninjau. Lakes Ranau, Di Bawah, Di Atas, Kerinci and Singkarak are drained by rivers that flow to the shallow east coast and so contain no eels (Delsman 1929).

Effects of the Catchment Area on Rivers and Lakes

The geology, geography and human use of a catchment area are extremely important to considerations of freshwater ecosystems because they influence the chemical composition of the water and the rate of water input. These in their turn will influence the composition of the biota.

Water Input

The major input to most rivers and lakes is from ground water discharge (figure 4.2). The water table – the top of the ground water zone – rises until it is exposed in the bottom of the deepest depression in the area, and this usually forms a river. Water entering a river may have fallen as rain days or even weeks beforehand. Even in Sumatra, rivers can run dry, particularly in the headwater regions, and the resident biota are forced to bury themselves in the moist substrate, to find permanent pools, or, for smaller forms, to form a cyst or some other resistant body that can begin to develop again when conditions become favourable.

At the other extreme, floods occur. Heavy rain does not usually cause serious flooding where the ground is covered with forest and most of the rain can permeate the ground to join the ground water or the interflow (water that flows beneath the ground but above the water table). A small proportion of rain water normally enters rivers and lakes via the surface runoff. However, where forest cover is removed, unobstructed rainfall cause the ground to compact, water is not held in the trees' root mat, and very little water enters the transpiration stream of plants so that most of the rain water flows along the soil surface and directly into streams. This causes erratic streamflow in the headwaters. At times of high water flow, the substratum of the streambed is likely to be scoured out, and many animals and plants are removed with it, thus also depleting the biological resources.

Water Chemistry

Lakes and rivers contain chemicals which affect the nature of the biological communities within them. All plants, both micro- and macroscopic, require a large number of elements, but particularly nitrogen, phosphorous and potassium. Aquatic animals also have specific chemical requirements; for

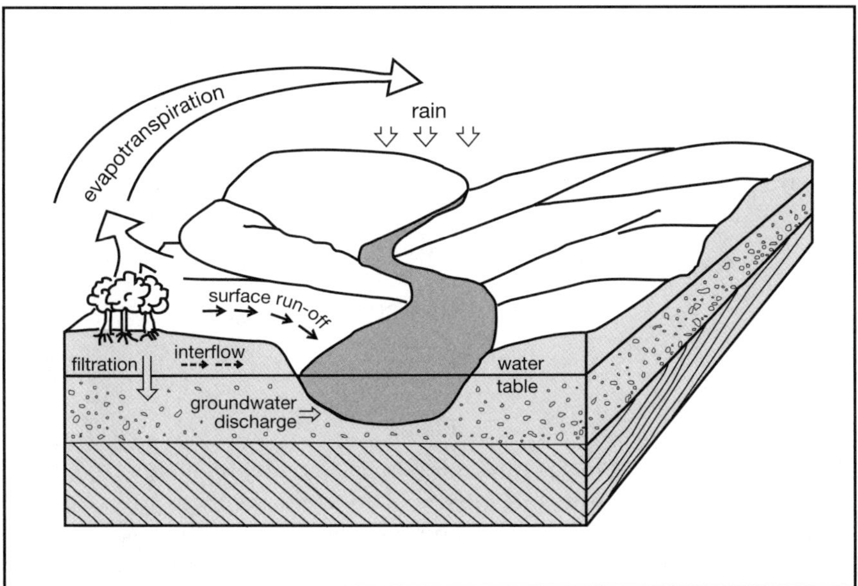

Figure 4.2. Diagram of the various pathways of water after arrival in a catchment area.

After Townsend 1980

instance, molluscs are dependent on sufficient supplies of calcium for shell growth (Berry 1963). The effects of high levels of phenolic compounds in water draining from nutrient-poor sites are described on page 177.

The concentrations of dissolved oxygen and carbon dioxide have important effects on all aquatic organisms. A proportion of these gases enter the water through diffusion from the atmosphere or, in fast-flowing rivers, by natural aeration in turbulent water. However, photosynthesis and respiration within the stream itself also have significant effects on gases in solution. For example, the marked diurnal pattern of photosynthesis may be reflected by considerable changes in oxygen and carbon dioxide concentrations, and in situations where large amounts of organic waste enters a stream (e.g., near habitation or a palm oil processing factory), their decomposition by microorganisms causes severe depletion of dissolved oxygen.

Minerals can also enter rivers and lakes from rain water – which is by no means pure and may contain quite appreciable quantities of many inorganic compounds – and from the ground water. Much of the latter will have been in intimate contact with both soil and unweathered parent rock and as a result will have incorporated inorganic and organic substances in solution or suspension. Thus the chemistry of river water is largely determined by the mineral nature of the catchment area. If this consists of easily eroded sedimentary rock, the concentration of minerals will be high; conversely, if the rock is resistant to weathering, the river water will be relatively low in minerals (Townsend 1980).

Biotic Components

Plants

The major groups of macrophytes, as the larger aquatic plants are known, are the flowering plants and mosses. Those macrophytes which do not float or have most of their leaves above water survive by being adapted to low oxygen levels, slow rates of diffusion and low light intensity of their environment. A good supply of large-celled aerenchyma tissue (through which gases can pass easily) is often found in these plants, thus facilitating adequate internal aeration. Adaptations allowing relatively high rates of photosynthesis under conditions of low light intensity include the absence of cuticle from stems and high concentrations of chloroplasts in the epidermal layer. Many plants, however, such as water lilies *Nymphoides,* duckweed *Lemna,* and the smallest of all flowering plants *Wolffia*[1], avoid these problems by having floating leaves which are in contact with the atmosphere. Some macrophytes, such as mosses, may be found attached to rocks while others, such as water hyacinth *Eichhornia crassipes,* are free-floating. Most of the flowering plants, however, are rooted in the substrate.

The macrophytes of four of Sumatra's lakes have been described in detail by van Steenis and Ruttner (1933). In the lakes (Toba, Di Atas, Singkarak and Ranau) *Hydrilla verticillata,* a submerged plant with whorls of three to eight leaves along the stem, was common. In Lake Singkarak it was found 8 m below the lake surface. Other plant species were common to different degrees in the four lakes.

Algae are the major primary producers in lakes and rivers. Attached algae are found on stones or on macrophytes and although they are generally microscopic in size, larger attached filamentous algae (such as *Spirogyra*) are also found. The other main group of algae is the phytoplankton, which consists exclusively of microscopic forms. Phytoplankton is rarely found in rivers, and any free-floating microscopic algae found in upstream regions are almost certainly not phytoplankton but attached algae which have become detached (Townsend 1980). Examples of phytoplankton are

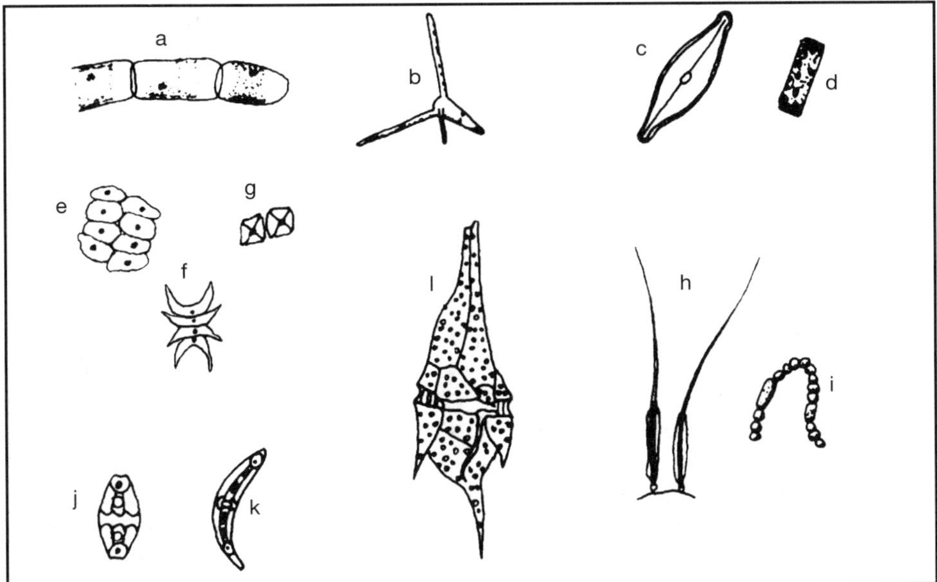

Figure 4.3. Examples of aquatic microorganisms. a - filamentous algae; b - fungal spores; c, d - diatoms; e, f, g - green algae; h, i - blue-green algae; j, k - desmid; l - dinoflagellate.

After Townsend 1980

shown in figure 4.3, and a description of the plankton in the Pekanbaru area is given by Anwar et al. (1980). A methodology for determining the population size of phytoplankton in ponds, where it is much more common, is given by Humner (1980). Studies of chlorophyll-*a* and chlorophyll-*b* in phytoplankton are extremely useful in assessing the primary productivity of aquatic ecosytems (Relevante and Gilmartin 1982).

Fungi and Bacteria

The major role of fungi and bacteria is as decomposers of dead organic matter (corpses, faeces, dead plant material, etc.). The fungi and actinomycetes (a common group of bacteria) are best suited to coping with solid matter because they adhere to and penetrate the surface, and often produce extracellular degrading enzymes. The smaller molecules released during this metabolism are rapidly consumed by bacteria. The ubiquitous nature of fungal spores and bacteria in most fresh water means that dead material is rapidly colonised. Many aquatic fungal spores have four long straight 'arms' diverging from the centre and these probably act as anchors to allow efficient attachment even in turbulent water (fig. 4.3b). These spores are sometimes concentrated by bubbles and the persistent foam below waterfalls or rapids is a good source of them. If this foam is collected, the spores can easily be examined under a microscope (Townsend 1980).

Animals

Animals of lakes and rivers can be roughly categorised by habitat. A few species, such as water skaters (Gerridae) (Cheng 1965), live on the water surface supported by surface tension, and this group of animals is known as the neuston. The species living in the mid-water are divided into two groups; those which are capable of swimming are called the nekton, the others swim weakly if at all and are called the zooplankton. The commonest zooplankton are the Rotifers (Karunakaran and Johnson 1978), and the small crustaceans Cladocera and Copepoda (Johnson 1962). Like the phytoplankton, zooplankton are concentrated in lakes or downstream sections of rivers. Plankton can be useful subjects in environmental studies, but identification, particularly of zooplankton, is often extremely difficult. Keys with pictures for the major plankton groups in Sumatra are needed. There are considerable (and not always easily explicable) fluctuations in the abundance of plankton at different levels in a body of water from hour to hour, day to day and season to season, and so the best method of sampling is by vertical pulls of a plankton net.

Most species of invertebrates are associated with river- and lakebeds and are known as the benthos. The majority of this benthos comprises insects but, in certain habitats, oligochaete worms, leeches, molluscs or crustaceans may predominate.

Food Webs

Constructing a food web for an aquatic or any other ecosystem demands more than simply compiling an inventory and deciding on the probable connections. It may take considerable detective work to elucidate some of the links in the web. For example, the most successful ponds raising Chinese carp *Cyprinus carpio flavipennis* are generally those in which the waters have been heavily fertilised with pig manure. Under such conditions the dominant organism is often the red protozoan *Euglena sanguinia* (fig. 4.4), the numbers of which are sometimes so dense that they form an orange-red film on the water surface. Since carp usually feed on macrophytes, it is not immediately obvious what role *Euglena* plays in the food web (Prowse 1958b). An example of a carefully produced food web is shown on page 148, and the methods used to produce it are by Green et al. (1976).

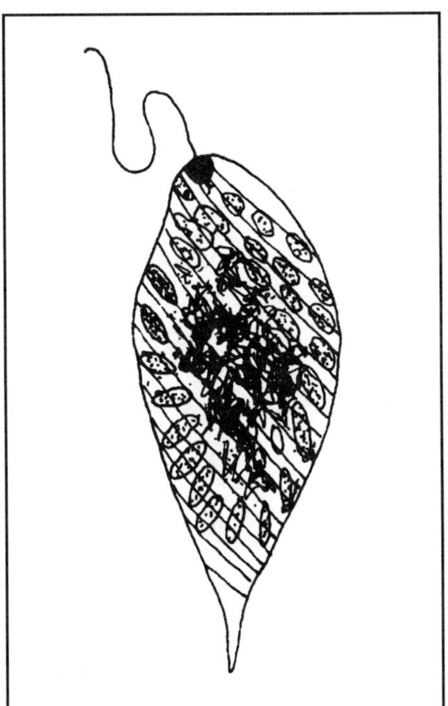

Figure 4.4. *Euglena sanguinia* – a protozoan common in some fish ponds.

Physical Patterns in Lakes

Lakes may be viewed as very slow moving rivers, in which the riverbed has become very wide and very deep. Many of the same species of animals and plants live in both lakes and rivers, and many of the adaptations they require are also the same. However, so great are some of the structural differences of lakes and rivers that their behaviour and characteristics require separate attention. Lakes are dealt with first because they are simpler than rivers.

The German Sunda Expedition studied five lakes in Sumatra:

- Lake Di Atas,
- Lake Ranau,
- Lake Maninjau,
- Lake Singkarak, and
- Lake Toba (Ruttner 1931).

In addition to these, Lake Kawar and Lake Mardingding were studied briefly by a team from CRES in order to collect data to illustrate topics discussed in this section (fig. 4.5).

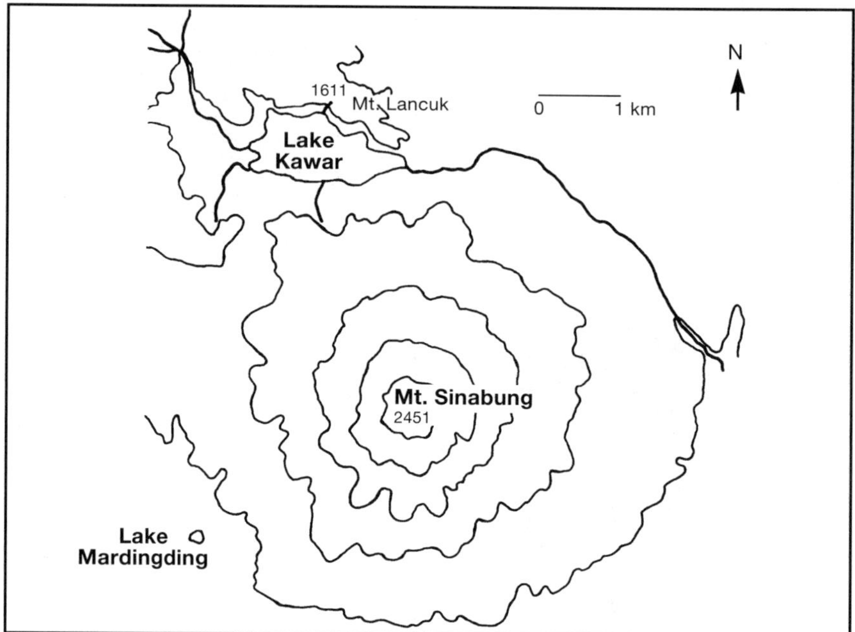

Figure 4.5. Locations of Lake Kawar and Lake Mardingding relative to Mt. Sinabung.

Lake Kawar and Lake Mardingding

Lake Kawar in the Karo highlands was formed several thousand years ago when a lava flow from the volcano Mount Sinabung blocked the River Tupin (Verstappen 1973). A small dam was built at the outflow in about 1985 for an irrigation scheme and this raised the level by about 2 m. Its present area is 116 ha and this is only slightly greater than before the dam was built because the lake sides slope steeply. North of the lake is undisturbed forest and to the south, there is low-intensity agricultural activity.

The CRES team recorded depth, gradients of temperature, conductivity and dissolved oxygen, benthos, and concentrations of nitrogen, potassium and phosphorus at 25 locations on the lake. From the depth soundings a map showing the approximate location of depth contours was produced (fig. 4.6). From these data the capacity of the lake could be calculated (Myers and Shelton 1980) and was found to be 41.5 million m^3 (equivalent to a cube with sides 346 m long). The outflow over the dam was measured and was found to be 25.8 m^3 per minute. Assuming, for the sake of simplicity, that this is an average flow rate, it is clear that the volume of the

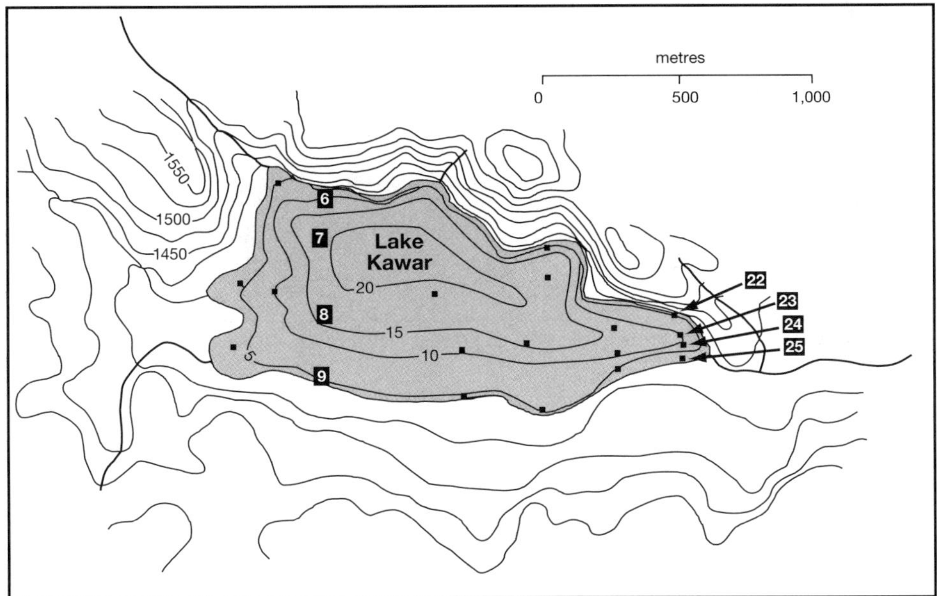

Figure 4.6. Height contours (every 25 m) around, and depth contours (every 5 m) within Lake Kawar, with the locations of the 25 sampling points.

lake is 'replaced' every three years. This fails to take account of see page from the lake into the ground water or of the fact that the deeper water is replaced more slowly than surface water, but it provides some indication of the time taken for water to pass through the lake.

Lake Mardingding lies on the southwest slope of Mount Sinabung. Its formation has not been described, but it may have formed in a small hole left by a minor crater. Its edges were reinforced with cement some years ago, presumably to aid irrigation. The same parameters studied on Lake Kawar were examined at 10 locations on Lake Mardingding. Its area is 0.4 ha and it has a volume of 4,200 m^3. Outflow was not measured but was estimated to be 1 m^3 per minute. Thus the lake volume would be exchanged in about three days (again, this does not take account of seepage into the ground water).

Temperature

The sun's rays penetrate to a certain depth in all lakes, and the surface water is therefore warmed. Warm water is less dense than colder water and so a layer of warm water 'floats' on top of the colder water. The warm layer

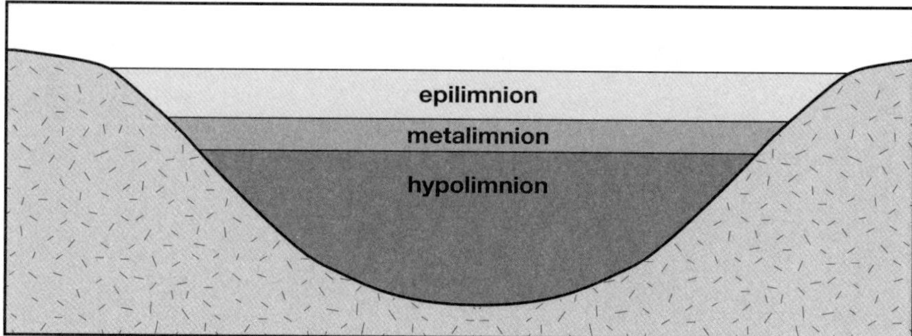

Figure 4.7. Hypothetical vertical section of a lake showing the layers.

is called the 'epilimnion', and the cold layer is the 'hypolimnion'. Separating the two is a fairly thin layer called the 'metalimnion,' across which there is a sharp temperature gradient called the 'thermocline' (fig. 4.7). Dark-coloured water absorbs more heat than light-coloured water. Thus the CRES team investigating the black-water lake of Lake Pulau Besar in the peat swamps of eastern Riau found the surface temperature to be 30.5°C, one or two degrees higher than would normally be expected. Even within a lake, surface temperatures can vary. In Lake Ranau, for example, the western end is 4°C warmer than the eastern end because of the influence of a hot spring (Forbes 1885).

The temperature profile from one of the deepest parts of Lake Kawar is shown in figure 4.8. The thermocline occurred at about 8.5 m, and although the temperature probe was only 12 m long, the hypolimnion appears to have begun 10 m below the lake surface and to have had an even temperature of 21°C. No thermocline or any other stratification was observed in the shallow Lake Mardingding. Sections through the middle and outflow end of Lake Kawar (fig. 4.9) show that the increased flow near the dam caused a certain amount of disturbance.

In none of the 15 lakes studied by the German Sunda Expedition did the temperature difference between top and bottom exceed 5.5°C. At maximum depth in the hypolimnion, the temperature was between 20.1°C and 27.0°C, being lowest in the mountain lakes, and increasing 0.4-1.4°C for every 100 m decrease in altitude (Ruttner 1931) (see p. 277).

Dissolved Oxygen

Thermal stratification has interesting consequences for the hypolimnion. Photosynthetic organisms thrive in the epilimnion near the light, thus

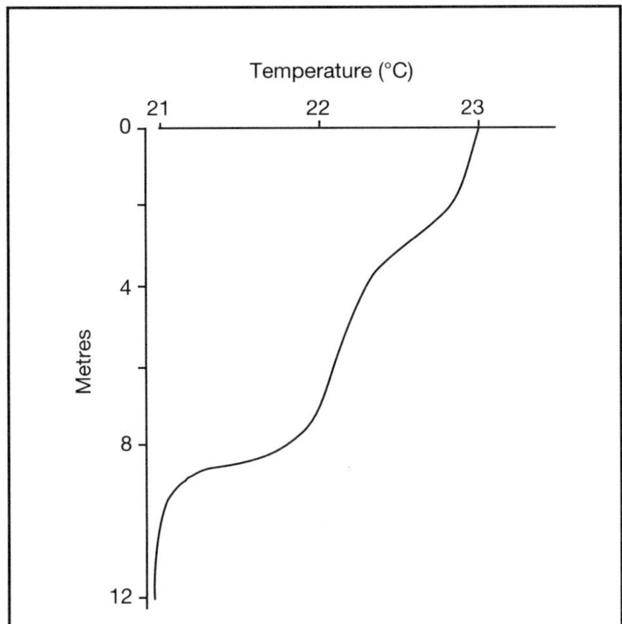

Figure 4.8. Temperature profile near the middle of Lake Kawar showing the thermocline at about 8.5 m.

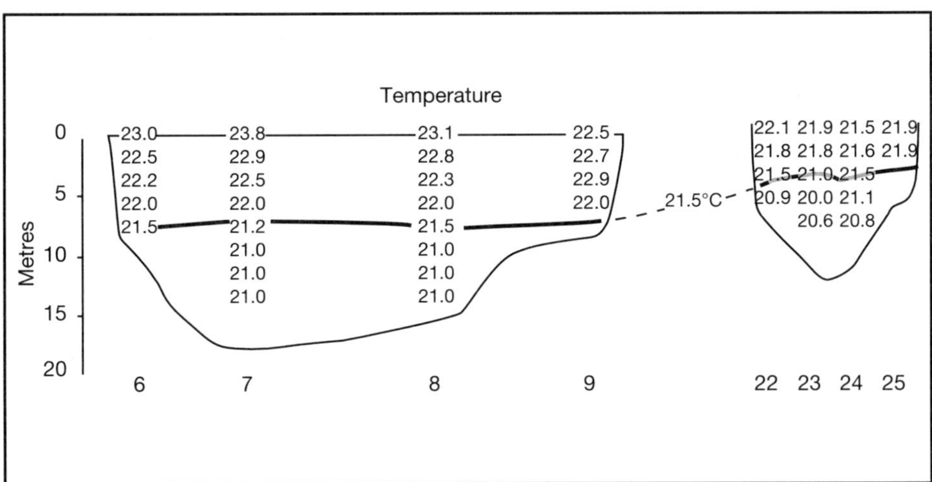

Figure 4.9. Sections through Lake Kawar showing temperature readings every 2 m.

keeping the epilimnion well supplied with oxygen. In the darker waters of the hypolimnion, however, there may be almost no photosynthesis and so almost no oxygen is produced. A few animals live in the sediments at the bottom of the lake, however, and remove the oxygen by their respiration. This oxygen deficit is made worse by the continual action of decomposers; the biota of the lighted surface will be continually dropping faeces and other debris into the hypolimnion, and these will be processed by bacteria as they fall. Bacterial respiration can quickly reduce the oxygen dissolved in the hypolimnion to virtually zero, and there is no way that oxygen can reach these layers unless the layers overturn (p. 145).

The dissolved oxygen profile from one of the deepest parts of Lake Kawar is shown in figure 4.10. This shows the boundary between oxygenated and deoxygenated water to be between 6.5 and 7 m. The dissolved oxygen probe was only 8 m long but the readings suggested the water would have been anoxic at about 10 m.

Sections through the middle and outflow end of Lake Kawar (fig. 4.11) show the boundary between the oxygenated and almost deoxygenated water to be at nearly the same depth across the body of the lake, and about 2 m shallower at the outflow end. It is interesting that the readings for dissolved oxygen 10 cm below the surface were an average of 7.1 ppm on the northern edge and an average of 7.5 ppm on the southern edge. Unlike the northern side, the south is not protected from the wind by projecting ridges and coves and the resultant surface disturbance takes up more oxygen.

Nutrients and Conductivity

As described above, much of the living matter in the epilimnion eventually finds its way to the hypolimnion when the animals or plants die, or when their consumers defecate their remains into the water. Thus, the epilimnion experiences a net loss of nutrient minerals while the hypolimnion experiences a net gain. The profile of conductivity (concentration of electrolytes) for one of the deepest parts of Lake Kawar is shown in figure 4.12. This shows that the boundary between the low concentration of dissolved minerals in the surface water and high concentration of dissolved minerals in the deeper water occurred at 7.5 m. Figure 4.13 reveals that this boundary was more or less constant across the body of the lake and slightly shallower and more disturbed at the outflow end.

Oxygen consumption (rate of respiration) is about 4-9 times faster in Sumatran lakes than in lakes in the temperate regions because the temperatures of the Sumatran hypolimnions are 15-20° higher. Thus carbon dioxide and other solutes are released very fast, and in deeper lakes much of the settling organic matter is mineralised before it even reaches the bottom. In the deepest lakes examined by the German Sunda Expedition, the deep bottom sediment consisted of almost pure silica in the form of

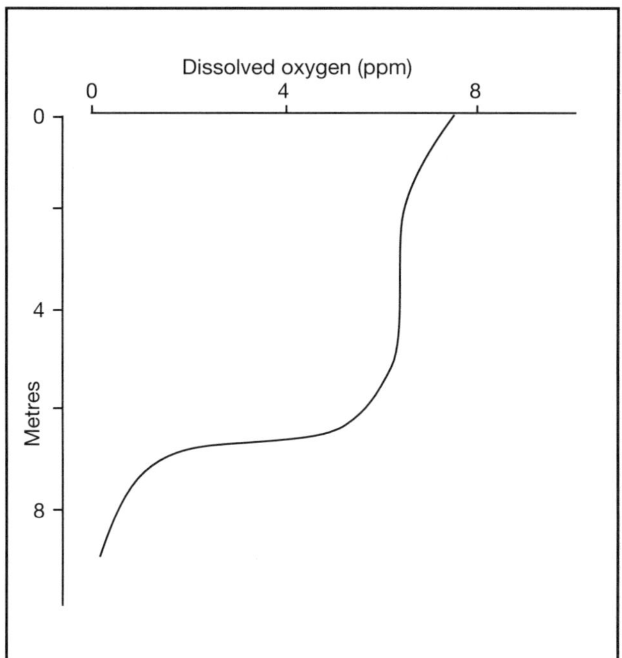

Figure 4.10. Dissolved oxygen profile near the middle of Lake Kawar showing a boundary at about 7 m.

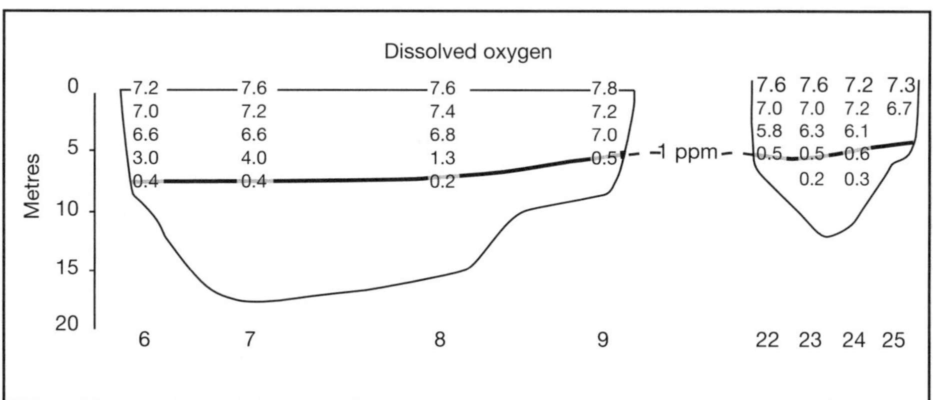

Figure 4.11. Sections through Lake Kawar (between locations 6-9 and 22-25) showing dissolved oxygen readings every 2 m.

Figure 4.12. Conductivity profile near the middle of Lake Kawar showing a boundary at about 7.5 m.

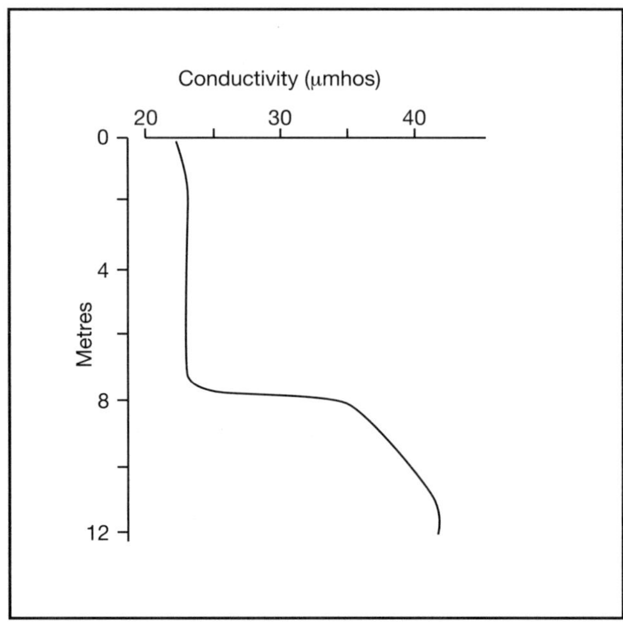

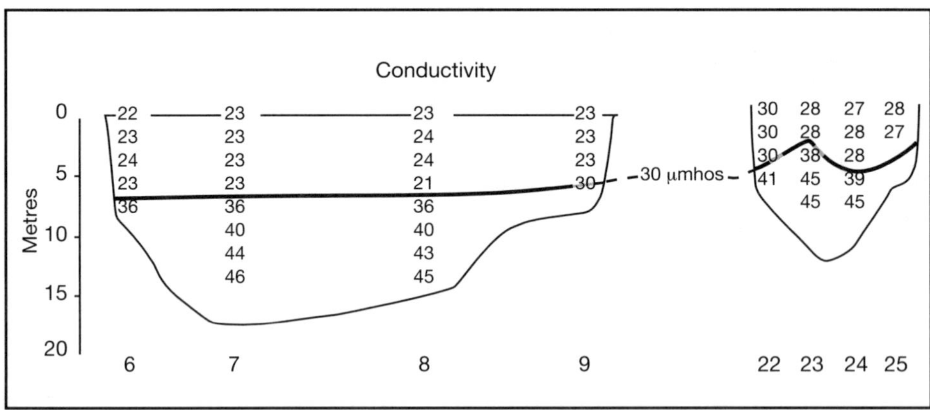

Figure 4.13. Sections through Lake Kawar (between locations 6-9 and 22-25) showing conductivity readings every 2 m.

diatom shells and spongilla needles (Ruttner 1931). Ruttner also found very high levels of phosphate and ammonium in some hypolimnions and suggested that water from deeper layers rather than the surface should be siphoned and used for rice field irrigation to save using artificial fertilisers. Algal growth in the euphotic zone (see below) is limited by the levels of nitrates and phosphorus, and the relative importance of these differs, depending on the catchment area.

Ecologists classify lakes between two extremes based on their nutrient content and organic productivity. 'Oligotrophic' lakes have low concentrations of nutrient minerals and harbour relatively little plant and animal life. The water in oligotrophic lakes is generally clear and unproductive. 'Eutrophic' lakes are rich in nutrient minerals and support an abundant fauna and flora. As a result of the dense plankton populations, water clarity is reduced.

Light Penetration

Light entering a water body is absorbed by the water itself, by dissolved substances and particles in suspension, mainly phytoplankton. In theory, light is never *totally* absorbed at any depth but where it is reduced to about 1% of the surface intensity, this represents the approximate depth at which the energy used by algae in respiration is balanced by the energy they gain by photosynthesis. This is known as the 'compensation point', and the water above the compensation point is known as the 'euphotic' zone. The depth of this zone depends to a large extent on the nutrient status of the lake because high nutrient levels lead to high phytoplankton populations which cause surface shading and so reduce light penetration. If a black-and-white disc (a Secchi disc) is lowered into water, the mean of the depth where it disappeared from view and where it reappeared is equivalent to 0.3-1.5 of the depth of the euphotic zone. Secchi disc readings at Lake Kawar were between 1.9 and 2 m. Dissolved oxygen was found down to almost 8 m, but this suggests that there was some mixing within the epilimnion rather than that photosynthesis occurred down to that depth. Ruttner (1931) found that the euphotic zone was generally the same as the epilimnion and varied from 20-30 m in Lake Toba to 2-5 m in smaller lakes.

Stability

In temperate regions, lake water is mixed in the autumn when the top layer starts to cool and strong winds exert a force on the lake water sufficient to overturn the layers. In most tropical lakes, however, the layering of water is more or less stable although overturns are known (see the paper by Green et al. [1976] for a lake in Java). When tropical overturns occur, the low oxygen concentration of the new surface water can kill fish and other animals.

Most of the 15 lakes investigated by the German Sunda Expedition were stratified and Ruttner (1931) found that their stabilities (resistance to mixing) and depth of thermoclines were related to surface area roughly according to the following proportions:

area	1 :	100 :	1,000
depth of thermocline	1 :	3 :	6
stability (0-20 m)	50 :	10 :	1

Thus the thermocline of Lake Toba (over 1,000 km^2) is over six times the depth of the thermocline found in Lake Kawar (1.16 km^2) and needs less than one-fiftieth the wind force to mix the top 20 m. So large is Lake Toba and so strong the winds that blow across it, that oxygen was found down to 425 m in the northern basin (Ondara 1969; Ruttner 1931).

Biotic Patterns in Lakes

Layers – Light and Oxygen

The distribution of plankton in lakes (and other bodies of water) is governed by a wide range of variables: water density and viscosity, night-time cooling, turbulence, temperature, light intensity and time of day. In addition, the form of feeding or photosynthesis has an effect (Davis 1955). Ruttner (1943) found that differences in plankton abundance did not only occur between the epi-, meta-, and hypolimnion but that considerable variation also existed within the epilimnion; those differences were not easy to explain. Phytoplankton were not usually found below the thermocline but in the very clear Lake Toba, for instance, some phytoplankton were even present in the hypolimnion.

Many fish obviously depend on the plankton but would be unable to feed on them if they were in the hypolimnion because of the low levels of oxygen found there. In Lake Kawar, for example, more than half the volume of the lake would not support fish because of a virtual absence of oxygen, and about 70% of the lake bed would not support benthos-eating fish because the lake bed is below the thermocline. Benthic animals either have to be able to cope with very little oxygen (such as the red, haemoglobin-filled chironomid fly larvae) or with no oxygen (and usually no light), such as anaerobic saprophytic[2] fungi and bacteria. In the low-oxygen, dark environment these organisms have few predators. Some animals make vertical migrations, coming to the surface epilimnion for oxygen and descending to feed. At Lake Kawar a total of 78 hauls of lake-bed samples resulted in just one snail *Melanoides tuberculata* being found. Even when sampling in shallower water closer to the edge, only an occasional snail could be found. This scarcity was probably caused by the low nutrient levels in the lake, which were also reflected in the absence of submerged macrophytes around those parts of the lake that were surveyed.

The high altitude (and therefore low temperature), deep water, and low nutrients of Lake Kawar do not recommend it for fisheries. Carp have been introduced with limited success but introductions of tilapia *Oreochromis mossambica* and gourami *Osphronemus goramy* failed. See the report by Ondara (1969) for an excellent account, in Indonesia, of the introductions into Lake Toba, which is also nutrient-poor, and the paper by Schuster (1950) for a discussion of the successes and failures of fish introductions into Indonesia.

At Lake Mardingding, the water was almost choked with water weeds and floating on the surface were quantities of duckweed *Lemna*. The nutrient levels at Lake Mardingding were much higher than at Lake Kawar and the oxygen levels at the bottom (1 m) of this well-mixed small lake were ± 7 ppm. This resulted in a considerably richer invertebrate (and probably fish) fauna, particularly amongst the water plants.

Water Hyacinth Community Ecology

One of the best known of Sumatra's lake plants is the water hyacinth *Eichhornia crassipes*. It is distributed throughout the tropics and subtropics (although it originated in Brazil) and has caused serious and expensive problems in many water management schemes because of its rapid growth. At the new Asahan River hydroelectric site, water hyacinth floating out of Lake Toba has to be held back by floating pontoons and then collected to prevent the plants entering the various sections of the hydro-scheme. Water hyacinth used to cover most of Lake Kerinci but a control program has reduced it considerably.

The above examples give the impression that water hyacinth is nothing but a pest. In fact, it can be a useful plant: Indonesian conservation groups have used it to make rough paper and efforts have been made to use it as cattle feed. It can also reduce levels of inorganic pollutants (Sato and Kondo 1981), and in moderate quantities plays an important role in lake fisheries.

At Lake Lamongan, one of the crater lakes on Mount Lamongan, East Java, water hyacinth forms a girdle around three sides of the lake up to 30 m wide, held in position by bamboo stakes. Around the fourth side, at the eastern end, there is a much more extensive mat. In the study by Green et al. (1976) it was found that there was fairly free movement by epilimnion water beneath the girdle of water hyacinth and thus the temperature, oxygen and pH up to 2 m beneath the mat was more or less the

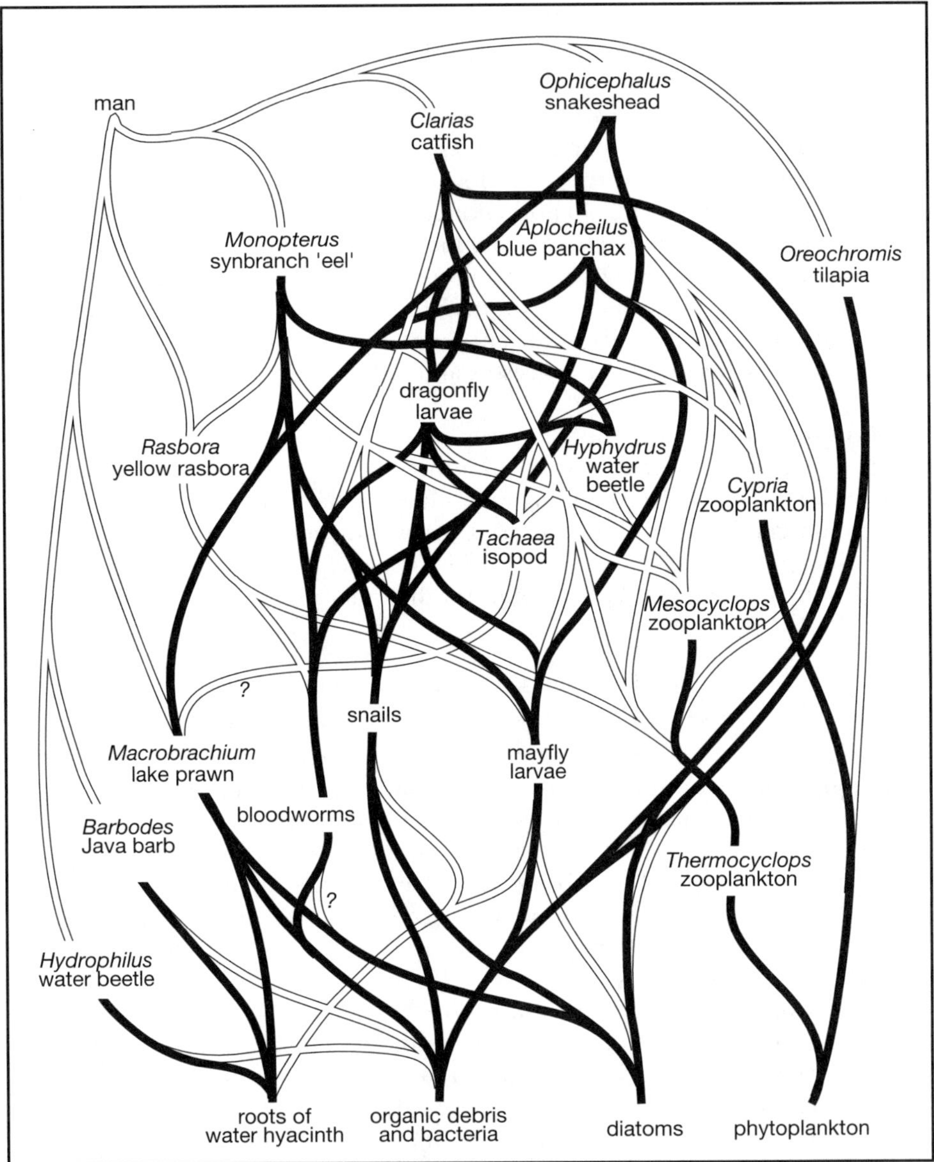

Figure 4.14. A simplified food web for part of the water hyacinth root-mat community in Lake Lamongan. The diagram is based on the gut contents of animals caught in or near the water hyacinth, or known to have fed there because of fragments of water hyacinth roots in the gut; or, in the case of *Ophiocephalus,* because the gut contained so many species from the water hyacinth community. The link between a consumer and its food is dark if the food item was ever recorded as the main gut contents of the consumer. Links based on indirect evidence are marked with a question mark.

After Green et al. 1976

same as that in the open water. At the eastern end, however, there was evidence of oxygen depletion. This was presumably caused by shading which prevented phytoplankton and macrophytes from photosynthesising.

The loose meshwork of water hyacinth roots extending 20-30 cm below the lake surface supported an abundance of aquatic organisms. The leaves reaching up to 1 m above the surface formed a habitat that was essentially terrestrial and was occupied by such air-breathing animals as could swim or fly to it. An analysis of animal stomach contents showed that few animal species seemed to feed directly on the tissue of the water hyacinth roots, but many fed on the organic debris and diatoms that covered the roots. Microinvertebrates were abundant, having been attached to, crawling on, or swimming among the roots, but unfortunately small protozoans and some rotifers are rarely recognisable in gut contents and so their role in the food web cold be not be completely determined (fig. 4.14). The water hyacinth root community forms the basis for an important trap-fishery for the people living around Lake Lamongan.

Physical Patterns in Rivers

Longitudinal patterns in the intensity or concentration of physical factors occur down the length of almost all rivers, but not necessarily in a smooth, continuous fashion. Large waterfalls, for instance, càn cause a major increase in the concentration of dissolved oxygen. Variation also exists across a stream in depth, substrate composition and current velocity, and these also have their effects on the biological communities.

Current Velocity

Despite what one may intuitively believe, the mean velocity of water increases from the headwaters to the lower stretches of a river. Less surprisingly, the discharge of a river (water volume per unit-time) also increases downstream as a product of the cross-sectional area and the mean velocity. This is due partly to additional water entering from tributaries and a reduction of friction on the riverbed caused by the change from large, irregular boulders to fine silt.

Current velocity also varies across the width of a river – being greatest near the middle and greatly reduced among macrophytes – and varies along the length of short stretches. In small streams, for example, shallow sections with fast water alternate with deeper pools where the water flows more sluggishly.

Figure 4.15. Relationship between current velocity and depth in an open channel.

After Townsend 1980

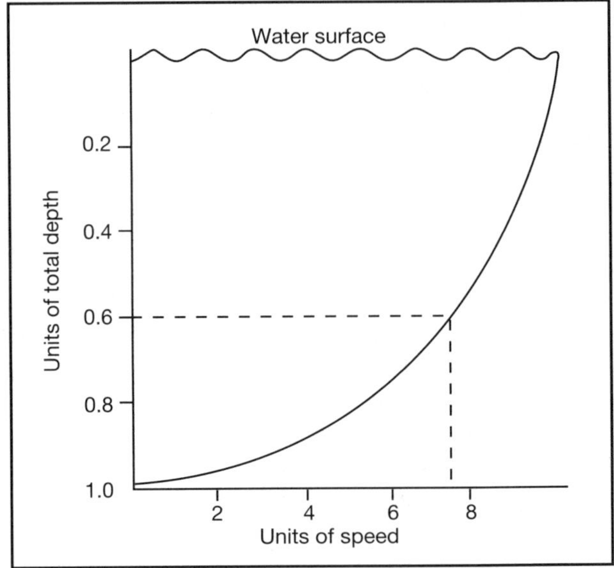

Shear Stress on the Riverbed

Current velocity decreases in a logarithmic manner from water surface to riverbed (fig. 4.15). The average velocity of the whole column of water is equivalent to that occurring at about 60% depth, and this may be an important factor in investigations of plankton and nekton which live in that region. This average velocity is, of course, of little relevance to animals living on the riverbed. They are subject to the force of 'shear stress' (t_o) on the bed. The greater the shear stress on the riverbed, the greater the chance that a benthic organism will be dislodged and washed downstream. Shear stress is calculated as follows:

$$t_o = yds$$

where 'y' is the specific weight of water, 'd' is depth and 's' is slope. Slope declines more quickly than depth increases (there could easily be a hundredfold decrease in slope but only a tenfold increase in depth) and thus for a given discharge frequency, t_o decreases downstream. It is clear then that the shear stress experienced by benthic organisms is likely to be a more important factor nearer the headwaters, despite the lower average velocity there (Townsend 1980).

Riverbed Particle Size

Shear stress affects not only animals and plants but also inorganic particles which can be dislodged. The particles which remain in a location will be those that are larger than the flowing water is able to carry away. The smallest ones will be carried in the water as 'suspended load' while the larger particles roll along the bottom as 'bed load'. As shear stress decreases downstream, so particle size also decreases downstream.

Temperature

Soil is well buffered against the effects of changing air temperature and solar radiation and so its temperature is much more stable than that of air. Headwater streams fed primarily by ground water will be at temperatures close to that of the surrounding soil. The water is gradually warmed as it moves downstream by contact with the air and by being warmed directly by sunlight. The rise is approximately proportional to the logarithm of the distance travelled (Townsend 1980).

Although water temperature is frequently measured in river or lake research programmes, full use is rarely made of the data, and its usefulness to environmental science is often underestimated. A survey of midday temperatures was conducted in three areas in western Peninsular Malaysia (Crowther 1982) and this highlighted the wide range of thermal regimes encountered in quite a small area and showed that river temperatures are strongly influenced by the environmental characteristics of the catchment area. As would have been predicted from the theoretical discussion of temperature in rivers above, water temperatures were lower where there was major input from ground water and where the river was shaded, than they were in open plains where input from surface water was important and there was no obstacle to solar radiation. The thermal regime of a river might provide a useful measure of the environmental impact of land use changes. Forest clearance, for example, would lead both directly and indirectly to increased temperatures and greater temperature fluctuations. The direct effects would be to reduce the amount of shade along the river and to increase surface and soil temperatures by allowing more sunlight to reach the ground. Indirectly, felling would increase surface runoff, thereby reducing the ground water component which has a low temperature and low variability. Also, by favouring high rates of soil erosion, the sediment load would be increased, thus enhancing the rate at which the river absorbs solar radiation (Crowther 1982).

The ecological consequences of the above are discussed on page 157.

Dissolved Oxygen

The concentration of dissolved oxygen in rivers is determined by a number of physical and biological factors. A rise in temperature reduces the solubility of gases, so the increase in temperature downstream causes a decrease in dissolved oxygen concentration. This pattern is reinforced because in the headwaters the water is turbulent and well mixed, whereas downstream it is calmer and less mixed. There is also likely to be more decomposition of organic debris (consumption of oxygen) downstream, and if the water is turbid the rate of photosynthesis (production of oxygen) is lower. Striking daily patterns can be found because photosynthesis can produce oxygen-saturated water during the day, while decomposition of organic matter and plant respiration at night greatly deplete oxygen levels (Townsend 1980).

Mineral Nutrients

The concentration of ecologically important mineral nutrients or dissolved salts generally increases downstream. Rocks in the headwaters are generally resistant to weathering and the concentration of minerals will thus be low. Downstream, in areas of alluvium and sedimentary rock, the inflow of minerals, and therefore the conductivity of the water, will increase. This trend is amplified by the inputs of mineral nutrients from the wastes of human activities (Townsend 1980).

Biotic Patterns in Rivers

Longitudinal sequences in the distribution of many riverine animal and plant species are commonly found – presumably at least partly related to the changes in the abiotic factors discussed above. One of the few detailed studies on this topic in the Sunda Region was conducted by Bishop (1973) along the Gombak River. His data are extremely complex but the longitudinal distribution of just 12 fish species is shown in figure 4.16. Note the change in species at the forest edge and where pollution sources become frequent. Bishop also found a significant correlation between fish diversity and stream order (fig. 4.17), indicating the greater number and complexity of spatial and feeding niches for fish in the larger channels.

The method for assigning stream orders in shown in figure 4.18.

Current

Current velocity increases in a downstream direction but shear stress is greatest in the shallow, turbulent headwaters (p. 149), and so those organisms possessing adaptations enabling them to resist being swept away are usually found upstream. Examples of animals living in headwater rivers are shown in figure 4.19.

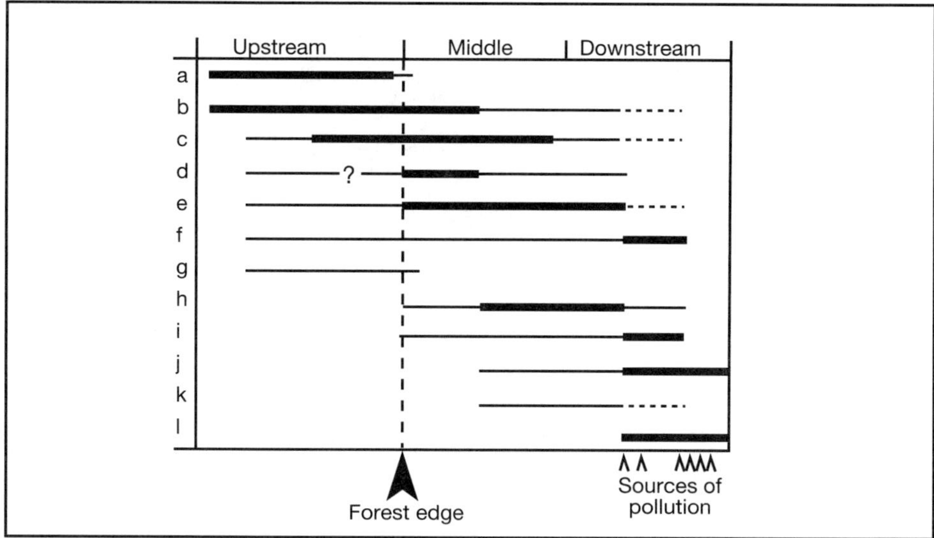

Figure 4.16. Longitudinal distribution of 12 species of fish in Gombak River. Horizontal dotted lines denote rare occurrence; narrow lines denote constant species; wide lines denote 1.0% or more of the fish catch. a – *Tor soro;* b – *Acrossocheilus deauratus*; c – spiny eel *Mastacembelus maculatus*; d – catfish *Silurichthys hasseltii*; e – *Glyptothorax major*; f – barb *Puntius binotatus*; g – snakeshead *Channa gachua*; h – *Mystacoleueus marginatus*; i – rasbora *Rasbora sumatrana*; j – halfbeak *Dermogenys pusilla*; k – *Acrossocheilus* sp.; l – guppy *Poecilia reticulata.*

Adapted from Bishop 1973

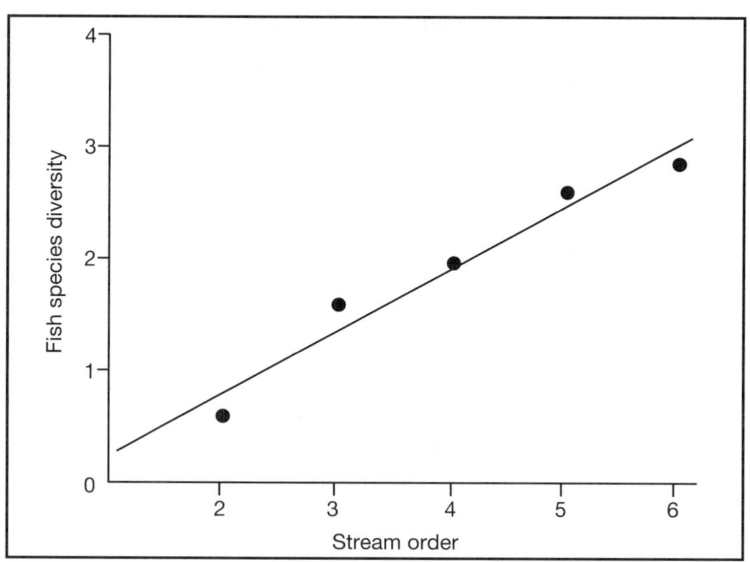

Figure 4.17. Relationship between fish species diversity and stream order.

After Bishop 1973

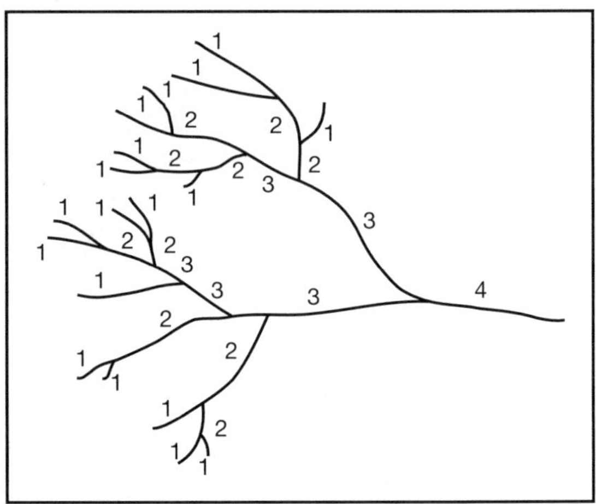

Figure 4.18. Hypothetical river basin showing system of assigning stream orders. Thus, 1+1=2, 2+2=3, but 2+1=2, 3+2=3 and so forth (see Strahler 1957).

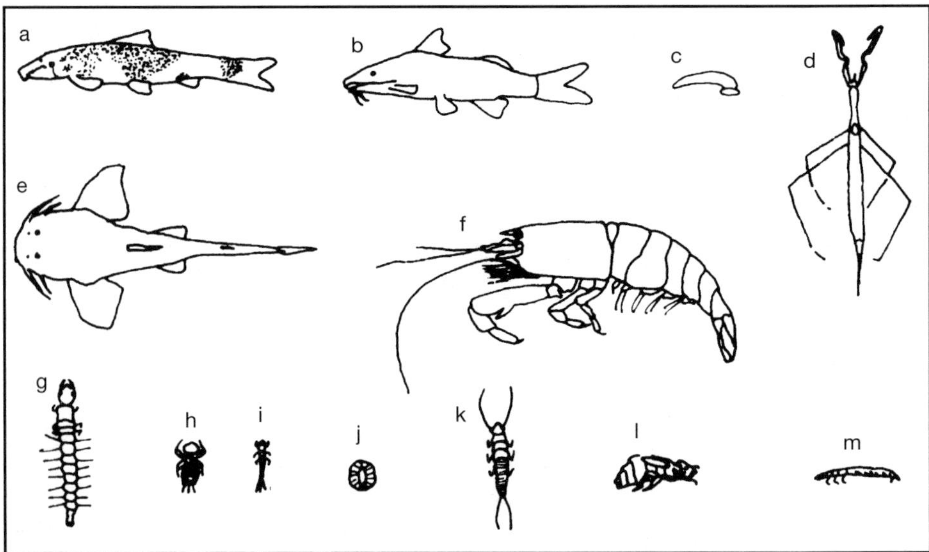

Figure 4.19. A selection of animals from torrent rivers in the Sunda Region. Relative sizes correct; about 1/3 natural size. a-pied suckerloach *Homaloptera ocellata*; b-catfish *Glyptothorax major*; c-leech (Hirudinea); d-giant water stick insect *Cercometus* sp.; e-catfish *Acrochordonichthys melanogaster*; f-torrent prawn *Atya spinipes*; g-hellgrammite larva (Corydalidae); h-mayfly nymph (Ecdyonuridae); i-mayfly nymph *Baetis* sp.; j-waterpenny larva *Eubrianax* sp.; k-stonefly nymph *Perta* sp.; l-dragonfly nymph (Gomphidae); m-caddis worm (Rhyacophilidae).

After Johnson 1957

Plants

The only plants found in very fast-flowing water are encrusting algae, unbranched filamentous algae, and mosses. These are usually squat species which resist shear stress, but this habit makes them more susceptible to smothering by sand. Despite these adaptations, they can, of course, be washed away along with the stones that they are attached to if the riverbed is unstable.

The adaptations which enable higher plants to live in fast-flowing water include a low resistance to water flow, high anchoring strength, and high resistance to abrasion. The ability to propagate from small fragments is also a common feature of plants of these zones. Plants especially adapted to living on the sides of rivers are called rheophytes and are typically shrubs with narrow leaves and brightly coloured fruit which are dispersed either by water or by fish (van Steenis 1952). Clearly, plants growing at the sides of fast streams (or growing amongst resistant species) are subject to less battering, and so variation in plant density occurs across the stream. Since discharge varies with time, plants at any point will, at least in part, show adaptations to the periods of maximum spate (Townsend 1980). Dudgeon (1982b), working on a river in a Hong Kong forest, found that floods caused a great reduction in epilithic algae (algae growing on stones or rock), but not in detritus, because detritus washed downstream was replaced by more detritus washed down from further upstream. Dependence on detritus as an energy source thus leads to a more stable existence than dependence on attached algae.

Invertebrates

There are many ways in which invertebrates have adapted to the conditions in turbulent headwater rivers. Many species have extremely flat bodies which allow them to live and move about in the relatively motionless layer just above the riverbed, or to live under stones, thereby avoiding the current. Other species hang on by means of hooks or suckers, or have streamlined shapes offering very little resistance to stream flow. Lieftinck (1950), writing about a lowland stream in West Java, states that at a meander in a river, the slow-flowing inside of the bend has an almost lake-like invertebrate fauna, whereas the fast-flowing outer bend has species with adaptations for the torrents of headwaters. The same paper also describes the many and varied adaptations of dragonfly larvae to life on the riverbed.

Fish

Many species of fish are adapted to life in fast-flowing water in ways which are similar to those evolved by invertebrates. Most are streamlined and round in cross-section, so offering little resistance to flow, and are frequently strong swimmers. These fish are components of the nekton but many fish of headwaters are effectively part of the benthos, spending most of their time on the riverbed. These fish are not such strong swimmers, are frequently flattened, live amongst stones and rocks, often have their eyes rather close to the dorsal surface and their month located ventrally. Many of these fish have suckers or friction pads to prevent them being washed away. Most of these fish also have the swim-bladder, an air-filled organ acting as a buoyancy control, much reduced in size.

Substratum

The current determines the occurrence of boulders in the headwaters and of silt near the river mouth (p. 150). Such differences obviously influence the distribution of invertebrates: species which are adapted to live under stones or in crevices only occur where the bed is stony, those that rely on their own hooks or nets for anchorage can only occur where the substratum is relatively stable, and the burrowing aquatic larvae of flies require a riverbed of fine particles. In the light of this, it is interesting to consider the impact on the river biota of removing middle-sized boulders from riverbeds for use as hard core for roads and other constructions.

In regions of a river where there is sufficient silt, rooted macrophytes can occur. These have a direct influence on the distribution of many invertebrates because more species and larger numbers tend to occur on plants than on the nearby mineral substratum, and some species or groups are confined to macrophytes. Most invertebrates on aquatic macrophytes do not in fact feed on the plant tissue. Instead, some graze on epiphytic algae growing on the macrophytes, others use the plant as an anchorage from which to filter food from the passing water, and others are predators. Artificial aquatic plants made from plastic string would probably be colonised by the same invertebrate community that colonise the plants and this method can be used for sampling (Macan and Kitching 1972). This emphasises the role of macrophytes as a living substrate rather than as a food (Townsend 1980).

The importance of particle size on a riverbed in determining species distribution was shown in a study of two species of freshwater crayfish *Orconectes* that inhabit the same river in the U.S.A. The rocky parts of the river were inhabited by *O. virilis* and the stretches with finer particles were inhabited by *O. immunis*; both species were found in the middle stretches of the river. Various factors such as dissolved oxygen levels were investigated to explain this distribution, but most important was the nature of the riverbed. Figure 4.20 shows the results of laboratory experiments in which both species were offered three types of riverbed independently and then with both species present. When separate it is clear that both species preferred a rocky riverbed and avoided a muddy riverbed. When both species

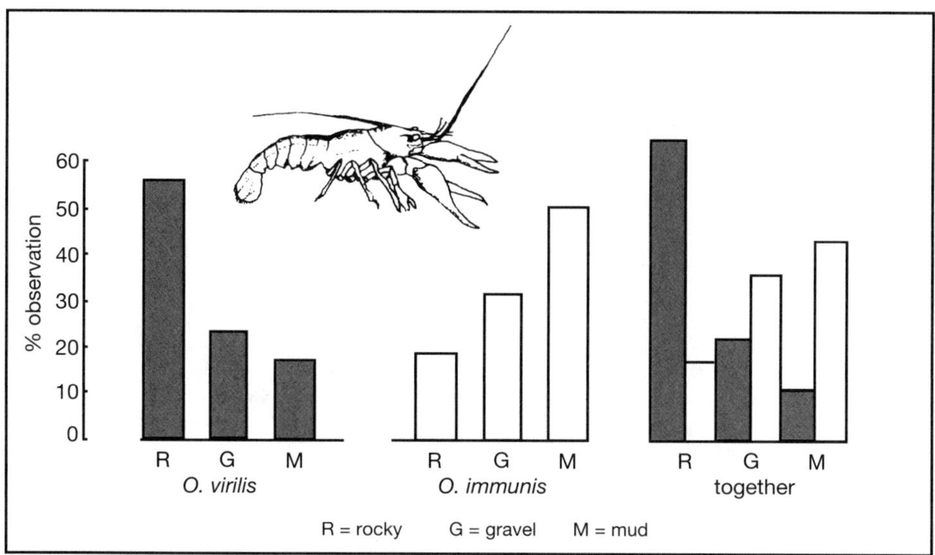

Figure 4.20. Substratum preferences for the crayfish species *Orconectes virilis* and *O. immunis* when alone and when together, expressed as a percentage of total positions recorded.

After Bovbjerg 1970

were present, however, the 'preference' of *O. immunis,* but not of *O. virilis* apparently reversed, because *O. virilis* consistently drove *O. immunis* away from the prime, safe, rocky sites (Bovbjerg 1970).

Temperature and Dissolved Oxygen

It has been shown that river temperature increases with distance from the river source (p. 151). Distributions of animals that do not seem to be related to current or nature of the riverbed might be explicable in terms of temperature, but there is such an intimate relationship between temperature and dissolved oxygen that these are better considered together.

The solubility of oxygen decreases with increases in temperature and so dissolved oxygen levels are generally lower downstream. This effect is exaggerated because water is less turbulent downstream so that the oxygen removed from the water by respiration of organisms is less easily replaced than in the aerated headwaters. This is especially important for organisms inhabiting the sediment where most of the decomposition of organic matter occurs. In this respect, a slow-moving river resembles a lake. Special adaptations to these conditions include the high haemoglobin levels in the bodies of bloodworms (larvae of chironomid flies) which enables them to live at very low oxygen levels, and movements of other animals which

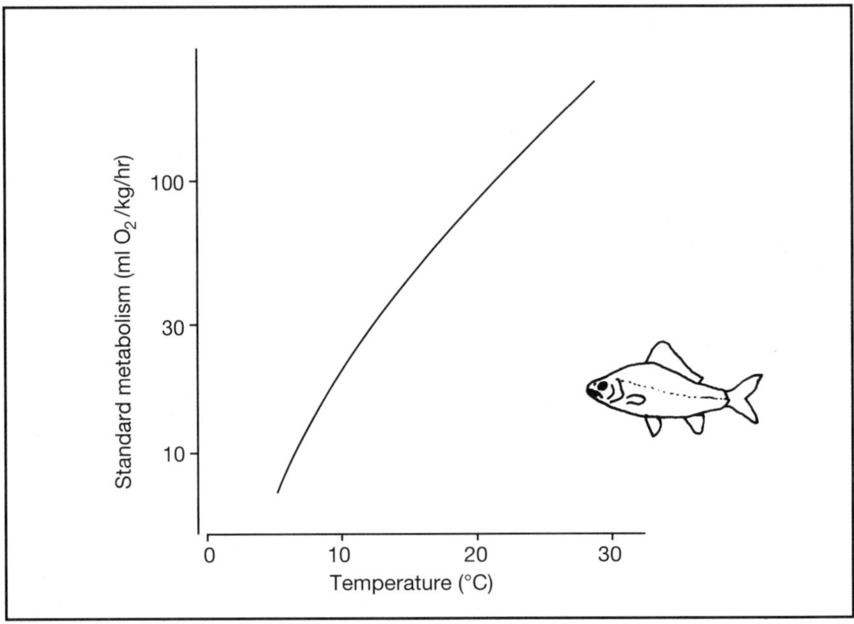

Figure 4.21. The relationship between standard metabolism and temperature for the goldfish.

After Varley 1967 in Townsend 1980

come up to the surface to take in gaseous air.

Most studies of the environmental tolerances and preferences of fish and other aquatic organisms have focused chiefly upon chemical parameters (Mohsin and Law 1980; Palmieri et al. 1980), including pollutants, and the possibility of thermal controls has generally been overlooked. It is well established that the metabolic rates of fish (and some other aquatic animals), and hence their demand for oxygen, increase in proportion to temperature, whilst haemoglobin has a lower affinity for oxygen, and levels of dissolved oxygen in water diminish at higher temperature (Townsend 1980). Some fish, such as goldfish[3], have a wide tolerance of temperature; the same species that lives in ornamental tanks in Sumatran homes can overwinter successfully in European, ice-covered ponds (fig. 4.21). Most fish and other aquatic organisms, however, have a narrow tolerance of temperature, although optimum temperatures will vary between species.

In addition, aquatic life is poorly adapted to rapid and marked temperature changes. Thus, irrespective of other factors, differences in temperature may result in short rivers arising out of mountains and with a significant groundwater component (such as the River Anai flowing through Padang) and long rivers arising from lowland lakes (such as the River

Indragiri arising from Lake Singkarak), supporting different numbers of different species.

As a result of forest clearance on the plains and the development of, for example, mining activities, many mean river temperatures have probably increased from 25° to 30°C. This leads to a 9.5% reduction in dissolved oxygen at saturation, and the effect would probably be compounded because the percentage saturation might well also fall. This, combined with a marked increase in temperature variability, is likely to have considerable impact on many species of aquatic animals and plants (Crowther 1982).

Mineral Nutrients

It would not be surprising if the distribution of certain species were to some extent determined by the changes in concentration of dissolved salts along a river. Virtually all aquatic molluscs, for example, secrete shells of calcium carbonate, and the majority of species are not found in water with less than 20 mg calcium/litre. Thus rivers running off volcanic areas are unlikely to have many, if any, molluscs. Also, plants which have adapted to growing in nutrient-poor (oligotrophic) conditions in mountain headwaters may often not be able to compete against species downstream which are not so adapted (Townsend 1980).

Biotic Factors

There is more to understanding the distribution of an organism than simply determining physical factors such as temperature and water flow. Every organism fares better within a certain range for each of a set of physical variables and the areas where these conditions are met may be called the organism's 'fundamental niche.' Unfortunately, another species may also be happiest in more or less the same conditions and this leads to competition between the species and the establishment of a species' 'realised niche'. Thus, taking the example of crayfish on page 156, the realised niche of *Orconectes immunis* was quite different from its fundamental niche because of the aggression of *O. virilis,* whose fundamental and realised niches were more or less the same. This type of competition is called 'interference competition'. Another type of competition, 'exploitation competition', is more subtle and refers to the indirect competition that takes place between species exploiting the same limited resource. Consumption of that resource by one species will reduce the amount available to the other species, and the one which is less able to convert the resource into reproductive output will either be forced to seek an alternative place to live or else perish. There are few detailed studies of this in any fresh waters but Pattee et al. (1973) and Lock and Reynoldson (1976) describe a series of experiments and observations of three species of flatworms (Tricladida).

These papers show the importance of competition but also stress that abiotic and biotic factors often interact in very complex ways and that distribution patterns may be influenced by different factors in different rivers. An account of resource partitioning among the fishes of rainforest rivers in Sri Lanka has recently been published (Moyle and Senanayake 1984).

Energy Flow in Rivers

At the level of communities or whole ecosystems, the study of ecology can be broadened to include the flow of energy through an ecosystem or part of an ecosystem. In this approach, organisms are therefore regarded as transformers of energy and this energy flow can be traced through the intricate webs of interaction. The biotic community relies on an energy base for the successive trophic levels and in most ecosystems this is provided by plants converting solar radiation through photosynthesis into high-energy organic molecules. Exceptions are caves (p. 315) where there is not enough light for green plants, and rivers in which a substantial proportion of the energy base is provided by dead organic matter. This organic matter can be divided into two components: 'allochthonous' (originating outside the system) and 'autochthonous' (originating within the system). The latter is a relatively minor component. The available organic matter, living and dead, is processed by a wide range of organisms, which include fungi, invertebrates and fish, all interacting in a highly complex manner.

The living organisms of a river can be classified according to their role in energy transfer. Thus, there are:

autotrophs – photosynthesising plants which represent the starting point for energy in most ecosystems;

microorganisms – fungi and bacteria whose major role is in the decomposition of organic material;

and four groups of invertebrates and vertebrates:

grazers – herbivores feeding on attached algae;

shredders – animals feeding on large units of plant material;

collectors – animals feeding on loose organic particles either on the riverbed or free in the water;

predators – animals which (usually) kill and eat other animals (Cummins 1974).

There are also three classes of organic materials:

- dissolved organic matter (DOM), arbitrarily defined as smaller than 0.00045 mm diameter,
- fine particulate matter (FPOM), less than 1 mm diameter, and

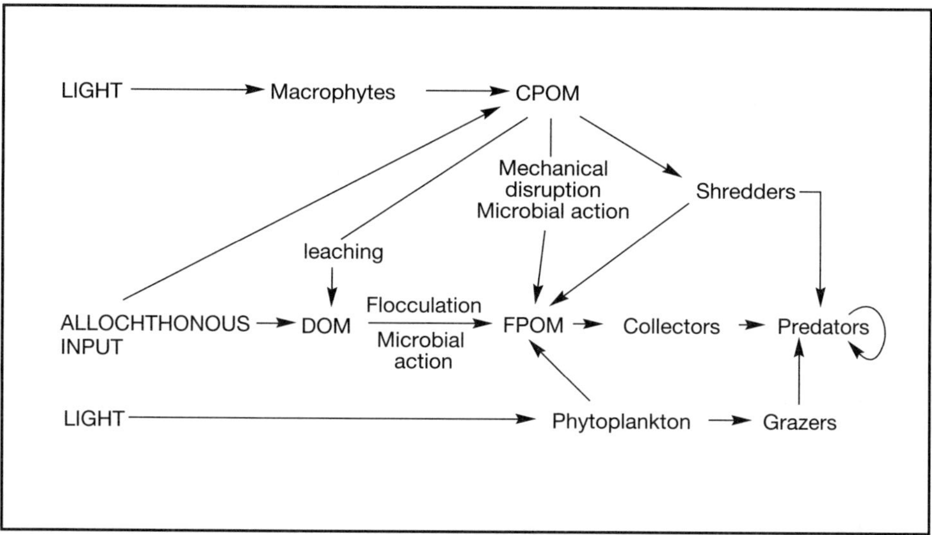

Figure 4.22. A simplified model of energy flow in a river ecosystem. To preserve clarity some arrows have been omitted. For example, all the animals contribute to FPOM in the form of faeces, dead bodies, etc.; some of the allochthonous input contributes directly to FPOM; the principal food of many fish in rivers consists of invertebrates, so the 'predator' category includes fish. However, some fish feed on macrophytes and detritus.

- coarse particulate organic matter (CPOM), more than 1 mm diameter and including whole leaves, twigs, etc.

The FPOM and CPOM components include the microorganisms associated with them. For more detailed discussion of and information relating to energy flow in rivers, see the papers by Petersen and Cummins (1974), Barlocher and Kendrick (1975), and Mann (1975).

Interactions between these groups and the three classes of organic materials are shown in figure 4.22.

Longitudinal Patterns

The actual details of the workings of a particular community depend greatly on the nature of its energy base. A hypothetical model is shown in figure 4.23 that represents the relative importance of the allochthonous and various types of autochthonous inputs. The majority of river headwaters, particularly if undisturbed by man, flow through forested catchment areas and receive a substantial allochthonous input from material

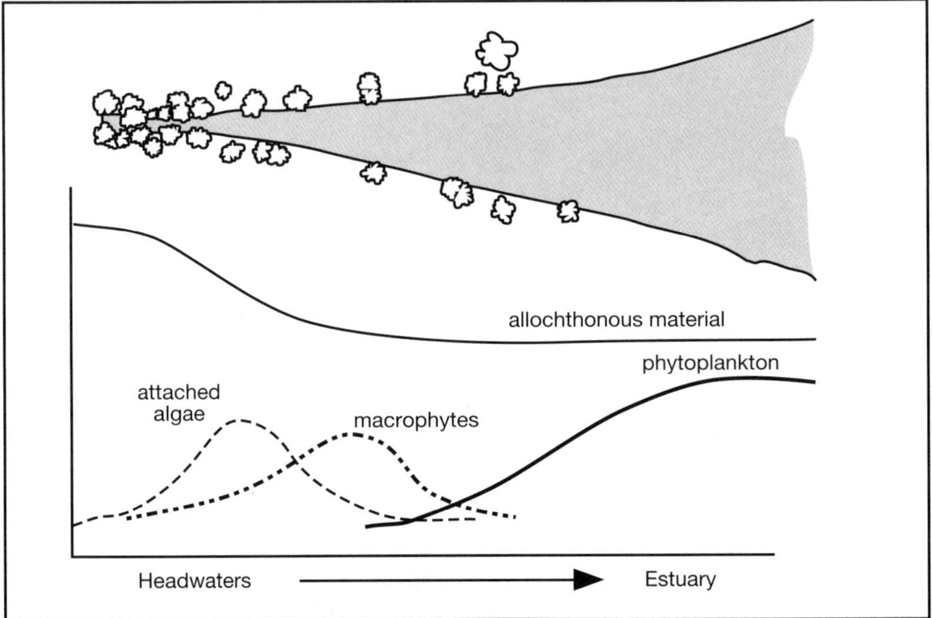

Figure 4.23. Hypothetical representation of the relative contributions of potential energy inputs to a river.
After Townsend 1980

that simply falls in from the forest canopy. The low level of light reaching the river naturally prevents or at least hinders the growth of attached algae or macrophytes. Lower down the river the shading only occurs at the margins and the autotrophic component increases. The attached algal component, which is generally better adapted to extreme flows and less dependent on a substrate of sediment, would be expected to reach its maximum role nearer the headwaters than the macrophytes. Both macrophytes and attached algae should continue to make significant contributions to the river energy budget until the depth and turbidity are such that light can no longer reach the riverbed, and then these plants will be restricted to the margins (Townsend 1980). Phytoplankton will usually only make a significant contribution where the river is long enough for this component to build up. The generation time for phytoplankton is one or two days (at least three or four days and frequently more for zooplankton), and since rivers generally flow between 20-60 km per day, it is clear that only long rivers, those on the east of Sumatra, will commonly have well-developed populations of phytoplankton.

Such longitudinal patterns have not been studied in Southeast Asia. The scheme illustrated is, it is repeated, hypothetical but it seems to fit the known facts from other regions. If it applies to Sumatran systems it can be seen that forest clearance in headwater areas can seriously disturb a river's energy input (and therefore the life that depends on it).

Benthos Dynamics in Rivers

When a heavy fall of rain causes increased flow, the shear stress exerted on the riverbed intensifies and the substrate is scoured, often with the loss of organisms on or in it, and as the flow subsides, so organisms from upstream will be deposited in their place. When the river flow is normal, however, benthic organisms are still moving, either on the riverbed or in the flowing water, and examination of the contents of a net which has been placed in a river and held above the riverbed confirms this. This phenomenon is called 'invertebrate drift'.

Invertebrate Drift

Bishop (1973) found that an average of 222,800 individual invertebrates would drift past a transect in a major headwater stream in just 24 hours. This is equivalent to about 160 individuals per 100 m^3 of river water. The drift varies not just with river flow but also through the day. Studies from various parts of the world, including Peninsular Malaysia, have shown that drift is highest at night, particularly just after sunset (Bishop 1973). This appears to be related to light levels rather than to chemical changes. Many invertebrates spend much of the day hiding under stones, etc., and only forage when darkness falls. It is logical that when they start to move, they are more susceptible to being swept away (Chaston 1969).

An index has been devised that expresses the percentage of the benthos drifting over a unit area of riverbed at an instant in time. Various studies in a number of rivers have shown that this value varies from 0.01% to 0.5% and Bishop (1973) found values of 0.003%-0.018% for headwater streams. These figures may appear low, but for a river flowing at 1 m per second, a value of 0.01% means that in one day at least 86,400 (seconds) x 0.01 = 864 times as many benthic organisms flow over a 1 m^2 area of riverbed as were in that area of riverbed. A study by Townsend and Hildrew (1976) showed that 2.6% of benthic invertebrates shifted their position each day by drifting. It is important to note that McLay (1970) showed that 60% of drifting invertebrates travel for less than 10 m before regaining a foothold.

Whether or not losing contact with the riverbed and drifting downstream is accidental, there may be adaptive significance in doing so. A riverbed, as with most habitats, is composed of 'patches,' some favourable for a particular organism and some unfavourable. The patch may be a food resource, a form of substrate, an area experiencing a certain set of biotic and/or abiotic conditions, etc. In some cases a patch may change its suitability, such as when a food resource is depleted or when a flood occurs (Bishop 1973; Taniuchi 1979). Drifting, although it involves certain risks, is an energy-efficient way of moving from an unfavourable to a possibly favourable patch, for a journey of 10 m along a riverbed is not inconsiderable given the size of many river invertebrates. It has been shown for some invertebrates that if the substrate landed upon is not a favourable one, there is a high probability of the animal re-entering the drift within a short time (5-30 minutes) (Walton 1978), suggesting that invertebrate drift is not entirely passive.

Colonisation Cycles

Sections above have stressed the importance of the downstream movement of organisms. It seems very reasonable, therefore, to ask how the upstream regions remain populated. Do upstream movements by some organisms compensate for the downstream losses?

Firstly, a downstream displacement of organisms does not necessarily lead to the extinction of those species in upstream stretches. One way to view drift is as a dispersal mechanism for removing animals (possibly as eggs or as larvae) which, had they stayed in the headwaters, would have exceeded the habitat's carrying capacity. It is obvious that not all young invertebrates could remain in the area where their eggs were laid because they would soon exhaust the initial food resource (Peckarsky 1979).

The above explanation is not the whole story, however, because an organism drifting downstream is likely to leave its zone of suitable environmental conditions. It would therefore be reasonable to suggest that adult invertebrates that managed somehow to reach regions upstream of their optimum habitat to breed would have an evolutionary advantage because their young would have a greater chance of developing in that optimum habitat (Townsend 1980). A 'colonisation cycle' is thus envisaged with eggs being laid in the headwaters, dispersal of larvae occurring downstream and an upstream flight or other movement of adults to the headwaters to complete the cycle. This is commonly known as Müller's hypothesis.

The first two stages of this hypothesis are irrefutable, but evidence for the upstream movement of adults is less convincing and a review of early work on this subject has been written by Bishop and Hynes (1969). Twin traps set to catch insects flying upstream and those flying downstream along the Gombak River, revealed that the predominant direction of flight was in fact downstream (Bishop 1973). As a rule, winged adults of invertebrate species with aquatic larval forms are not strong fliers and their flight direction might simply reflect the prevailing wind direction. Strong winds occur most frequently in rainy seasons and these are the periods when insect dispersal is most common (Fernando 1963). Adults of invertebrate species which spend their entire life cycle in fresh water are not usually strong swimmers or walkers, but they may travel near the river edge where shear stress is least so that the upstream journey requires the least possible energy. Most studies have found that upstream movements represent only about 7%-10% of the individuals that move downstream (Moss 1980; Williams 1981). It must be remembered, however, that if only a single female reaches the upstream regions, she may lay hundreds or thousands of eggs.

The study of invertebrate drift deserves more attention. If a factory is to be sited in the middle stretches of a river, one of the many problems that should be considered by an environmental impact study is the impact on the recolonisation of upper stretches of the river by invertebrates (or, for that matter, fish). Is the effluent going to be poisonous or debilitating to adults moving upstream? Which months are the most critical? Since invertebrates are common food for fish, and fish are common food for humans, the problem of the mechanism of invertebrate drift is wider than that of esoteric biology.

Chapter Five

Peatswamp Forests

INTRODUCTION

Peat is a soil type with a very high organic content. It is defined, for the purposes of this chapter, as a soil at least 50 cm deep which, when dried and burned, will lose more than 65% of its mass (Driessen 1977), and is therefore more than 65% organic matter. Certain international agricultural-oriented criteria now define peat as a soil with more than 30% organic matter[1] (Driessen and Soepraptohardjo 1974), but peatswamp forests do not develop on such soils (Anderson 1964). It is estimated that the natural area of peat soils in Sumatra was once 7.3-9.7 million ha (Andriesse 1974), or about one-quarter of all tropical peat lands (Driessen 1977).

Peat deposits can be of two forms:

- ombrogenous peat: the most common type which has its surface above the surrounding land. Plants that grow on this peat only obtain nutrients from within the peat, directly from the rain, or from the plants themselves. There are no nutrients entering the system from the mineral soil below the peat or from rain water running into it. This type of peat is generally found near the east coast behind the mangroves and is often very deep (± 20 m). The peat and its drainage water are very acid and poor in nutrients (oligotrophic), particularly calcium.
- topogenous peat: a less common type formed in topographic depressions. Plants growing on this peat extract their nutrients from mineral subsoil, river water, plant remains and rain. Topogenous peat can be found behind coastal sand bars and inland where free drainage is hindered, such as in mountain depressions and extinct craters. The peat is usually found in a relatively thin layer (± 4 m), but a 9 m core was taken from the bottom of a pond on the peaty plateau west of Siborong-borong, south of Lake Toba (p. 18). The peat and drainage water are slightly acid (pH 5.0) with relatively abundant nutrients (mesotrophic). The surface of the peat is a fibrous crust, either hard and compacted or soft, overlying a semi-liquid interior which contains large pieces of wood and other vegetable remains.

Much of the fundamental work on peat soils by Polak (1933) was conducted in Sumatra, and early vegetation studies concentrated on the potential of peatswamp forests for forestry (van Bodegom 1929; Boon 1936; Endert 1920; Rakoen 1955; Sewandono 1938). More recently, studies have been conducted on vegetation ecology and soils of peatswamp forest in Riau (Anderson 1976; Driessen 1977); Ahmad (1978) classified coastal swamps (probably topogenous) in West Sumatra, Yamada and Soekardjo (1980) studied peatswamp forest in South Sumatra, and Whitten (1982c) investigated a topogenous peatswamp on Siberut Island, West Sumatra. Teams from the Environmental Studies Centre at IPB (Bogor) have conducted various studies on peat in the Upang delta and Banyuasin area of South Sumatra (PSL-IPB 1977). In addition, Morley et al. (1973) examined small peaty lakes in central Sumatra.

A CRES team visited the peatswamp forest around Lake Pulau Besar, Bengkalis, in the Caltex oil concession in order to collect some data for this chapter.

Peatswamp Formation

Formation of Ombrogenous Peatswamp

It was shown on page 84 that mangrove forest colonises areas of recently deposited alluvium along the coast. As the coast, and therefore the mangrove forest, advances seawards, so other plant communities develop in its wake (Anderson 1964). These are no longer inundated by tides and so coastal decomposers such as crabs are rare or absent. Decomposer microorganisms are unable to thrive in the high sulphide and salt conditions of the soil and so a layer of undecomposed vegetable matter (peat) begins to form over the mangrove clay soil.

This general scenario has been confirmed by analyses of pollen from cores of peat taken from a swamp in Sarawak (Wilford 1960). At 13 m below the present surface, a mangrove clay was found overlain by remains of mangrove vegetation. Above this, pollen from the various vegetation types typical of peatswamp forest were found in order of succession. The whole successional sequence had taken 4,500 years (Wilford 1960), indicating the period over which coastlines have been advancing by deposition of alluvium rather than being flooded by the rising sea (p. 12) (fig. 5.1). Although the average rate of peat accumulation was 0.3 m/100 years, the rate during the early stages was much higher (0.475 m/100 years for 12-10 m below the present surface). More recently the rate of accumulation has fallen (0.223 m/100 years for the top 5 m) (Anderson 1964) as nutrients became progressively more scarce. This difference explains the convex shape of the

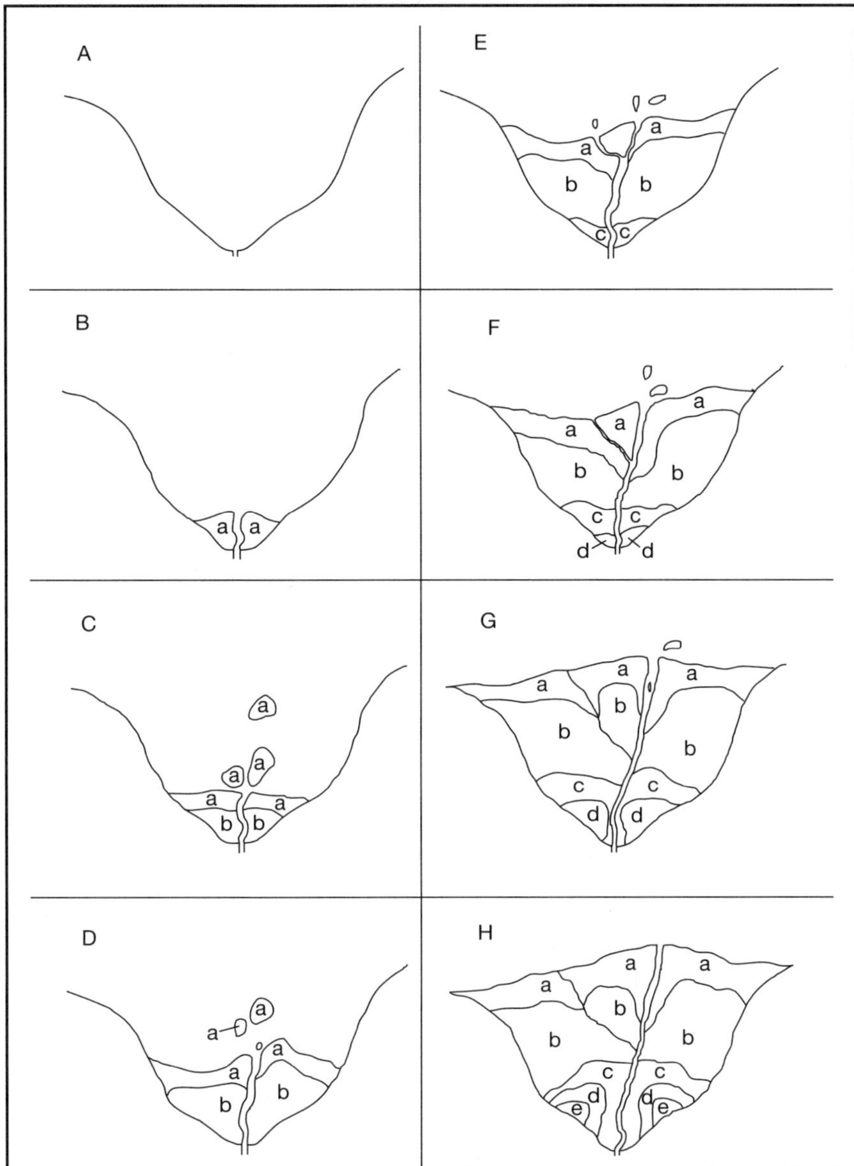

Figure 5.1. Hypothetical formation of a coastal peatswamp forest. A – an estuary after the final rise in sea level (about 8,000 years ago). B-H – deposition of alluvium, colonisation by mangrove forests. a – mangrove pioneers (e.g., *Avicennia*), b – late mangrove species (e.g., *Bruguiera*), c – peatswamp forest pioneers on thin peat and slightly brackish soils, d – mixed peatswamp forest on thicker peat soils above the level of adjacent rivers, e – dwarfed 'padang' forest on thick peat.

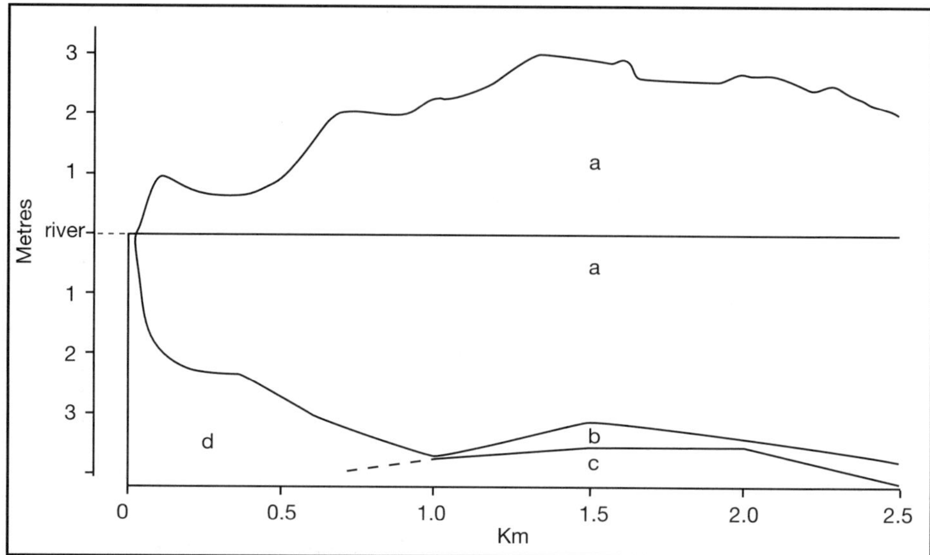

Figure 5.2. Vertical section through a peatswamp near Muara Tolam, near the River Kampar, Riau. Note that the vertical and horizontal scales are different. a – peat layer (a matted crust above, looser below); b – peat remains and clay deposited beneath mangrove forest during an earlier time; c – clay deposited beneath the sea; d – clay deposited beneath river.

After Driessen 1977

peatswamp surface. As swamps replace the mangrove and as the distance to the sea increases, so rivers deposit alluvium along their banks, forming levees which are raised above the level of the original swamp. This is the cause of the saucer-shaped base of peatswamps. A cross-section of one edge of a peatswamp in Riau is shown in figure 5.2.

Ombrogenous peat can form over freshwater swamp if conditions are suitable (Morley 1981) but this seems to be relatively unusual.

Formation of Topogenous Peatswamp

Topogenous peatswamp is formed in a small lake or behind a barrier to drainage such as a sand ridge, which results in waterlogged conditions. The soil pH is not entirely unfavourable to decomposer microorganisms, and many of the plants are able to root in the mineral silt and clay below the peat. This means that peat accumulation is slow, great depths of peat are not formed, and that the brown humic acids characteristic of ombrogenous peat are not so pronounced.

The area of topogenous peatswamp in Sumatra is not large, and small lakes only occur in coastal western Sumatra and certain locations inland

(Hansell 1981). In the classification by Ahmad (1978), topogenous peatswamp is described as shallow freshwater swamp and fringe brackish swamp. There is virtually nothing known about the vegetation or ecology of topogenous peatswamps in Sumatra.

Drainage and Drainage Water

Although the surface of an ombrogenous peatswamp is raised above the surrounding areas, rivulets form on its surface. This is probably so because the heavily compacted mass of woody peat, and the saucer-shaped mineral base of the swamp, prevent lateral drainage (Anderson 1964). Water flowing out of such a peatswamp is generally clear but appears tea-coloured by transmitted light, and opaque black by reflected light in the same way as does Coca-Cola. Such waters are known as blackwater rivers and are generally very acidic (pH 3-4.5). They contain many fewer inorganic ions than do clear, white or muddy waters – even in the same drainage basin – have low concentrations of dissolved oxygen, and high concentrations of humic acids (Janzen 1974a; Mizuno and Mori 1970; Mohr and van Baren 1954).

The low nutrient content of blackwater rivers is partially explained by the nature of the soil, for which the only source of nutrients is rain water (p. 167). The humic acids are chelating agents for inorganic ions, binding them into larger molecules, thus preventing their uptake by plants. The generally low oxygen levels are probably due to the scarcity of aquatic plants. The high acidity of the water is due to the humic acids.

Humic acids are phenolic compounds, which are one of the many groups of plant defence (or secondary[2]) compounds and have been found to be very important in plant-animal interactions. Phenolic compounds are generally toxic to animals and decomposers because they bind with protein (pp. 106 and 231) and are difficult to degrade. They therefore persist to a greater degree than other chemicals in plant debris. Humic acids would therefore be expected to have detrimental effects on organisms attempting to live in blackwater rivers.

It seems that the water draining into the blackwater rivers has high concentrations of phenolic compounds (and possibly other toxic compounds) because:

- the leachate from the living vegetation and decomposing litter in the peatswamp soils is exceptionally rich in phenols and other defence compounds (but see the paper by St. John and Anderson [1982]), and because
- the soil leads indirectly, and high input of phenols leads directly, to a soil and litter decomposition community (of bacteria, fungi, etc.) which is unable to break down these compounds (Janzen 1974a).

The effect of phenolic compounds on the soils, aquatic and forest communities, will be discussed further on page 177 after a description of peatswamp vegetation.

VEGETATION

Composition

Because the surface of an ombrogenous peatswamp is out of the reach of flood water and because the only nutrient input comes from the nutrient-poor rain, there is a decrease towards the centre of a swamp in the amounts of mineral nutrients in the soil, and this is particularly marked for phosphorus and potassium (Muller 1972). The top 15 cm of a peatswamp soil, where a mat of roots is formed, generally has more nutrients than deeper down. This trend of increasing infertility towards the centre of a peatswamp seems to be reflected in the vegetation by the:

- decreasing canopy height,
- decreasing total biomass per unit area,
- increasing leaf-thickness (an adaptation to poor soils) (see pp. 257 and 286), and
- decreasing average girth of certain tree species (Whitmore 1984).

Thus there is no single type of peatswamp forest. Instead, in response to the changes in nutrients, depth of peat, etc., a series of forest types develop which intergrade with each other. Anderson (1963) has described the composition of forest types from certain peatswamp areas in Sarawak in great detail but considerable differences exist in species composition of equivalent forest types (1976), and not all forest types were found in each area.

In general, however, the sequential pattern of forest types (fig. 5.3) represents a change from:

a) a high forest with an uneven canopy which is similar in general appearance to other Sumatran lowland forests, but with fewer species and stems per unit area, and a lower canopy (36-42 m); to:
b) a dense forest with a relatively even canopy, but stunted and with various xeromorphic characters; to:
c) a very dense 'pole forest' with a canopy barely 20 m high and trunks with a diameter rarely wider than 30 cm; to, in some areas:
d) an open, savanna forest (padang) with few trees over 15 m high (Anderson 1964; Whitmore 1984).

Anderson (1976) examined trees of 30 cm girth (± 10 cm diameter) at breast height and over in three areas of peatswamp in Riau at Teluk Kiambang, Muara Tolam and near the Siak Kecil River.

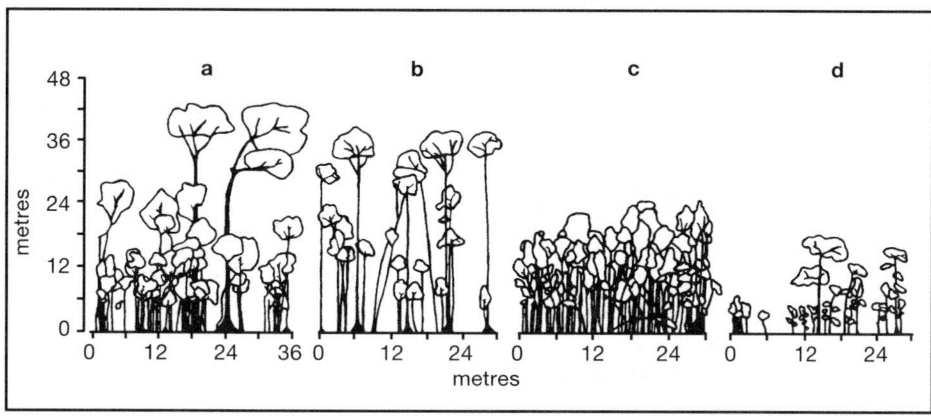

Figure 5.3. Sections through four types of peatswamp forest. a – high-canopy mixed forest; b – lower canopy forest; c – pole forest; d – padang forest.
After Anderson 1961

Although he had recognised six forest types in Sarawak forest, he divided these areas of Sumatran forest into just two[3]. The three locations had markedly different species compositions (table 5.1), indicating that no generalisations should be made about the species present from this small sample. Further variation is indicated by the dwarf forest dominated by *Tristania* found by Polak (1933) in the central region of the Pareh Peninsula, and the dominance of somewhat stunted *Tristania obovata* and *Ploiarium alternifolium* found in peatswamp forest in the Musi delta region of South Sumatra by Endert (1920). *Tristania* is easily recognised by its smooth, peeling bark which varies from pale-grey to orange in colour. It is

Table 5.1. Tree species found in two types of peatswamp forest at three locations in Sumatra.

	Teluk Kiambang	**Muara Tolam**	**R. Siak Kecil**
Mixed swamp forest	*Durio carinatus**	*Artocarpus rigidus*	*Strombosia javanica**
	Palaquium burkii	*Gonystylus bancanus*	*Mezzettia leptopoda*
	Dyera lowii	*Palaquium burkii*	*Palaquium walsuraefolium*
	Shorea platycarpa		
	Shorea uliginosa		*Koompassia malaccensis*
	Shorea teysmanniana		
Pole forest	*Campnosperma coriacea*	*Eugenia elliptifolia*	*Palaquium burckii**
		Shorea teysmanniana	*Blumeodendron kurzii*
		Mangifera havilandii	*Palaquium walsuraefolium*
			Campnosperma coriacea

* = dominant species
From Anderson 1976

a member of the family Myrtaceae, and like its relatives *Eucalyptus*, *Melaleuca* (paperbark tree) and *Eugenia* (e.g., rose apple, clove), its leaves have an aromatic smell when crushed.

Yamada and Soekardjo (1980) found that in their 0.01 ha central peatswamp plot, *Polyalthia glauca* was the most common tree. The topogenous peatswamp examined on Siberut Island was dominated by *Stemonurus secundiflorus* (Icacinaceae) and *Radermachera gigantea* (Bignoniaceae), which together made up nearly 50% of the trees (Whitten 1982c).

Palms are not common in peatswamp forest but a few species are more or less confined to this ecosystem. Both Endert (1920) and Yamada and Soekardjo (1980) mention *Salacca conferta,* a close relative of the edible salak *Salacca zallaca.* Salak is a stemless palm with long (6 m) leaves armed along the midrib with fans of long white spines (Whitmore 1977). Its habit of growing in dense clumps makes salak one of the most formidable natural barriers in Sumatran forest. The tall serdang palms *Livistona hasseltii* of Sumatran peatswamp forest often emerge above the forest canopy (Dransfield 1974), and their crowns of fan-shaped leaves can be seen clearly from the air. A total of 22 palm species have been found in the peatswamp Berbak Reserve in Jambi, making it the most palm-rich peatswamp forest yet known (Dransfield 1974). The forest around Lake Pulau Besar examined by a CRES team had large numbers of the bright-red sealing-wax palm *Cyrtostachys lakka.*

The only epiphytes found in the pole forest and peat padang are those with some means of obtaining mineral nutrients other than just from mineral uptake from their substrate and the rain. Almost all these epiphytes provide shelter for small, non-aggressive ants *Iriodomyrmex,* and in return receive nutrients in the form of discarded food, dead ant remains and ant waste products (Huxley 1978; Janzen 1974b).

Muller (1972) investigated fossil pollen in coal (fossilised peat), and marine and coastal sediments in Sarawak. He found that some of the tree genera present in peatswamps there today have been present since the Oligocene (26 million years ago) and that most have been present since at least the mid-Miocene (± 18 million years ago). The striking similarities between the ancient peatswamp and those of today indicates that the climatic changes of the Pleistocene did not influence the composition of the peatswamp communities to any great degree.

Structure

Structural and other data taken from peatswamps by Anderson are shown in table 5.2. For some variables, such as number of trees, there is considerable variation between plots in the same forest type and so the averages should be treated with caution. The number of species per plot is very similar between the forest types. This is not the pattern generally found in

Kalimantan (Anderson 1964, 1976), where pole forest is much poorer in species and may indicate that the soils of the Sumatran pole forests are not as nutrient-poor as those in Kalimantan. Other areas of Sumatran pole forest in peatswamp may well have a lower species diversity.

The number of trees, as expected, increases from mixed swamp forest to pole forest, but the average girth decreases. The total basal areas[4] (of the trees measured) were generally higher in mixed swamp forest than in pole forest, as was the average basal area (total basal area divided by number of trees). The single 0.01 ha plot investigated by Yamada and Soekardjo (1980) had a total basal area of 1.56 m^2 (3.1 m^2/0.2 ha), one-third of which was contributed by a single large *Alstonia angustiloba.*

Ecological Consequences of Low Nutrient Levels

The low nutrient levels of the soils in mature peatswamp forest (p. 171) almost certainly limit primary productivity. Although studies have yet to be conducted to investigate this, the low total basal area of pole forest (table 5.2) and the doubtless even lower total basal area of stunted peat padang, indicate that their primary productivity is low. In such a habitat, but where climatic conditions are favourable for animal life, plants would be expected

Table 5.2. Characteristics of two types of peatswamp forest in 0.2 ha quadrats at three locations in Sumatra.

	Teluk Kiambang	Muara Tolam	R. Siak Kecil	Average
No. species				
Mixed swamp forest	37, 37, 31	49, 44	29, 32	37
Pole forest	25, 49	46	32, 31	37
No. trees				
Mixed swamp forest	163, 159, 110	137, 110	89, 89	122
Pole forest	220, 123	176	169, 149	167
Average girth (cm)				
Mixed swamp forest	80, 85, 95	76, 85	86, 97	86
Pole forest	69, 67	70	68, 69	69
Total basal area (m^2)				
Mixed swamp forest	8.3, 9.3, 8.0	6.3, 6.3	5.3, 6.7	7.2
Pole forest	8.3, 4.4	6.8	6.3, 5.7	6.3
Average basal area (m^2)				
Mixed swamp forest	0.051, 0.059, 0.072	0.046, 0.058	0.060, 0.076	0.060
Pole forest	0.038, 0.047	0.039	0.038, 0.038	0.036

From Anderson 1976

to defend their leaves and other edible parts as fiercely as possible against potential herbivores. The three main reasons for this are:

- The loss of a leaf or other part to a herbivore would be proportionately more serious for a plant growing in peatswamp than for a plant growing in a more fertile habitat. This is because of the greater 'cost' of replacing the eaten or damaged leaf part in infertile habitats. Thus it would be expected that proportionately more of the energy and nutrient resources of a peatswamp forest plant were used for defence.
- In a habitat with relatively low productivity, the plant should produce better-protected leaves in order to increase the life of each leaf. Leaves have a finite life, usually related to the amount of damage they sustain, and so it is advantageous to a plant to ensure that this damage occurs as slowly as possible. It is possible that the relatively thick cuticle may reduce the leaching of nutrients from living leaves.
- In habitats with low productivity and containing vegetation whose seeding strategies involve anti-predator[5] measures such as extreme toxicity or gregarious fruiting (p. 221), there is usually a low species richness of trees and noticeable grouping of species (Janzen 1974a).

The trees over 30 cm girth at breast height found by Anderson (1961, 1963) in Sarawak peatswamp belong to families known to be particularly rich in defence compounds such as latex, essential oils, resins, tannins, and other phenolic or terpenoid compounds (table 5.3). This indicates that most leaf-eating animals from lowland forest on good soils would find it difficult to live in peatswamp forest.

As stated above (p. 172), many of the plants growing in the inner part of peatswamp forests have thick leaves. When these are crushed there is usually a resinous, acrid or aromatic odour or taste. This is most atypical of tropical vegetation but common for plants growing in poor conditions (Janzen 1974a), particularly where nitrogen and phosphorus are in short

Table 5.3. Tree families in Sarawak peatswamp forest. The number of genera are shown in parentheses.

Anacardiaceae (3)	Ebenaceae (1)	Oleaceae (1)
Anisophylleaceae (1)	Euphorbiaceae (1)	Rubiaceae (1)
Annonaceae (5)	Fagaceae (1)	Rutaceae (1)
Apocynaceae (1)	Guttiferae (6)	Sapindaceae (1)
Aquifoliaceae (1)	Leguminosae (1)	Sapotaceae (2)
Burseraceae (1)	Meliaceae (1)	Sterculiaceae (3)
Crypterioniaceae (1)	Myristicaceae (1)	Thymelaceae (1)
Dipterocarpaceae (5)	Myrtaceae (6)	Xanthophyllaceae (2)

From Anderson 1963

supply (Beadle 1966). Thick leaves are probably not only a defence against harsh environmental conditions but also a means of deterring herbivores (p. 300).

The lack of ant-free epiphytes mentioned on page 174 has been interpreted by Janzen (1974a) as a consequence of low productivity due to a low rate of inorganic ion input to tree crowns from bird faeces, insects, and falling leaves and twigs.

Ecological Consequences of High Levels of Secondary Compounds

Microorganisms

Some litter microorganisms can degrade phenolic compounds in soil and water (Alexander 1964; McConnell 1968) but they do so slowly, requiring a well-oxygenated substrate and accessory energy sources (Burges 1965). It is not surprising that microorganisms have difficulty in degrading phenolic compounds because such chemical defences are the result of selection specifically favouring compounds which are difficult for microorganisms and higher animals to digest. As a result, a leaf from a peatswamp forest will probably lie on the ground for many weeks, unattacked by decomposers, until most of the defence compounds have been leached out by rain water. It is not unexpected, then, to find that phenolic compounds have negative effects on mycorrhizae, other fungi, bacteria, roots, vertebrates, insects and worms (for references see the paper by Janzen [1974a]).

The effects of these defence compounds on microorganisms will be enhanced in the pole forest and peat padang soil by:

- the initial low nutrient quality of the soil (p. 171) making the microorganisms even more dependent on the litter itself for nutrients;
- the high acidity of the soil;
- the low nutrient quality of the litter because of selection favouring withdrawal of as many nutrients as possible from the leaves before they drop, especially on low-nutrient soils;
- a low rate of litter input because of low primary productivity.

Aquatic Animals

Data from Peninsular Malaysia suggest that blackwater rivers have an impoverished and distinct fauna. Malayan blackwater rivers have only about 10% of the fish fauna of other rivers (Johnson 1967a). Cladocera (water fleas), annelid worms, rotifers, nematodes and protozoans are rare;

algae are generally rare except for a few species which are locally abundant, and macrophytes may be absent (Johnson 1968). Of the 15 fish species found in blackwater rivers, nine were air-breathers or lived near the surface (Johnson 1967b) and almost all the insects were air-breathers (Johnson 1967a). This may be due solely to the difficulties of existing at low oxygen levels but it could also be that the protein-binding defence compounds, such as tannins, deleteriously affect animals' gills. Thus the aquatic larvae of biting insects, such as mosquitoes, are rare in peatswamp forest. The productivity of animals and plants in these rivers is low and the biomass of fish may be only 0.5 g/m^2. Bishop estimated the fish biomass of a 'normal' small Malayan river to be about 18 g/m^2 (Bishop 1973).

What factor is responsible for this distinct aquatic ecosystem? Low nutrient levels have been suggested, but Johnson (1967b) has shown that calcium levels (for example) in blackwater rivers are generally higher than in other types of river. The low pH cannot be the only factor because those blackwater rivers in Malaya which later flow over limestone have a pH over 6.0; these streams have snails, more insects and fish, but the fish are restricted to blackwater rivers, rarely occurring in other rivers with a 'normal' pH (Johnson 1968). The blackness of the water and hence low light penetration might be thought to be an important factor but this would not explain why there are so few animals and plants at the water surface, nor why such animals and plants exist in very muddy water where light penetration is even less. So it would seem that the main factor influencing this ecosystem is the high level of phenolic compounds.

Terrestrial Animals

Peatswamp forests do not support an abundance of terrestrial wildlife. Merton (1962) said of an inland peatswamp in Peninsular Malaysia "it was the absence of animals rather than their presence which was so striking". Janzen (1974a) described an animal collecting trip in peatswamp forest in Sarawak as "generally a waste of time". An investigation of primate densities in peatswamps of Peninsular Malaysia found that away from rivers, in the species-poor forests, few primate species existed and those species present lived at densities of less than three groups per km^2 (Marsh 1981). Typical figures for lowland forest would be 10 groups per km^2 (Marsh and Wilson 1981). Similarly, in a study of one group of Mentawai gibbons on Siberut Island, it was found that these animals consistently under-used the area of inland peatswamp in their home range in comparison with the other forest types available to them (Whitten 1982c,d). This behaviour is understandable because primates depend on tree and vine fruit for most of their diet and the generally low productivity of peatswamp forests would result in fruit being available only rarely.

Chapter Six

Freshwater-Swamp Forests

Introduction

Freshwater-swamp forest and related vegetation grow on soils where there is occasional inundation of mineral-rich fresh water of pH 6 and above and where the water level fluctuates so that periodic drying of the soil surface occurs. The floods may originate from rain or from river water backing up in response to high tides. The major physical differences between this and peatswamp forest are the lack of deep peat (a few centimetres thickness is sometimes found but this seems to have little effect on tree species composition), and the sources of water being rivers and rain water (Whitmore 1984). Although freshwater-swamp forest normally forms on riverine alluvium, it can also be expected, under natural conditions, to form on alluvium deposited by larger lakes where wave action prevents the formation of peat, such as in 'Lake' Bento, Kerinci.

The high agricultural potential of their soils has meant that the freshwater-swamp forests of Sumatra have suffered considerably from human disturbance. In 1982 only 22% remained of the original area of freshwater-swamp forest in Sumatra and very little of this was in an undisturbed state. The area now is even less.

Soils

Freshwater-swamp forest grows on recent freshwater alluvial soils. Such soils are usually more fertile than the soils on the surrounding slopes and, after draining, have obvious potential for agriculture, but they are less fertile than some of the soils on recent marine alluvium or volcanic ash (Burnham 1975). Recent alluvium from rivers and lakes has had relatively little time for soil formation, and the absence of clear soil horizons is one of its characteristics. It is further characterized by its low position, close to the water table. The actual content of the soil varies greatly with the nature of the parent material but it is usually composed of relatively fine particles, the last to be deposited by a flooding river. Recent, waterlogged alluvial soils are usually grey (or mottled-grey and orange where the soil occasionally dries) as a result of 'gleying'. The main processes of gleying are the reduction of iron compounds to their ferrous forms and their partial re-oxidation and precipitation (Young 1976). The grey colour of these soils seems to arise from the lack of mixing of the often peaty soil surface with the lower layers. This is because soil animals that would normally take organic matter into the soil (such as termites and earthworms) are not tolerant of the frequently anaerobic conditions caused by waterlogging such as occur in swamp soils. Gley soils are amongst the most poorly understood yet agriculturally most important tropical soils (Young 1976).

VEGETATION

Composition and Structure

The vegetation of freshwater swamps varies considerably in response to the wide variation in its soils. In some areas grassy marshes may be the natural vegetation, in others some form of palm- or pandan-dominated forest, and in yet others forest which is similar in structure and species composition to lowland rainforest. Under certain conditions, long, winding tree buttresses, stilt roots and pneumatophores are common. Few plant species are restricted to the freshwater-swamp ecosystem but the tree species found there tend to be gregarious and to form species-poor associations.

Very little information appears to be available on the floristics of freshwater-swamp forests in Sumatra (Kartawinata 1974) but much can be learned from the highly readable account by Corner (1978) of freshwater-swamp forests in South Johore and Singapore. In an enumeration of various forest types on Siberut Island, it was found that freshwater-swamp forest was more similar in species composition to neighbouring flat and non-swampy areas of lowland forest than it was to nearby peatswamp forest (Whitten 1982c). The total basal area, median basal area and height were not exceptional in comparison with other forest types but the trees tended to be smaller. The area examined was dominated by the extremely gregarious *Alangium ridleyi*, a medium-sized tree with sweet fruit, which accounted for 20% of the trees. Lake Bento, which lies south of Mt. Tujuh and northwest of Lake Kerinci, is surrounded by the highest (in elevation) forest-clad marsh in West Malesia (FAO/de Wulf et al. 1981).

Unfortunately, the primary freshwater-swamp forest in Way Kambas Reserve, set up to protect the swamp ecosystems, has been extensively logged (FAO/Wind et al. 1979). Fragments remain elsewhere but it is likely that it will never now be possible to know exactly how such forests functioned in Sumatra.

Adaptations to Floods

Trees of freshwater-swamp forest often have to endure prolonged periods of flooding. Inundation with water will flood air spaces in the soil which, in a normal, moist topsoil, can account for about 60% of the soil volume (Bridges 1970). With these spaces filled, the soil soon becomes anaerobic and plants have difficulty in obtaining sufficient oxygen through their roots.

Very little is known about trees' tolerance of, and responses to, flooding. Where logging roads have dammed small rivers in forests, it is clear from the many surrounding tree skeletons that flood tolerance is not inherent to all trees. Information that does exist concerning the effects of flooding in tropical forests comes from South America. The flood tolerance of the swamp tupelo *Nyssa sylvatica*, for instance, depended on its capacity to accelerate anaerobic respiration and to oxidise the 'rhizosphere' (the immediate surroundings of the roots) (Hook et al. 1971; Keeley 1978, 1979; Keeley and Franz 1979). Another tree, *Sebastiana klotzchyana*, commonly found in South American freshwater-swamp forests, was found to continue growing during flood conditions, whereas growth of shoots from young trees of other species found in dry areas was inhibited during artificial flood conditions. After one month of flooded conditions the roots of *S. klotzchyana* showed a significant decrease in uptake of oxygen (indicating they had partially 'switched-over' to anaerobic respiration), whereas roots of trees from dry areas showed a significant increase (as though they were making up for a period of reduced respiration). Rates of anaerobic respiration and the levels of metabolic products such as ethanol, malate, lactate and succlanate, varied between trees from flood areas and dry areas, and showed that a range of responses are to be found even among flood-tolerant trees (Joly and Crawford 1982).

If the roots of a plant are under water, oxygen must reach them from the exposed parts of the plant. Some trees in freshwater swamps have strongly lenticellate bark; the bark has large lenticels or holes through which diffusion of gases can occur. The submerged roots and pneumatophores of aquatic or semi-aquatic plants are usually very aerenchymatous (they have large channels for air movement). To test whether a plant has aerenchymatous tissue, blow down one end of a cut section of root and watch for bubbles at the other end while it is held under water (Corner 1978). Bubbles will be seen if the root is aerenchymatous. It should be emphasised that since the above characters are not common to all swamp species, there is clearly more than one way to adapt to anaerobic soil conditions.

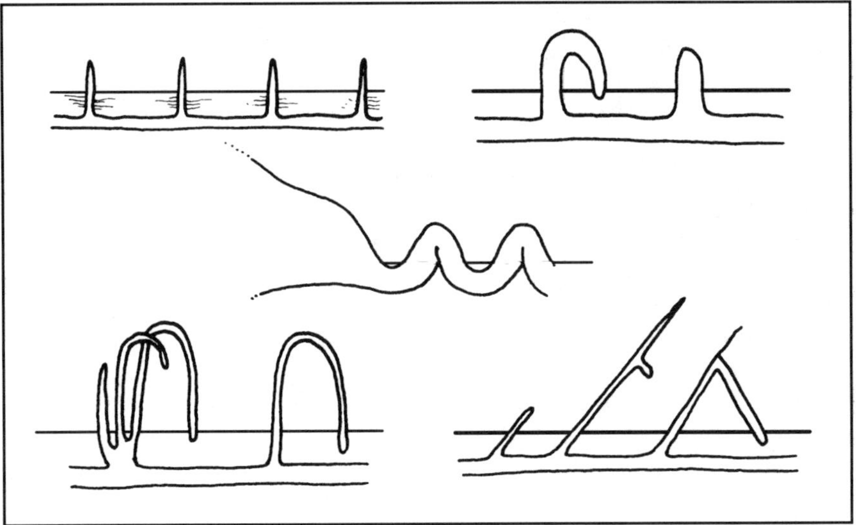

Figure 6.1. Five types of pneumatophores found in freshwater-swamp forest. *After Corner 1978*

In a freshwater-swamp forest in Johore, Peninsular Malaysia, Corner (1978) found 36 species (in 19 families) with pneumatophores and these pneumatophores were of five forms (fig. 6.1):

- erect conical pegs (similar to *Sonneratia,* p. 77);
- erect plank buttresses degenerating into squat knee-roots where there was shallow tidal flooding (similar to *Bruguiera,* p. 77);
- slender loop-roots formed by erect pneumatophores bending down into the soil;
- thick loop roots formed in the same way as the previous form;
- roots formed by a descending root growing laterally from a slanting ascending root. The top section may later die.

Fauna

The occasional high floods caused by heavy rain can create rather peculiar conditions. Corner (1978) described graphically the state of the animal life in freshwater-swamp forest during a 6 m flood of the Sedili Besar River in Johore, Peninsular Malaysia.

> During the three days at Danau we paddled in a canoe through the flooded forest. The force of the flood was lost among the trees...Whenever we touched leaf, twig, trunk of floating log, showers

> of insects tumbled into the canoe. Everything that could, had climbed above the water. Ants ran over everything that could, had climbed above the water, even scorpions, centipedes and frogs... Lizards clung to the trunks; earthworms wriggled in the water... I realised the importance of the hillocks in and around the swamp forest to animal life, for anything that could escape the flood must have fled there. We met no corpses. Pig, deer, tapir, rat, porcupine, leopard, tiger, monitor lizard and snakes must have congregated on those hillocks in disquieting proximity.

The fauna of freshwater-swamp forests is both much more diverse and more dense than the fauna of peatswamp forests, and is similar to that in lowland dry-land forests. Nothing is known in detail about the species inhabiting Sumatran (or even other Southeast Asian) freshwater swamps but Wind et al. (1979) report siamang, dark-handed gibbons, monkeys, rusa, barking deer, pigs, elephant, tapir and tiger from Way Kambas Reserve, Lampung. The areas of swamp grassland also appear to be utilised by most of these species.

The swamp grassland of the southern part of Way Kambas is used by herons, egrets (see fig. 2.30), terns, bitterns, pond-herons, whistling ducks, pygmy geese, lesser adjudants, milky storks (fig. 6.2) and occasionally the rare white-winged wood duck (Wind et al. 1979). This species (fig. 6.3) is considered 'endangered' by the International Union for the Conservation of Nature and Natural Resources (IUCN) and has been reported in Sumatra only from the eastern side of Lampung in and around freshwater-swamp and lowland forest (Holmes 1976, 1977a). It may possibly live in peatswamp areas but no information is available. Its range used to include the swampy areas adjacent to lowland forest from Java to Burma but it seems to be extinct from all that area except for parts of Assam (northeast India), possibly northern Burma, and southern Sumatra. In the northern parts of its range it appears to require dense forest, but Holmes (1977a) found the ducks in the much more open and disturbed forest of Lampung. This ability to adapt to disturbed areas encourages some optimism regarding the security of its future.

During the wetter seasons these ducks can live almost entirely within forest, but when the slow-running streams they frequent begin to dry up they venture into areas of swamp or even rice fields. They fly to these areas just before dusk and, if not disturbed, may remain there until two or three hours after dawn. White-winged wood ducks are omnivorous, eating floating aquatic plants, seeds, insects, worms, molluscs, frogs and even small snakes and fish (MacKenzie and Kear 1976). They are usually seen in pairs but when the young are old enough to fly, family groups may be seen. One of the best ways to locate these birds is to listen for the evening light call which is described by MacKenzie and Kear (1976) as a "prolonged, vibrant, wailing honk sometimes breaking to a nasal whistle at the end".

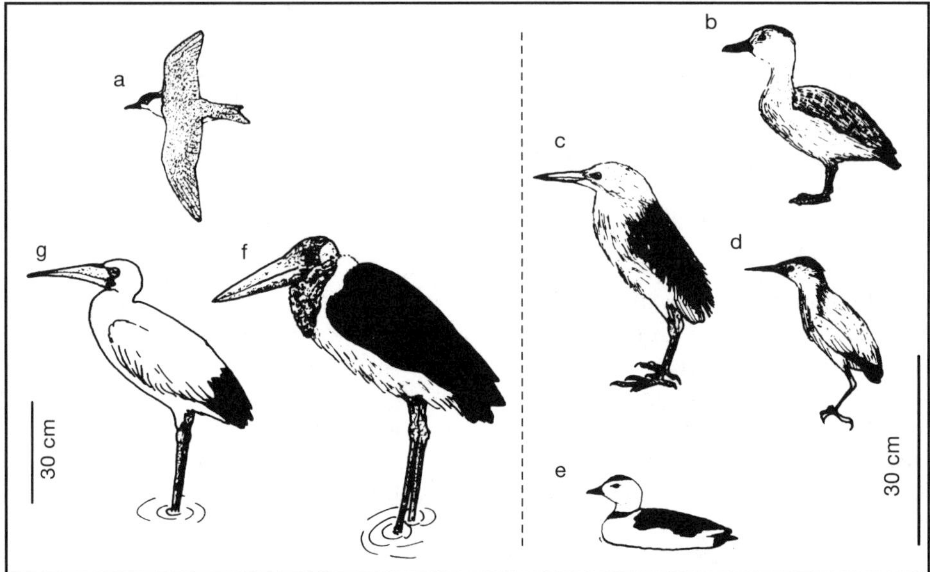

Figure 6.2. Birds commonly found in grassy swamp a - tern *Sterna hirundo*, b - Java tree duck *Dendrocygna javanica*, c - Java heron *Ardeola speciosa*, d - Yellow bittern *Ixobrychus sinensis*, e - pygmy goose *Nettapus coromandelianus*, f - long-billed stork *Leptoptilus javanicus*, g - milky stork *Ibis cinereus*.

After King et al. 1975

Figure 6.3. White-winged wood duck *Cairina scutulata*.

After MacKenzie and Kear 1976

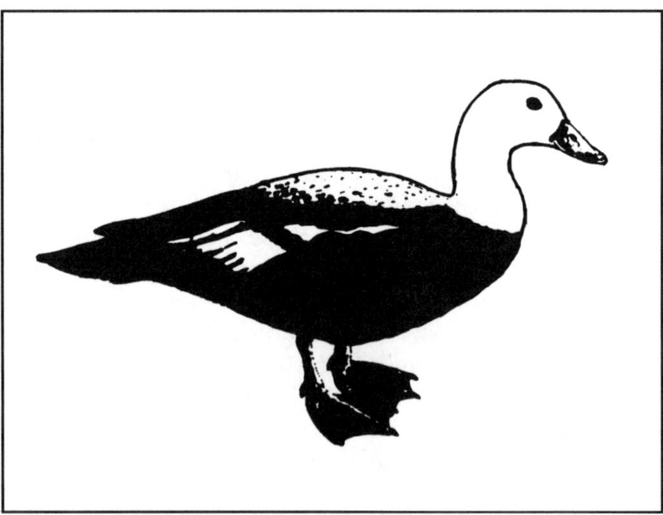

This call is never, it seems, given by birds flying alone. White-winged wood ducks nest in the holes of swamp trees, particularly 'rengas' (probably *Melanorrhoea*), and their breeding season lasts from December to March or April (Holmes 1977a).

The freshwater-swamp areas used to have large populations of the estuarine crocodile *Crocodylus porosus* and of the smaller false ghavial *Tomistomus schlegeli.* The false ghavial, which never exceeds 5 m in length, resembles the true ghavials in having a snout which is much longer and more slender than the snouts of most crocodiles. The crocodiles are the last remaining members of the Archosaurs, the group of reptiles including dinosaurs that dominated the earth during the Mesozoic. They are the largest inhabitants of freshwater swamp and their movements inhibit the encroachment of aquatic plants into waterways. In areas with a prolonged dry season they maintain residual water holes which serve as restocking reservoirs for smaller aquatic organisms which would otherwise perish. Crocodiles enrich the nutrient content of the water by converting prey into waterborne faecal particles which are food for a host of invertebrates and fish. Crocodiles have been mercilessly hunted, partly through fear (unjustified in the case of the false ghavial which rarely eats anything but fish) and unbridled greed for their skins. Luckily, the remaining small populations of estuarine crocodiles are now protected throughout Indonesia.

Like all animals that can live in both seawater and fresh water, the estuarine crocodile has physiological adaptations which enable it to control the concentration of its plasma. In a marine environment, where fresh water is scarce, the crocodile passes very concentrated urine, reabsorbing water in the kidneys, and very dry faeces by reabsorbing water in its intestines. These two measures enable it to conserve water. Calcium, magnesium, potassium, ammonium, uric acid, and bicarbonates are excreted in the urine. Sodium chloride, however, does not pass through into the urine (Grigg 1981), but is excreted through the crocodile's external nasal glands near the nostrils and lachrymal glands in the corners of the eyes.

When a female crocodile is ready to lay eggs, she seeks a shady location on land where she builds up a dome-shaped nest of leaves or tall grass (Greer 1971). Up to 50 or more eggs are laid in the middle of this nest, where they remain damp and are protected from direct sunlight. The heat generated by the decomposition of vegetable matter may help the incubation but if the mother senses that the eggs are becoming too hot she will spray urine over the nest to cool it down.

Just before the eggs hatch, the young crocodiles make high-pitched croaks which are audible even outside the nest. The mother responds by scratching away the now-hardened surface of nest material and as the young crocodiles wrestle their way out of their shells, she (and sometimes the father too) picks them up gently in her mouth and carries them away to a secluded 'nursery' area in a swampy bank. They stay there for a month

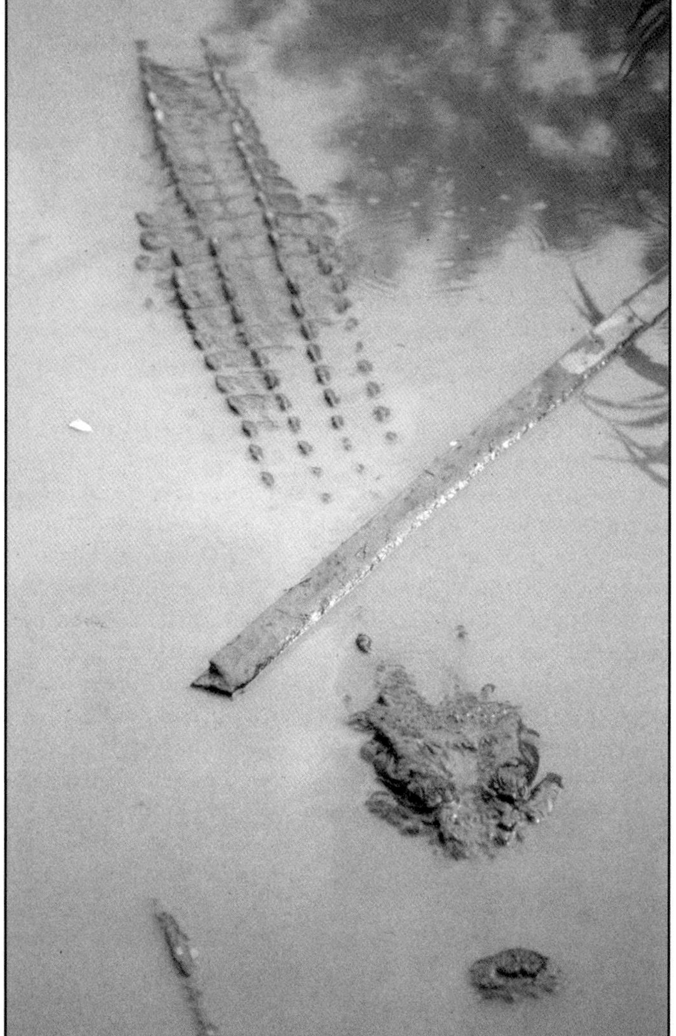

Figure 6.4. Large estuarine crocodiles are now extremely rare in Sumatra, but reports of smaller individuals in fresh- and peat-swamp areas persist.
A.J. Whitten

or two, catching large insects and small vertebrates such as fish and frogs, while their parents guard them closely.

Perhaps the most fascinating aspect of crocodile nesting behaviour is that the sex of the young animals is determined by the temperature of incubation. This has been known for various groups of reptiles (Vogt and Bull 1982) and has recently been determined for the Mississippi alligator of northern America (Ferguson and Joanen 1982), although experiments have yet to be conducted on Asian crocodiles. Field and laboratory experiments have shown that alligator sex is fully determined and irreversible by the time of hatching and that incubation temperatures of up to and including 30°C produced all females, whereas 34°C and above produced all males. Since nests on exposed levees or other dry areas receive more sunlight and are hotter than those nearer wet, shaded swamp, the sex ratios of hatchlings will be different between these habitats (Ferguson and Joanen 1982). It is likely that the sexes of Sumatran crocodiles are also determined by incubation temperature, and this is important for management of the species. Felling of swamp trees, particularly those near riverbanks, will reduce the amount of shading and may increase the average ambient temperature of any area. This in turn may increase the number of male crocodiles. However, the efficiency of nest temperature regulation by the female is unknown as are the social constraints to any alterations in sex-ratio.

Chapter Seven

Lowland Forests

INTRODUCTION

The lowland forests of Sumatra are among the most diverse, awe-inspiring, complex and exciting ecosystems on Earth. They are probably the best-studied forested natural ecosystem on the island, but for information on how these forests actually function it is frequently necessary to turn elsewhere in the Sunda Region. In many respects, the Sunda Region lags behind tropical America in unravelling and understanding ecological interactions. For this reason, occasional examples are taken from lowland forests in countries such as Costa Rica, when illustrations of particular aspects of ecology are required.

Diversity of Plants

The diversity of tree species in Sumatran lowland forest is extremely high. In a valley area around the River Ranun in North Sumatra, the Simpson Index of Diversity[1] for trees of 15 cm diameter and over at breast height was 0.96, and in the neighbouring hills it was 0.93 (from data in MacKinnon [1974]). In a hill forest on Bangka Island investigated by a CRES team, the index of diversity was 0.94. Both studies used local vernacular names which always underestimates actual species number, as explained on page 198. Corresponding figures for a forest in Europe would be much lower, at about 0.4-0.6.

The problem of explaining the very high diversity of plants in tropical forests, particularly lowland forests, has taxed many minds. Before detailed investigations of fossil pollen began about 15 years ago, it was thought that tropical forests had experienced climatic stability for millions of years and this had allowed time for the evolution of many species. This has now been shown to be false as tropical vegetation has experienced considerable changes over time (p. 17). Other writers have suggested that the formation of isolated forest blocks or refuges during the peaks of glacial activity, when climates were cooler and drier, would have resulted in the species of each block following a course of evolution different from species in other blocks. When the climate ameliorated and the forest blocks were reunited,

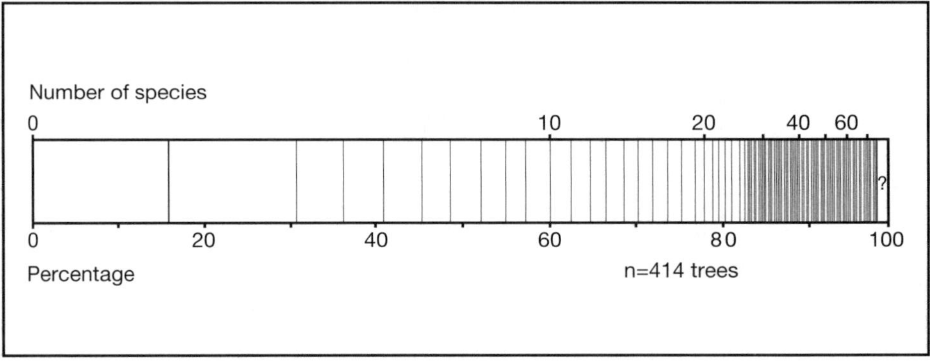

Figure 7.1. Relative abundance of tree species near the River Ranun, Dairi. Note that 73% of the species contribute only about 20% of the trees.
After MacKinnon 1977

some of the closely related species would be incompatible for reproduction and thus a greater degree of diversity would exist than before the forest blocks were separated. This is generally known as the refuge theory (Haffer 1982) and has been useful in explaining diversity in certain taxa in Amazonia and Africa. It is less satisfying as a theory in Southeast Asia, however, where it is suggested that the whole of Peninsular Malaysia, Sumatra, Borneo and West Java formed one 'refuge' and the majority of New Guinea the other. The refuge theory does not explain all species distributions (Endler 1982) and it is clear that more detailed surveys are required (Meijer 1982a).

An ecosystem with high diversity has, by corollary, a large number of rare species (fig. 7.1) so, as Flenley (1980) suggests, instead of asking "why is lowland forest so diverse", we could equally well ask "why are there so many rare species?". Reproductive strategies adopted by living organisms can be described as falling within a spectrum between two extremes: r-selected strategies in which as many offspring are produced as possible, and K-selected strategies in which few offspring are produced but each has a great deal of care, attention and material resources devoted to it in order to ensure its survival and success. Thus elephants, cows and bats are K-selected, and rats, pigs, and many fish and insects are r-selected. If an organism adopts the K-strategy in a lowland forest, it will almost certainly be relatively rare. It is then necessary to ask "under what circumstances can rarity be an advantage?". It is generally true that the herbivores on a plant species in lowland forest are restricted to a few species which have evolved as the plant species evolved and found ways to circumvent whatever forms of physical or chemical defence the plant has adopted (see p. 231). The chances of a pest finding an individual of a particular tree species are greater if that tree is common, and so the pressure exerted by herbivores on an abundant or otherwise dominant tree species puts it at a disadvantage and permits subordinate species to coexist (Janzen 1970).

A hypothesis has been suggested by Ricklefs (1977) that greater heterogeneity in soil properties and surface microclimate in the tropics might be the basis of the trend of increasing tree species diversity from temperate to tropical regions. He proposes that ranges of angle and intensity of incident solar radiation, precipitation, temperature, and the partitioning of nutrients between soil and vegetation are wider in a tropical forest clearing than in a similar clearing in temperate regions. The differences are because:

a) the large biomass of a tropical forest modifies factors such as light levels, humidity, temperature, and environmental constancy to a greater extent than does a temperate forest of smaller biomass;
b) more rapid decomposition of leaf litter and other organic detritus in tropical forests accelerates the release of mineral nutrients and organic detritus from trees and increases the flush of nutrients into the soil (see also Anderson and Swift [1983]);
c) even if nutrient levels in soils under temperate and tropical forests were similar, the humus content of tropical forest soils is lower. Humus content influences the retention of soil moisture and the stability of other soil properties when exposed to more intense physical forces within forest gaps;
d) the higher rainfall of tropical areas increases the leaching of certain ions from exposed soils. This differential leaching would further increase variation across a forest gap;
e) for much of the day the sun is more or less overhead and so the light hits the soil in a tropical forest gap more directly and for a longer time each day than in a temperate forest.

These differences create a greater heterogeneity of environmental conditions for seedling establishment in tropical forests and it is this heterogeneity that provides the basis for resource partitioning and the coexistence of competing species.

The natural rarity of some plant species is compounded because some also are dioecious (bearing flowers of only one sex on one tree) and because many plants are endemic to quite restricted areas. One consequence of this is that reserves hoping to conserve these species have to be extremely large (Ng 1983).

Diversity of Animals

The large number of animal species in lowland forest is generally associated with structural and taxonomic heterogeneity of the plants. The number of animal species is high in Sumatra and other parts of the Sunda Region, but not as high as in other tropical regions. For example, the total number of birds in about 15 ha of primary lowland forest in seven sites is shown in table 7.1 and it is clear that those for sites in the Sunda Region are considerably lower than for South America (Pearson 1982). If exactly equivalent habitats are compared, however, counts of Sunda forest birds are higher than counts in Africa and not far behind South America (D. Wells, pers. comm.).

Significance was achieved, however, when the geological history of the localities was considered in terms of separations from, and connections to, other landmasses. Since islands theoretically have higher extinction rates and lower immigration rates than similar continental areas (p. 46), sites in Borneo, Sumatra, and New Guinea would be expected to have fewer species than sites in tropical continental South America.

Inger (1980) compared the densities of frogs and lizards in lowland forests in Borneo (the data are probably applicable to Sumatra) and Central America. In Borneo about 1.5 individuals were found per 100 m^2 but in Central America the figures ranged from 15 to 45. He proposed (as Janzen [1974, 1977a] had done earlier) that the low abundance was due to many major tree species in the Sunda Region exhibiting the phenomenon of gregarious fruiting at long intervals (p. 221). This in turn would reduce the numbers of seed-eating insects available as frog and lizard food. The same argument would apply to other insect-eating animals.

Table 7.1. The number of bird species found in 15 ha of primary lowland forest at seven locations.

	No. of species/15 ha primary lowland forest
Ecuador	232
Peru	204
Bolivia	181
Gabon	158
Sumatra	151
Borneo	142
New Guinea	114

Based on Pearson 1982; figure for Sumatra from Rijksen 1978 and R. Eve and A.M. Gigue, pers. comm., for rather more than 15 ha at the Ketambe Research Station.

VEGETATION

Characteristics

Tropical lowland forests are characterised by the huge amount of plant material they contain, and this can be measured in terms of the amount of carbon present. The tropical forests of the world (i.e., not just lowland forests but also montane, swamp and dry forests) cover 1,838 million ha or 11.5% of the earth's land surface, yet they contain 46% of the living terrestrial carbon (Brown and Lugo 1982). The type of lowland forest found in Sumatra would certainly contain an even more disproportionate percentage of the carbon. Of the total of organic matter in a lowland forest, only about 1% is accounted for by the litter, about 40% by the soil, but about 60% is in the vegetation (Brown and Lugo 1982). In temperate regions the corresponding biomass figures are about 10%-20%, 35% and 50%.

Lowland forests are characterised by the conspicuous presence of thick climbers, large buttressed trees, and the prevalence of trees with tall, smooth-barked trunks. Although the canopy may occasionally be dominated by small-leaved leguminous trees (such as the magnificent *Koompasia excelsa* and *K. malaccensis*), the vast majority of trees have simple mesophyll-sized leaves (p. 283) between 8 and 24 cm long. The leaves of understorey trees, including the immature individuals of large trees, often have larger leaves (Parkhurst and Loucks 1972).

Leaves with smooth edges and 'drip-tips' (fig. 7.2) are particularly common in lowland forest and there is considerable convergence between plant families in leaf shape which, together with the high diversity of species, frequently makes identification somewhat difficult. Even so, characters such as sap, type of bark, size of buttresses, leaf vein arrangement and arrangement of leaves on twigs give ample information to allow identification of the majority of specimens, even without flowers. For a convenient key see Wyatt-Smith and Kochummen (1979) but see page 198 for cautions. Since plant taxonomy is based on flower anatomy, final identification relies on the examination of flowers.

The drip-tips are most prevalent in understorey saplings. Since one theory of the function of drip-tips is that they allow water to run easily off the leaf surface and thus prevent or hinder the growth of epiphylls (p. 210), it is reasonable that these should be most common where the relative humidity is highest (i.e., rates of evaporation lowest).

Forest Growth Cycle

Different types of tropical rain forests, and indeed forests throughout the world, have many fundamental similarities because the processes of forest

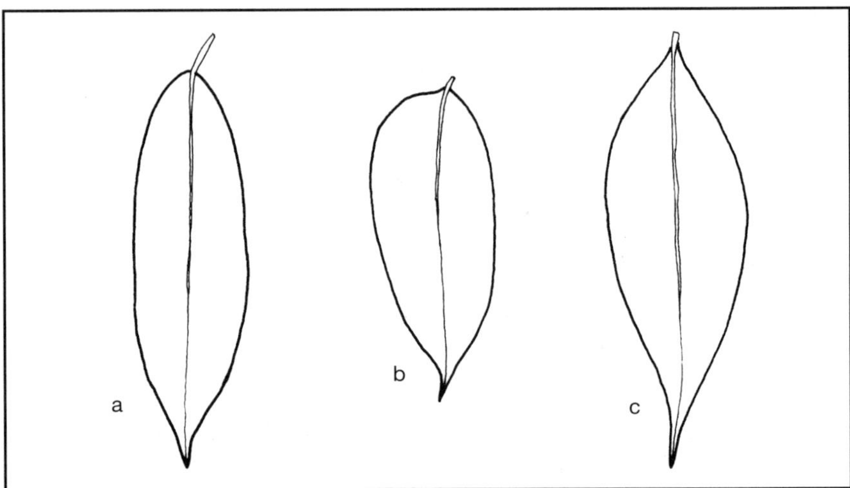

Figure 7.2. Leaves of a) *Dialium platysepalum*, b) *Dracontomelon mangiferum*, and c) *Sterculia foetida* to show drip-tips characteristic of many lowland forest trees.

succession and the range of ecological strategies of tree species are more or less the same (Whitmore 1982a, 1984). These strategies form what is known as the forest growth cycle; that is, the events of a large mature tree falling over in a closed forest, forming a gap, the gap filling with a succession of plant species until another large mature tree falls over. Vigorous, light-tolerant (shade-intolerant) species grow up in the gap and these create favourable conditions for seeds of shade-tolerant (but not necessarily light-intolerant) tree species to germinate and for the seedlings to grow and eventually to supersede the initial gap fillers. The process leading from a gap to mature forest is known as secondary succession. Whitmore divides the cycle into three phases – the gap, building and mature phases – which together form a mosaic throughout the forest which is continually changing in state and shape (Whitmore 1982a, 1984) (figs. 7.3. and 7.4).

For example, a patch of forest in building phase may return to gap phase should a tree in a neighbouring mature patch fall across it. The size of a gap depends on its cause – from the falling of a single dead tree, to a landslide, a fire, or even a widespread drought. The floristic composition of a forest will depend on the size of former gaps because a large gap will be filled initially with tree species requiring light, whereas small gaps will be filled by shade-tolerant tree species. Many of the latter may have grown slowly for many years to become part of the understorey but have been unable to form part of the main canopy until a small gap was formed. It

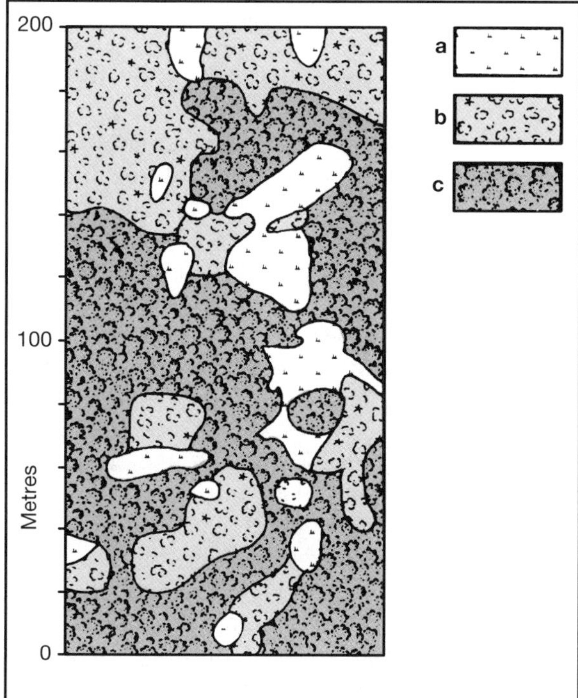

Figure 7.3. Mosaic of a) gap, b) building and c) mature phases at S. Menyala, Peninsular Malaysia. The extensive area of building phase was caused by partial clearance in 1917.

After Whitmore 1984

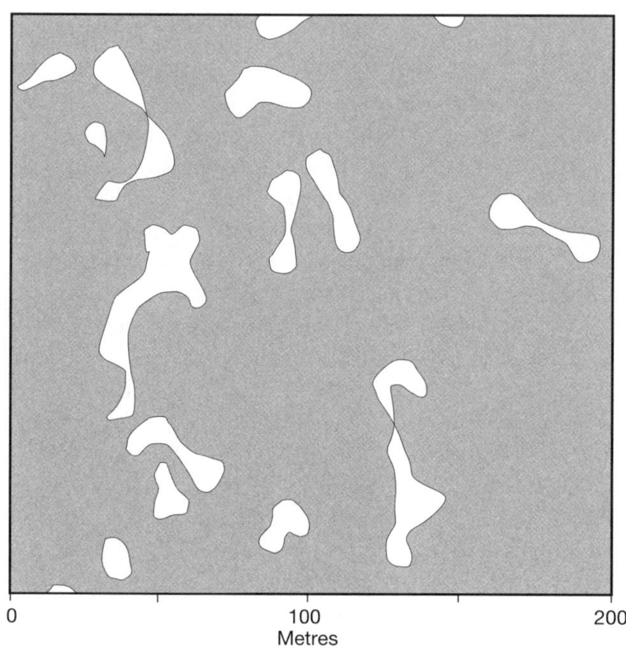

Figure 7.4. Gaps near the River Pesisip, Batang Hari, formed within 10 years preceding the study of Huc and Rosalina (1981b).

should be noted that since a single tree can contribute to gap, building, and mature phases, it is incorrect to refer to 'gap-phase' or any other phase species (Whitmore 1982a). Terms which describe the behaviour of a species, such as pioneer, shade-tolerant or light-intolerant, are more useful. Since forest comprises a mosaic of gap, building and mature areas, it is a matter of debate whether the term 'climax' vegetation has any meaning in the field.

Gaps and their plants have been studied in Sumatra at various locations in West Sumatra, Jambi, South Sumatra and Lampung by Huc (1981) and Huc and Rosalina (1981b). They concluded that the size of gaps influences the duration of the forest growth cycle. Small gaps created by dead standing trees produced short cycles without a gap phase. That is, immature trees beneath the dead tree were able to grow up as soon as the light intensity increased. Large gaps created by the fall of several adjacent large trees allowed the full growth cycle to occur (fig. 7.5). Huc and Rosalina's findings confirmed those of Kramer (1933) on the slopes of Mt Gede. Kramer made artificial clearings of 1,000, 2,000, and 3,000 m^2. The first was soon filled by regenerating mature-forest trees but the larger clearings were swamped by vigorous pioneer species. This has also been observed in Mount Leuser National Park by Rijksen (1978). Huc and Rosalina (1981) calculated that the average duration of the growth cycle in the forests they examined was about 117 years, and at Ketambe Research Station an approximate duration of 108 years has been estimated by van Noordwijk and van Schaik (pers. comm.). These results are comparable to estimates ranging from 80 to 138 years in Costa Rica (Hartshorn 1978) but rather less than the 250 to 375 years estimated by Poore (1968) for part of Peninsular Malaysia.

Within each of the different stages of the forest growth cycle the trees which are found have a number of ecological characteristics in common. For example, the first species that grow up in a gap (often called pioneer species) generally produce large numbers of small seeds frequently, if not continuously. These seeds tend to be easily dispersed (often by wind, squirrels or birds) and, probably because of this, these species tend to have wide geographical distributions (Huc 1981). They have high viability even in full light, and the seedlings grow rapidly. In addition, these species are often dioecious (each plant bearing flowers of only one sex), which leads to considerable genetic variability. An index of the pioneering ability of different species has been devised (Sakai et al. 1979).

The trees of later stages in the succession are shade-bearers (that is, they tolerate growing in shade, but if more light is available they will grow faster) and can regenerate beneath their living parents. These species often have large seeds with rather specialised means of dispersal and a relatively restricted geographical distribution. Between these two extremes are large numbers of species of 'late-secondary growth' which are unable

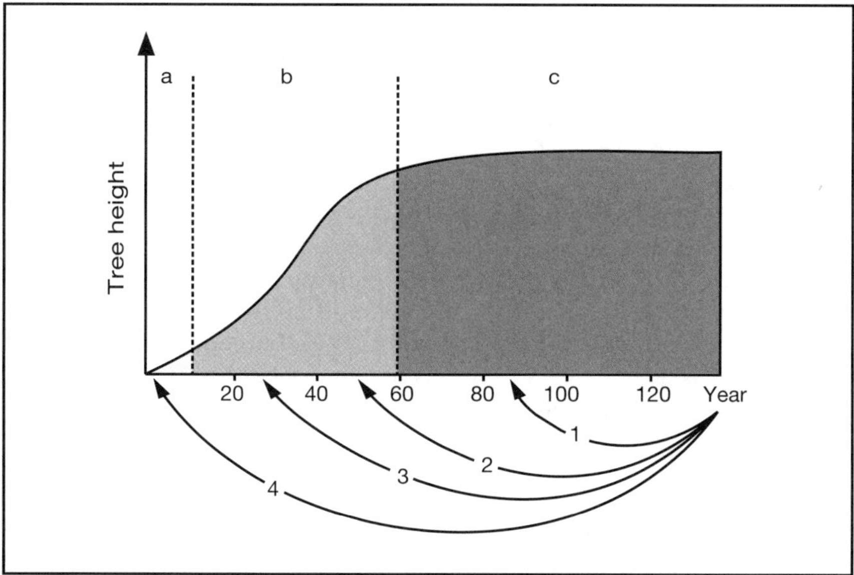

Figure 7.5. Silvigenetic cycles. 1, 2 and 3 are short cycles due to the replacement of dead trees by trees below, with no gap formation. 4 is a long cycle due to the formation of a gap in which pioneer vegetation grows. a, b and c represent pioneer-, building-, and mature-phase forest.

After Huc and Rosalina 1981

either to colonise open sites or to regenerate beneath their living parents (Whitmore 1982a, 1984).

Changes in species composition during succession may be due to availability of nutrients, availability of seeds, suppression of some species by chemicals produced by the same or other species (allelopathy), crown competition or root competition (Huc 1981), but there is still very little information available on how regeneration actually occurs. There are three ways in which trees can grow: from shoots growing out of roots, stumps or fallen trunks, from established seedlings, or from seeds. Beneath a mature forest there are clearly no seedlings of light-demanding species and, conversely, in gaps light-intolerant species will die. Seed may lie dormant in or on the soil surface (seed bank), or it can be brought into an area in the 'seed rain' (seeds dispersed from their parents). The existence of seed banks in rain forest in northern Thailand has been demonstrated by Cheke et al. (1979). They found dormant, viable seeds of pioneer species such as *Trema, Mallotus, Macaranga, Melastoma* and grasses in locations where those species were not represented. Similar findings are reported from Peninsular Malaysia (Liew 1973; Ng 1980; Symington 1933). Some of the seeds that were found 20 cm below the soil surface were still able to germinate although they must have arrived many years before. In addition, a long period of dormancy for these seeds is suggested by the low rate of seed input and the large numbers of seeds found.

Whereas trees of the gap and early building phases may have seeds which can lie dormant for many years, this is not so for many of the trees of the mature phase. These tend to fruit infrequently (p. 219), to have poor dispersal and little or no potential for extended dormancy. Their continued existence in an area therefore depends on fruit from mature trees falling into a spot with the correct conditions for immediate germination. Dipterocarpaceae trees fall into this category. Secondary succession is discussed further on page 340.

Floristic Composition and Variation

As was stated in the introduction to Part B, the lowland forests of West Malesia (p. 44) are amongst the most diverse and species-rich ecosystems in the world. Vernacular names allow differentiation of some of the trees and represent a common and useful means of conducting a general forest survey (MacKinnon 1974; Rijksen 1978; Whitten 1980b). It should, however, be realised that vernacular or commercial names generally greatly underestimate the number of species present. For example, 'meranti' describes the genus *Shorea* but the 'specific' names such as 'meranti merah', 'meranti putih' and 'meranti damar hitam' each represent a group of species. For example, in Sumatra there are over 26 true species of 'meranti merah' (W. Meijer, pers. comm.).

Most types of forest formation in West Malesia are dominated by a single family of tree called Dipterocarpaceae (fig. 7.6). These are the only forests in the world in which a single family has such a high density of genera, species and individuals (Whitmore 1982a, 1984). This domination is usually, but not always, particularly marked in lowland forests. Sumatra has 112 species of dipterocarps including 11 endemic species (Ashton 1982).

The emergent trees in Sumatran lowland forest, some of which can reach 70 m tall, are generally of the families Dipterocarpaceae (e.g., *Dipterocarpus, Parashorea, Shorea, Dryobalanops*) and Caesalpiniaceae (formerly part of Leguminosae) (e.g., *Koompasia, Sindora* and *Dialium*). An extensive and interesting enumeration of large trees in part of Peninsular Malaysia was conducted by Whitmore (1973). For identification of large and medium-sized trees see Meijer (1974), Symington (1943) and Wyatt-Smith and Kochummen (1979). It must be remembered that the last two of these books were written for Peninsular Malaysia, and although most Sumatran tree species are included, an unknown tree may be one of the ones that is not. It is generally quite satisfying to identify a tree to its genus.

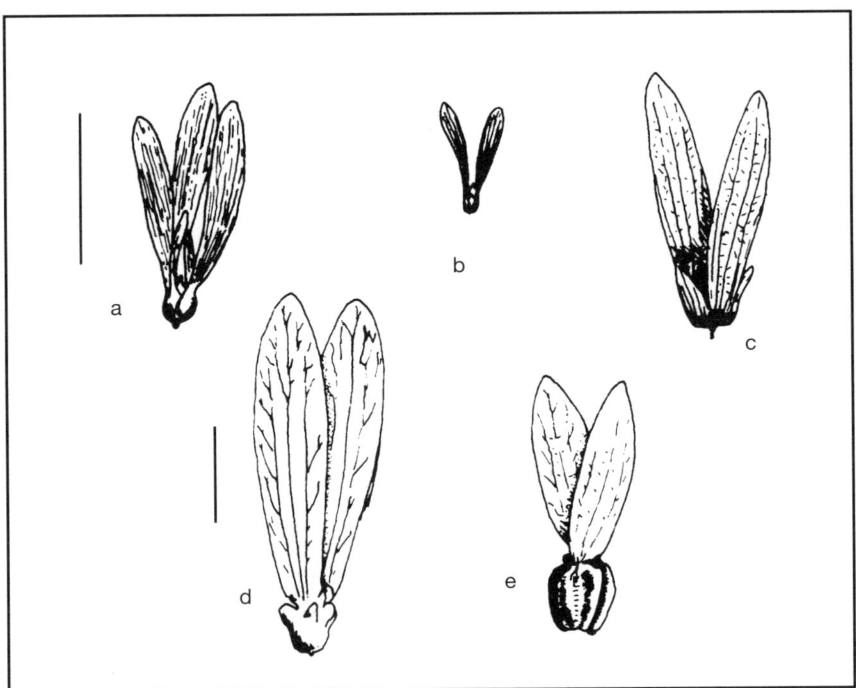

Figure 7.6. Fruits of live Sumatran dipterocarp trees: a) *Shorea assamica*, b) *Hopea ferruginea*, c) *Vatica maingayi*, d) *Dipterocarpus cornutus*, and e) *Dipterocarpus palembanicus*.
After Symington 1943

Common families for the smaller trees are Burseraceae, Sapotaceae, Euphorbiaceae, Rubiaceae, Annonaceae, Lauraceae, Myristicaceae and smaller species of Dipterocarpaceae, but some will also be immature emergents.

Dipterocarp trees dominate much of the lowland forests of Sumatra. For example, on the hill ridges examined on Siberut Island, between 20% and 59% of all trees over 15 cm diameter at breast height were dipterocarps; Myristicaceae, Euphorbiaceae and Sapotaceae were the next most common families. On level ground between the ridges, Euphorbiaceae, Dipterocarpaceae and Myristicaceae ranked more or less first equal (Whitten 1982e). Other families may be dominant, however, such as at the Ketambe Research Station where Meliaceae accounts for 17% of the trees over 10 cm diameter and dipterocarps are not particularly common, accounting for only 4% of the trees (LBN 1983; Rijksen 1978). Near the Pesisip River, Batang Hari, the dominant families are Dipterocarpaceae and Olaceae (particularly *Scorodocarpus borneensis*) (Franken and Roos 1981).

Numerous surveys of Sumatra's lowland forests have been conducted for forestry evaluations but these are of limited use for ecological work because:

- they enumerate only those trees providing profitable timber, and
- they concentrate on areas where timber trees are likely to be most common.

Variation in the floristic composition of a given phase in the growth cycle of lowland forest can be quite considerable, but because the variation is often continuous (that is, there are no sharp boundaries, forest of one composition changes gradually into forest with another composition), it is very hard to study. Poore (1968) was of the opinion that floristic composition was largely determined by chance factors, particularly at the times of fruit dispersal and seedling establishment. Within a given area, however, composition is related to large-scale habitat features such as soils and topography (Ashton 1976; Baillie and Ashton 1983; Franken and Roos 1981). A detailed study of species composition and various soil and other site variables in Sarawak showed that soil texture, levels of iron and aluminium oxides, and the acidity of the soil parent material were the most important factors determining species composition (Baillie and Ashton 1983). These effects are very subtle and occur more as a variation in the relative abundance of different species rather than as the presence or absence of one or more species. This is partly because a particular soil type will not always exclude ill-adapted species, but adults of such species will be relatively rare and have reduced vigour. The minor variations between the forest type studied on Siberut Island were sufficiently clear to the gibbons occupying the area for it to be possible to detect differences in their behaviour and activity between the various forest types (Whitten 1982c,d).

A particularly noticeable variation in lowland forest occurs along streams, and certain tree species tend to be restricted to these areas. Below these trees, alongside fast-flowing rivers, rheophytes (river-plants) can be found. These are usually shrubs which, although diverse taxonomically, have narrow leaves and often brightly coloured fruits dispersed by either water or fish (Jacobs 1981; van Steenis 1952, 1978; Whitmore 1984).

Layering

The canopies of forest trees are often said to form several (usually three) 'layers'. While this is a convenient theoretical concept it has limited application in the field. An emergent layer is often quite clear but below that the supposed layers may be difficult to distinguish. Some trees may clearly belong to a particular layer but many more are generally indefinable. Part of the difficulty in applying the concept of layering lies in the existence of the mosaic of gap, building and mature phases in the forest (p. 193), each of which is in a dynamic flux. In an analysis of various forest types on

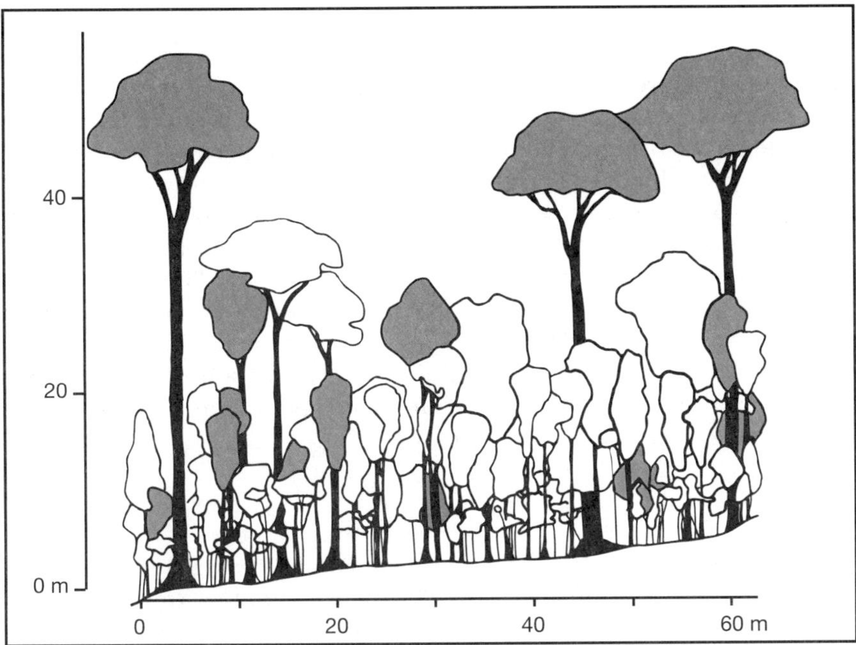

Figure 7.7. Mature and building phases at Belalong, Brunei. Dipterocarps are shaded – note that in the building phase these have the tall, monopodial crown of youth.

After Ashton 1964, in Whitmore 1984, in which the identification of each tree is given

Siberut Island, each of the 2,459 trees along 11.5 ha of 20 m wide transects was scored for the position of its crown relative to the neighbouring canopy. 'Emergence' (defined as 0%-25% of the crown in contact or enclosed by other crowns) did not correlate with tree height because, within a large area in building phase, a medium-sized tree can be emergent, and in mature forest a 40 m high tree can be more or less completely enclosed by taller trees (Whitten 1982c).

The shape of tree crowns changes as the trees grow and thus a certain amount of layering of crown shape occurs. Young trees tend to have monopodial stems (that is, a single main stem) and the crown is tall and narrow around that stem. Mature trees, however, tend to be sympodial (that is, without a single main stem) at the canopy level. This difference is shown in figure 7.7. Layering is discussed more fully by Whitmore (1984) and Richards (1983).

Basal Area

The basal area for trees of 15 cm diameter and over at breast height in seven lowland forest types on Siberut Island ranged from 16 to 42 m^2/ha. The tree palm *Oncosperma horridum* was most common in the forest types with the lowest basal area of trees, and when this species was added, the range decreased to 23-42 m^2/ha. It was suggested that the inclusion of all other woody plants (e.g., small trees, other palms, shrubs and climbers) in the basal area calculation would have resulted in the basal area of the forest types becoming very similar (Whitten 1982c). The basal area of trees in 17 ¼ -ha plots at Ketambe Research Station was 16-45 m/ha (van Noordwijk and van Schaik, pers. comm.) and near the Pesisip River, Batang Hari, was 27.9 m^2/ha (Huc and Rosalina 1981). Two adjacent forest areas of 0.5 ha examined by a CRES team on Bangka Island had basal areas equivalent to 24.7 and 25.7 m^2/ha.

Biomass

The above-ground biomass of two areas of Pasoh Forest Reserve in Peninsular Malaysia was estimated using destructive sampling (felling, cutting and weighing) by Kato et al. (1978). The results were 475 and 664 t/ha. Ogawa et al. (1965) have estimated root biomass as one-tenth of above-ground biomass and so Kato et al. (1978) suggest that average total plant biomass for the Pasoh Forest Reserve was 500-550 t/ha. The changes in biomass with height above ground are shown in figure 7.8. The total plant biomass increment was found to be about 7 t/ha/yr. The ratio between leaf area and leaf dry weight (sometimes known as mean specific leaf area) varied with height above the ground; it decreased steadily from 13 m^2/kg in ground-layer seedlings to 7 m^2/kg in the tallest trees, showing that leaves are generally thicker in the higher parts of the canopy.

Leaf Area Index

Leaf area index (total leaf area perpendicularly above a unit area of ground) was measured at Pasoh Forest Reserve in different height categories (fig. 7.9) (Kato et al. 1978). It was relatively constant between 10 and 35 m and the increase below this was caused by the plants in the shrub and herb layers. There were relatively few leaves between 5 and 10 m, and leaves of climbers maintained a surprisingly constant contribution (average 10%). The total leaf area index in the two destructive plots was 7.15 and 7.99 m/ha.

Litter Production

Litter production has been investigated at Ketambe Research Station and by teams from the National Biological Institute (LBN 1983) and by van

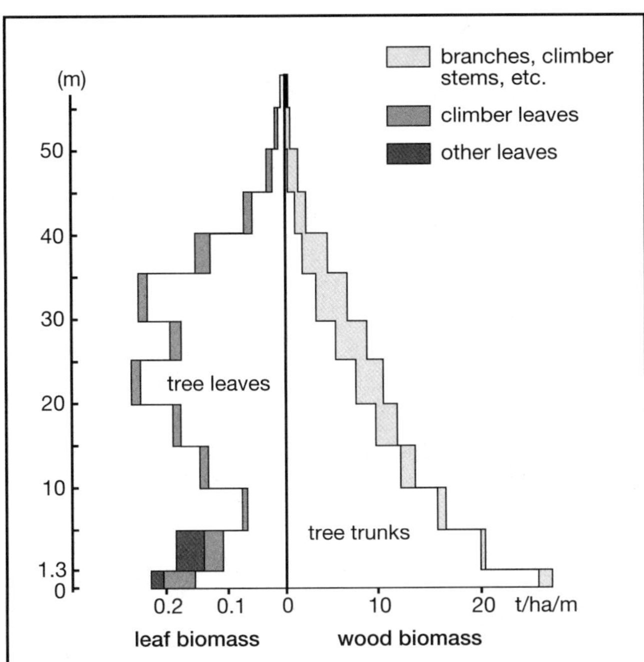

Figure 7.8. Vertical distribution of leaf and wood biomass density in a plot at Pasoh Forest Reserve, Peninsular Malaysia.

After Kato et al. 1978

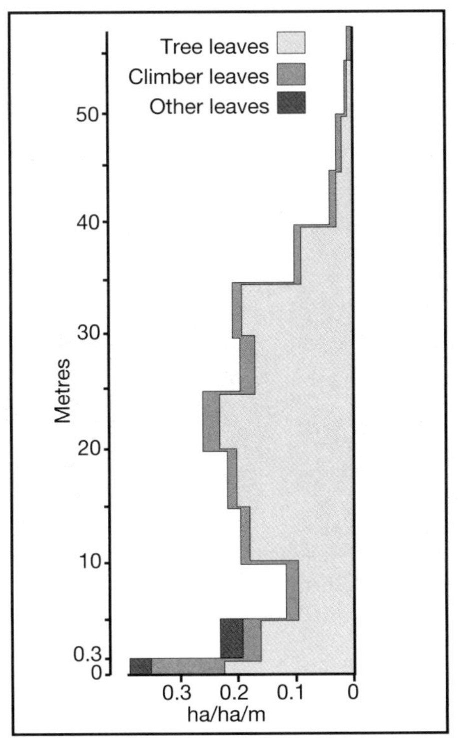

Figure 7.9. Vertical distribution of leaf area index in a plot at Pasoh Forest Reserve, Peninsular Malaysia.

After Kato et al. 1978

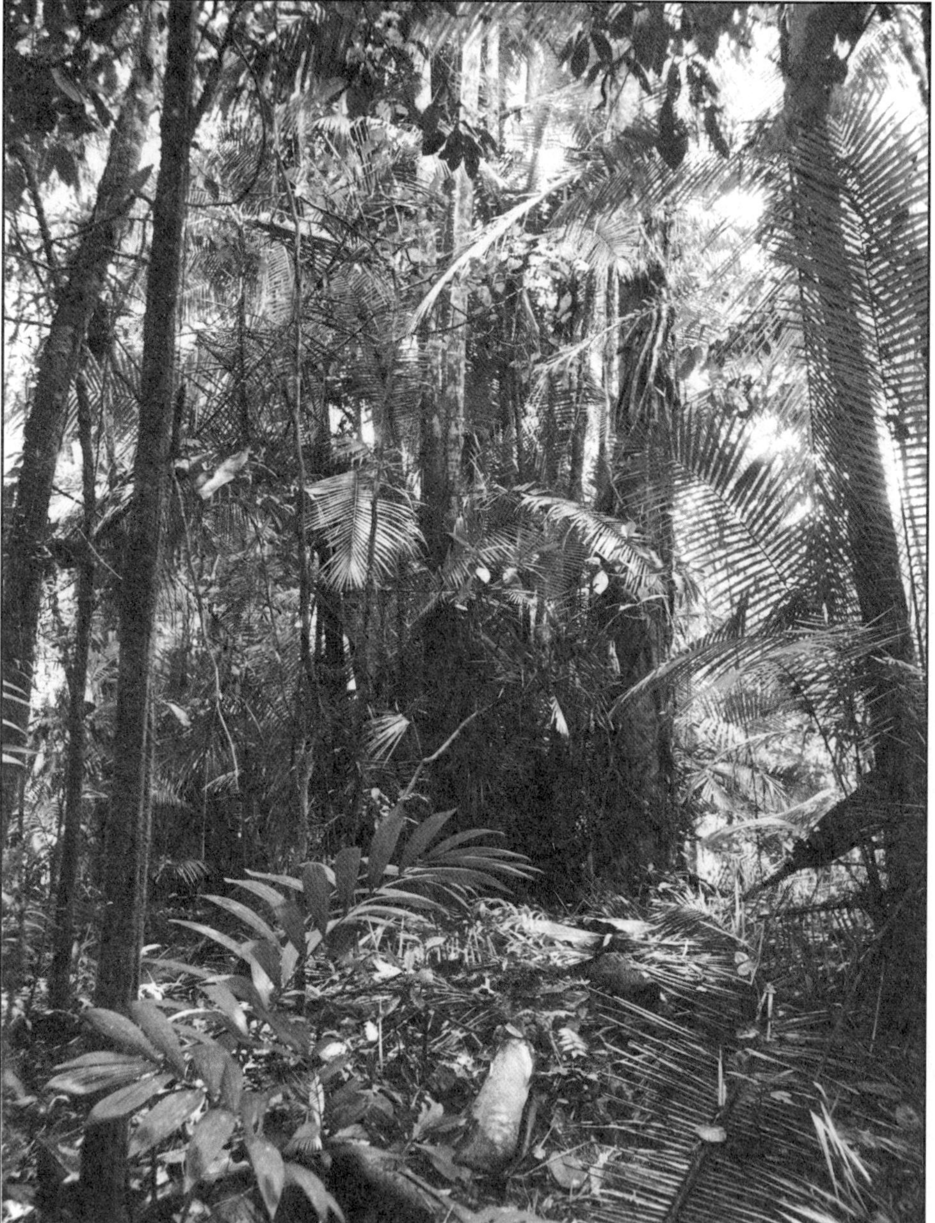

Figure 7.10. Lowland forest on Siberut Island, West Sumatra, showing many young rattans and the abundance of the spiny tree palm *Oncosperma horridum.*

Noordwijk and van Schaik. The second study indicates that 4.7-8.3 t/ha/yr of leaves and 2.2-5.9 t/ha/yr of other litterfall. Litter production was lower in forests with lower pH (lower fertility) (van Noordwijk and van Schaik 1983, in prep). Yearly production of all small litter (twigs, branches and leaves) at Pasoh Forest Reserve was found to be between 7.5 and 10.6 t/ha/yr (Lim 1978; Ogawa 1978) with a likely average of 9.0 t/ha/yr. Similar values were obtained in lowland forest in Sarawak (Proctor et al. 1983) and East Kalimantan (Kartawinata et al. 1981). The proportion of the litter in different forms at Pasoh was as follows: leaves 72%, twigs 17%, fruit and flower parts 6%, other 5% (Lim 1978). Ogawa (1978) found that the annual leaf fall (6.5 t/ha) was less than the leaf biomass (8 t/ha) and thus calculated the mean life of the leaves to be 15.2 months. Litter usually accounts for only 1% of the organic carbon in a lowland forest compared with about 20% in a European forest (Brown and Lugo 1982).

Total Primary Production

Very few studies of total forest primary production (p. 86) have been conducted anywhere in the Sunda Region, largely because of the many methodological difficulties involved (Whitmore 1984). At Pasoh Forest Reserve three methods were used (Kira 1978; Koyama 1978; Yoda 1978a; Yoda and Kira 1982). The results suggested a net primary productivity of 25-50 t/ha/yr and a gross primary productivity of 70-100 t/ha/yr. Kato et al. (1978) remark that the net primary productivity is lower than for rubber and oil palm plantations. As was described on page 86, however, a mature forest would be expected to have a lower productivity than a managed, growing forest or an agricultural plantation.

Mineral Cycling

Cycling of minerals through a lowland forest has not been investigated in its entirety in the Sunda Region, but relevant papers have been written on various stages of the cycle (Anderson et al. 1983; Lim 1978; Manokaran 1978, 1980; Ogawa 1978; Yoda 1978a, 1978b; Yoda and Kira 1982; Yoneda et al. 1978). A review of mineral cycling in tropical forests has recently been published (Vitousek 1984). See page 296 for details of a study of mineral cycling in a lower montane forest.

Ecology of Some Forest Components

Roots

A relatively small proportion of the organic carbon and other nutrients in a lowland forest is in the soil. Organisms such as fungi, bacteria, soil animals and roots therefore compete for what are probably limiting resources. It appears that almost all trees in lowland forests and almost all non-grain crops can develop a mutualistic relationship between their roots and fungi (Janos 1980a). This is called a mycorrhiza. The most common form found in the tropics has the hyphae ('root' strands) of the fungus penetrating the plant's roots, where they take up carbohydrates from the plant. For its part the fungus channels minerals into the roots (Janos 1980b, 1983). Many plant species with mycorrhiza appear not to possess root hairs because these structures are improved upon by the mycorrhizae. Mycorrhizae are probably common among trees in heath forest (p. 255), but are poorly developed among plants of early successional stages which normally grow on soil enriched from fire ash, deposits left by flood water, or the decomposition products of fallen trees (Janzen 1975). In such situations a plant would be at a disadvantage if it had to wait for its roots to be infected with fungus before it could start growing and competing (Janos 1980b, 1983). Some species cannot grow without mycorrhizae, whereas others grow better without mycorrhizae where mineral nutrients are not limited, but better with mycorrhizae in poor soils (Janos 1980a, 1983). For trees that depend on mycorrhizae it is advantageous for their fruit to have large seed reserves because they provide a source of food for the seedling before infection with fungus occurs. It is not surprising, then, that seedlings kept experimentally without mycorrhizae, stopped or slowed their growth after attaining a size correlated with the average dry weight of a seed for that species (Janos 1980b).

Buttresses and Trunks

Buttresses are common features of trees in lowland forest. Setten (1953) found that 41% of 18,067 large ('timber-sized') trees in a forest in Peninsular Malaysia had buttresses reaching more than 1.35 m up the trunk. Different species are relatively constant in the presence, shape and surface characteristics of buttresses and these characters can be helpful in identification (Wyatt-Smith and 1979).

Buttresses are sometimes said to act as structural supports for trees whose roots are relatively shallow and where the substrate affords little anchorage (Henwood 1973; Smith 1972). They are also said to be composed of 'tension wood' to reduce the pulling strain on the roots and so would be expected to form on the uphill side of a trunk, on the side of a trunk opposite a major congregation of heavy climbers, or on the side the wind blows from – that is, on the 'tension' side. From subjective impressions, however, it is clear that some trees do not 'obey' the rules and instead form buttresses on the 'compression' side of the trunk; see, for example, photographs by Corner (1978). In addition, of course, many trees grow well in similar conditions without buttresses.

Recent hypotheses on buttress formation have been unrelated to structural problems. It has been suggested, for example, than since buttresses increase the space of forest floor occupied by an individual tree, they could be viewed as a competitive mechanism (Black and Harper 1979). The physical presence of a buttress prevents or hinders the establishment of neighbouring (i.e., competing) trees. If this hypothesis is correct then it should be possible to observe that:

- the density of trees near a buttressed tree are less than near a tree of similar bole size, canopy form and age without buttresses;
- the distance to the nearest neighbours of buttressed trees are greater on the average than from a non-buttressed tree to its nearest neighbours;
- the density of trees around different individuals of a buttress-forming species bear a relationship to the age of the individual trees. Older buttressed trees should have both more distant and fewer neighbours in the smaller size classes;
- species diversity of trees around a buttressed tree are lower than around a non-buttressed tree because some species are likely to be less competitive against buttresses than others (Black and Harper 1979).

These hypotheses need testing, and collecting appropriate data does not require in-depth botanical knowledge.

Trees with rotten, empty trunks are surprisingly common in lowland forest. This may appear to be unfortunate for the tree but it has been suggested that it is in fact a strategy of distinct advantage. Empty cores of large trees are inhabited by bats, porcupines and rats in addition to a host of invertebrates. Faeces and other products from these animals are broken down by microbes and thus provide an exclusive supply of nutrients for the tree in a generally nutrient-poor environment (Fisher 1976; Janzen 1976a).

It is not generally known that tree trunks exhibit a daily fluctuation in girth, being largest in the early morning and smallest in the late afternoon. These changes are presumably caused by water being drawn out of the plant by evapotranspiration during the day. Although measurable, the changes are very small, being in the region of 0.1 mm around an average timber tree (Yoda and Sato 1975).

Climbing Plants

Climbing plants abound in many types of lowland forest. The crown of a large climber may be as large as that of a tree but because (with the exception of rattan) they do not possess any real commercial value, they have been little studied. Identifying climbers is very difficult but the coiled or convoluted ones with flattened stem sides are generally leguminous species, and the ones with regular hoops around the stem are members of the Gnetaceae, relatives of the conifers. The best-known member of the Gnetaceae in Sumatra is the tree *Gnetum gnemon* from which *emping* crisps are made, but almost all members of the family are climbers.

When young, many species of tree-hanging climbers in the forest understorey can look very similar to young trees as they grow slowly, waiting for an opportunity to rise to the canopy. Such opportunities are afforded when a gap is formed, and it is then that a young climber can attach itself to a rapidly growing tree and be carried upwards. During a similar period of apparent dormancy, some species are growing a large tuber below ground. In response to a particular cue the energy available in the tuber is suddenly used up in a rapid burst of upward growth, whether or not a gap has formed (Janzen 1975). Some climbers, such as rattans, have elaborate hooks by which they can climb their own way to the light at the top of the canopy. Other climbers begin to coil their way up their chosen host as soon as they have germinated. Some species have populations which coil in a left-handed manner in the northern hemisphere, and in a right-handed manner in the southern hemisphere so as to make most efficient use of the sun's rays (Janzen 1975). It would be interesting to know whether this phenomenon exists in Sumatra, which is more or less equally divided by the equator.

The stems of climbers can sometimes be seen to loop down towards the forest floor and back again, indicating that the major bough or the whole tree supporting the climber may have fallen and the climber then grew up again. Mature climbers are rarely supported by a single tree but rather grow horizontally through the canopy, sewing together the tree crowns. This may help to prevent trees from falling over in strong winds but, if a tree is felled, the climbers in its crown often pull over other trees as well (p. 366).

A considerable portion of the apparent 'seedlings' of climbers (and a certain proportion for some forest trees) are in fact no more than shoots growing up from a horizontal root of an established climber (Janzen 1975). This is a means by which a plant can increase the size of its crown, and since these shoots are obviously genetically identical to their 'parent', the actual number of genetically distinct individuals in a given area may be extremely low.

Climbers compete with trees for light, nutrients and water and can cause mechanical damage to trees. In well-managed production forest this

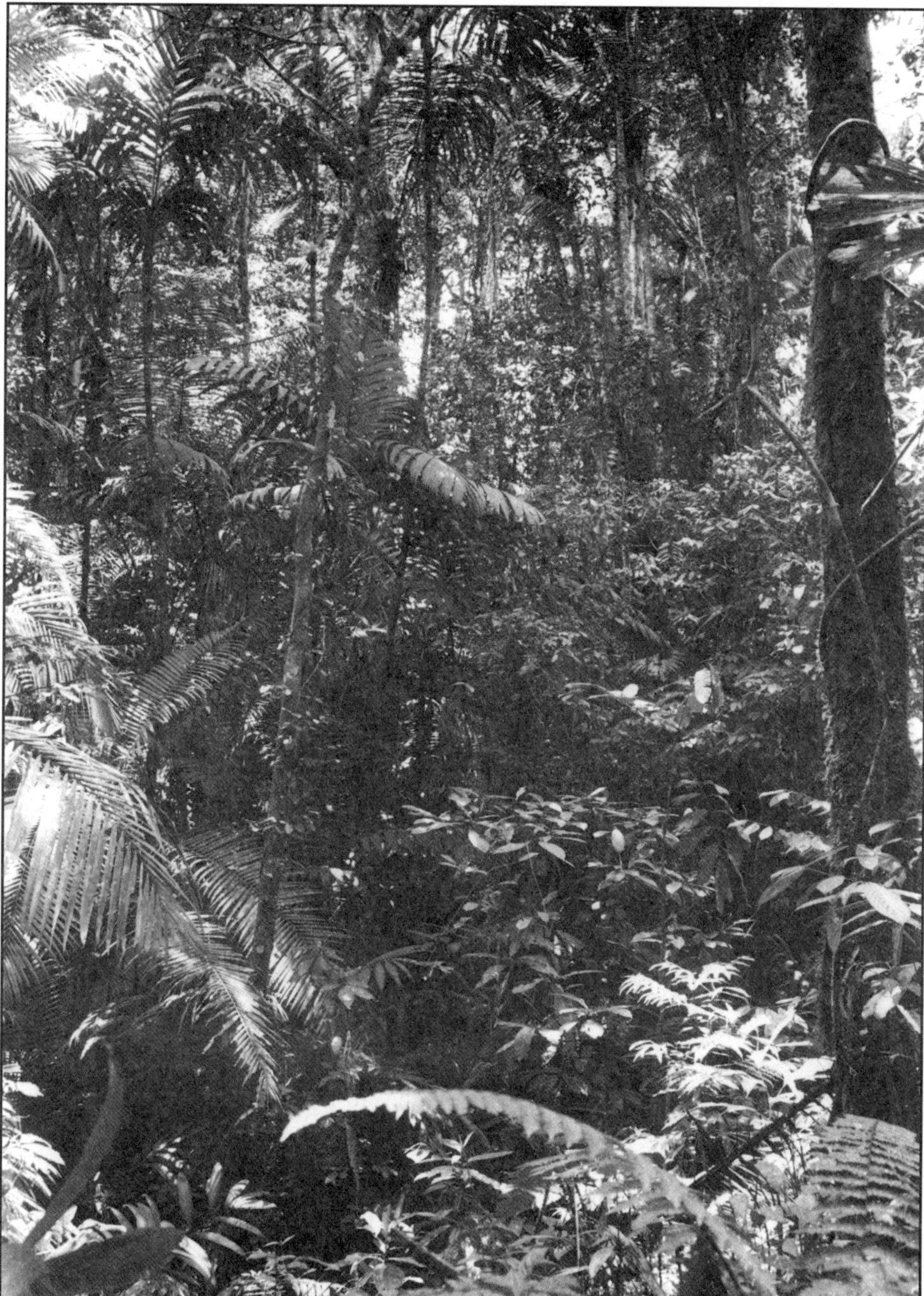

Figure 7.11. Lowland forest near Kebamse, Mount Leuser National Park.
A.J. Whitten

is the reason climber stems are cut (see p. 366). In addition, bole diameter is often greater than average for trees with heavy climbers in their crown, indicating that trees have to divert resources away from vertical growth and reproduction in order to increase the strength of the supporting structure (Whitten, unpubl.). Where climbers are abundant, regeneration of trees can be seriously delayed, and so for all these reasons it is an advantage to a tree to shed or to avoid climbers.

Fast-growing trees are better able to avoid climbers than are slow-growing trees. Trees which are able to rapidly increase their bole diameter as well as their height are also more likely to escape certain climbers which coil up the stem, because such climbers are limited in the diameter of support they can ascend. Rapid increase in girth can also break potentially strangling plants. Some trees exhibiting a symbiotic relationship with ants (such as *Macaranga*, p. 376) may benefit from the ants removing vines from the branches as they have been shown to do in Central America (Janzen 1973a). Other strategies used by trees to avoid climbers are suggested by Putz (1980).

Epiphytes and Epiphylls

An epiphyte is a perennial plant rooted upon, not in, a larger host, and one which does not have to produce or maintain massive woody stems and branches to live above the forest floor. Epiphytes are common in lowland forest, but even more common in certain montane and heath forests (pp. 257 and 284), and they are thought to make a significant contribution to the totals of biomass and species-richness in a forest (Benzing 1983). The major higher plant families of epiphytes in Sumatra are Gesneriaceae, Melastomaceae, Rubiaceae (including the ant-plants *Myrmecodia* and *Hydnophytum*), Asclepiadaceae and Orchidaceae. In addition to these are numerous epiphytic ferns, lichens and bryophytes (including mosses and liverworts).

Bryophytes are small but common epiphytes in lowland forest. Their small size makes them less restricted than other plants in the range of microhabitats they can occupy, and these are shown in figure 7.12. The most luxuriant bryophyte communities in wet lowland forest are generally found on the bases of large trees (see Pócs [1982] for detailed descriptions).

Epiphytes might seem to have an easy life but they live on a substrate which is extremely poor in nutrients. Most of them rely upon nutrients dissolved in rain, a small amount of litterfall, and occasional mineral inputs from animals. Epiphytes also have to contend with a very erratic water supply – being drenched when it rains, and parched when the sun is shining and the wind is blowing (Janzen 1975).

Epiphytes cope with growing in nutrient-poor conditions in various ways (although few species will exhibit all these features):

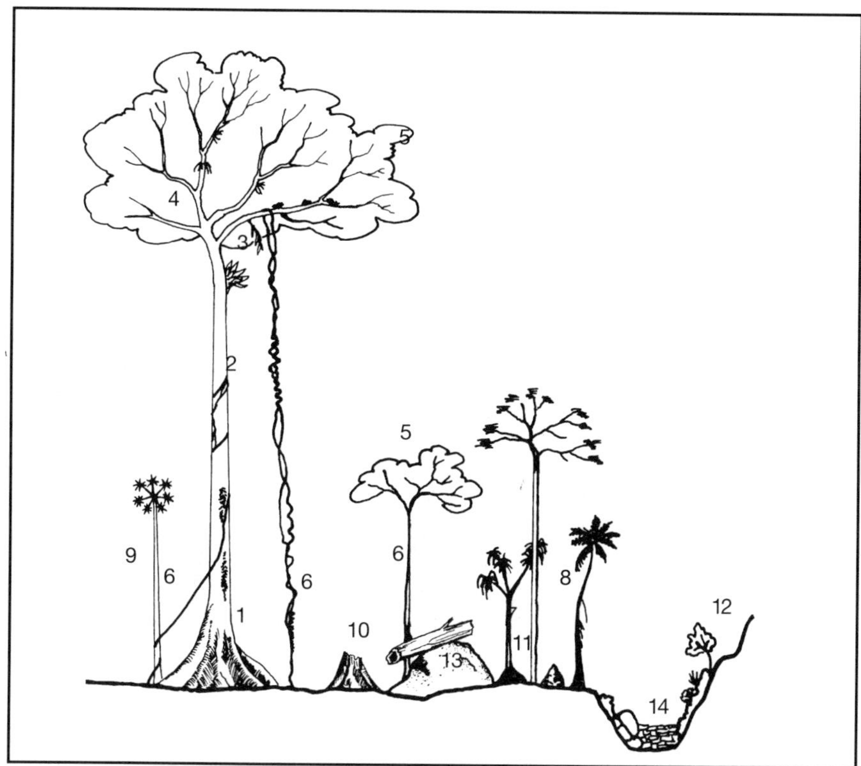

Figure 7.12. Bryophyte microhabitats in a lowland forest. 1. bases of large trees, 2. upper parts of trunks, 3. macro-epiphyte 'nests', 4. bark of main branches, 5. terminal twigs and leaves, 6. bark of climbers, shrub branches and thin trunks, 7. pandan stem, 8. tree-fern stems, 9. palm trunks, 10. rotting trunks, 11. soil surface and termite mounds, 12. roadside cuttings, 13. rocks and stones, 14. submerged or emergent rocks in streams.
After Pócs 1982

- their own tissues may have very low nutrient concentrations;
- juvenile stages are prolonged even though little growth is required to become mature;
- their vegetative parts are reduced in size;
- they may obtain nutrients from unconventional sources such as ants;
- their leaves tend to be long-lasting (due perhaps to some form of defence against herbivores) so that the replacement of nutrient-expensive leaves is required only infrequently; and

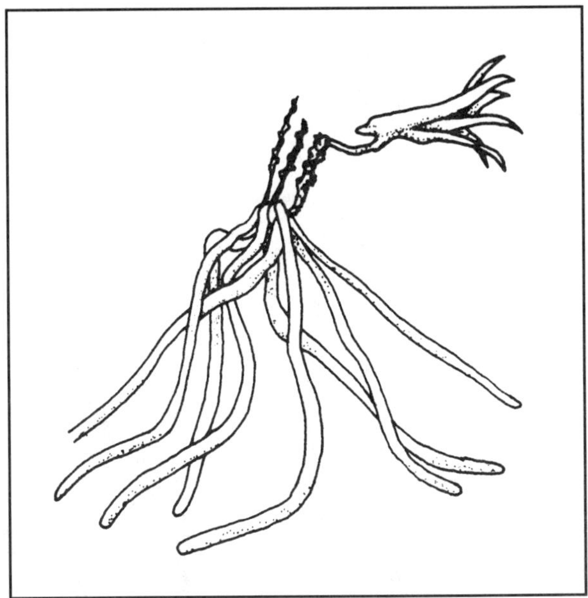

Figure 7.13. *Taeniophyllum*, an example of a leafless orchid.

- their flowers are pollinated by animals because wind dispersal of the pollen would result in 'expensive' losses (Benzing 1983).

In addition, it has been observed that some epiphytes concentrate their roots on the underside of boughs where water and nutrients would remain longest (Whitten 1981).

The third strategy, vegetative reduction, has been studied by Benzing and Ott (1981). They investigated three species of shoot-less orchid; that is, orchids with no leaves and a shoot which merely initiates roots and produces occasional flowers (fig. 7.13). It was found that these orchid roots were green in colour and that they were capable of fixing carbon in their tissues, whereas roots of normal, leafy orchids, while able to photosynthesise, were unable to exhibit a net gain of carbon. Benzing and Ott consider that the shootless habit, by reducing the 'cost' of the vegetative body, has evolved for reasons of nutrient economy so that maximum nutrient resources can be directed towards reproductive effort.

As Janzen (1975) has pointed out, epiphyte leaves seem to suffer very little damage from herbivores, and he further suggests that these valuable leaves may be defended chemically to prevent their loss. There appears to be no data yet to confirm this. Some epiphytes, notably *Myrmecodia* (p. 257), harbour ants in their stems and these may well protect their leaves from caterpillars and other arthropod herbivores. It seems unlikely that the succulent leaves of *Myrmecodia* contain concentrations of chemical defences because on Siberut Island, Mentawai gibbons, whose diet indicated a marked avoidance of potential sources of defence compounds, ate *Myrmecodia* leaves (Whitten 1982b,e). *Myrmecodia* on Siberut Island appeared to grow further away from the tree trunks than other species and it may have been helped in such extreme habitats by the nutrients provided through its relationship with ants (Whitten 1981). Although epiphytes (by definition) do not take nutrients directly from the host plant, they have been called 'nutritional pirates' (Benzing 1981, 1983).

This is because they can tie up nutrients in their own biomass from dust, leaf leachates and rain which would otherwise have fallen onto the soil to be utilised by the hosts. While this is true, Nadkarni (1981) has found that host trees can turn the tables at least partially back in their favour by producing small roots beneath the mats of dead and living epiphyte tissue. These roots thus give the tree access to canopy nutrient sources. The main boughs of some tree species are almost invariably covered with epiphytes but it would appear that other species actively deter epiphyte growth, presumably because the relationship is disadvantageous. Thus, on Siberut Island, *Sterculia macrophylla* were always laden with epiphytes but *Endospermum malaccensis* bore none (Whitten 1981).

Epiphylls are mosses, liverworts, algae and lichens which grow on the living surface of leaves in shady situations where the air is almost continually saturated. Thus they are generally restricted to the leaves of young trees, shrubs and long-lived herbs of the lowest layer of forest up to about 2-3 m and also near streams. Over 10 species of epiphylls can occur on a single leaf and some of these are obligate; that is, they are found only on leaves (Pócs 1982). Epiphylls are restricted to certain types of leaves and are possibly more common on leaves lacking drip-tips (p. 194).

Gaps

Pioneer plants (p. 196) are generally thought to be low in defensive compounds; their growth strategy is to grow and reproduce quickly because, in the nature of succession, they are limited in both space and time. Resources are not diverted into possibly unnecessary defence. Plants growing in gaps, the location of which is unpredictable, are able to avoid herbivores partly through that unpredictability of occurrence and partly through relatively short life cycles (Janzen 1975). Maiorana (1981) noticed that some species growing in shaded habitats suffered more from being eaten than did the same species in open habitats. Strangely, when leaves from both habitats were presented to snails, generalist herbivores frequently used in palatability experiments (Edwards and Wratten 1982; Wratten et al. 1981), it was found that the plants from open areas were preferred. It is suggested that the 'defence' used by pioneer plants might in fact be the hot, unshaded, low humidity of the gap environment itself.

The fast-growing and often succulent plants of lowland forest gaps are food for the larger herbivorous mammals such as deer, tapir and elephant (p. 373). Studies of birds in forest gaps indicated that the number of species caught in nets in gaps was greater than those caught in surrounding forest (Schemske and Brokaw 1981), perhaps because forest birds cross gaps to reach areas of forest but birds of gaps rarely enter forest.

As described on page 194, gaps experience considerable differences in microclimate compared with mature forest (Ricklefs 1977) and this is reflected in the behaviour of some animals which are found in both habitats. For example, the orientation of webs of one spider species was found to be north/south in closed habitats and east/west in open habitats. The longer a spider spends on the web, the more prey it captures (some prey manage to wriggle loose before being killed by the spider), and so presumably an orientation is chosen to allow maximum time on the web. Web orientation seems to be related to the body temperature of the spiders because both too little and too much heat can affect spider activity. Thus, in open habitats, the spiders' webs face east/west to reduce heat load and in closed habitats face north/south to increase body temperature (Biere and Uetz 1981).

Figs

Fig plants exhibit numerous interesting characteristics (e.g., the strangling habit of some species – figure 7.14) and represent an extremely important source of food for many forest animals. Fig plants are, however, probably most interesting for the way in which pollination is effected.

According to botanical terminology, figs are syconia, not fruit, but ecologically they *are* fruit and so that noun is retained (Janzen 1979d). The wall of flowers arches over and around so that the flowers are inside a small, almost entirely enclosed, cup (fig. 7.15). The hole at the end of a fig is partially closed by scales formed by bracts.

Fig flowers are pollinated exclusively by tiny (±1 mm) fig-wasps (Agaonidae) (fig. 7.16), and both figs and wasps are entirely dependent on each other for their survival. In addition, one species of fig plant (and there are about 100 species in Sumatra [Corner 1962]) will generally have only one species of pollinator wasp. The female flowers develop first and female wasps fly to the figs, probably attracted by smell. One or more females climb inside the fig by forcing their way through the scales and in the process lose their wings and antennae. Once inside, a female pushes her ovipositor down into the ovary of one of the female flowers and lays an egg. It used to be thought (Corner 1952; Galil 1977) that females oviposited in ovaries of flowers specifically intended for developing larvae but various authors have shown that this is not so (Janzen 1979d). During oviposition and as she moves around inside of the fig, the female deposits pollen

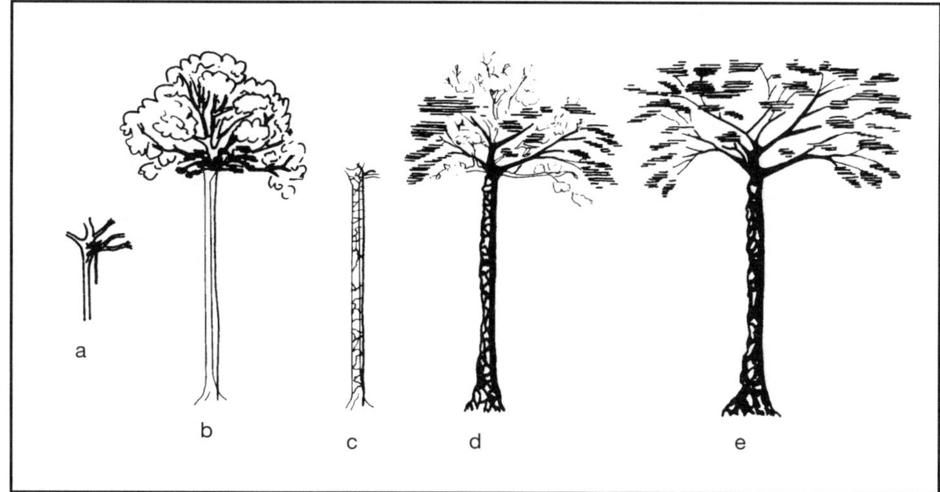

Figure 7.14. One means by which a strangling fig forms. A bird, squirrel or monkey deposits a seed in the crown of a tree. A small bush grows, from which a long root descends to the ground. Side roots then form and grow around the trunk, joining where they meet. The supporting tree eventually becomes encased in a basket of roots which eventually strangle the host to death. In time the fig will stand like a tree with a hollow trunk.

After Corner 1952

from the flowers where she hatched onto other flowers and so effects pollination. Thus although the wasps pollinate the flowers, they are also seed 'predators'. Janzen (1978b, 1979b) found that in a sample of 160 figs from four species, 98% of the figs had more than 30% of their potential seeds killed by pollinating wasps. The larvae develop and pupate and the wingless males emerge first. They search for female pupae and fertilise them before they emerge.

Janzen (1979c) found that most figs are only entered by one or two females. This means that there is a very high probability of brothers and sisters mating, a situation avoided in most other organisms. The males' role is not yet completed because they then make a tunnel through the fig to the outside. The carbon dioxide level inside the fig is about 10%, but when the hole is made the concentration drops to atmospheric levels (i.e., 0.03%). This change appears to stimulate the development of the male flowers, the emergence of the females and the process of ripening. In some species the females fly out through the tunnel made by the males and in other species both males and females eat away the scales at the fig

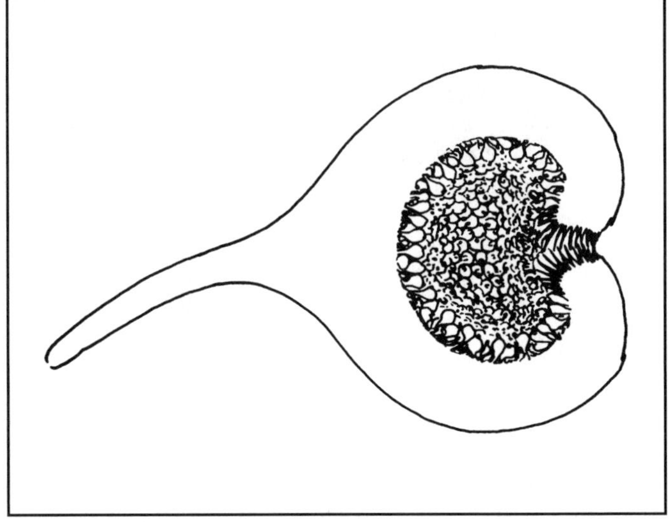

Figure 7.15. A cross-section through a fig fruit showing the ring of male flowers (arrowed) around the end, and the bract scales partially closing the opening.
After Corner 1952

Figure 7.16. Fig-wasps; female above, male below.

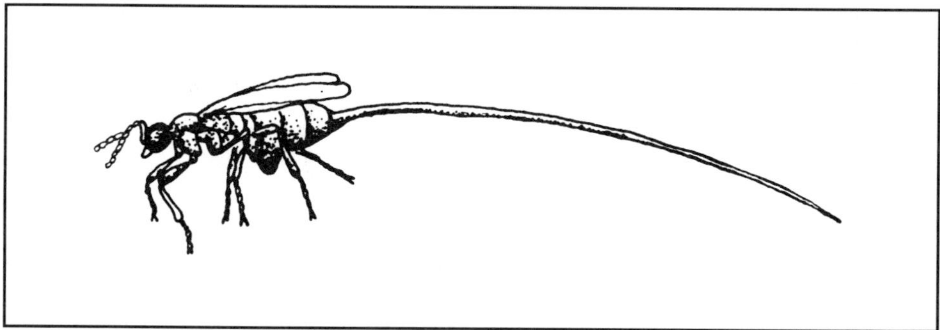

Figure 7.17. Fig-wasps' wasp parasite.
After Corner 1952

entrance. Either way, the females fly off carrying pollen in search of a developing fig; and so the cycle continues. The males die within the fig.

Not all the insects emerging from a fig are fig-wasps because small parasitic wasps oviposit from the outside of the fig into developing fig-wasp larva. How they locate a larva without seeing it is unknown. The story is further complicated by some of the parasites being parasitised themselves by other wasps (fig. 7.17).

Rafflesia

One of the best-known wild flowers in Sumatra is *Rafflesia,* although it is rarely seen by most people except in photographs. It is a member of the family Rafflesiaceae which includes two other peculiar flowers found in Sumatra, *Rhizanthes* and *Mitrastemon* (Palm 1934; Meijer 1997). Sumatra has six species of *Rafflesia*: *R. arnoldii* (the largest flower in the world) throughout (Brown 1822, 1845; Justessen 1922), *R. gadutensis,* endemic, found in west central Sumatra, *R. hasseltii* found in central Sumatra, *R. micropylora,* endemic in Aceh, *R. patma* in southern Sumatra, and *R. rochussenii* in northern Sumatra (Meijer 1997). *Rafflesia* occurs mainly in lowland forest but van Steenis (1972) reports finding one at 1,800 m on Mount Lembuh, Aceh.

Rafflesia is exceptional not just because of its size but because it is the most specialised of all parasitic plants — it has no stem, leaves or roots. Apart from the flower, the plant consists of only strands of tissue growing inside the stem and larger roots of a few species of *Tetrastigma* climbers (family Vitaceae). Flower buds form on the stem and, for *R. arnoldii,* take

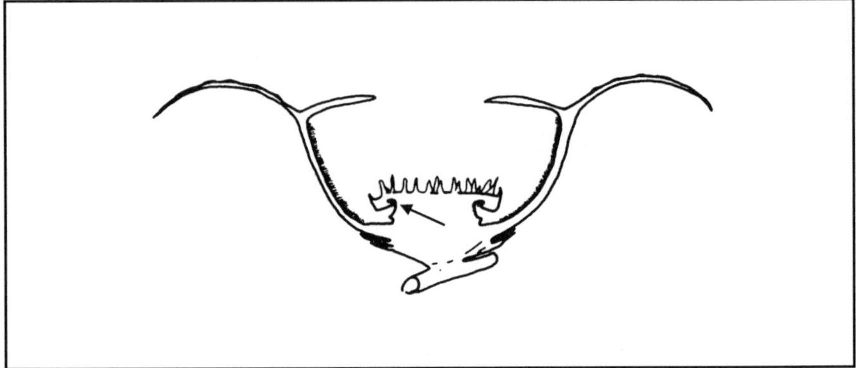

Figure 7.18. Cross-section of *Rafflesia arnoldii* on a *Tetrastigma* root. Arrow indicates location of sexual parts.
After Meijer 1958

about 19-21 months before they open as flowers. *R. arnoldii* buds of 10 cm diameter will take about five months to flower, of 15 cm diameter two to three months, and of 25 cm 20 to 30 days (Meijer 1958). There appears to be no evidence for seasonal flowering but this has not been investigated in detail.

When *Rafflesia* flowers open (fig. 7.18), a smell of decaying meat arises from inside the flower and this attracts flies which presumably effect pollination. Male and female flowers are separate, and for pollination to occur both need to be open simultaneously. Flowers may remain open for a week and then wither and decay. The fruits inside the flower are very small and their means of dispersal to other *Tetrastigma* is unknown (Ernst and Schmidt 1983; Justessen 1922). It seems feasible, however, that pigs, squirrels, rats and tree shrews may get *Rafflesia* fruits stuck in their feet while searching for food in the forest undergrowth. If they disturb the ground around another *Tetrastigma* climber, the seeds may be rubbed off onto the bark where they can germinate.

The name 'bunga bangkai' is sometimes given to *Rafflesia* but this name is also used for *Amorphophallus titanum*, the tallest flower in the world which was featured on the old Rp 500 banknote. *Amorphophallus* plants are quite easily recognised by their characteristic blotchy stems and unusual leaves (fig. 7.19), but those seen in disturbed areas are generally smaller species. The flower of *Amorphophallus titanum* has a foetid odour which attracts small beetles and these effect pollination (van der Pijl 1937).

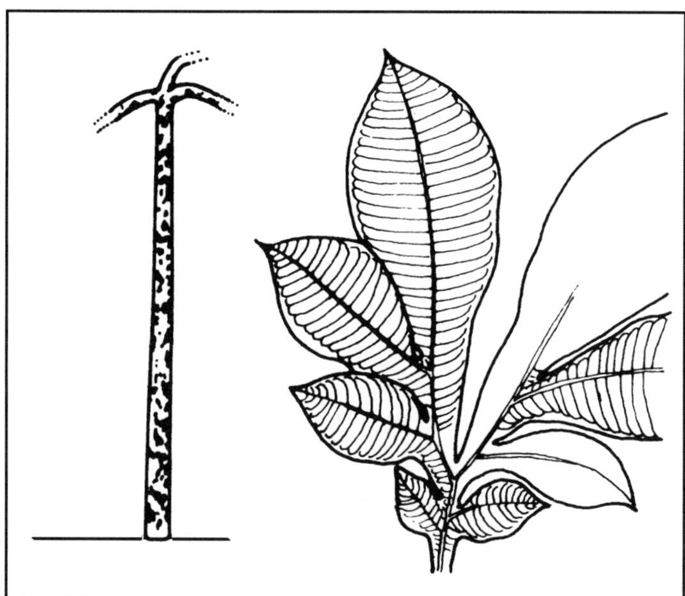

Figure 7.19. The form of the stalk of *Amorphophallus* and detail from part of a leaf.

Cycles of Flower, Fruit and Leaf Production

As far as is known, cycles of flower, fruit and leaf production have been studied in Sumatra only at the Ketambe Research Station (Rijksen 1978), and on Siberut Island (Whitten 1980b). None of the published accounts cover a period of more than two years. There is considerable variation within and between species, between months and between years, and the patterns become clearer if the cycles are studied for longer periods such as the six years achieved at Ulu Gombak (Medway 1972a) and the long-term but intermittent studies at Kuala Lompat (Raemaekers et al. 1980), both in Peninsular Malaysia.

At Ulu Gombak 61 mature canopy trees representing 45 species were monitored. Only one of these flowered and fruited at consistent non-annual intervals, and that was the large strangling fig *Ficus sumatrana* which had a cycle of four to five months. Strangely, that rhythm was unconnected with the weather and yet it always produced a heavy crop. Ten species were represented by two or more trees in the sample and only five of these synchronised their flowering and fruiting. In only 10 species was

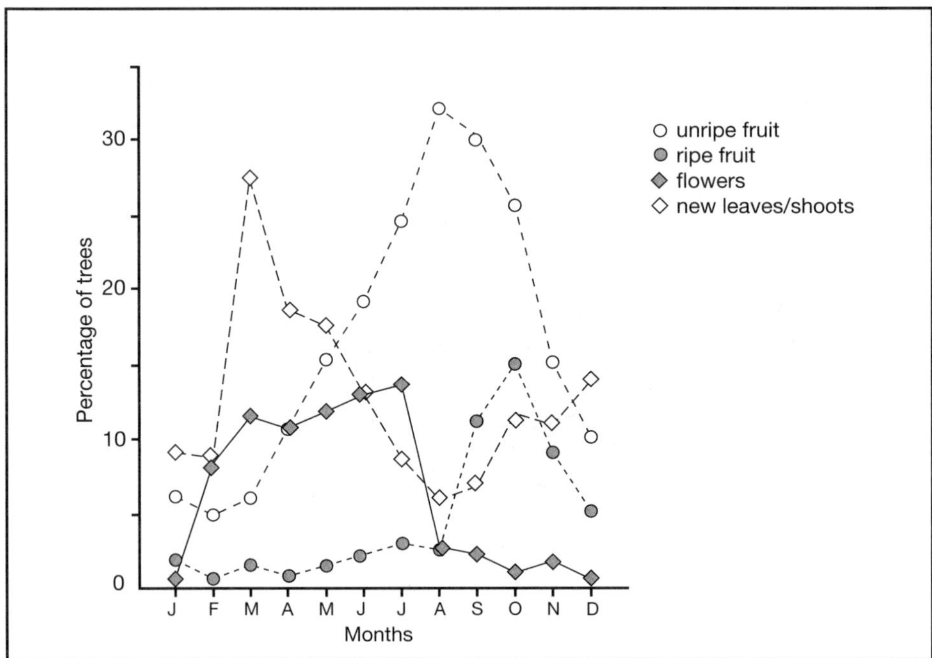

Figure 7.20. Monthly percentages of trees with unripe fruit, with ripe fruit, with flowers and with leaf flush at Ulu Gombak, Peninsular Malaysia. Figures are averaged over six years.

Data from Medway 1972a

flowering and fruiting consistent with annual cycles; the remaining 34 species flowered and fruited at irregular intervals usually exceeding one year and irrespective of weather. Ten species produced flowers each year, but only six of these produced fruit every year. This illustrates the fact that flowering does not necessarily lead to fruiting. This can be because of a lack of pollination, fruit and seed predation, or adverse weather conditions. In a review of this subject it is suggested that abortion of flowers or young fruit is often caused by a lack of necessary resources such as inorganic nutrients, water and products of photosynthesis (Stephenson 1981).

Leaf flush was also variable between species. Three species produced new leaves almost all of the time and 21 species showed consistent annual patterns. Five other species exhibited irregular biannual patterns. Eight species were deciduous (lost all their leaves), four on an annual basis and four at intervals greater than a year.

Despite these variations, the tree community does show clear patterns of flower, fruit and leaf production (fig. 7.20). At the Ulu Gombak forest new leaves were produced twice each year with a major peak in March to June (just after the driest time of year) and a minor peak from October to December beginning just before and extending into the wettest time of year (Medway 1972a). Flowering at Ulu Gombak peaked after the driest time of year, that is from March to July, and, not surprisingly, fruiting peaked in August to September just before the wettest time of year. The flowering of many lowland forest trees seems to be initiated by water stress (Whitmore 1984). The apparent higher percentage of fruiting trees than flowering trees is caused by the fact that fruit remains on a tree longer than flowers and is thus counted in more months. For this reason the percentage of trees with ripe fruit is also shown.

The rainfall patterns in much of northern Sumatra are very similar to those of Ulu Gombak and so similar cycles of flower, fruit and leaf production would be expected. In the south of the island, however, where the driest months are usually June/July (Oldeman et al. 1979), the cycles are presumably adjusted accordingly.

Gregarious Fruiting

Gregarious (or mast) fruiting is a term applied to the simultaneous mass fruiting of certain trees over wide areas at intervals in excess of one year and often in excess of five years. In the tropics, this phenomenon is only found in Southeast Asia and virtually only among the Dipterocarpaceae. On pages 88, 177 and 259, mention is made of the strategy of chemical defence of leaves by trees growing with low productivity on poor soils in climates favourable for animal life, leading to low numbers of animals. Janzen (1974a) has proposed that gregarious fruiting is an indirect consequence of low productivity and, possibly, relatively low animal biomass (p. 192).

The primary purpose of gregarious fruiting is to escape seed predation. The trees produce so much seed that, after satiating the appetites of seed-predators which find the fruit, some seed still remains for germination. The synchrony also means that seed predators are unable to specialise on the particular fruit and so the population of potential seed predators is kept at a low level until gregarious fruiting begins. If only a single tree species adopted the strategy of gregarious fruiting there would be no chance of satiating the seed predators and the strategy only works because most of the family members fruit at the same time. It was shown on page 198 that dipterocarp trees often make up a considerable proportion of the trees in a lowland forest.

Dipterocarp seeds are eaten by squirrels, rats and probably also by larger animals such as pigs, tapir, deer, rhino and elephant (Payne 1979; Poore 1968; Woods 1956). They do not seem to be heavily defended with chemicals. Janzen (1974a) suggests that if such oil-rich, defence-poor seeds lay on the floor of most African, Central or South America forests, none would escape predation. It was shown on page 192 that there are fewer species of animals in the Sunda Region than in Central and South America and it is the subjective view that animal biomass is probably also less (Janzen 1974a, 1977a). This is due, at least in part, to:

- the long gaps between periods of fruiting which reduce the amount of regularly available food needed to sustain denser populations; and
- too much of the forest floor being occupied by immature dipterocarp seedlings and saplings which 'steal' resources from species which flower and fruit (produce animal food) at comparable heights.

If it is presumed that a fruiting interval for dipterocarps of only two years is sufficient to avoid most of the insects that might specialise on dipterocarp seeds (Medway 1972) (p. 225), and to reduce the populations of vertebrate seed predators, it is necessary to ask why they wait so much longer than two years. It is probably because the longer interval between fruiting seasons enables more products of photosynthesis to be stored, and thus more fruit to be produced. The longer the interval, the smaller is the percentage of the total seed crop consumed or lost. It appears that intervals of six years are common but the interval appears to be longer on poorer soil (Janzen 1974a).

It is interesting that although fruiting time is synchronised between species, flowering time is not. This disparity is probably because pollinating insects are shared between species. However, fruit of late-flowering trees develop quickly and fruit of early-flowering trees develop more slowly. It will have been noticed by those accustomed to working in forests, that mast fruiting does not operate perfectly and trees sometimes make 'mistakes' (McClure 1966; Medway 1972; Wood 1956). Janzen (1974a) suggests that since flowering appears to be triggered by climatic events (probably occasional, severe droughts) within a generally very uniform climate, minor localised droughts may sometimes cause flowering in a small population of a species. Trees sometimes deal with such mistakes by absorbing flower buds or flowers, or by producing sterile fruit (Medway 1972; Wood 1956). None of the seeds produced in fertile fruit by a single tree or small population are likely to survive because the crop would be insufficient to satiate all the potential seed predators (Wood 1956).

Effects of Flower, Fruit and Leaf Production Cycles on Animals

Since many of the trees in lowland forest are pollinated by insects, it is not surprising that peaks in abundance of certain insect species coincide with peaks of flower production. These peaks occur at the drier times of year and usually coincide with, or come slightly after, peaks of leaf production which are exploited by butterfly and moth caterpillars. Other insect species, particularly those associated with rotting wood, or those dependent on

All photos in this section by Tony Whitten

The Mentawai snub-nosed monkey *Simias concolor* is peculiar in having both single-female and multi-female groups.

The Mentawai "joja" *Presbytis potenziani* is the world's only leaf monkey found only in single-female groups.

This striking frog *Rana siberu* is known only from Siberut Island.

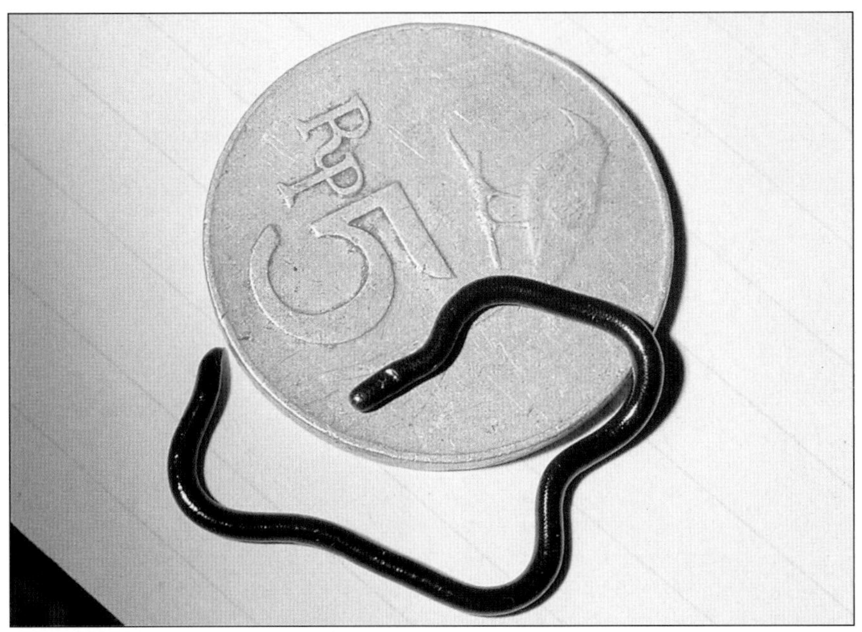

The soil dwelling burrowing snake *Ramnotyphlops braminus* is quite common but rarely seen. It grows to just 15 cm, and only females are known.

Pythons *Python reticulatus* are quite common and are avidly sought by farmers to reduce predation on their small livestock and to sell to skin traders.

Distribution of four species and twenty one subspecies of leaf monkey to illustrate the zoo-geographic zones.

Based on Chasen (12), Chasen and Kloss (13), Medway (75), Miller (85), Pocock (135), Whitten (125) and Wilson and Wilson (127, 128). From top left: *Ptn - Prebytis thomasi nubilis; Ptt - P.t. thomasi; Pmm - P. melalophos margae; Pfpa - P. femoralis paenulata; Pfpe - P.f. percura; Pfn - P.f. natuna; Pfr - P.f. rhionis; Pm? - P. melalophos (subspecies not yet named); Pfc - P. femoralis canus; Pm? - P. melalophos (subspecies not yet named); Pmfu1 - P.m. fuscomurina (lowland form); Pmfu2 - P.m. fuscomurina (highland form); Pmfl - P.m. fluviatilis; Pmme - P.m. melalophos; Pmm - P.m. nobilis; Ppp - P. potenziani; Pps - P.p. siberu; Pmf - P. melalophos; Pmm - P.m. nobilis; Ppp - P. potenziani; Pps - P.p. siberu; Pmf - P. melalophos ferruginea; Pmb - P.m. batuana; Pma - P.m. aurata; Pms - P.m. sumatrana.* The only leaf monkey that lives in the stippled area is the silvered leaf monkey *Presbytis cristata*, which is also common throughout the rest of Sumatra in coastal and riverine areas.

A megalith from Pasemah showing a man wrestling with an elephant.

One of the Hindu-Buddhist temples near Padang Lawas, South Tapanuli.

A long Islamic grave of an early Indian nobleman at Barus, Central Tapanuli.

Pneumatophore roots radiating from *Sonneratia alba*; near Bubun, Langkat.

The yellow-ringed cat snake *Boiga dendrophila* is quite common in mangrove forest. Its venom has little effect on people.

One of the *Ophiusa serva* caterpillars which defoliated areas of *Excoecaria* mangrove trees near Medan in 1983.

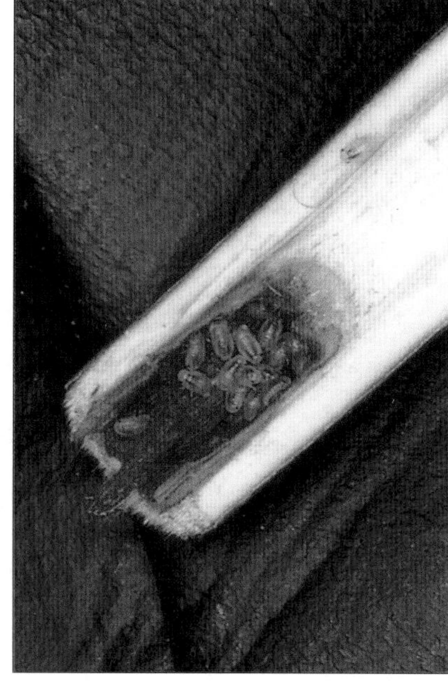

Isopod crustaceans in a *Rhizophora* root tip.

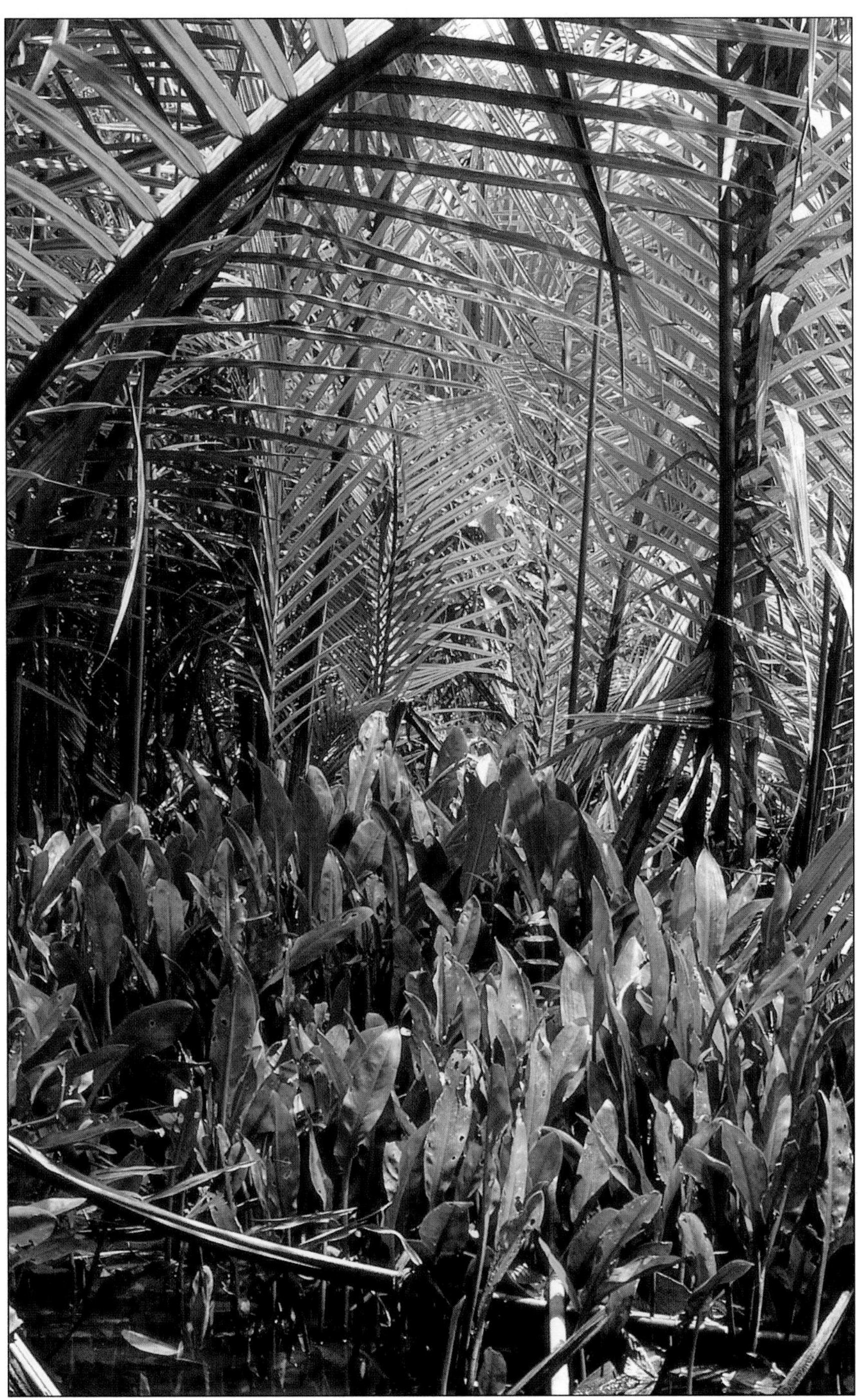

Nypa palm swamp near Sicanggang.

Typical pes-caprae vegetation, north of Padang.

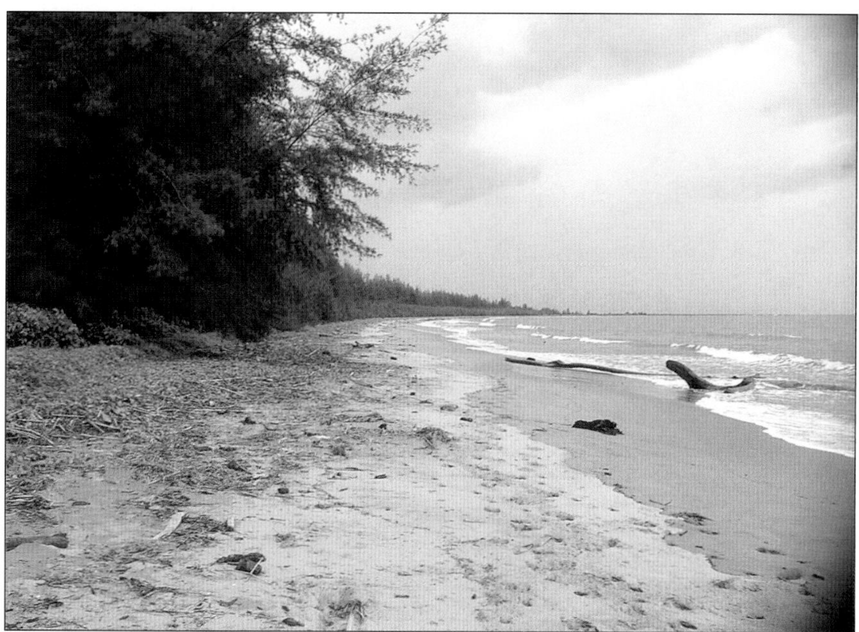

Casuarina forest near Singkil.

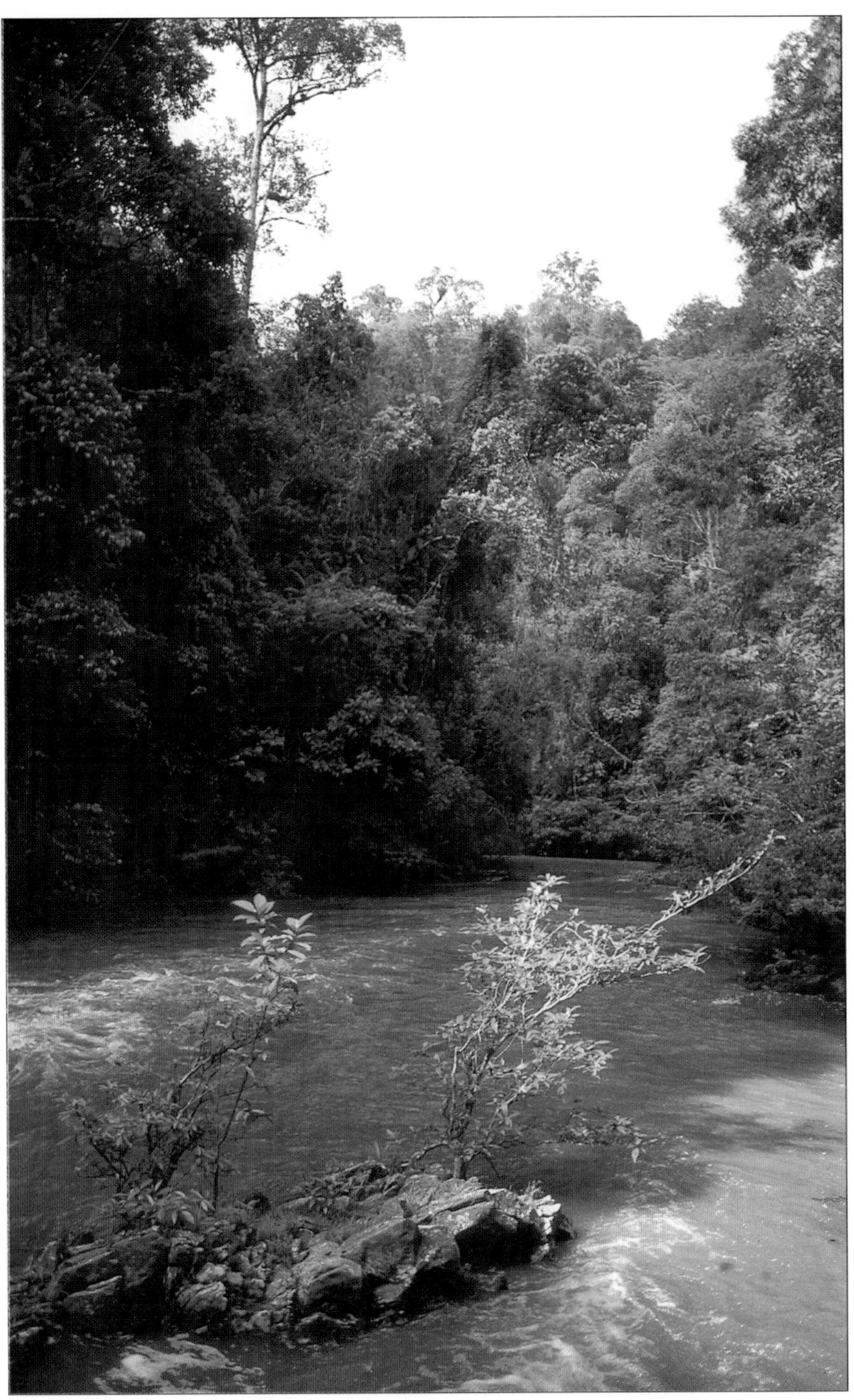

A river in Mt. Leuser National Park. In such forested rivers there is a high diversity of fish and other aquatic organisms.

Lake Dusun Besar, Bengkulu, is fringed by lance-like *Hanguana* plants.

Shallow, fast-moving streams such as this have specially adapted animals and plants.

Aerial view of a blackwater lake in peatswamp forest on Padang Island, Riau.

Stinkhorn fungus *Dictyophora indusiata.*

Agamid lizard *Gonocephalus grandis.*

The fish-eating false ghavial *Tomistomus schlegeli* is becoming increasingly rare as its swamp forest homes are diminished by fire and clearing.

A view from Mt. Kerinci to Lake Tujuh across the highest freshwater swamp in Sumatra.

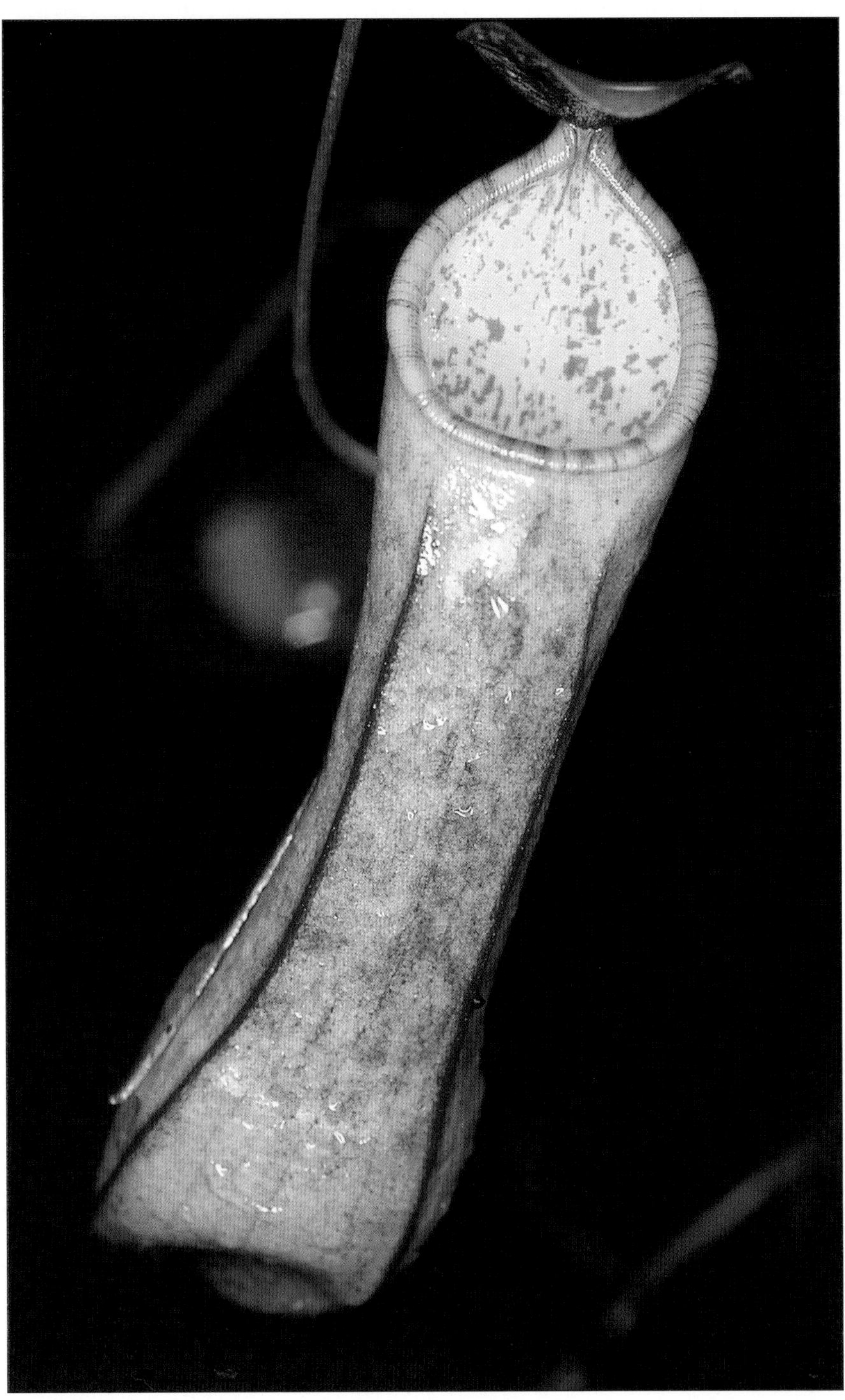

On Bangka pitcher plants are common on the infertile sandy soils.

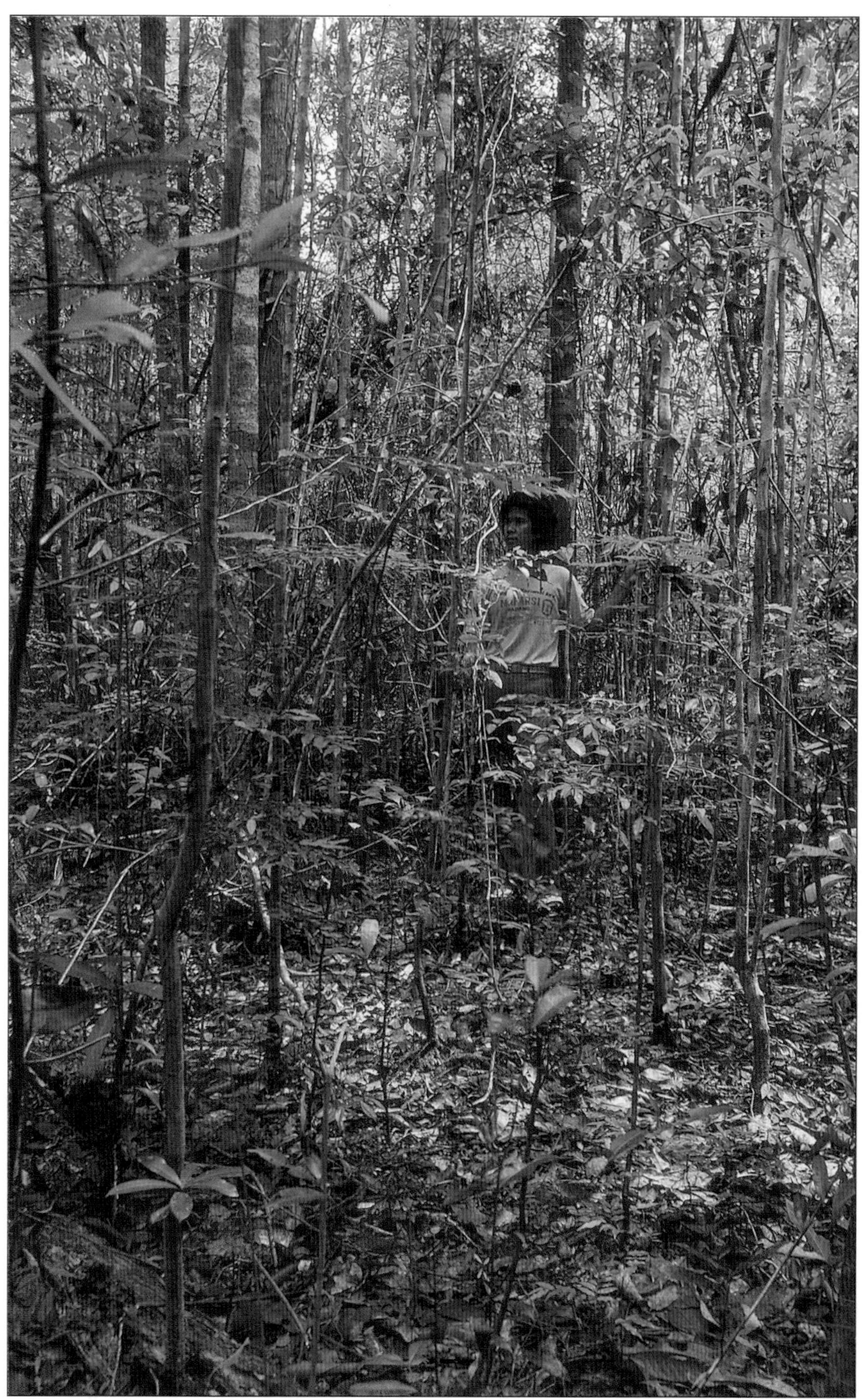

Typical small trees in a remnant heath forest, Bangka.

Padang forest, Bangka.

The convex leaves of *Dischidia* on Bangka shield colonies of biting ants.

Tin tailings on Bangka Island with dying heath forest at the back.

Insectivorous sundews *Drosera* in padang forest, Bangka.

Forested limestone hills, Payakumbuh

The world's largest flower *Raffesia arnoldii*, West Sumatra.

Scutigerids are relatives of millipedes and are found among leaf litter and in caves.

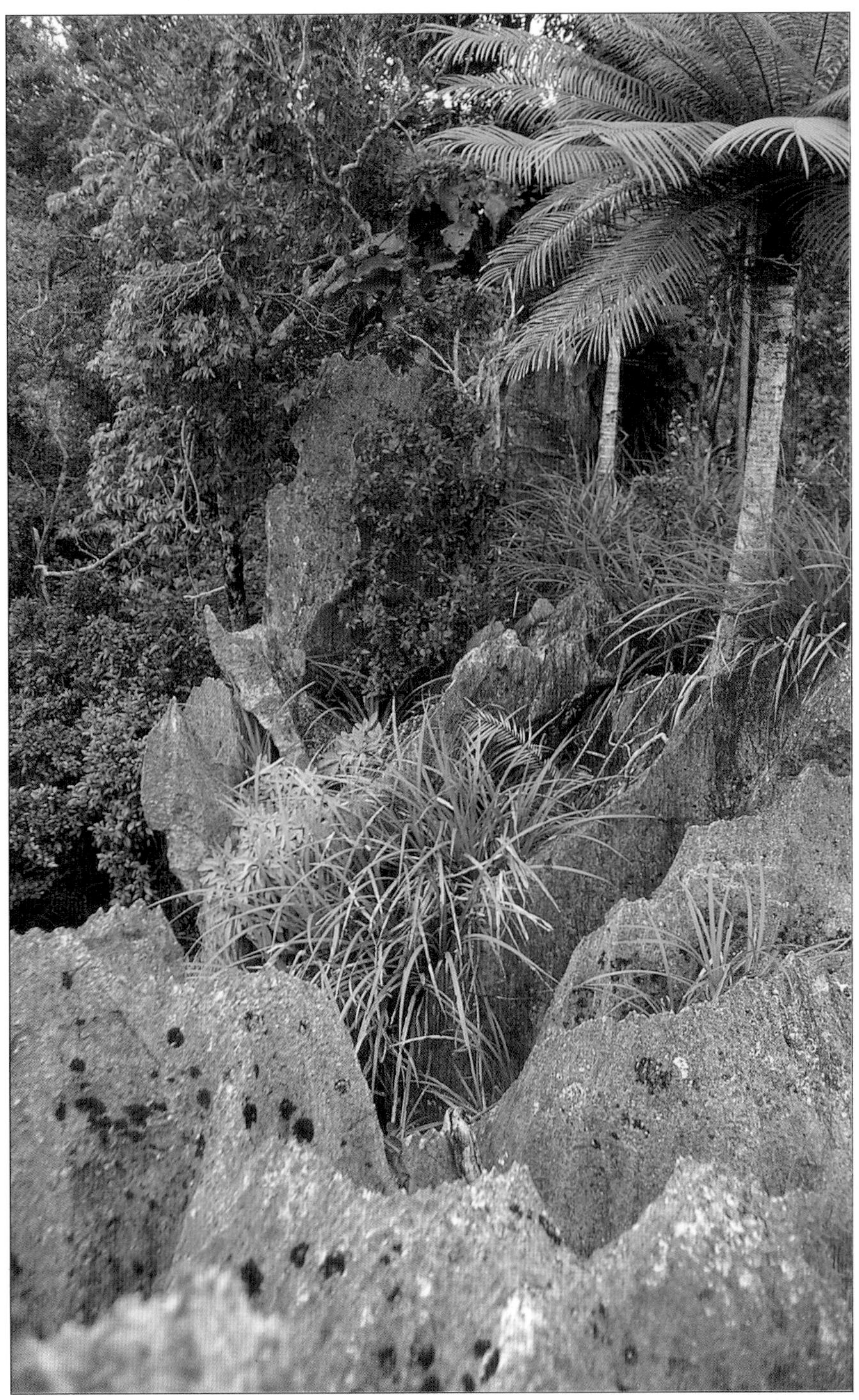

Limestone hills have characteristic plants adapted to long periods of drought, as well as many highly restricted species of snails, Lho'Nga.

The ironwood forests of Jambi are among the very few natural tropical forests which exist in virtually single-species stands.

The seeds of ironwood are large, produced continuously, and probably toxic.

Ironwood provides an excellent timber resistant to rot and insects.

The tree heather *Vaccinium varingaefolium* is found from about 2,000 m to the highest vegetation belts in Sumatra.

Upper montane forest, Mt. Kemiri, 2,700 m.

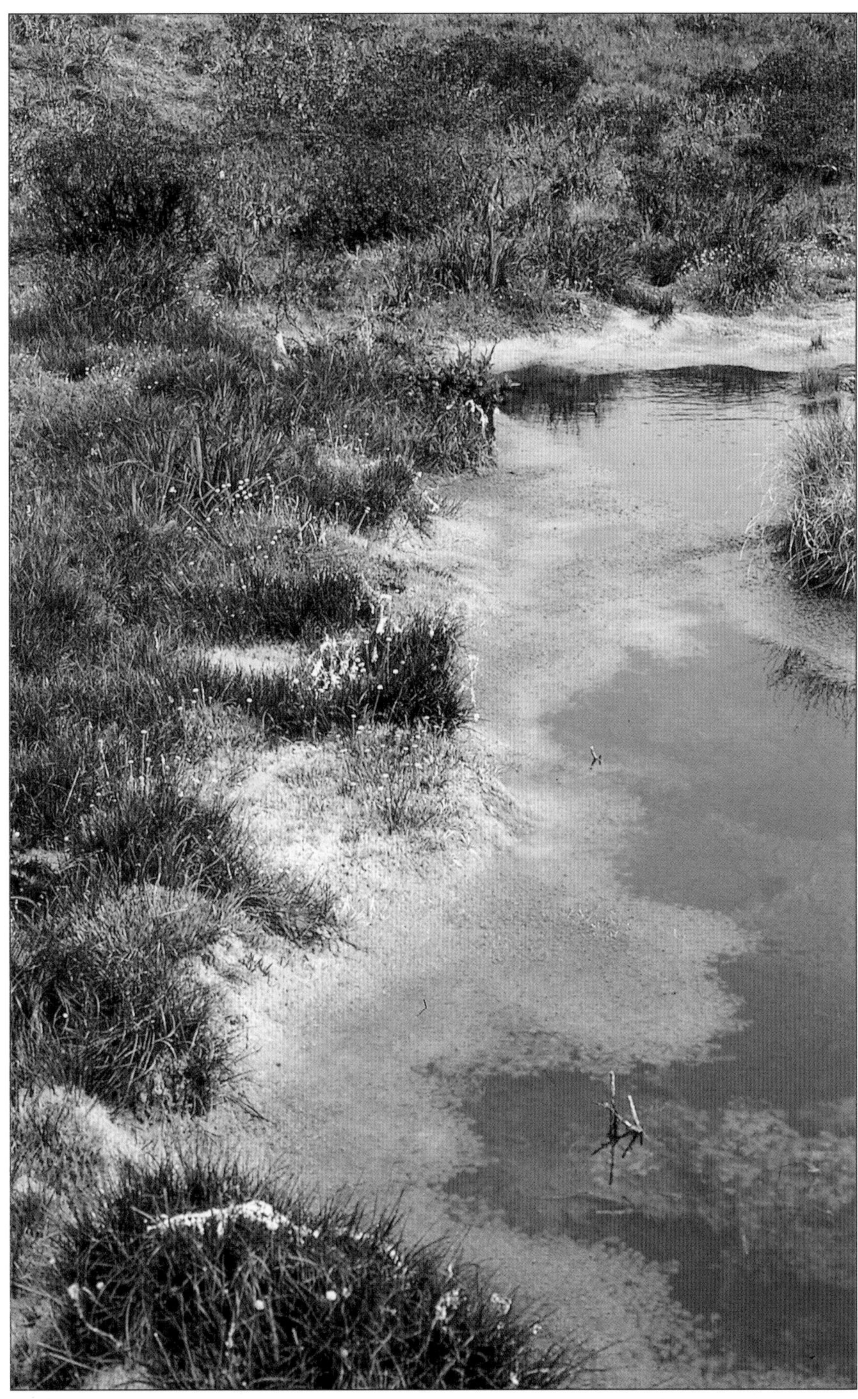

Sphagnum lake in a hollow formed by glacial action near the 3,200 m summit of Mt. Kemiri.

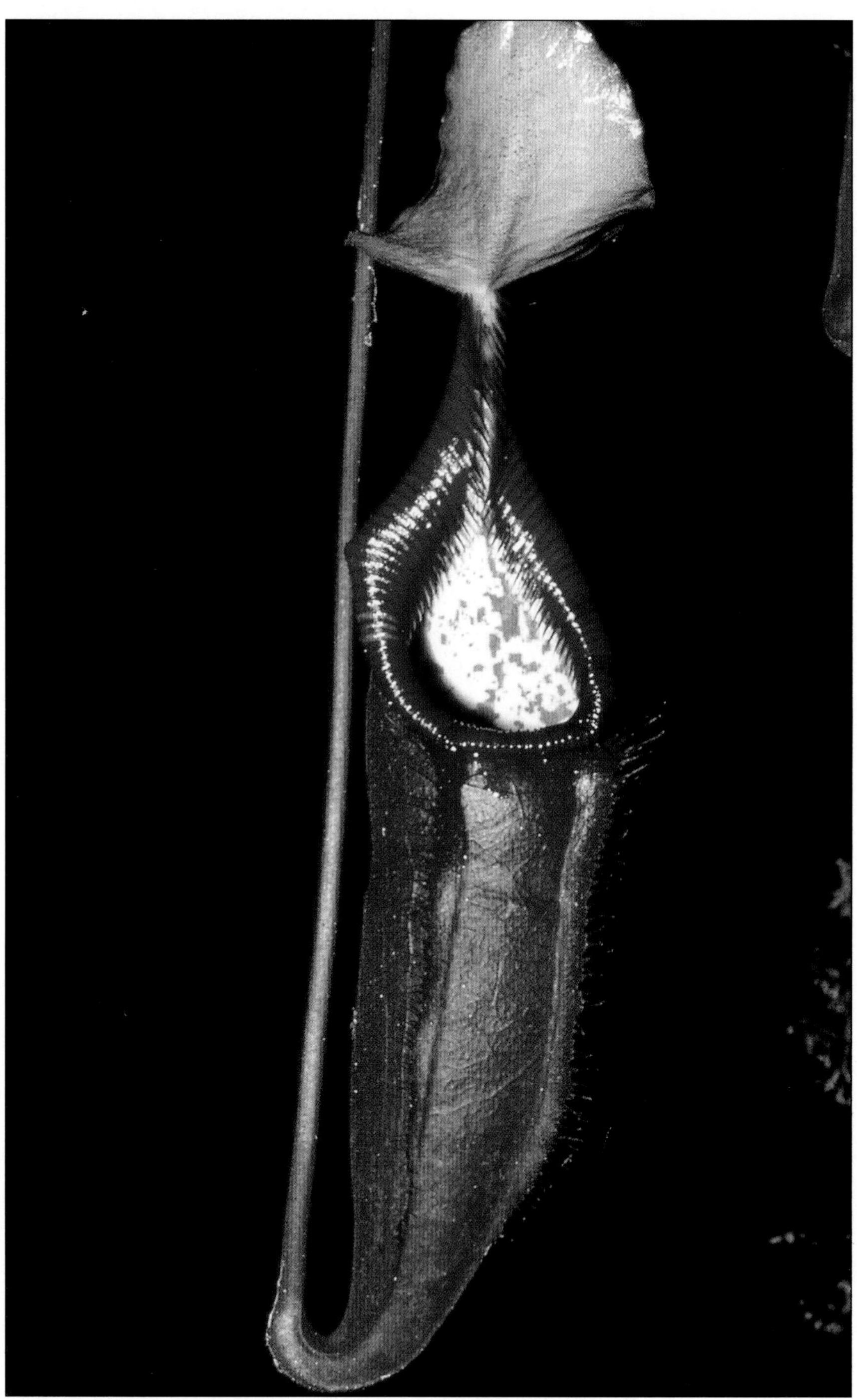

Nephenthes gymnamphora pitcher plant, 2,500 m, Mt. Kemiri.

Potentilla borneensis, 2,900 m, Mt. Kemiri.

The hairy undersides of some montane herbs, such as this *Potentilla,* may be a protection against high temperatures, intense ultraviolet radiation or frost.

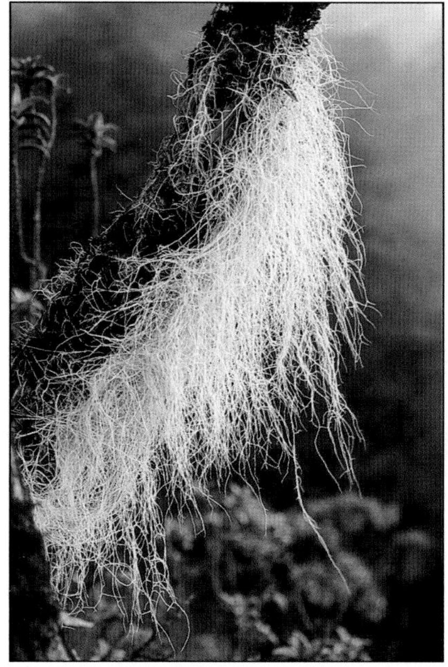

Usnea lichen growing in misty upper montane forest, Mt. Kemiri.

Lobelia sumatrana, 2,900 m, Mt. Kemiri.

Epigenium pulchellum orchids, 3,300 m, Mt. Kemiri.

Viola biflora, 3,250 m, Mt. Kemiri.

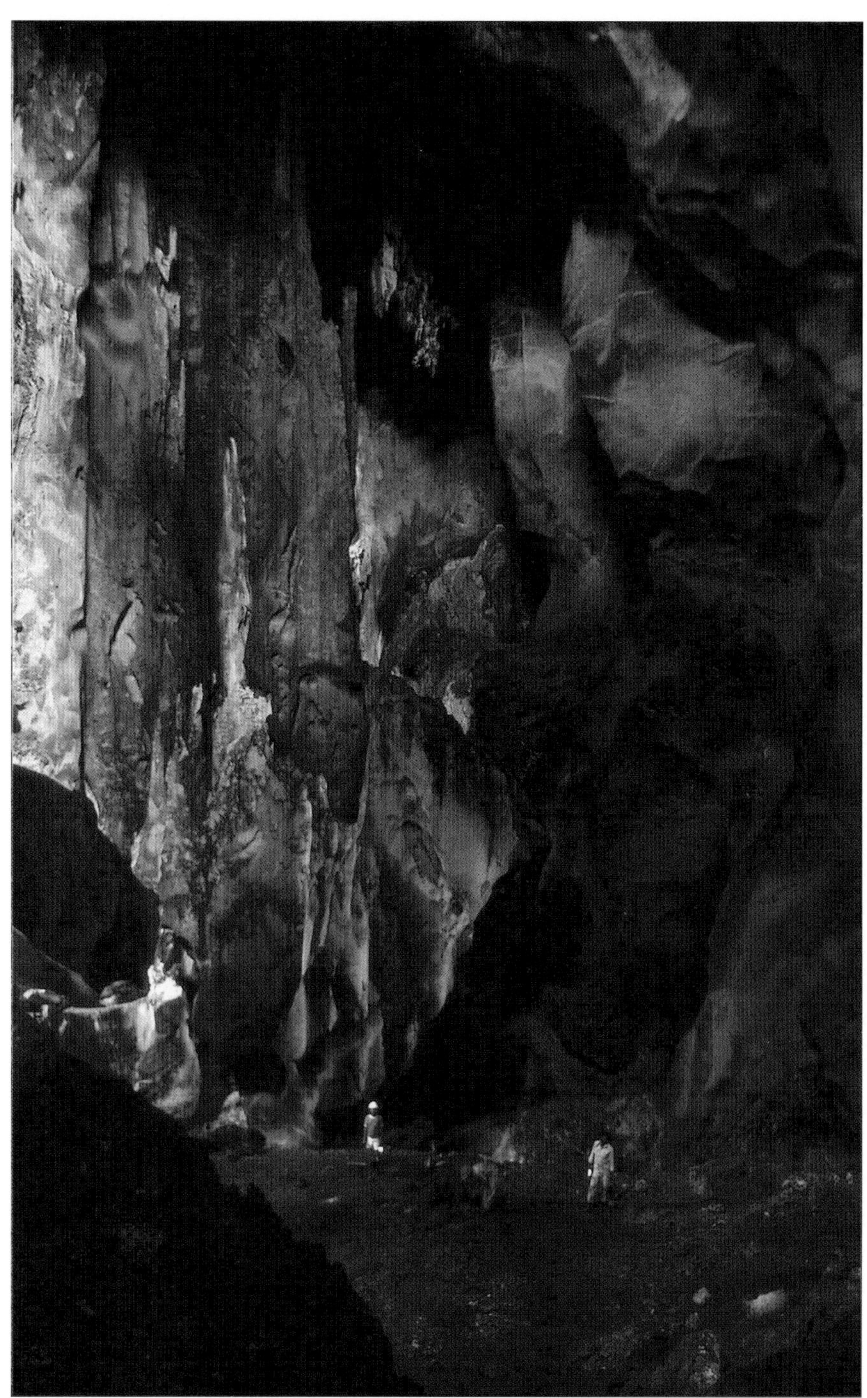

A large cave near Lho'Nga.

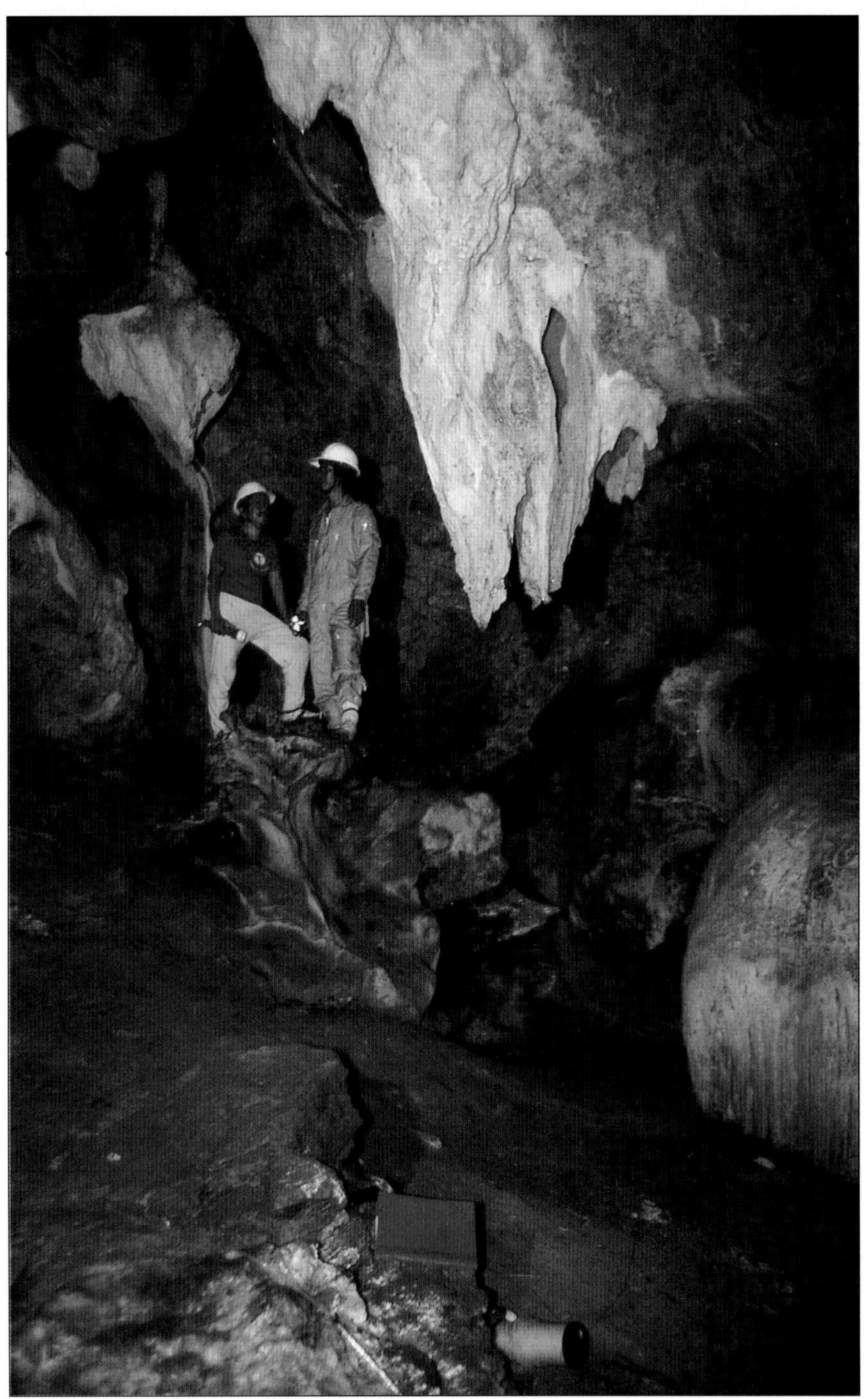

One of the authors (Jazanul Anwar, left) researching in a dry cave near Bohorok.

The cave fruit bat *Eonycteris spelaea* is a major pollinator of commercial fruit trees.

False vampire bats *Megaderma spasma* feed on lizards, frogs and insects picked off the ground and branches.

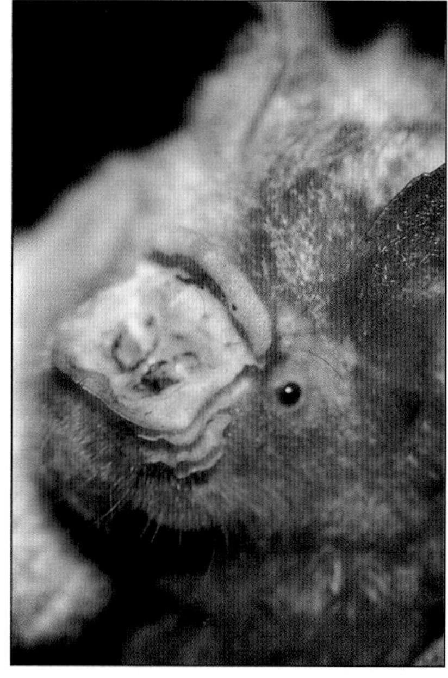

The roundleaf horseshoe bat *Hipposideros diadema* is one of the largest insectivorous bats living in caves.

Many saw mills waste a great deal of wood in saw dust and mis-sized planks.

Canals dug in peatswamp areas allow its water to drain away, drying the peat soil and making it more susceptible to fire.

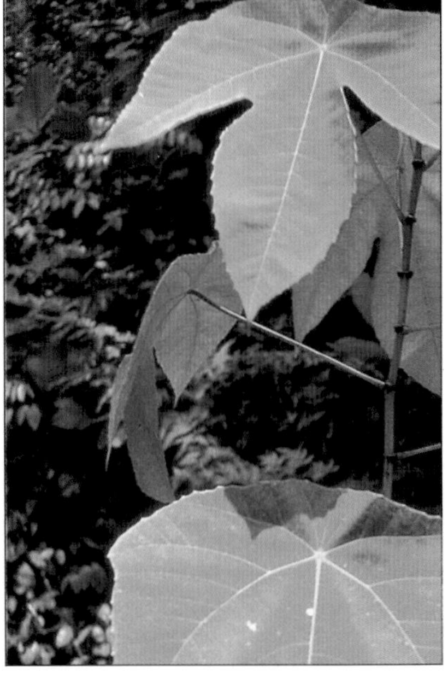

Species of *Macaranga* are among the most characteristic trees of secondary growth.

Sustainable swidden cultivation is practiced by very few in Sumatra today. The exceptions include groups of Talang Mamaq near Bukit Tigapuluh National Park.

Prints of beleaguered tigers can be seen around remnant areas of forest.

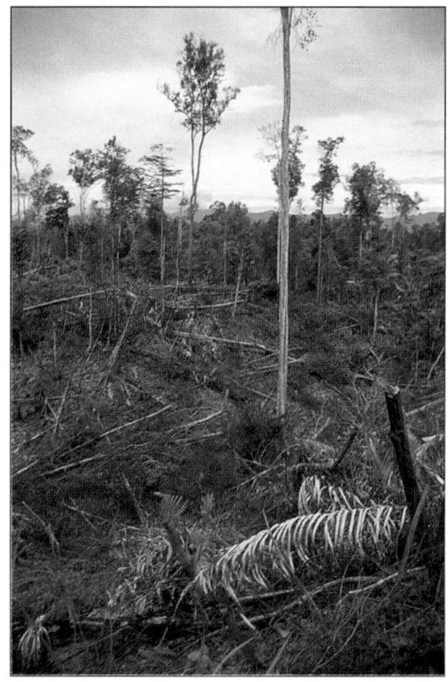

Legal logging begets illegal logging begets clearance for agriculture.

Rubber, both in commercial plantations and in rubber "forests" is a mainstay of many of Sumatra's rural population.

There has been a massive increase in the area under oil palm in the last few years, with a concomitant decrease in the area of lowland forest.

The tomb bat *Taphozous longimanus* is a common roof-roosting bat in Sumatra's cities.

Bronchocela cristatella is a common lizard in leafy gardens.

Flying foxes *Pteropus vampyrus* visit fruit trees such as this jambu to dring nectar.

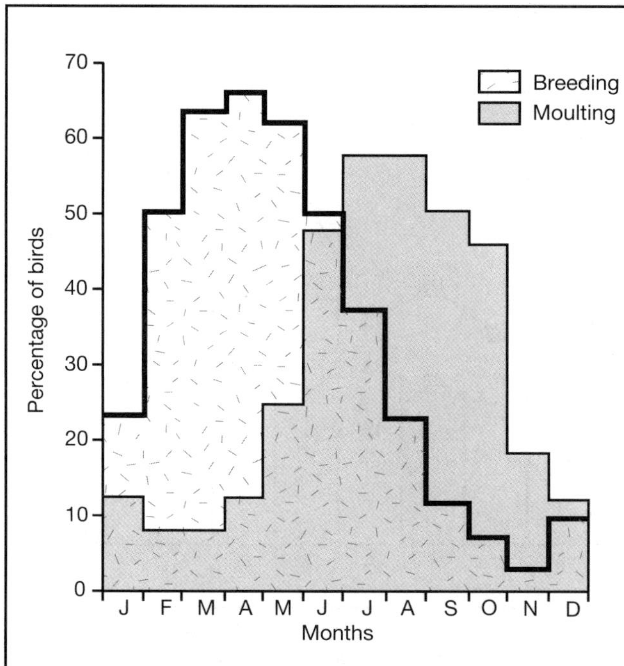

Figure 7.21. Percentage incidence of breeding and moulting birds of insectivorous of partially-insectivorous habit in the Malay Peninsula. The heavy line describes the percentage of breeding birds and the light line the percentage of moulting birds.
After Wells 1974

pools of water for breeding, become most abundant during the wetter months (McClure 1978; Hails 1982).

These variations in insect abundance are probably reflected in the behaviour of insectivorous or partially-insectivorous birds. The second category is included because, for birds that feed largely but not entirely on fruit, seeds or nectar, insects represent a protein-rich source of food essential for the energy-expensive tasks of feeding their young and for moulting. The insectivorous and partially-insectivorous birds of Peninsular Malaysia breed and moult all year round but there are major peaks in March-May and July-August, respectively (fig. 7.21). It is believed that these peaks depend on food availability (Wells 1974) and similar results have been found for insectivorous birds in Sarawak lowland forest (Fogden 1972), and insectivorous bats (Gould 1978b).

Frugivorous mammals and birds differ in their response to changing fruit abundance depending on the degree to which they are restricted to particular areas (p. 237). For example, animals such as bearded pigs, certain hornbills (p. 245), mynahs, broadbills, small parrots and, to some

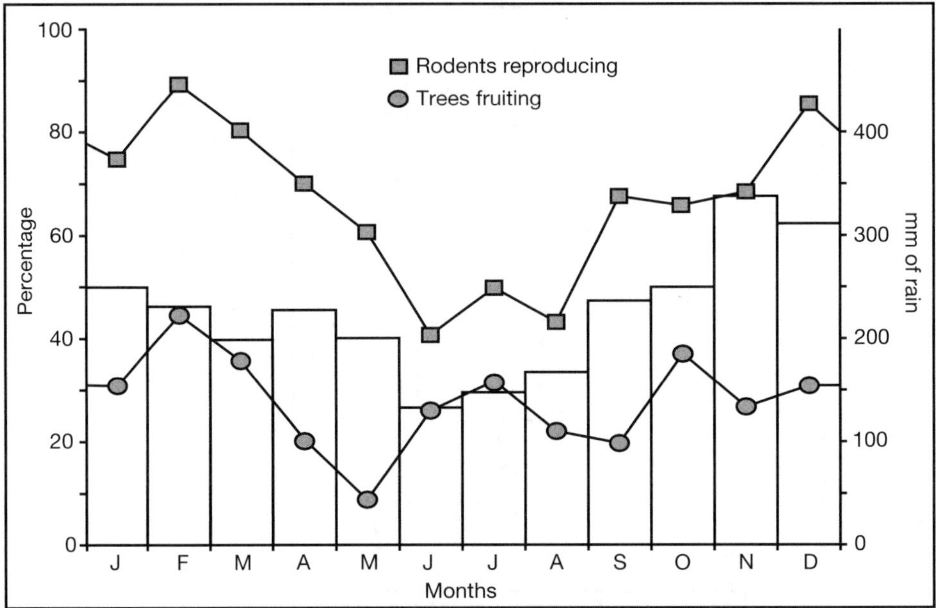

Figure 7.22. Relationship between rainfall (columns), reproductive activity as a percentage of 13 rodent species (upper graph), and percentage of trees in fruit (lower graph) in a lowland forest in Zaire. Note the similar shapes.

After Dieterlen 1982

extent, orangutans are free to move away from an area when food is scarce because they are either nomadic, migratory or have huge home ranges. Frugivores with fixed home ranges, however, such as squirrels, gibbons, some monkeys and certain hornbills, have to eat more non-fruit foods or take aseasonal fruit. Important among these aseasonal fruits are figs (Leighton and Leighton 1983).

Variations in fruit abundance have been shown to affect the behaviour of gibbons (Raemaekers 1979). During months when fruit was scarce the animals did not travel very far and stayed close to a few good fruit sources, but when fruit was abundant they travelled widely in their home range and ate a wide variety of foods. The peaks of pregnancy in Malaysian forest rats coincided with the seasonal peak in fruiting (Harrison 1955; Medway 1972), and this has been shown in more detail for rodents in a lowland forest in Zaire (Dieterlen 1982) (fig. 7.22).

Seed Dispersal

The dispersal of a seed is not simply the event of a mature fruit being released by its parent. Dispersal implies the carrying of a seed by some agent to a place where the seed will, perhaps, germinate, grow and reproduce. Most seeds die because the places they are dropped are unsuitable for growth; that is, they are not successfully dispersed. Seeds found directly below the parent tree are probably undispersed seeds, that is, they have simply dropped from the branches. Very few seeds and seedlings beneath their parent tree have any chance of survival because they will experience severe competition from others of their species. In addition, the parent tree and its offspring would represent a clumped food resource for seed predators. Thus dispersal of seeds can be envisaged as a means of escape from seed predators (Howe and Smallwood 1982) as well as a method of colonising widely separated available sites. In the species-rich lowland forest where, as shown on pages 189 and 198, most species are rare, a seed dispersed away from its parent will stand a relatively low chance of being adjacent to a tree of the same species.

A further advantage of dispersal away from the parent is that the biological and physical environment of the location where the parent germinated has in almost all cases changed by the time it has matured (Janzen 1975). In general, this means that the environment below the parent tree is now unsuitable for its own seeds. Gap, building and mature forest areas (p. 194) occur in more or less predictable proportions and patterns in a forest, and the means of seed dispersal will aim to maximise the colonisation of suitable areas with seeds.

The area over which seeds come to rest on the forest floor is known as a 'seed shadow'. The seed shadow of a tree is generally most dense close to the parent tree and much less dense overall when dispersed by an animal than when dispersed by wind. Whereas a wind-generated seed shadow is quite homogeneous, a seed shadow generated by animals will usually be heterogeneous. That is, animal-dispersed seeds will occur in dung piles, along animal paths, especially in certain vegetation types, etc. Exceptions are fig trees whose fruit is fed upon by many species of animals whose modes of living are so varied that a relatively homogeneous seed shadow can result. There is also variation in seed shadows for neighbouring individuals of the same species because access for the dispersing animals may be easier into one tree than into another, and because other attractive food sources may be close to one but not to the other (Janzen 1975; Janzen et al. 1976). These seed shadows are not simply of academic interest but are of major importance in determining future forest structure and composition (Flemming and Heithaus 1981).

Wind-dispersed fruits are generally borne by tall trees – the only forest trees to be affected by wind to any extent. Wind-dispersed fruits are either

Figure 7.23. Lowland forest on Siberut Island, West Sumatra. The most common family of trees is Dipterocarpaceae followed by Myristicaceae, Euphorbiaceae, and Sapotaceae.

A.J. Whitten

very light or have wings – in both cases their descent to the ground is delayed. Although wind does not disperse tree seeds particularly far from the parent, wind can at least be relied upon, whereas animals cannot.

The majority of forest tree seeds are dispersed by animals which eat the fruit and carry away the still-living seeds in their guts to be deposited some distance away when the animal defaecates. Seeds can be dispersed by animals in at least six different ways and the fruit can be classified accordingly (Payne 1979, 1980). They are:

- bird fruits (divisible into opportunistic fruit, seed-eating bird fruits, specialist bird fruits and mimetic fruit),
- opportunistic fruits,
- arboreal mammal fruits,
- bat fruits,
- terrestrial rodent fruits, and
- large mammal fruits.

Thus fruits can usually be identified as being adapted for dispersal by a particular group of animals. It is worth considering what effect hunting and forest reduction has had on those species of tree whose fruits are adapted for dispersal by large mammals such as Sumatran rhino (van Strien 1974) and elephants (Olivier 1978a).

The dispersal agent for a particular fruit depends on the nutritional requirements of the animals, the accessibility of the fruit, its shape and its size. The fruits produced by a tree vary in size and, in some species, by the number of seeds they contain. Thus different portions of the fruit crop may be dispersed by different animals over different distances (Howe and Smallwood 1982; Howe and Vande Kerckhove 1981; Janzen 1982a).

It is obviously to a plant's advantage to ensure that its fruit are dispersed to the right type of location, and numerous means are used to selectively advertise the presence of fruit and to encourage certain dispersers while discouraging others. These characteristics have been shaped over evolutionary time by reciprocal interactions between animals and plants but, unlike the process of pollination, few species-specific relationships exist (Janzen 1983a; Wheelwright and Orians 1982). Edible fruits are basically seeds covered with a bait of food, and Janzen (1982b) has hypothesised that grasses and other herbs also encourage active dispersion by surrounding dry, small seeds with a bait of nutritious foliage. When the foliage is eaten, the seeds are, too. The fruit bait can vary greatly in nutritional value and plants tend to adopt one of two strategies. They can produce large numbers of 'cheap' fruit (small, sugary), most of which may be wasted or killed by the many species of frugivores attracted to them. Alternatively, a plant can produce smaller numbers of 'expensive' fruit (large, lipid-rich, oily) that have to be searched for by those few specialised species of frugivores which gain a balanced diet almost entirely from fruit (Howe 1980, 1981; Howe and Smallwood 1982; Howe and Vande Kerckhove 1980).

Fruits advertise their ripeness to dispersers using colour, texture, taste, conspicuous shapes, and odour. However, between ripening and being taken by a disperser, the fruit is also exposed to destructive animals and microorganisms.

The destructive animals may be termed 'seed predators'. It should be noted that a disperser animal can become a predator if it eats the fruit before the seeds are mature. Thus a fruit needs to be as unattractive as possible before the seed is viable and this is commonly achieved by the presence of defence compounds whose concentration is reduced as ripening progresses. For example, on Siberut Island the endemic gibbons eat the fruit of wild sugar palms *Arenga obtfusifolia*. These fruits contain water-soluble sodium oxalate which, if it comes into contact with mucous membranes (such as lips, mouth and throat), absorbs calcium and becomes water-insoluble needle-sharp crystals of calcium oxalate. The pain and the loss of calcium can kill. As the fruit ripens, however, the sodium oxalate is broken down and when fully ripe the fruit can be eaten safely (Whitten 1980a).

Some fruit do not lose all their defence compounds when they are ripe and this is thought to be a means of protecting fruit from being eaten or destroyed by non-disperser animals or microorganisms (Herrera 1982; Janzen 1983b). Defence compounds can be divided into two groups: toxins which can debilitate or kill, and quantitative defences or digestion inhibitors such as resins, gums, volatile oils and phenols which deter potential seed predators or cause some mildly adverse physiological effect (such as feeling sick) (Maiorana 1979; Waterman 1983; Waterman and Choo 1981). A review of these defended ripe fruits concluded that the species producing them were at a competitive disadvantage for dispersal relative to species without defended ripe fruit. If the defences are viewed as means by which undesirable seed predators can be avoided, then extreme toxicity may be a 'last resort' (Herrera 1982).

Orangutans are probably the sole fruit disperser for some plants amongst which is the climber *Strychnos ignatii* (Rijksen 1978). The large fruit contains the 'deadly' alkaloid strychnine but it appears to have no effect on orangutans except for an excessive production of saliva. Indeed, strychnine appears to be dangerous only for carnivorous animals and for omnivores such as humans (Janzen 1978a). Orangutans also eat the fruits of the ipoh tree *Antiaris toxicaria* which contain another poisonous alkaloid (Rijksen 1978). Ipoh latex is used by some peoples in the preparation of poisoned arrows and darts but it is actually not as deadly as popularly believed (Burkill 1966; Corner 1952). It may be that these plants use their toxins to deter the 'wrong' dispersers.

Seed predators are commonly larvae of small beetles and these show a high level of specificity. Of over 975 species of shrubs and trees in an area of Costa Rica, at least 100 species regularly had beetle larvae in their mature or nearly-mature seeds. Three-quarters of the 110 beetle species found

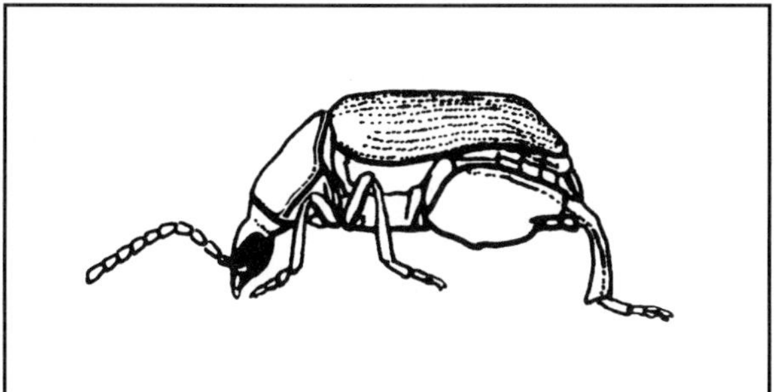

Figure 7.24. A bruchid beetle.

(primarily from the family Bruchidae) were confined to a single species of plant; if a beetle species was found on more than one plant species, the plants were closely related. Of the 100 species of plants, 63 were legumes and 11 were Convolvulaceae (Janzen 1980). This 'preference' for legumes might be because most plant families have had strong defences against attacks from bruchid beetles (fig. 7.24), whereas a few families have not. It seems more likely, however, that bruchid beetles became legume seed predators many millions of years ago and were able to counter whatever forms of defence the plant used. As the legumes evolved and diversified so the beetles evolved – a case of coevolution. Since plant defences appear to be so complex, the specialisations required to predate on a particular species or genus are such that a beetle species would be unlikely to be able to also become a predator on seeds in another plant family.

In general, the larger the seed the longer the time between ingestion and excretion. In experiments with a tapir, it was found that a considerable proportion of large, unchewed seeds were killed in the animal's gut because they germinated before being excreted and thus were acted upon by digestive juices (Janzen 1981b).

Destructive microorganisms (fungi and bacterial) represent a form of competition for frugivores. Fungi are not generally envisaged in this role but Janzen (1977b, 1979a) has suggested that this interpretation is valid. To a microbe, a ripe fruit is a considerable food resource which can be digested and converted into more microbes. It is obviously preferable, having started, to exploit the whole of the fruit and so competitors have to be dissuaded from using it as food. The microbe produces antibiotic substances to prevent other microorganisms from colonising the fruit, and toxic and unpleasant-tasting substances to deter larger frugivores. Producing tasteless toxins alone would not be a successful strategy because an animal may still eat the fruit, kill the microbe and later die itself. In that case neither party wins. Fruits invaded by microbes have to look, smell and taste different from untainted fruit or else the microbes have lost the battle.

Some fruits are known to produce their own antibiotics against decomposer microorganisms (Janzen 1978a) but this field of study has received little attention.

Leaves and Bark as Sources of Food

Leaves

About 50% of at leaf consists of cellulose, a complex sugar molecule which makes up the outer cell walls. Most animals lack the necessary enzymes to break down the cellulose into easily digestible, volatile fatty acids but some employ certain bacteria (or fungi – p. 233) to conduct this first stage of digestion on their behalf. The possession of these bacteria allows access to a widely distributed and abundant food source. The bacteria themselves form an important source of protein for the host and can synthesise most vitamins except A and D. In vertebrates there are two major forms of bacteria-assisted digestion: foregut and hindgut fermentation. Foregut fermentation is found in, for example, cattle, deer and leaf monkeys, and hindgut fermentation in, for example, rabbits, horses, koalas, rodents and flying lemurs (Bauchop 1978; Muul and Lim 1978). Leaves are eaten by a wide range of mammals and larval and adult insects, but not by amphibians and by very few reptiles or birds (Morton 1978; Rand 1978). It is estimated that 7%-12.5% of leaf production is eaten by insects (Leigh 1975; Wint 1983) and only 2.4% by vertebrates (Leigh 1975). However, as Janzen (1978a) points out, these figures are usually calculated from the remains of fallen leaves and so do not necessarily include leaves or shoots eaten in their entirety.

The actual amount of leaf material eaten is not the most important ecological measurement because different parts have different values to the plant and hence different 'costs' for replacement. For example, the loss of a small shoot from a seedling is considerably more costly to the individual plant than the loss of a great many leaves on a mature plant. Leaves pay back their cost by photosynthesising and contributing the products of photosynthesis to the plant. So, once a leaf has paid back its 'debt', its value to the plant decreases with time. It is clear, then, that for advanced ecological analysis of an ecosystem, a list of plant species eaten by a particular animal species is not nearly so useful as knowledge of what parts of each species are eaten and the likely cost to individual plants of the material lost.

To the eyes of a human observer the leaves of a lowland forest, or indeed other tropical vegetation types, vary in shape and shade of green. To an animal which depends on leaves as a food source, however, they are 'coloured' "nicotine, tannin, lectin, strychnine, cannabinol, sterculic acid, cannavanine, lignin, etc., and every bite contains a horrible mix of these" (Janzen 1981a). These phenolic and alkaloid compounds (Walker 1975) are just some of the toxic and digestion-inhibiting chemicals used by plants to defend their leaves against herbivores (Edwards and Wratten 1980). Some of the ecological consequences of these chemicals are described on pages 177 and 259. See also page 213 for a discussion of the defence strategies of plants growing in gaps.

A recent laboratory study in North America has added a further dimension to the study of defence compounds. The response of tree seedlings was monitored before and after 7% of their leaf area was removed. Within two days the seedlings had increased the concentration of phenolic compounds in their remaining leaves – perhaps as a defence against further loss of leaves. Strangely, however, nearby undamaged seedlings of the same species behaved similarly. This suggests that the damaged plants were able to communicate a warning to the other plants. How this is effected is as yet unknown (Baldwin and Schultz 1983).

Some animals may be able to eat leaves because they have coevolved with the plant – that is, changes in a plant's chemical defences have been met by changes in the animal's digestive capability (Edwards and Wratten 1980; Wratten et al. 1981). It could be, however, that an animal has evolved the ability to cope with the defences with no corresponding innovation in the plant's chemistry. Conversely, a plant may produce defences which cause the partial 'defeat' of the herbivore with no response by the animal.

There is probably no plant whose defences preclude attack by all herbivores, and there is certainly no animal which can eat all types of leaves. A particular moth caterpillar in Costa Rica is able to eat leaves containing 40% by dry weight of phenolic compounds. This caterpillar eats almost only phenol-rich leaves from nearly 20 species, but it grows at only half the rate of moth caterpillars eating phenol-poor leaves from just a few species of plants (Janzen 1983c). Similar results were obtained in experiments with locusts (Bernays 1978; Bernays and Chamberlain 1980). When eating a leaf rich in defence compounds, a herbivore gains less usable food per unit effort (per mouthful) than when eating a leaf low in defence compounds, but it probably has more species to use as food and experiences relatively less competition from other herbivores unable to cope with the chemicals.

Experiments on the selection of leaves by a captive South American tapir showed that out of 381 species offered, only 55% were eaten, and only one of the 55 leguminous species was accepted. The tapir of the Sunda Region is also a very selective feeder in the wild, eating from saplings of only a fraction of the available tree species (Medway 1974; Williams and Petrides 1980).

Monkeys are also very selective in the plants they eat, choosing particular species, particular ages and parts of leaves, and quantities. Recent studies of food selection by leaf monkeys in Peninsular Malaysia and Sabah have shown that leaves selected as food are consistently low in lignin (fibre), and usually relatively high in protein; that is, the leaves are highly digestible. Tannin levels did not show consistent significant correlations with food selection (Bennett 1984; Davies 1984). Similar results have been obtained in Africa and India (Oats et al. 1977, 1980) and it is now generally felt that the role of tannins in food choice has been overemphasised (Waterman 1983). They do, however, probably remain a potent force in ecosystems where poor soil conditions cause some form of stress in the trees growing on them.

Tannins and lignins are classed as digestion-inhibiting phenolic compounds (although lignin is 10-20 times less effective than tannins), whereas others, such as the alkaloid strychnine, are classed as toxins. The term 'toxic' is really a measure of the energy expended by an animal in utilising a food relative to the energy value of usable materials gained from the digestion process. Toxicity is therefore an outcome rather than an inherent property of a chemical, although some chemicals are more likely to be toxic than others. Whether a certain percentage composition of a potentially dangerous chemical is toxic or not will depend on the food value of the material eaten and on the efficiency of the animal's detoxification system. Thus, a leaf may be more toxic than a seed with the same weight and concentration of defence compounds, because the seed has a greater nutritive value (Freeland and Janzen 1974; Janzen 1978a; McKey 1978). Defence compounds should not be thought of as purely disadvantageous from the herbivore's point of view because they can have decidedly beneficial effects, as described in the next section.

Bark

Bark is, perhaps, an unlikely food but it has been recorded as a constituent of the diet of orangutans (Rijksen 1978), gibbons (Whitten 1980b), elephants (Olivier 1978a), Sumatran rhino (van Strein 1974) and squirrels (Payne 1979; Whitten 1979, 1980). Bark is not a rich food source and it may be eaten as a supplementary food to provide dietary fibre or trace elements.

On Siberut Island the small squirrel *Sundasciurus lowii* eats bark as a major item of its diet (p. 244), and an analysis of the bark and bark-eating was conducted by J. Whitten (1979). Physical characteristics of trees (size, roughness of bark, ease of access for squirrels, etc.), food value and concentration of phenolic compounds in the bark of trees chosen as food sources were compared with similar data from trees not chosen as food sources. Trees on which the squirrels fed tended to be large, smooth-barked and relatively free of climbers, and certain tannins were absent in the barks eaten. These barks were not noticeably richer in food value than those which were not eaten.

Defence compounds in bark, and indeed other food items, need not necessarily be selected against by animals; indeed, Janzen (1978a) suggests that certain compounds may be actively sought for specific purposes. For example, elephants have been reported to feed extensively on a certain leguminous creeper before embarking on a long period of travel, and it may be that the plant contains some stimulating or sustaining chemical (Hubback 1941). The high levels of tannins in some barks may be eaten to combat heavy parasite loads or upset stomachs. For example, Asian buffalo eat a plant with an informative scientific name *Holarrhena antidysenterica* (Ogilvie 1929) and Sumatran rhino have been observed to eat so much of the tannin-rich bark of the mangrove tree *Ceriops tagal* that their urine is stained bright orange (Hubback 1939). It is worth noting, however, that the once-popular drug for diarrhoea 'Enterovioform' contains 50% (dry weight) tannin, a concentration not unknown in nature, but its role in combating diarrhoea has now been largely discredited.

Soil and Its Animals

The soil under the majority of lowland forest in Sumatra is red-yellow podzolic, and this has been described by Burnham (1975), Sepraptohardjo et al. (1979) and Young (1976).

The majority of soil fauna in the Sunda Region is known very poorly and the only quantitative studies seem to have been conducted in Peninsular Malaysia and Sarawak (Collins 1980, 1983). In 1 m^2 at Pasoh an average of 2,000 ants, 300 termites, 30 earthworms, 30 fly larvae, 50 spiders and 100 beetles and their larvae were found. The total biomass of these invertebrates was about 3 g/m^2 (3,000 kg/km^2) but this and the proportions of different groups varied greatly from month to month. Similar results with more termites were obtained in Sarawak where the biomass was 4-6 g/m^2 (Collins 1980). To compare these figures with vertebrate biomass see page 250. Among the smaller animals at Pasoh, over 10,000 mites (Acarina) and about 56,000 nematode worms were found per m^2. These totals are not particularly high and some temperate areas would have greater densities but species diversity in the tropics is almost undoubtedly greater.

Termites

The soil animals about which about most is known are termites. Termites may look superficially like ants but they are classified in a completely different order which is more closely related to cockroaches. Termites are Isopterans, ants (together with bees and wasps) are Hymenopterans (fig. 7.25). Termites are like ants, however, in that they form enormous colonies with (at least in some parts of the world) possibly a million or more members. A colony is not a simple community of distinct individuals because all except two of the colony members are siblings. The other two, the 'royal pair', are the parents. The parents originated in other termite nests from which they flew along with thousands of others. In these swarms some manage to escape the hordes of ants, amphibians, reptiles, birds and mammals for which such a swarm is a food bonanza. After landing, their wings drop off and if a male finds a female, they will look for a suitable crack in the ground or a tree (depending on the species) and here they build their 'royal cell'. They copulate and the female begins laying eggs. The larvae, unlike the helpless larvae of ants, bees and wasps, are fully able to move around and undergo several moults (like cockroaches) rather than a single metamorphosis before becoming adults. These first larvae have to be fed by their parents but as soon as they are large enough to forage for food and to build walls for the nest, the royal pair devote themselves entirely to the production of eggs. The abdomen of the female, or queen, grows to grotesque proportions and a production rate of thousands of eggs per day is common.

A termite colony can be likened to a single, although sometimes disparate, organism because none of its components is capable of independent life. The workers are blind and sterile, and the soldiers which protect the columns of foraging workers and guard entrances to the colony's nest have jaws so large that they can no longer feed themselves and have to be fed by the workers. The king and queen are also fed by the workers. In the same way that communication between different organs or parts of a single body is effected by chemicals flowing through a body's tissues, so chemicals (pheromones[2]) link the different members of a termite colony into a coordinated and organised whole. All of the colony members continually exchange food and saliva which contain pheromones. Workers pass these materials from mouth to mouth and also consume one another's faeces to reprocess whatever partially digested food remains. Workers feed the soldiers, larvae and royal pair as well as collect the queen's faeces. There is thus a continual interchange of chemicals through the colony and this coordinates the operations of the colony. For example, larvae are potentially fertile termites of either sex but the food with which the workers feed them contains quantities of pheromone from the queen which inhibits the development of larvae and produces workers. The soldiers also produce a pheromone which prevents the larvae from developing into soldiers. When, for some reason, the number of soldiers falls, the level of 'soldier repression pheromone' circulating through the colony members falls and some eggs develop into soldiers. On other occasions the queen will reduce her repressive pheromones and her eggs will develop into fertile winged termites which, when the air is moist, will leave the nest in a swarm and the cycle of colony formation begins again.

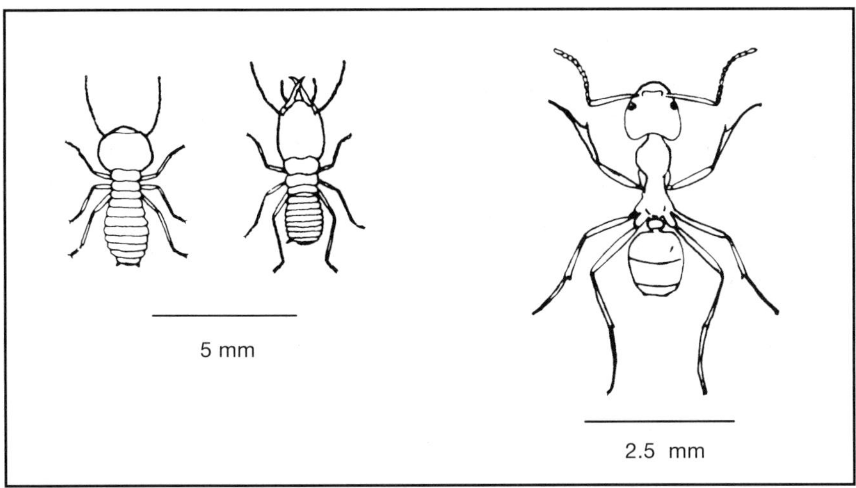

Figure 7.25. Typical termites (worker on left, soldier on right) and an ant.

Most invertebrate decomposers of the lowland forest soil depend on free-living saprophytic fungi and bacteria to break down indigestible plant material into a digestible form before they can play their role, and termites are no exception. One subfamily of termites, the Macrotermitinae, represented in Sumatra by *Macrotermes, Odontotermes* and *Microtermes,* maintain a fungus in 'gardens' in their nests which releases them from competing for partially decomposed material outside their nest. The workers of the above genera build complex frameworks, or combs, from their own round faecal pellets (not, as was once thought, from chewed food) – other groups of termites use faeces for building the fabric of their nests. (An identification key to genus level based on the head shape of soldiers has been written by Harris [1957]). Since the main food of the Macrotermitinae is wood and leaf litter, the combs look like thin, fragile pieces of moist, spongy, rotten wood. These combs are the food of a form of fungus called *Termitomyces* which is unknown outside termite nests. Small white dots, which are clumps of asexual spores, can sometimes be seen on the combs if a nest is cut open. It takes up to two months for the fungus to process the combs (Collins 1982).

This relationship with the fungus confers several advantages on the termites. No animal has the necessary digestive enzymes to digest lignin, a largely inert compound which is the main component of wood and thus one of the major items in a termite diet, and so lignin is excreted unchanged in the termite faeces. The fungus, however, can digest lignin.

The termites are also unable to digest cellulose – the main component of plant cell walls – but the fungus does produce the necessary enzymes for breaking down cellulose. When the termites eat the fungus-permeated comb, these cellulose-digesting enzymes persist in the termite gut, thereby improving the efficiency of its own digestion (Collins 1982). The fungus removes carbon from the comb during the digestion process and releases carbon dioxide as a result of its respiration. This has the advantage to the termite that the proportion of nitrogen in the combs increases roughly four-fold when processed by the fungus (Matsumoto 1976, 1978b). The tremendous contribution the fungus makes to Macrotermitinae nutrition has resulted in these termites being unable to live without *Termitomyces* – an example of obligate symbiosis. It has made them so successful that they have been reported to be responsible for the removal of 32% of leaf litter in Pasoh Forest Reserve in Peninsular Malaysia (Matsumoto 1978a). In the same forest Abe (1978, 1979) studied the role of termites in the decomposition of fallen trees. He found that the rate of wood removal was higher for small branches (3-6 cm diameter) than for trunk wood (30-50 cm diameter) which is harder and more difficult to cut and remove. In a species of meranti *Shorea parviflora*, 81% of the tree's dry weight of small branches was removed in the first 18 months after it fell compared with 18% of the trunk. Termites thus play a major role in the breakdown and cycling of plant material (Wood 1978).

There are at least 55 species of termite at Pasoh (Abe and Matsumoto 1978, 1979) and for population studies Matsumoto (1976, 1978b) chose the four most abundant species with conspicuous mound nests, part of which projected above the soil surface. Approximately 100 nests were found per ha for three of the species but there were only 15 large nests of *Macrotermes*. The population size per nest for *Macrotermes* (about 88,000) was, however, at least twice that for the other species. The number of nests built by other genera in trees averaged four per species per ha. Nests and the queen or queens they contain can survive for many years but some colonies may last less than a year (Abe and Matsumoto 1978, 1979). Colonies die for various reasons: because of extensive parasitism by small wasps, because the queen or king dies, or because the nest is destroyed by pangolins. Harrison (1962) examined the stomach of a pangolin *Manis javanica* and found approximately 200,000 ant workers and pupae within it. If this is taken as an average daily diet (probably an underestimate), then a single pangolin could eat 73,000,000 ants or termites per year.

Spacing of Vertebrate Animals

Different species organise the distribution and grouping of their members in a variety of different ways. It is beyond the scope of this book to cover the theoretical reasons behind different kinds of social structure but the major patterns found in Sumatran lowland forest animals are described below.

Social Systems

Adult male orangutans, slow loris, pangolins, shrews, tigers, bears, tapirs and Sumatran rhinos all live in more or less exclusive ranges within which single female/offspring groups live in somewhat smaller, overlapping ranges. The male attempts to maintain exclusive breeding rights over these females and the extent of his claimed area is communicated to other males by loud calls, regularly visited dung or urine sites or a combination of these two (Borner 1979; Lekagul and McNeely 1977; MacKinnon 1977; Rijksen 1978; van Strein 1974; Williams 1979). Male pigs also live a largely solitary existence but several females and their offspring form groups of 10 to 20 animals. When a female pig is ready to give birth she will go off on her own and remain alone until the piglets are old enough to run after her.

Mouse deer live singly for most of the year but form stable pairs for breeding (Davison 1980). Elephant males and females also live apart but the males will sometimes form bachelor groups and the females form herds with their young. Even when one or more males are attending the females and offspring, that group is still led by a matriarch (Olivier 1978a).

Wild forest dogs *Cuon alpinus* live in mixed-sex packs of 10 or more, led by a dominant male. They travel and hunt together and are quite capable of killing a large deer even though its greater size, antlers and speed would make it unlikely prey in a one-to-one contest.

Almost all monkeys live in single- or multi-adult-male groups with more adult females than adult males (polygynous). The exception, unique among the world's monkeys, is the joja leaf monkey *Presbytis potenziani* of the Mentawai Islands. It lives, like gibbons, siamang and tarsiers, in permanent, monogamous (one adult male and one adult female) groups within home ranges (the areas normally used by individuals or groups) containing territories which are defended against others of the same species (Aldrich-Blake 1980; Curtin 1980; Gittins 1980; Gittins and Raemaekers 1980; Niemitz 1979; Raemaekers 1979; Tilson and Tenaza 1976; Whitten 1982b, 1982d; Whitten and Whitten 1982; Wilson and Wilson 1973). The other Mentawai leaf monkey or simakobu *Simias concolor* lives in both monogamous and polygynous groups (Tilson 1977; Watanabe 1981; Whitten 1982b).

Most species of forest birds are more or less solitary for most of the year. Amongst these, the great argus pheasant *Argusianus argus* (fig. 7.26) is of

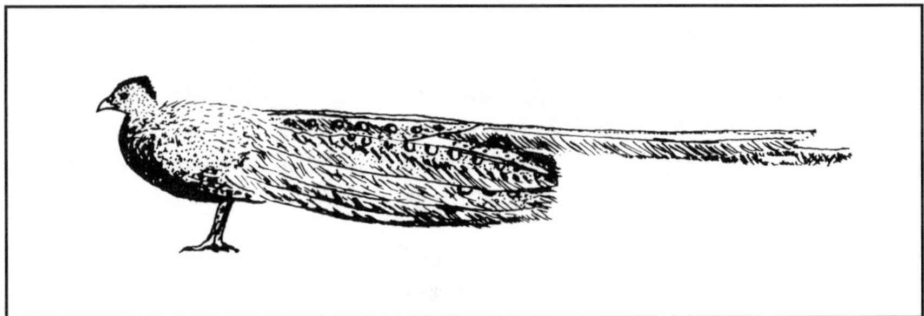

Figure 7.26. Great Argus Pheasant *Argusianus argus*.
After King et al. 1975

particular interest. The male scratches conspicuous, cleared 'dancing grounds' about 12 m^2 in area on the forest floor on the top of hills or knolls. He utters his well-known 'ki-au' call from the dancing grounds to attract females for breeding and when they arrive he performs a dramatic dance displaying his ornate, 1.5 m long tail. After mating he plays no further role in the development of the chicks (Davison 1981a).

Most pigeons (Columbidae), small parrots (*Loriculus* and *Psittinus*), ioras and leafbirds (*Aegithina* and *Chloropsis* – see p. 42), minivets (*Pericrocotus*), some hornbills[3] (Bucerotidae) and some other birds are gregarious and form flocks for most of the year but separate into pairs for breeding. Drongos (Dicruridae) and some hornbills spend most of their adult life in pairs and these tend to be territorial. Other hornbills are territorial, communal breeders living in more or less permanent flocks of up to 10 (Leighton 1982). In addition, mixed-species flocks are found (Croxall 1976; Leighton 1982). These have been interpreted as a means by which relatively specialised insect-eating birds can increase their food supply, particularly during times of food shortage (p. 222), because of the insects flushed by the activity of the other species in the flock. The different responses of insects to disturbance and the different specialisations of the species could ensure benefit for all the flock members (Croxall 1976).

Niches

The type of niche discussed below is the realised niche (p. 156), or that part of the total environment which a species actually exploits. A species' niche has been likened to its profession and this is a useful way to understand the term.

The enormous structural complexity and the high plant species diversity of lowland forest provides an uncountable number and often

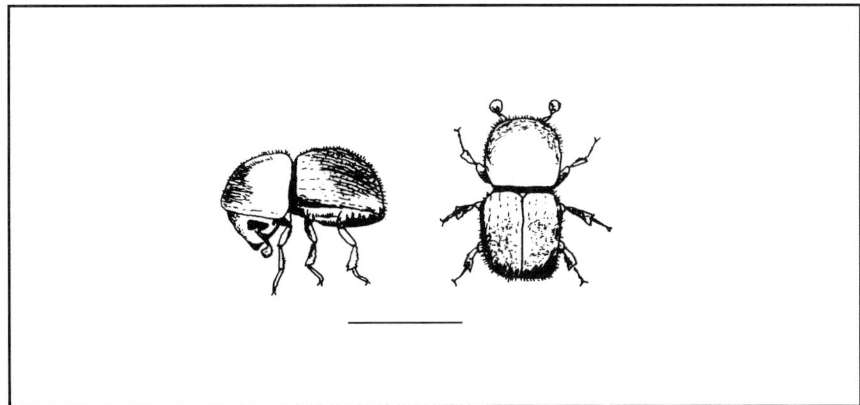

Figure 7.27. Side and dorsal views of a scolytid beetle.

unexpected range of niches for possible exploitation by animals. For example, certain bark beetles (Scolytidae) (fig. 7.27) breed only in the leaf stalks of certain species of trees (Beaver 1979a,b). The large number of animal species in turn provide a wealth of exploitation opportunities for other animals. For example, when spiders move about they leave behind them trails of fine 'silk' which may be on a substrate or suspended between two objects. Certain flies, wasps and moths habitually hang on these abandoned threads at night and are thus out of danger of walking predators (e.g., ants) and can effect a rapid escape from danger (Lahmann and Zuniga 1981). In addition, certain wasps have developed the ability to take flies out of spiders' webs before the fly is killed by the spider and without getting caught themselves (Bristowe 1976).

Three Sumatran reptiles (the geckos *Cnemaspis kandianus* and *Peropus mutilatus* and the skink *Sphenomorphus cyanolaemus*) are 'specialists' on buttress in lowland forests. That is, when reptiles were collected for six weeks at Bukit Lawang, Langkat, those three species were only caught on buttresses. There were even microhabitat preferences within the 'buttress-niche' depending on the amount of litter present between the arms of the buttresses (Voris 1977).

As stated elsewhere (p. 326), although the niches of two species cannot be exactly the same, one species frequently shares part of its niche with another species. This can cause competition between two species such that the removal of one species will increase the abundance of the other. This has been shown for frogs found on the banks of lowland forest streams in Sarawak. Three species (two of which are found in Sumatra) were studied along three streams and each was found to range from the stream bank different distances into the forest from the stream: thus *Rana blythi*, the largest of the three, was found up to 10 m from the stream, *R. ibanorum* up to 7 m, and *R. macrodon* up to 3 m. The last two species were almost exactly the same size and only slightly smaller than *R. blythi*. The prey of the three species also showed considerable overlap. After a complete censuses of the frogs had been made along the three streams, *R. ibanorum* was removed from along the stream where *R. blythi* was least common, *R. blythi* from the stream where *R. ibanorum* was least common, and the third stream was left as a control. The results are shown in figure 7.28. Where *R. ibanorum* was removed, *R. blythi* increased its population, but where *R. blythi* was removed, the *R. ibanorum* population fluctuated but did not increase. Thus *R. ibanorum* seemed to have been restricting the realised niche of *R. blythi* but not vice versa. Exactly how the competition was effected is not known (Inger 1969; Inger and Greenberg 1966).

Niche Differentiation

The manner in which similar species are able to coexist in the same area is termed 'niche differentiation'. This can be effected by spatial separation, dietary separation, temporal separation, or, more usually, a combination of all three. To illustrate this, the niche differentiation within three groups of lowland forest animal communities will be described: primates, squirrels, and hornbills.

Primates

The most species of primates in an area of mainland Sumatran lowland forest is eight[4]: orangutan *Pongo pygmaeus* or tarsier *Tarsier bancanus*[5], siamang *Hylobates syndactylus*, white-handed gibbon *H. lar* or dark-handed gibbon *H. agilis*, silvered leaf monkey *Presbytis cristata*, Thomas' leaf monkey *P. thomasi* or banded leaf monkey *P. melalophos* or eastern leaf monkey *P. femoralis*, long-tailed macaque *Macaca fascicularis*, pig-tailed macaque *M. nemestrina*, and slow loris *Nycticebus coucang*. The tarsier and slow loris have a very different niche from the others: they are both nocturnal, the slow loris eating fruit and insects, the tarsier mainly insects. The niche separation of the other six is not so obvious.

Spatial separation. The actual areas of lowland forest used by the diurnal primate species are more or less identical, although long-tailed macaques and, to some extent, silvered leaf monkeys are more common in riverine forest (Wilson and Wilson 1973). Wilson and Wilson also report that silvered leaf monkeys rarely enter deep into lowland forest but, on Bangka Island where banded and eastern leaf monkeys are absent, they appear to

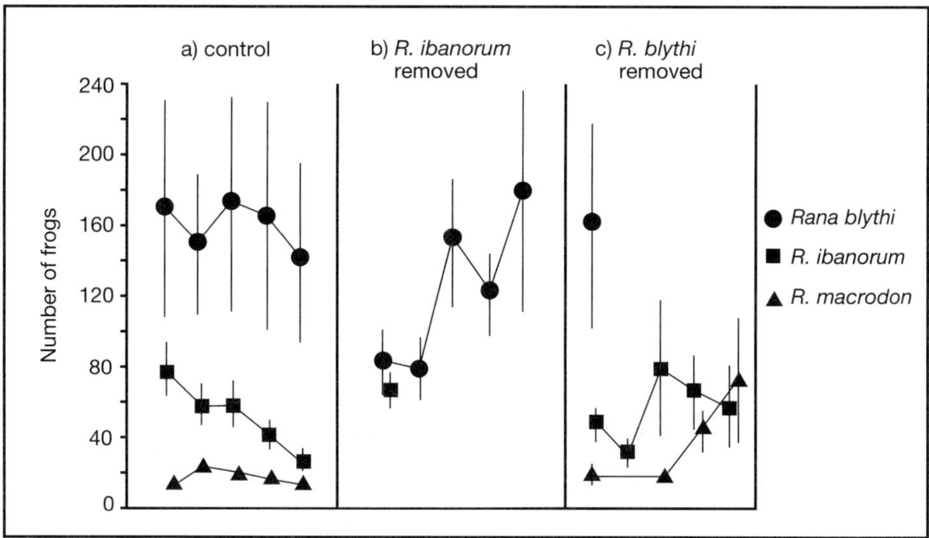

Figure 7.28. Estimates of frog population size on three streams in Sarawak at five census periods distributed over one year, showing mean and one standard deviation either side of the mean: a) untampered with to act as a control, b) streams with *R. ibanorum* removed, after the first census, c) streams with *R. blythi* removed after the first census.

After Inger and Greenberg 1966

be quite common inland (Whitten, unpubl.). Each species has its own altitudinal range (p. 305). Marked separation also occurs in the use each species makes of the canopy (fig. 7.29). The difference between the two macaques is the most marked, with pig-tailed macaques being much more terrestrial (Crockett and Wilson 1980). The two species do use the canopy in a similar manner for feeding but pig-tailed macaques are anatomically better suited to walking on the ground than to jumping or leaping. The siamang and gibbon are very similar in their use of the canopy (MacKinnon and MacKinnon 1980) but the orangutan uses the middle canopy more than either of them (Rijksen 1978). Orangutans never leap between trees and often rock back and forth on trees in the middle canopy of the forest until they can reach a neighbouring tree.

Dietary separation. Representative diets of some of the diurnal primates are shown in figure 7.30. These figures disguise considerable variations between days, months and seasons, but illustrate the major differences in

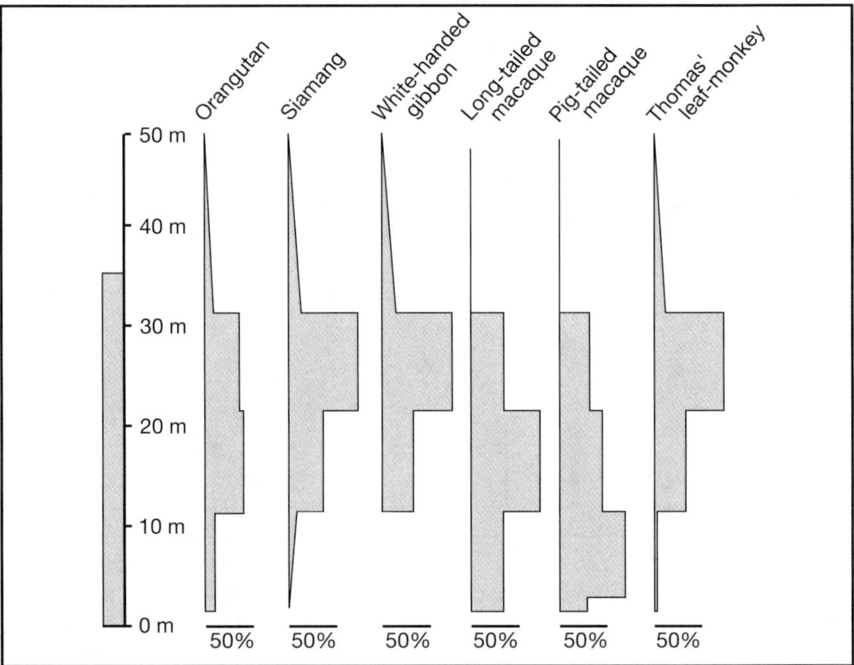

Figure 7.29. Vertical use of the canopy by six primates at Ketambe Research Station, Mount Leuser National Park. Bars beneath each column represent 50% of total observations.

After Rijksen 1978

the relative importance of fruit and leaves. The differences are clearer still if the actual food items are considered. In other words, different plant species and different parts of the same plant species are counted separately; this is shown in figure 7.31. Thus, whereas in figure 7.30 siamang and banded leaf monkeys were apparently quite similar in their diet, the actual overlap between the food items they eat is very small. Insufficient data are available for a complete analysis, but in the Ranun River area of North Sumatra pig-tailed macaques have a high dietary overlap of 48% with their close relative, the long-tailed macaques (MacKinnon and MacKinnon 1980), and the diets of orangutan and white-handed gibbons overlap by 40% (MacKinnon 1977). The orangutan diet probably overlaps more with that of gibbons and siamang than with the diets of the other primates.

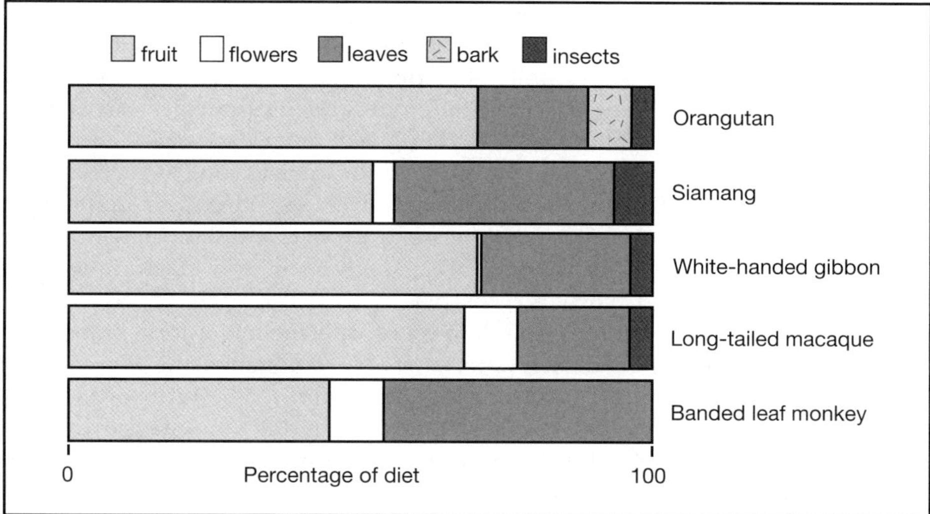

Figure 7.30. Average diets for five Sumatran primates. The proportions are typical but disguise a wide variation between days and between months. Thomas' and Mentawai leaf monkeys probably have similar diets to banded leaf monkeys. Silvered leaf monkeys probably eat more leaves than fruit.

From data in MacKinnon 1977; Rijksen 1978; MacKinnon and MacKinnon 1980 and from E.L. Bennett, pers. comm. and A.G. Davies, pers. comm.

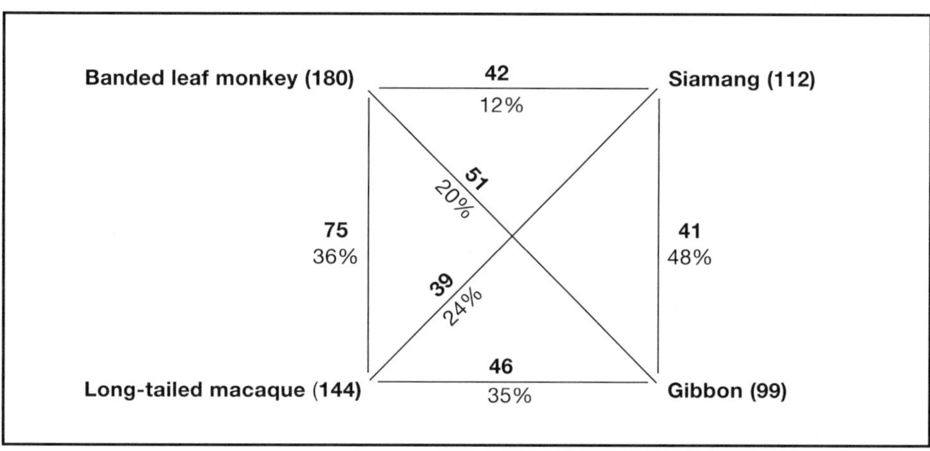

Figure 7.31. Number of different food items eaten by four primate species (in brackets), the number of shared items, and the percentage of dietary overlap between each pair of species. E.L. Bennett (pers. comm.) believes the overlap between white-handed gibbons and banded leaf monkeys is only about 1%.

Based on MacKinnon and MacKinnon 1980

The greatest dietary overlap is between the siamang and gibbon, which also appear to have the most specialised diets (fewest types of items). The way in which they segregate food resources between them is a function of body size which influences their different foraging strategies. Siamang are twice as heavy as gibbons but travel only half as far each day. They feed at only half as many feeding sites, but feed for twice as long at each site. Siamang have to travel direct routes between trees with abundant food but gibbons are able to exploit smaller, more dispersed food sources (Gittins and Raemaekers 1980; MacKinnon 1977; MacKinnon and MacKinnon 1980; Raemaekers 1979). These two compete by depleting each other's food supplies, and exploitation competition is probably the major form of this competition. The larger siamang, however, is able to dominate the gibbon if they meet at a food source (disturbance competition). The rarity of meetings between the two species, even when they occupy the same areas of forest, is probably avoidance of this form of competition by the gibbons which are able to seek alternative food sources (Raemaekers 1978).

Temporal separation. The activity patterns of five Sumatran primates is shown in figure 7.32. Different activity patterns have probably evolved so that food is eaten at optimally spaced intervals through the day. Thus leaf monkeys, which eat seeds and leaves (both of which contain high levels of defence compounds), need to space their feeding more or less evenly through the day so that the effects of these compounds do not cause debilitating peaks. Conversely, gibbons and siamang eat mainly fruit and young leaves and feed more intensively during the early part of the day. Since leaves are most nutritious at the end of the day when the products of photosynthesis (carbohydrates and sugars) have accumulated, it would be expected that these are eaten most in the afternoon. This is found in leaf monkeys (Curtin 1980), siamang and gibbons (Gittins and Raemaekers 1980). Temporal separation also occurs in the times at which the different primate species call (fig. 7.33) (see also p. 248).

Squirrels. Ecological separation of three diurnal squirrels was investigated on Siberut Island by J. Whitten (1980). The species (relative sizes shown in fig. 7.34) were separated largely on their vertical-use of the forest (fig. 7.35), but in the areas of overlap in the lower levels they were separated by both activity (as indicated by calling) (fig. 7.36) and by diet. The large *Callosciurus melanogaster* ate mainly fruit and insects, the small *Sundasciurus lowii* ate bark, moss, lichens, insects and earthworms, and the largely ground-dwelling *Lariscus obscurus* ate mainly fruit (fig. 7.37).

Similar results were obtained in Peninsular Malaysia by Payne (1979). Peninsular Malaysia has the two very large diurnal squirrels *Ratufa affinis* and *Ratufa bicolor*, which are both also found in Sumatra. These two species are almost exactly the same size and weight (factors which usually influence

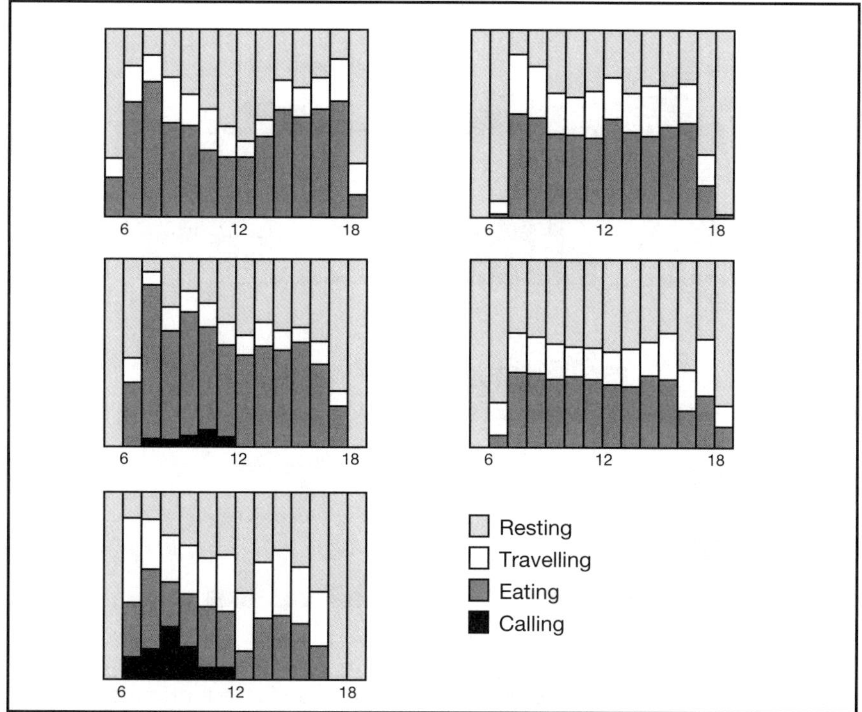

Figure 7.32. Daily activity patterns of five primate species. Note the relatively high proportion of travel for white-handed gibbons, the double peak of feeding and marked midday rest for orangutan, and the relatively late start for banded leaf monkeys and long-tailed macaques.

After MacKinnon 1977, MacKinnon and MacKinnon 1980

ecological separation) and exploit the forest in very similar ways. Their niche separation is thought to be due to minor differences in canopy use and diet (Payne 1979).

Hornbills. Sumatra has 10 species of hornbill and at least eight can be found in a single area of forest. The means by which available resources are divided between them so that these species can coexist have not been studied in Sumatra, but this subject was studied in East Kalimantan where seven species (all of which are found on Sumatra) live in the same forest (Leighton 1982). Two of the species are nomadic, flocking species, three live as territorial pairs and two live in territorial but communal groups. Fruit forms the major part of all their diets but they will all also eat small

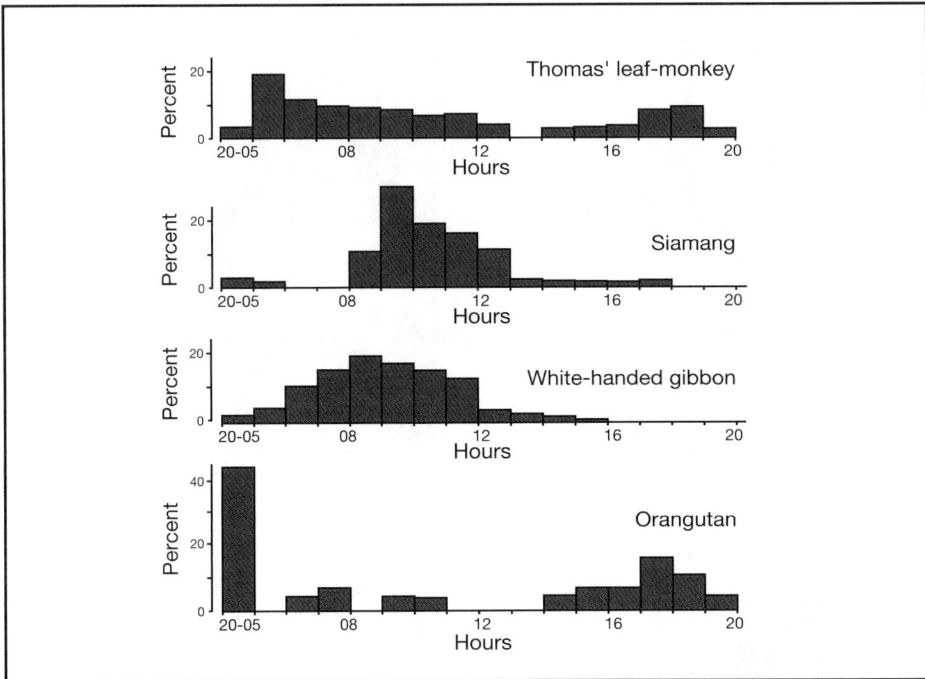

Figure 7.33. Percentage of calls occurring through the day for four primate species around the River Ranun, Dairi.

After MacKinnon 1977

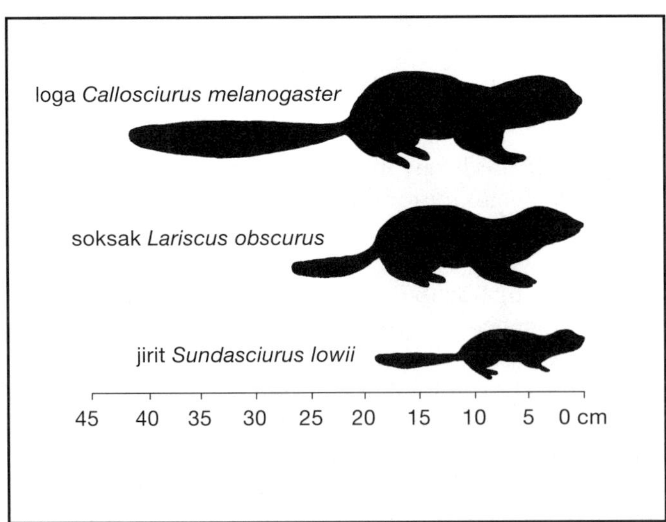

Figure 7.34. Relative sizes of the three diurnal squirrels on Siberut Island.

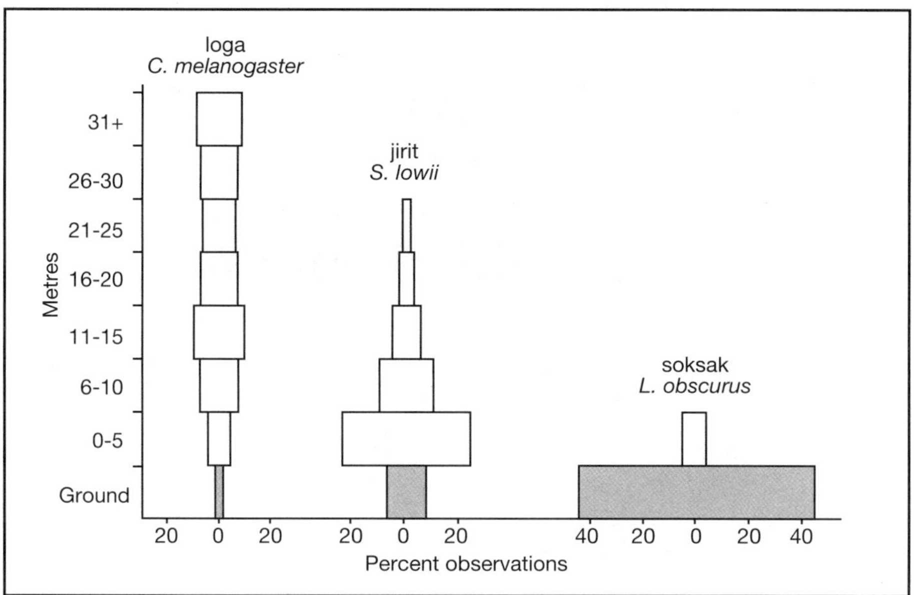

Figure 7.35. Vertical distribution of three species of squirrel on Siberut Island.
After J. Whitten 1980

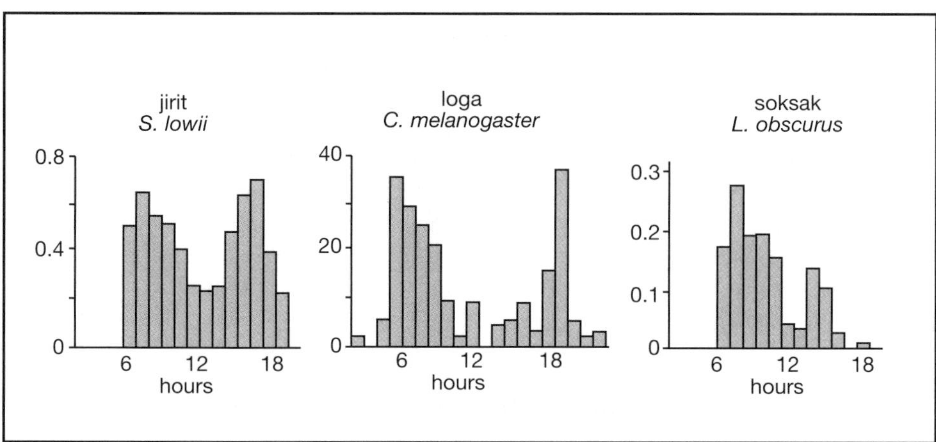

Figure 7.36. Distribution of calling through the day by three species of squirrel on Siberut Island. Note that the vertical axis for *S. lowii* and *L. obscurus* are calls per hour, whereas for *C. melanogaster* the vertical axis is total number of calls heard.
After J. Whitten 1980

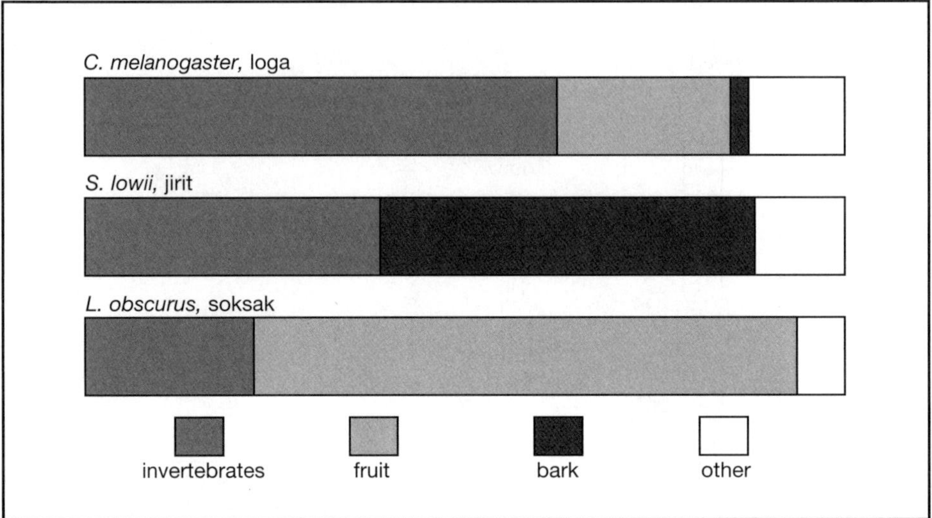

Figure 7.37. Diets of three squirrels on Siberut Island.
Simplified from data in J. Whitten 1980

animals. The two major classes of fruit eaten, sugary figs and lipid-rich fruits, differed in their abundance, distribution and profitability (net rates of energy gain for the hornbills). The economics of searching for these different foods together with the hornbills' body weight and competitiveness probably influence group size and the relative importance of figs or other fruit to the birds. The nomadic species travel large distances (i.e., expend considerable energy) but track temporary peaks in fruiting of lipid-rich fruits within localised habitats over a large area. The territorial species, living within fixed areas and thus with lower energy costs, supplement their diets with figs and animal prey between the periods of lipid-rich fruit availability.

Temporal Separation of Animal Calls

The way in which sound transmits through forest, particularly the structurally complex lowland forest, is quite different from the way sound moves through open air. Foliage, air turbulence, air temperature gradients and ground effects can rapidly degrade the vocal signals which animals use to communicate with others of the same species. In brief, the principles governing sound transmission in lowland forest are:

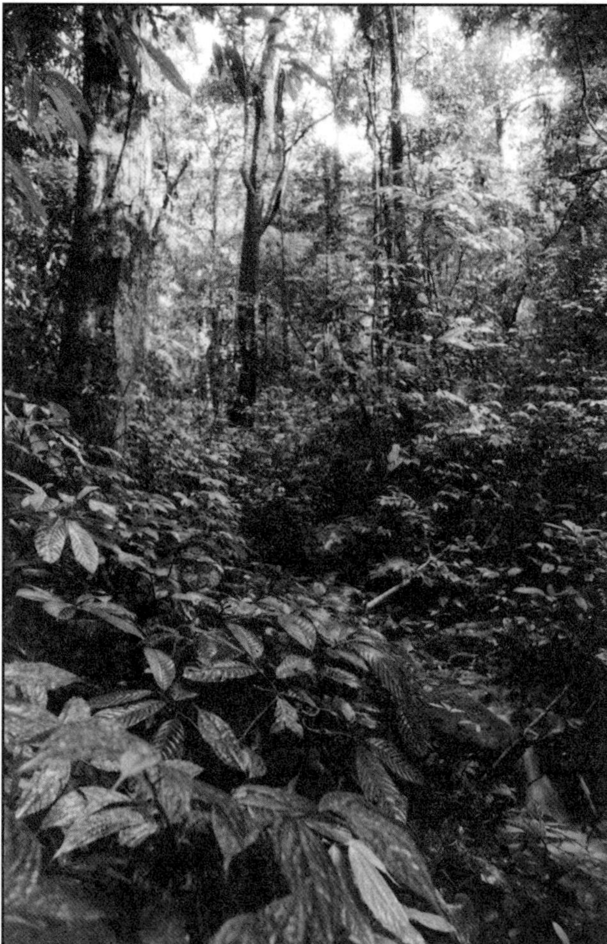

Figure 7.38. Lowland forest near a stream in Mount Leuser National Park, Southeast Aceh.
A.J. Whitten

a) sounds with wavelengths shorter than objects in their path will be reflected whereas longer wavelengths will not;
b) low-frequency sounds are absorbed less rapidly by humid air than high-frequency sounds;
c) animals calling from the forest canopy, above the range of ground effects and at times when other animals are not calling, increase the distance to which their calls will be carried;
d) complex structural properties of forests produce 'sound windows'

Table 7.2. Approximate densities of various large birds and mammals in Sumatran lowland forest. [1] long-tailed macaques are more or less restricted to riverine forest so figures given here are per km of riverbank (van Noordwijk and van Schaik 1983).

Species	Individuals /km²	Group size	Groups /km²	Biomass kg/km²
Great Argus pheasant *Argusianus argus*	3	-	-	?
Helmeted hornbill *Rhinoplax vigil*	0.4	2	0.2	?
Rhinoceros hornbill *Buceros rhinoceros*	1.1	2	0.4	?
Bushy-crested hornbill *Anorrhinus galeritus*	4	6	0.7	?
Black giant squirrel *Ratufa bicolor*	11	-	-	18
Banded leaf monkey *Presbytis melalophos*	26	13	2	150
Thomas' leaf monkey *Presbytis thomasi*	29	10	2.9	168
Long-tailed macaque *Macaca fascicularis*	30[1]	15	2[1]	120[1]
Pig-tailed macaque *Macaca nemestrina*	6	30	0.2	27
Gibbon *Hylobates lar/agilis*	6	3.5	1.5	22
Siamang *Hylobates syndactylus*	9	3.5	2.6	75
Orangutan *Pongo pygmaeus*	1.5	-	-	36.5
Tiger *Panthera tigris*	0.02	-	-	4
Elephant *Elephas maximus*	0.1	15	0.007	300
Tapir *Tapirus indicus*	0.1	-	-	20
Rhinoceros *Dicerorhinus sumatrensis*	0.01	-	-	8

After Chivers and Davies 1978; Davison 1981b; Flynn and Abdullah 1983; Lekagul and McNeely 1977; MacKinnon 1974, 1977; MacKinnon and MacKinnon 1980; Medway & Wells 1971; van Noordwijk and van Schaik 1983; Olivier 1978a; Payne 1979; Poniran 1974; Rijksen 1978; Robertson 1982; van Strien 1974; West 1979; World Wildlife Fund (Malaysia)/Payne and Davies 1981

through which certain frequencies can pass but others cannot;

e) temperature increases with height above the forest floor during the day and this causes sound to be trapped and attenuated within the forest (Richards and Wiley 1980; Whitten 1982f).

Given the above principles one would expect animals to call when conditions allow maximum transmission of their calls. Examination of the spacing of calls by cicadas (Young 1981), birds and other vertebrates (Henwood and Fabrick 1979; Medway 1969), and gibbons (Whitten 1980b) have shown that calls are indeed given at the times and frequency that provide for best transmission.

Density of Larger Animals

Knowledge of the home range occupied by a single group cannot necessarily be used to calculate a density, because for some species there are frequently gaps between home ranges or, for other species, home ranges overlap. The densities of selected large mammals and birds in Sumatran lowland forest shown in table 7.2 *must not* be used to calculate densities in a given patch of forest in, for example, an environment impact assessment. They are included in this book so that the reader can form an impression of the natural rarity of many of the larger animals and of the density that might be expected. Densities vary considerably between areas and between forest types (Marsh and Wilson 1981; Payne and Davies 1981) and results from one area should not be applied indiscriminately to another area. The methodology of faunal surveys can be found elsewhere (MacKinnon 1981; Marsh and Wilson 1981; Payne and Davies 1981). Data are available on numbers and biomass of insectivorous birds in lowland forest (Wells 1978), on approximate ranges of forest rats (Harrison 1958) and of other mammals (Gitins 1980; Medway and Wells 1971) and on the density and biomass of squirrels (Payne 1979).

Chapter Eight

Uncommon Lowland Forests

Introduction

Lowland forests vary quite considerably in composition (p. 198) and three types are discussed individually below because they are so strikingly different. They are: heath forest/padang vegetation, ironwood forest and forest on limestone. The first two types were investigated by a team from CRES and are discussed in detail below.

Heath Forest/Padang Vegetation

The only places where extensive areas of heath forest and padang vegetation can be found in Sumatra are on Bangka and Belitung Islands, but small areas also exist in parts of eastern Sumatra, the Lingga and Natuna Islands and Jemaja Island in the Anambas group (Frey-Wijssling 1933b; Kartawinata 1978b; van Steenis 1932, pers. comm.; Valeton 1908; Verstappen 1960). Heath forest is known by botanists as 'kerangas forest' from an Iban (Sarawak) word which refers to forested land which, if cleared, would not grow rice. It has nothing to do with 'rengas' trees (e.g., *Gluta, Melanorrhoea*), although these may occasionally be found within the forest. Padang vegetation is often associated with heath forest and although its status as a completely natural vegetation type is still open to some debate (p. 263), it is certainly a long-lasting and relatively stable secondary growth forming after heath forest is cut or burned.

Soil

Heath forest usually grows on soil derived from siliceous parent materials which are inherently poor in bases and commonly coarsely textured and free draining (Burham 1975). The soils in intact heath forest or under bushes in padangs have a covering of brownish-black, half-decomposed organic material. In the open padangs the surface is generally pure white sand about 0.5-5 cm thick above a darker layer (Burham 1975). In the ecological literature the soils of both heath forests and padangs are called white-sand soils.

White-sand soils usually originate from ancient eroded sandstone beaches which were stranded due either to uplift of land or fall of sea level. If conditions are suitable for soil development, a hard iron-rich layer (podzol pan) forms a metre or so below the soil surface (Burham 1975). One of the first descriptions of a tropical podzol was by Hardon (1937) from a padang soil on Bangka Island (table 8.1). The hard iron pan can cause temporary waterlogging of the soil after heavy rain before the water is able to drain away.

The 'A-horizon' soils have a very high percentage of sand and only about 1% of clay (Hardon 1937). This leads to high porosity, rapid leaching, and low ion retention properties so that white-sand soils are probably among the most nutrient-poor soils in the world (Mohr and van Baren 1954). Analyses of soils from a Sarawak heath forest are also given by Proctor et al. (1983) and Richards (1941).

Drainage Water

The water draining from heath forest is generally a black water (see p. 171). A water sample taken from a small stream flowing out of the heath forest visited by the CRES team was very acid (pH 4-4.5) and had low nutrient levels (2 ppm nitrogen, 1 ppm potassium and 1 ppm phosphorus). These results are characteristic of heath forest water (Whitmore 1984). A study of white-sand soils in the Rio Negro region of Amazonia concluded that, apart from iron, the concentration of nutrients in the river water draining from them was the same as the region's rain water. This suggests that there are no nutrients being weathered from the parent material

Table 8.1. Description of a podzol from Bangka Island.

Horizon	Depth	pH	Description
A_0	0-10 cm	2.7	Black cover of half-decomposed organic material intermixed with coarse quartz sand.
A_1	10-25 cm	3.9	Loose greyish-black humic quartz sandy layer.
A_2	25-40 cm	6.1	Loose greyish-white quartz sandy layer.
B_1	40-70 cm	3.9	Dark brown very compact quartz sandy hardpan.
B_2	70-100 cm	4.6	Loose light brown quartz sandy layer.

From Hardon 1937

(Ungemach 1969). The ecological aspects of black waters in peatswamp areas described on pages 176-177, are all relevant to drainage water for heath forests and padang vegetation.

Vegetation

Heath Forest. No detailed survey of heath forest ever seems to have been conducted on Bangka or Belitung Island and therefore much of the following is taken from studies conducted in Sarawak and Brunei. Heath forest can be strikingly different in its floral components and structure from usual lowland dipterocarp forest and it can itself show considerable variation (Brunig 1973; C.G.G.J. van Steenis, pers. comm.). In normal lowland forest, fresh, green-coloured vegetation evenly and loosely fills the space from the floor to the multi-layered canopy, whereas in heath forest a low, uniform, single-storey canopy seems to be formed predominantly by crowns of large saplings and small 'pole' trees (Ashton 1971). These sometimes form such dense vegetation that walking can be difficult. For example, in East Kalimantan heath forest, 454-750 trees of 10 cm diameter at breast height and over were found per ha (Riswan 1981, 1982). Brown and reddish colours are common in the foliage of the upper part of the canopy (Ashton 1971). Many of the tree species have thick leaves (p. 286). The medium leaf thickness of 12 tree species measured by the CRES team was 0.25 mm (range 0.10-0.45 mm).

Trees of large girth or with buttresses, and large woody climbers are very rare. Palms are also found and the CRES team found several giant mountain rattans *Plectocomia elongata*, a species which, as the name suggests, is normally found on mountains, where the soils are also poor (Whitmore 1977) (p. 281). Thin, small climbers may be quite common, as may epiphytes, several of which have associations with ants or catch small insects. The ground flora is sparse and frequently consists of mosses and liverworts. The number of tree species is relatively low. Proctor et al. (1983) found 123 species in a 1 ha plot in Sarawak compared with 214 species in a 1 ha plot in nearby dipterocarp forest. However, a total of 849 species from 428 genera are reported from the heath forests of Sarawak and Brunei (Brunig 1973).

Many of these features of heath forest vegetation have parallels in mountain and peatswamp vegetation, which also grow on poor, acid soils (pp. 171 and 281). The similarities with peatswamp forests are particularly striking as many of the tree species are the same. In figure 5.3 (p. 173), the four types of peatswamp forest in Sarawak have, from tallest to shortest respectively, 27%, 55%, 54%, and 70% of their species also recorded from heath forest (Whitmore 1984).

Various aspects of heath forest suggest it is restricted in its productivity by the low nutrient content of the soil. First, heath forest generally has a lower biomass than lowland forests on usual latosols. For example, Proctor et al. (1983) calculated the above-ground biomass of heath forest in Sarawak to be 470 t/ha (dry weight) compared with 650 t/ha in nearby dipterocarp forest. These are probably both above average for their type. The basal area of trees in an East Kalimantan heath forest was 6.4-16.9 m^2/ha (Riswan 1981, 1982) compared with 25-30 m^2/ha in a normal lowland forest. The heath forest examined by a CRES team at Padang Kekurai, Bangka, had a basal area of 13.5 m^2/ha for trees of 15 cm diameter and over. Second, plants with supplementary means of mineral nutrition are common (e.g., myrmecophytes and insectivorous plants) (Huxley 1978; Janzen 1974b). Third, heath forest is easily degraded into padang if burned or abandoned after brief cultivation; that is, heath forest does not seem to regenerate. This may be because the nutrient level of the soil has fallen below some level critical for full regeneration. While the heath forest is intact, nutrients from the detritus are recycled into the vegetation efficiently by the plants' roots.

The small litterfall (leaves and small wood) in the plot of heath forest examined by Proctor et al. (1983) was relatively low in its concentration of nitrogen and this may be due to the very low soil pH which could limit the mineralisation of the organic nitrogen. They suggest that the small leaves typical of heath forest trees are likely to be an adaptation to low nitrogen levels but the mechanism of this adaptation is not explained. The leaf thickening had earlier been explained as an adaptation to periodic drought as well as to low levels of soil nutrients (Brunig 1970; Whitmore 1984). Experiments by Peace and MacDonald (1981) have, however, shown that the heath forest species they examined had no special ability to avoid or resist desiccation. However, if features of heath forest such as the relatively even canopy and the small, often slanting and reflective leaves are seen as adaptations to reduce water loss, then the roots would not have to cope with such large volumes of water containing potentially toxic hydrogen ions (very acid soils have high concentrations of hydrogen ions) and phenolic compounds. Phenolic compounds can significantly inhibit the uptake of ions by certain crops (Glass 1973, 1974), and occur in high concentrations in heath forest leaves.

Padang Vegetation. Padang is a shrubby vegetation in which the tallest trees usually reach only about 5 m, but trees reaching 25 m are not unknown. Plants from padangs on Bangka and Belitung have been collected by several botanists (van Steenis 1932). In the padang (Padang Kekurai) examined by the CRES team, the dominant small trees were *Baeckia frutescens* and *Melaleuca cajuputih*, but *Calophyllum* sp., *Garcinia* sp. and *Syzygium* sp. were quite common. The most notable absences were of trees normally common as colonisers of open ground such as *Macaranga* spp., *Mallotus* spp. and *Melastoma* spp. Bare patches of white sand were frequent and these were

Figure 8.1.
Myrmecodia tuberosa.

often bordered by the sedges *Fimbristylus* sp., *Rhyncospora* sp., *Xyris microcephala,* and *Xyris* sp.; the lily *Dianella* and the bush *Rhodomyrtus tomentosa* were also locally common. Pitcher plants *Nepenthes* sp. were abundant and Valeton (1908) also records the presence in the damper parts of Bangka padangs of the small insectivorous plant *Drosera burmanii.* The climbing epiphyte *Dischidia* sp. covered the lower trunks of some of the larger trees. This plant has yellowish leaves shaped like wrinkled, inverted saucers. When the plant is pulled off a tree and the underside examined, numerous ants are found sheltering beneath the leaves, which also cover the plant's roots. This mutualistic association is similar to that found in *Myrmecodia* (fig. 8.1) and *Hydnophytum* (Huxley 1978; Janzen 1974b) (both of which are also found in heath forests), with the plant providing protection and the ants supplying nutrients in the form of waste food, faeces and dead ants.

The leaf thickness of 10 plant species were measured and the median value was 0.30 mm (range 0.25-0.70 mm), thicker than leaves in normal lowland forest, indicating an impoverished soil (p. 286).

Padang has been regarded as degraded heath forest caused by fire or felling (Janzen 1974a; Whitmore 1984), and as a natural ecosystem in its own right growing on extremely impoverished soils (A.J.G.H. Kostermans, pers. comm.; Specht and Womersley 1979). The former is certainly true but it does not necessarily follow from that that the latter is false. Padang is very slow growing (Riswan 1982). Janzen (1974a) mentions 1-2 m of plant growth in 30 years on the padang at Bako National Park, Sarawak, and it seems unable to regenerate back into heath forest (p. 377). Decomposition of dead plant material (burned stumps, etc.) is also very slow and it is possible that a padang formed many decades (or even centuries) ago might show no signs of its heath-forest past. Kartawinata (1978) reports having found charcoal 50 cm below the soil surface in a heath forest in East Kalimantan. This may indicate that heath forest regenerated from a padang vegetation formed by fire but the depth of the charcoal shows that the fire was clearly a very long time ago.

From the work of Brunig, Specht and Womersley (1979) counted 83 plant species that had been recorded from padangs in Sarawak and Brunei, and 48 of these were *not* found in heath forest.

Fauna

Animals in heath forest and padang vegetation face very similar problems to those in peatswamp forest. Poor soils cause low productivity which causes plants to defend their edible parts with toxic or digestion-inhibiting compounds to make themselves unattractive to herbivorous animals (Janzen 1974a). Not only that, but the nutritive value of the vegetation can be less on white-sand soils than on normal soil. Hardon (1937) analysed the chemical composition of leaves of two species of plant, one set from padangs on Bangka Island and another from specimens growing at the

Table 8.2. Nutrient levels in leaves from two plant species, one set grown on poor soils on Bangka Island, and the other set grown of fertile soils at the Botanic Gardens in Bogor.

	Dacrydium elatum		*Rhodomyrtus tomentosa*	
	Padang, Bangka Island	**Botanic Gardens, Bogor**	**Padang, Bangka Island**	**Botanic Gardens, Bogor**
Ash	2.16	7.59	2.58	3.70
CaO	0.59	3.43	0.13	0.58
MgO	0.33	0.69	0.21	0.35
K_2O	0.52	0.57	0.36	0.50
Na_2O	0.08	0.05	0.24	0.14
MnO	0.02	0.15	0.07	0.45
P_2O_5	0.10	0.15	0.07	0.45

From Hardon 1937

Botanic Gardens in Bogor on an andesitic laterite soil. The results are shown in table 8.2. Low nutrient levels were also found in leaves of heath forest trees by Pearce and MacDonald (1981).

In view of the above it is hardly surprising that animal communities are considerably reduced in heath forest and padang vegetation. Wallace (1869) describes a visit to the port of Muntok on Bangka in 1861 and writes, "A few walks into the country showed me that it was very hilly and full of granite and lateritic rocks with a dry and stunted vegetation; and I could find very few insects". Janzen (1974a) describes his intensive, but barely successful, attempt to find vertebrates and invertebrates using baits and direct observations in the heath forests/padangs of Bako National Park, Sarawak, and writes that "the absence of bird calls was deafening". In the same area, Harrison (1965) estimated there was only one mammal per 2 ha and one bird per 0.4 ha. The results of collecting frogs, lizards and snakes in three forest types is shown in table 8.3. Heath forest had less than half the number of species found in the other forest types (Lloyd et al. 1968). Similarly, in a comparison of a number of forest types at Mulu, Sarawak, heath forest had the smallest number of dung beetles (Scarabaeidae) (Hanski 1983).

From what was written about the effects of nutrient-poor soils on the chemical defences of tree leaves in peatswamp forest (p. 175), it would be reasonable to expect similar strategies to reduce the impact of herbivores in heath forest. Only two studies are known that have compared the levels of phenolic compounds in leaves between a forest growing on acid, nutrient-poor white-sand (heath forest) soil and a forest growing on more usual soil. In the first study, McKey showed that there was approximately twice the concentration of phenolic compounds in the leaves of trees growing on the white-sand soils. Monkeys living in forest on usual soils ate a wide range of leaves, whereas those living in the forest on white-sand soils avoided almost all the leaves expect the most nutritious ones where the benefit of eating a good source of food was greater than the cost of ingesting defence compounds (McKey 1978; McKey et al. 1978). Although no sweeping generalisations should be made from results of a single study, there is reason to suppose that the acute lack of animals in heath forest in

Table 8.3. The number of different types of reptiles in three types of forest.

	Frogs	Lizards	Snakes	Tortoises	Total
Heath forest	24	13	13	0	50
Alluvial forest	33	33	38	2	106
Hill forest	53	31	35	1	120

After Lloyd et al. 1968

Sumatra and adjacent regions is at least partly due to higher than normal concentrations of toxic phenolic and other defence compounds. The second study found that the mean value of total phenols in oven-dry heath forest leaf litter was 2.49 mg/100 g, compared with 1.68 mg/100 g in nearby dipterocarp forest (Proctor et al. 1983). It might be expected, then, that consumption of the heath forest leaves would be relatively low. In fact, the percent of leaf area missing from fallen leaves (22% with no damage, 50% with 20% damage, 15% with 20%-40% damage) was similar to that found in the other three forest types examined (Proctor et al. 1983). This does not, however, necessarily negate the hypotheses of Janzen (1974a) because although leaves may fall after the same approximate degree of destruction, the rate of destruction of heath forest leaves may be much lower.

Leaf samples from padang and heath forest were collected by the CRES team but the results of the chemical analyses are not yet available.

Fauna of Pitcher Plants. Pitcher plants provide an excellent example of an ecological microcosm. It is well known that insects are lured into the pitchers where they slip on the smooth, waxy lip and drown in the pool of water held in the bottom of the pitcher. Glands inside the pitcher secrete digestive enzymes and so the nutrients from the corpses are available to the plant.

It is less well known, however, that the water inside a pitcher supports a community of insect larvae and other organisms which are resistant to the digestive enzymes. Most of these animals feed directly on the drowned insects or feed indirectly on them by being predators on bacteria and other microorganisms which have themselves fed on the corpses. Since many of these species spend only the early part of their life cycle inside pitchers, they represent an export of nutrients which would otherwise have been used by the plant (Beaver 1979c,d).

Thienemann (1932, 1934) documented the fauna of pitcher plants in Sumatra, but the only study to consider the ecology of these unlikely communities is by Beaver, who worked on Penang Island. In the three species of pitcher plants examined (*Nepenthes albomarginata, N. ampullaria* and *N. gracilis*) he found 25 species of insect and three species of arachnid. The predominant family was Culicidae or true mosquitoes, but eight other families of flies (order Diptera) were found as well as representatives of the Hymenoptera (bees, wasps, ants), Lepidoptera (butterflies and moths), Araneae (spiders) and Acari (mites). Bacteria, protozoans, rotifers, nematode worms, oligochaete worms and crustaceans are also known from pitcher plants but these occur most frequently in old pitchers which have been deserted by insects and arachnids. These old pitchers no longer attract prey and so the input of nutrients is low and population levels of these microorganisms can never become high. Microorganisms are

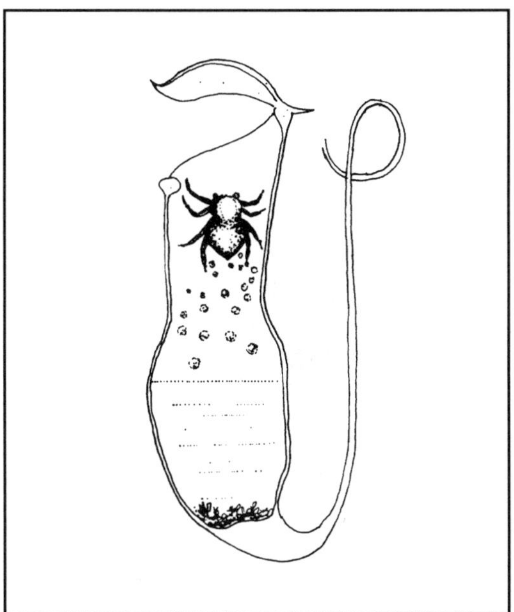

Figure 8.2. A vertical section of a pitcher showing a thomasiid spider waiting for its prey. The round marks below it are glands on the inside of the pitcher wall from which digestive enzymes are secreted.

probably infrequent in younger, attractive pitchers because of the filter-feeding activities of some mosquito larvae. A negative correlation between protozoa and mosquito larvae in such habitats in North America was shown by Addicott (1974).

The fauna of pitcher plants can be divided into three groups using terms coined by Thienemann (1932) according to the frequency of use of pitchers. Nepenthebiont species normally live only in pitchers and these account for 66% of the pitcher fauna found in Peninsular Malaysia and Singapore (Beaver 1979c). A few species are considered nepenthephils because they are common in pitchers but are frequently found in other habitats, and other species are called nepenthexene. This last group occur in pitcher plants but are far more often found elsewhere. One such nepenthexene mosquito, *Aedes albopictus,* is the only medically important insect to have been found in pitchers.

The fauna of pitchers occupy a range of feeding and spatial niches. The drowned animals in the pitcher fluid are fed upon directly by the larvae of certain *Diptera,* the larger species of which generally only tolerate a single individual because of aggressive competition for limited food supply. Mosquito larvae generally feed by filtering the microorganisms from the pitcher fluid, but some will also browse over the surface of the corpses,

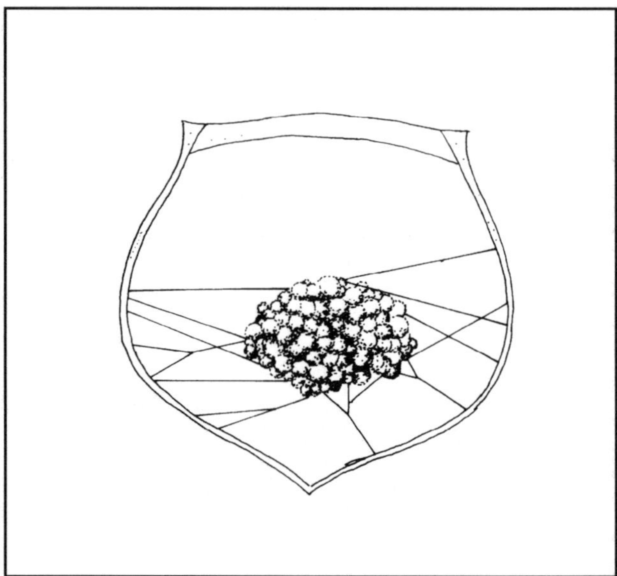

Figure 8.3. A horizontal cross-section of a pitcher showing the web of a mycetophilid fly larva.

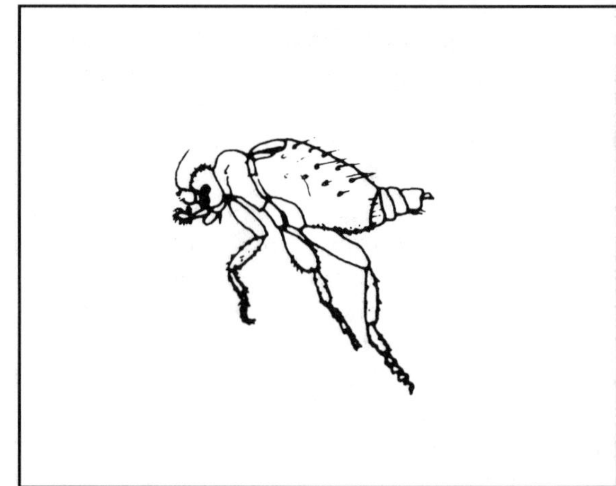

Figure 8.4. A phorid fly, larvae of which are parasitised by small wasps in pitchers. Phorids are peculiar flies in that most species are virtually wingless and most are less than 1 mm long.

After Mani 1972

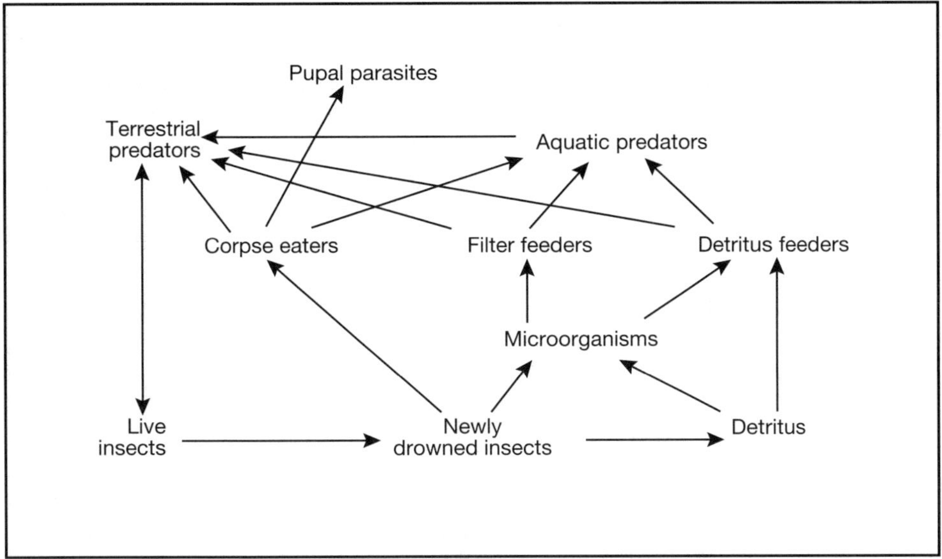

Figure 8.5. A simplified food web of the community living in a pitcher, showing the functional groups.

After Beaver 1979d, pers. comm.

ingesting loosened particles. At the very bottom of the pitcher, finely divided detritus from the remains of animals and plants accumulate and this is fed upon by the larvae of biting midges (Ceratopogonidae) and by mites. The pitcher fluid is also inhabited by insect predators such as a predatory mosquito larva which preys on other mosquito larvae; because of their large size, only one is found per pitcher. Above the fluid, a thomasiid spider (fig. 8.2) waits, and the predatory larva of a mycetophilid fly spins a web, to intercept insects falling from the pitcher rim as well as adult insects emerging from the pitcher after developing in the fluid (fig. 8.3). Finally, small encyrtid and diapriid parasitic wasps lay eggs in the pupae of phorid flies (fig. 8.4). A simplified food web for the community in two species of pitcher plants is shown in figure 8.5.

Distribution of Heath Forest on Bangka and Belitung Islands

It has been suggested or implied that the natural vegetation for the majority of Bangka and Belitung Islands would have been heath forest (Frankena and Roos 1981). This may not be the case. Berkhout (1895) and de Leeuw (1936) described the forests and general situation of Bangka Island but make no mention of a forest like heath forest, nor of species that would be especially expected from heath forest or padang. Instead they describe the conventional dipterocarp forests that today clothe only the hills. Descriptions of birds (Berkhout 1895), fishes (de Beaufort 1939) and snakes (Westermann 1942) from Bangka and Belitung Islands make no mention of an unusual forest type nor give any hint that the fauna was poor in species. Heath forests are notoriously poor in species (see p. 258). A.J.G.H. Kostermans (pers. comm.) spent six months collecting botanical specimens primarily in the south of Bangka Island and found areas of padang but no heath forests nor brown or blackwater streams, as would be expected to flow out of heath forests (p. 254). However, when flying from Pangkal Pinang to Jakarta it is possible to see blackwater rivers disgorging into the sea on the east-facing shore of Bangka. In addition, van Steenis (1932) has suggested that padang vegetation on Bangka Island is restricted to parts of the north and northeast and he makes no mention of heath forest.

After several days of travelling on Bangka Island, the CRES team finally found a belt of heath forest in the north of the island about 10 km north-east of Belinyu, on the landward side of 'Padang Kerurai'[1]. It is suspected that heath forest, on Bangka Island at least, is restricted in general to the soils called Brown Yellow Podzolic/Podzol Association by Soepraptohardjo and Barus (1974), and in particular to where the podzols are more pronounced, such as the old marine terraces near the coast. Hardon (1937), however, described a podzol from a site more than 12 km from the sea. The CRES team also found a white-sand area near Kurau, 25 km south of Pangkal Pinang, at the site of a large transmigration site for fishermen. All forest/scrub had been cleared from at least 200 ha and only a few paper-bark *Melaleuca* trees and coarse sedges *Gahnia* grew on the poor soil. Janzen dedicated his paper about the ecology of white-sand soils to "those unfortunate tropical farmers (who) attempt to survive on white sand soils... because they appear unexploited" (Janzen 1974a).

In inland areas of Bangka Island which are also shown on maps as having Brown Yellow Podzolic/Podzol Association soils and being over old sandstone parent material (e.g., around Poding and Petaling), small rubber plantations and fruit orchards are common. Large areas are used for pepper and this suggests that the capability of the land is limited but not critical. In the descriptions of typical heath forest/padang soils, however, it is suggested that they are virtually unusable for permanent agriculture. This suggests that heath forest would not necessarily have covered these flat or undulating inland areas.

It is interesting that in the form of the Melayu language spoken on Bangka Island, the word 'ngarangas' (similar to kerangas) is used to describe the dead and dying trees around the tin tailings. It is possible that ngarangas used to refer to heath forest.

The present area of intact or partially disturbed heath forest is not known but it is probably much less than that estimated by FAO/MacKinnon (1982). It is probably Sumatra's most endangered ecosystem, and a thorough survey should be undertaken as soon as possible.

Ironwood Forest

Introduction

Ironwood forest in Sumatra is of special interest because of its extremely low diversity of tree species, being dominated by the ironwood *Eusideroxylon zwageri.* The ecological interest centres on its apparently stable state despite its species-poor composition.

The Tree

Ironwood is a member of the laurel family (Lauraceae) and is distributed through Sumatra, Kalimantan, and the southern Philippines. There used to be two members of the genus *Eusideroxylon* but the one called *melagangai* has recently been moved to its own genus *Potoxylon* by Kostermans (1979). Ironwood is a large tree (50 m tall and 2.20 m diameter at breast height) with a warm red-brown bark, large leaves and heavy (± 300 g) fruits. Its timber is economically very valuable because of its strength and durability — it can resist rotting for up to 40 years when in constant contact with wet soil, or a century in dryer conditions (K. Kartawinata, pers. comm.; Riswan 1982; Suselo 1981). It is used for bridges, piers, road foundations, floors and, of course, the popular 'sirap' roofing tiles.

Ironwood seems to be restricted to southern Sumatra, having been found in Jambi, South Sumatra, and Bangka and Belitung Islands (De Wit 1949; Greeser 1919; Koopman and Verhoef 1938; Soedibja 1952), but its range has been greatly reduced by the heavy demand for its timber. It has been shown that ironwood could be an important plantation species, and guidelines for its cultivation have been proposed (Koopman and Verhoef 1938). Commercial germination of seeds and growth of seedlings can be quite successful if conducted in moist, shady places such as beneath secondary growth with a closed canopy (Koopman and Verhoef 1938; Tuyt 1939). The main threat appears to be a certain amount of predation of the seeds by porcupines (p. 270).

A characteristic feature of ironwood is that, when felled, numerous coppice shoots grow from its base. If cut carefully these can be used for plantation stock, and Beekman (1949) found that these shoots grew faster than seedlings.

The Forest

Ironwood forest may have ironwood as virtually the only species of large tree present, or ironwood may be one of many species but clearly the dominant one. An example of the former was investigated by a team from CRES in Rimbo Kulim[2] near the River Kahidupan, north of Muara Tembesi in Jambi. It was difficult to find a patch of unexploited forest but eventually sufficient was found to enumerate 0.5 ha (10 x 500 m) of undisturbed forest. A total of 84 trees of 15 cm diameter at breast height were found and 81 (96%) of these were ironwood (fig. 8.6). Two of the remainder were unidentified small trees, and the other was a large specimen of *Koompasia* sp. The Simpson Index of species diversity (p. 189) for this situation is only 0.07. As many as 33 ironwood trees of 20 cm diameter and over can be present per ha in Kalimantan (van der Laan 1925). The equivalent figure for the forest examined by the CRES team was 126, indicating a perhaps unusual richness.

When inside the forest at Rimbo Kulim it is not immediately obvious that the forest is species-poor because there is a wide range of tree sizes. It is quite unlike, for example, being in a mangrove forest where the vegetation within a zone is relatively homogenous. The distribution of trunk diameters for ironwood trees at Rimbo Kulim is shown in figure 8.7. The undergrowth consists of many ironwood seedlings as well as small palms (*Pinanga* sp. and *Licuala* sp.) and the tall palm *Livistona* sp.

The ironwood forest examined by Suselo (1981) was not far from Rimbo Kulim and was also dominated by ironwood, but ironwood was only one of 37 tree species in his plot of 0.2 ha; He found 65 trees of 10 cm diameter or more, but only 17 (26%) were ironwood (fig. 8.8). (A similar percentage was found in a nearby plot (Franken and Roos 1981). Ironwood, however, formed the main canopy of the forest. In ironwood forest examined by Greeser (1919), *Koompasia, Shorea* or *Intsia* were the main emergents and ironwood was a subdominant species.

In other areas where ironwood grows, such as Bangka Island, it can occur in mixed dipterocarp forests as just an occasional species.

Soils and Topography

Witkamps (1925) considered that ironwood was generally restricted to sandy soils of Tertiary origins. The ironwood forest examined by Suselo (1981) was, however, on a clay-loam (for the top 20 cm) on top of clay. The soil analysis for three samples taken at Rimbo Kulim is summarised in figure 8.9, and reveals a sandy silt-loam (but close to a clay-loam) top soil with clay below. The soil is moderately acid with quite high carbon but a moderate C/N ratio. The soil contains very low concentrations of phosphorus, sodium and magnesium but a fairly high concentration of potassium. The total exchangeable bases are very low and the cation

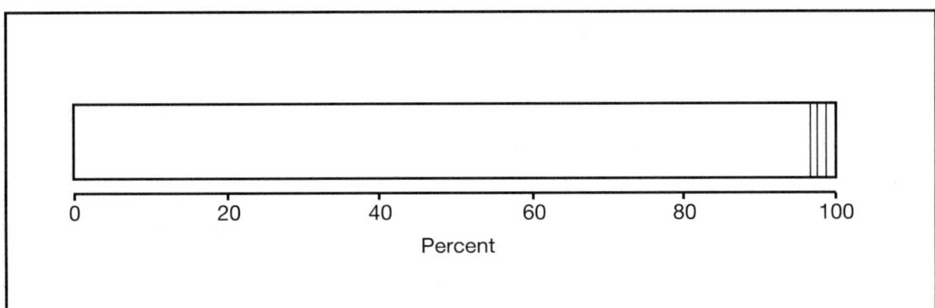

Figure 8.6. Relative abundance of the four tree species (n= 84) at Rimbo Kulim, Muara Tembesi, Jambi. Ninety-six percent of the sample comprises ironwood *Eusideroxylon zwageri*.

Data collected by a CRES team

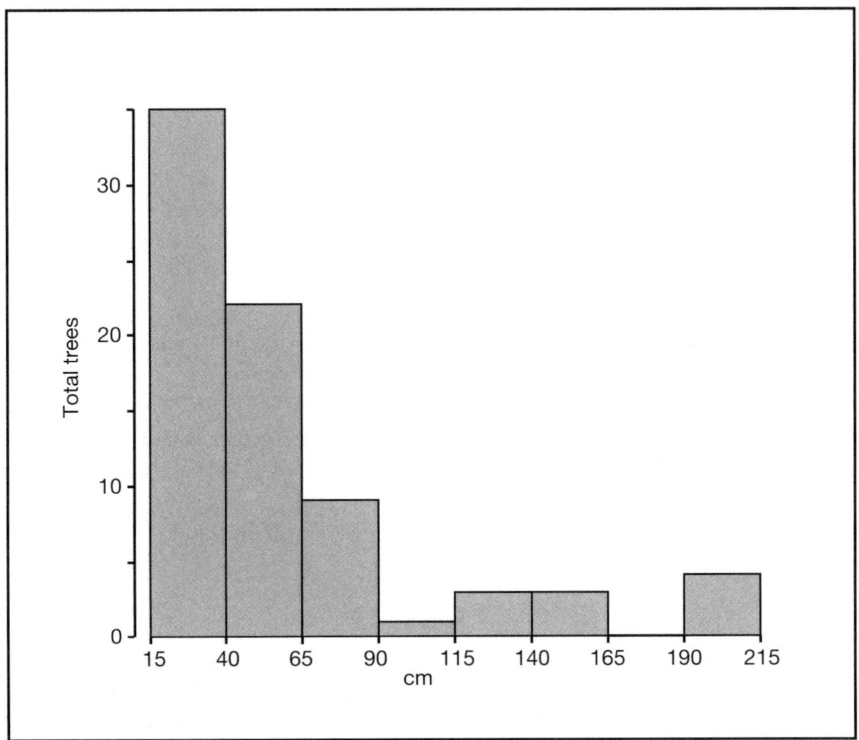

Figure 8.7. The distribution of different diameter classes among the ironwood trees at Rimbo Kulim, Muara Tembesi, Jambi.

Data collected by a CRES team

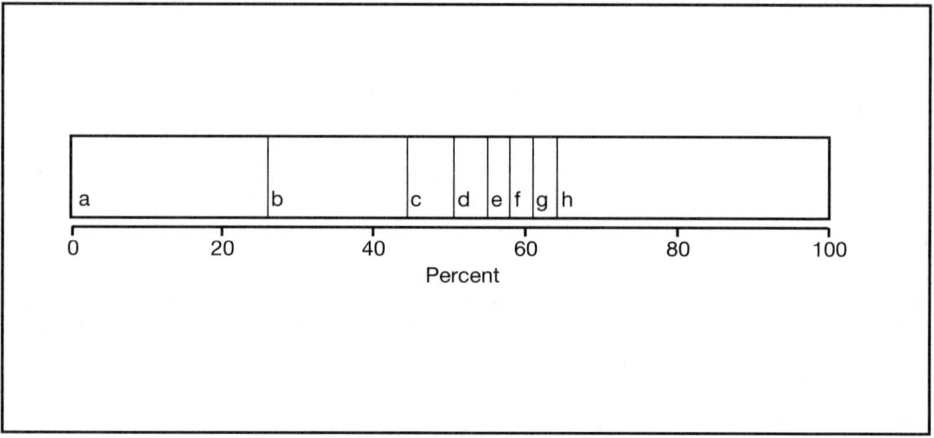

Figure 8.8. Relative abundance of tree species (n= 65) at the forest examined at Dusun Aro, Jambi. a- *Eusideroxylon zwageri,* b- *Hydnocarpus polypetala*, c- *Aglaia* sp., d- *Dysoxylon* sp., e- *Diospyros* sp., f- *Ixonanthes icosandra*, g- *Knema* sp., h- 23 other species.

Based on data from Suselo 1981

exchange capacity quite high.

The distribution of ironwood forests shown in figure 1.11 (p. 21) coincides with areas of red-yellow pozolic soils on folded sedimentary rock shown on the map in BPPP/Soepraptohardjo et al. (1979). There are, however, areas with the same soil characteristics in the same climatic zone where ironwood is known not to dominate the forests. Various authors have commented on ironwood's preference for slightly undulating topography but the cause of any limitation has not been established.

Fauna

Nothing seems to have been written on the fauna of ironwood forest. The birds seen during the CRES visit to Rimbo Kulim seemed similar in diversity and number to those in normal dipterocarp forest. Two groups of the unnamed black-and-white subspecies of leaf monkey *Presbytis melalophos* were seen but were not observed to feed. Where ironwood is virtually the only tree species, food may be a problem for many animals. The mature

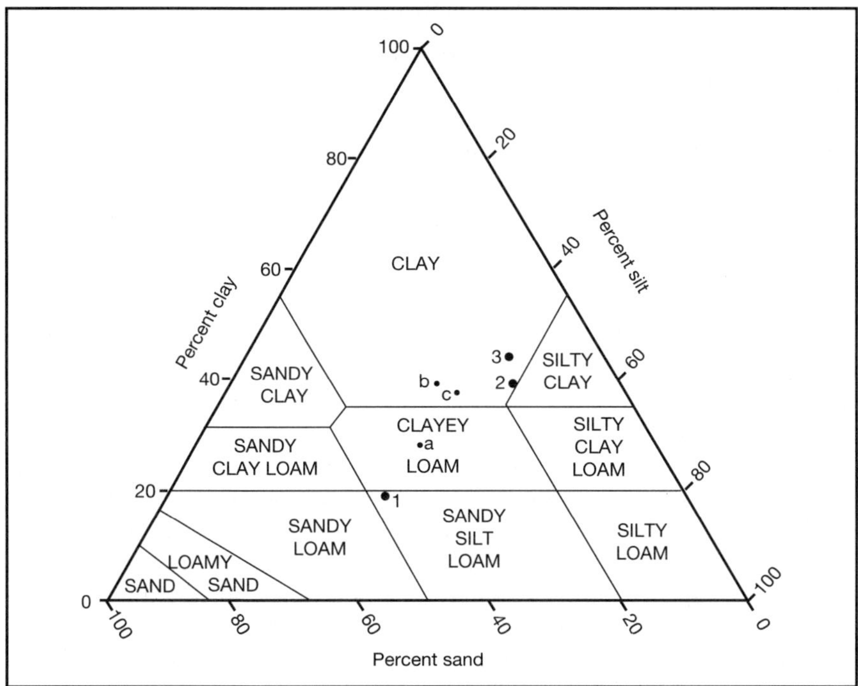

Figure 8.9. The structure of soils under two ironwood forests in Jambi. Rimbo Kulim soil: 1 - 1.3 cm, 2 - 5-10 cm, 3 - 10-15 cm (soil collected by CRES team). Dusun Aro soil: a - 1-5 cm, b - 20-30 cm.

Data from Suselo 1981

leaves are thick (0.35 mm) and coarse, and the fruit are very tough, large and heavy such that almost no arboreal animal would be able to eat them. An exception might be Prevost's squirrel *Callosciurus prevosti* which was seen several times at Rimbo Kulim. A few fruit were taken from Rimbo Kulim and small holes were observed in the fruit skin, caused probably by small beetles. Investigation showed that the seed had been slightly damaged but apparently not enough to halt germination.

In the forest investigated by Suselo (1981), 14 of the 37 species found in the plot produce fruit which would be suitable for frugivorous primates such as gibbons (Whitten 1982e). In the forest around Rimbo Kulim, dark-handed gibbons *Hylobates agilis* were heard singing, presumably where ironwood was less dominant.

Ecological Significance

There is a general and popular view that in tropical regions where climate and soils are favourable for plant growth, high species diversity in natural ecosystems is necessary for ecological stability. On page 79 it was shown that mangroves do not conform to this preconception, and it would seem that the same applies to certain ironwood forests. Indeed, the former ecological tenet of ecosystem complexity leading to stability has been called into question.

On page 221 the strategy of gregarious fruiting was described for the dominant, large dipterocarps which, at their canopy level at least, form species-poor stands. Janzen (1974a) suggests that a second type of strategy common in species which occur at relatively high densities, is the production of large and toxic seeds on a more or less continuous basis. Some mangrove species fall into this category and so, apparently, does ironwood. Koopman and Verhoef (1938) describe the flowering of ironwood as irregular but with a peak around the middle or end of the dry season. Ripe fruits are found approximately three months after flowering. The results of analyses of ironwood seeds collected by the CRES team are not yet available but the very large numbers of seeds found on the ground, and the very large numbers of seedlings under parent trees (Suselo [1981] found 920 saplings per ha) is indicative of very little seed predation. Porcupine damage of ironwood seed-beds has not been described in detail but it is conceivable that only parts of the seed were eaten and, like the monkeys described eating leaves from trees growing on white-sand soils (p. 259), the benefits of eating a highly nutritious item of food may exceed the costs of ingesting defence compounds. In any case, porcupine damage of seeds in natural situations does not seem to be important.

The CRES team also collected leaves, but again the analysis results are not yet available. However, their feel suggests a high concentration of fibre (lignin) and the timber itself must be defended very effectively against insect attack for it to be so exceptionally durable. The ecology of a leguminous tree called *Mora* was studied in Trinidad where it accounts for 40%-60% of trees in the forests where it is found. Most of what Janzen proposed as theory for such trees, as described above, fits for *Mora*. It suffers very little seed predation in comparison with other trees around it. Its major seed predator is a large rat-like rodent *Oryzomys* but the seeds are generally less than one-third eaten. Indeed, this damage is usually confined to the surface of the cotyledons and the seeds were still able to germinate (Rankin 1978). It is possible that *Mora* seeds, and those of ironwood, are protected by lectins, a group of glycoproteins which are sticky and act as digestion inhibitors. Lectins are very strongly selected against by rodents (Janzen 1981c) and this many be the reason porcupines do not do more damage to the abundant ironwood seed crop.

Species-poor ironwood forests would make ideal study areas to try to unravel the reasons why such abundant trees are not exploited or controlled by herbivores. Species-rich forests are incredibly complex and any ecological investigation confronts numerous confounding variables. Ironwood forests have some of these variables controlled and if an understanding of the ironwood ecosystem can be achieved, then attempts to understand species-rich forests are more likely to succeed.

Unfortunately, licences continue to be granted for the exploitation of ironwood trees and so it seems that the time remaining to make a thorough investigation of the exceptional forests of Sumatra is severely limited.

Forest on Limestone

Introduction

Sumatra has limestone areas although few exceed 10 square kilometres in size (fig. 8.10). Limestone forms a characteristic landscape called karst which occurs in two forms, cockpit (or labyrinth) karst and tower karst, both which are found in Sumatra. Cockpit karst such as at Lho'Nga, North Aceh, and in the hills near Payakumbuh, has a regular series of conical or hemispherical hills and hollows with moderately steep sides (30-40°). Tower karst such as is found in the river plants near Payakumbuh consists of low (± 300 m) isolated hills with precipitous sides (60-90°) and is often riddled with caves, separated by flat depressions (Jennings 1972; Verstappen 1960). The surface of the limestone may be either rocky or covered with intricately sculptured, razor-sharp pinnacles as at Lho'Nga.

Soils

On moderate limestone slopes and hollows, brownish-red latosols would be expected to be formed which are clay-rich and leached. As might be predicted from the parent material, the soils are often richer in bases, particularly calcium and magnesium, with a higher cation exchange capacity than soils in similar situations but on other parent materials. On limestone crests and shelves an acid, humus-rich, peat-like soil can develop directly on top of limestone rock, but this has not been reported (as far as could be determined) from Sumatra. On steep slopes, of course, soil cannot form except in very small quantities in fissures and cracks (Burham 1975). Soils over limestone can be very variable depending on the purity of the parent material and the topography (Crowther 1972a).

The soil beneath the forest on limestone examined by Proctor et al. (1983) in Mt. Mulu National Park, Sarawak, lay on a 25-30° slope and was

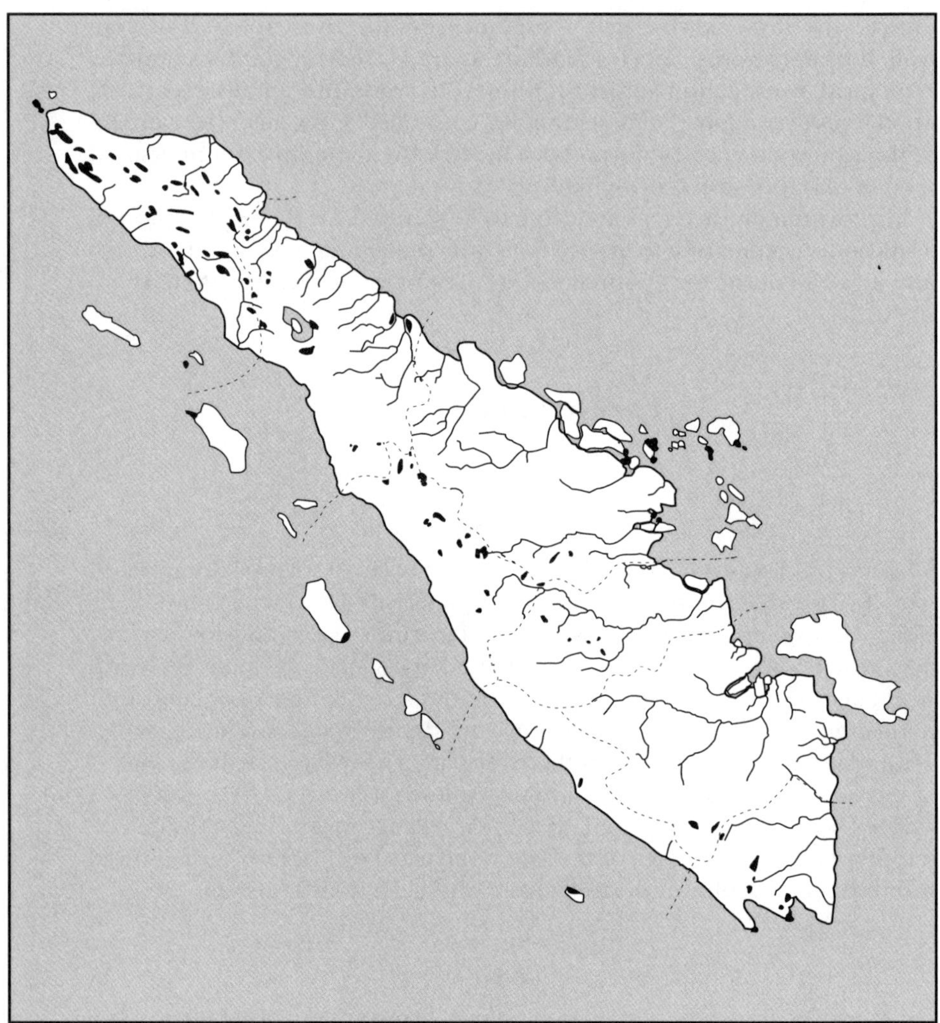

Figure 8.10. The distribution of limestone in Sumatra.

Adapted from Verstappen 1973

extremely shallow (average 11 cm, range 0-55 cm), and highly organic (80% weight lost on ignition). Its base saturation was 32% compared with 1.6% in dipterocarp forest and 2.9% in heath forest. Similarly, its cation exchange capacity was 210 m equivalents/100 g, compared with 37 in dipterocarp forest and 110 in heath forest. Despite the high organic content of the soil on limestone, its pH was 6.1, the highest recorded from the forests on Mt. Mulu.

Soil collected by a CRES team from small ledges on the rock walls outside Kutabuluh Cave, Tanah Karo, was basic and had very high concentrations of calcium and magnesium (table 8.4).

Vegetation

Anderson (1965) identified a range of habitats influenced by limestone in Sarawak.

a) At the base of limestone hills the soils, although derived from other parent material, are influenced by the water running off the limestone and by eroded limestone fragments. The relatively base-rich, fertile soils support a number of characteristics tree species but cultivation often extends to the very base of the hill.

b) Characteristic species can also be found at the base of limestone hills where the base itself is also over limestone.

c) The steeper slopes can support an untidy forest of trees whose roots cling to the rough surface or penetrate the rocks to emerge in caves below. Where the slope is too steep, small herbs thrive, many of which are not found in any other habitat. Most of these plants are subject to repeated water stress because the small volume of meagre soil in which they grow does not have sufficient water capacity. As a result they exhibit 'poikilohydry' — the ability to lose most of their water, except that in their protoplasm which resists drying out, and then to revive on rewetting (Whitmore 1984).

Table 8.4. Result of analyses made on soil samples from outside the cave at Kutabuluh. Soil collected by a CRES team.

FRACTION (%)			pH	C	N	C/N
Sand	Silt	Clay	H_2O	(%)	(%)	
52	31	17	7.2	6.43	0.87	7.4

P-av (ppm)	Exchangeable (m.e. /100 g)				T.E.B. (m.e./ 100 g)	C.E.C (m.e./ 100 g)	B.S. (%)
	K	Na	Ca	Mg			
177	1.82	0.34	51.01	5.32	58.49	50.64	116

d) Small scree slopes also support a characteristic flora.
e) Forest of the moderate slopes and summits of limestone hills grows in the peat-like humus described in the previous section, and has certain similarities in species with those found in heath forest and inner peatswamp forest (see Anderson [1965] for details).

This was one type of forest examined as part of the study by Proctor et al. (1983a,b). In their 1 ha plots they found only 74 tree species in forest on limestone, compared with 215 in dipterocarp forest and 122 in heath forest. Forest on limestone also has a relatively low density of trees, small total basal area and mean basal area per tree, low tree height and trunk diameters of 10-20 cm. Dipterocarpaceae was the most important plant family. The phenolic content of leaf litter from forest is not easily interpreted, possibly because leaf litter rather than leaves taken directly from the parent plant was examined.

Limestone hills are peculiar in their discontinuous distribution, their relatively base-rich soils, the high levels of calcium in the rock, and their very steep or sheer slopes, which can become periodically very dry. It is not surprising, then, that some plants are confined to them. For example, of 1,216 plant species recorded on limestone hills in Peninsular Malaysia, 254 species (21%) are found only on limestone hills and 130 of these species are only found in Peninsular Malaysia (Chin 1977). Considering the greater extent of karst limestone in Sumatra, it would be safe to suggest that an even greater number of plants are endemic to Sumatran limestone. About 10% of the plant species found on, but not restricted to, limestone in Peninsular Malaysia are specialised in living in crevices (Henderson 1939). Peninsular Malaysian limestone hills also represent the only known habitat of a genus of small serdang palms *Maxburretia* (J. Dransfield, pers. comm.; Whitmore 1977, 1984), and similar plants should be looked for in any ecological surveys of limestone hills in Sumatra. For example, the CRES team visiting the Lho'Nga limestone in North Aceh found a species of pigeon cane palm *Rhapis* which was, until then, a genus unknown in the Malesian region.

Fauna

The only fauna restricted to limestone habitats are certain molluscs. This was first recognised in the last century but the only thorough account of the phenomenon for the Sunda Region appears to be by Tweedie (1961). He examined the mollusc fauna of 28 more-or-less isolated limestone hills in Peninsular Malaysia and found a total of 108 species. The most remarkable finding was that 70 (66%) of these species were restricted to just one limestone hill. One hill even had seven mollusc species not found on any of the other hills (fig. 8.11). These endemic species must be extremely intolerant of living away from their limestone environment. Strangely, other

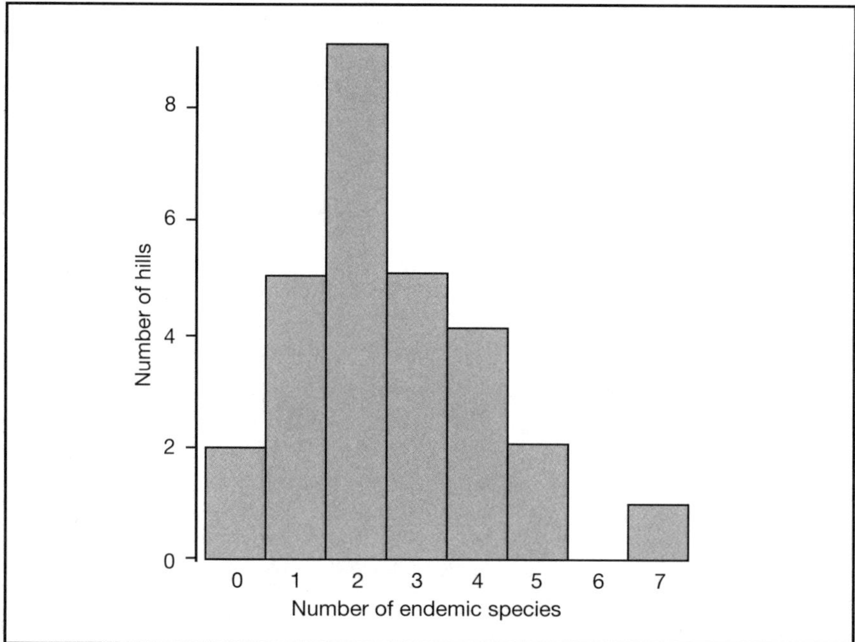

Figure 8.11. The distribution of endemic species of snails of different numbers of limestone hills. Thus, four hills each had a total of four endemic snail species, and five hills had only one endemic species.

Based on data from Tweedie 1961

species from the same genera as the endemic species were found on a wide range of hills; one such was collected from no less than 19 of the 28 hills. Some of these wider-ranging species have indeed been found in areas between the limestone hills, but others must be present in those inhospitable (for a mollusc) intervening habitats in such low numbers that transfers between hills are extremely rare occurrences.

Similar situations to the above doubtless exist for the isolated limestone outcrops in Sumatra. Obviously, those hills supporting more endemic species would be more worthy of protection from use as, for example, limestone quarries. Anyone interested in pursuing such a line of study should consult van Benthem Jutting (1959) as the first step.

Chapter Nine

Mountains

Introduction

Sumatra has a large number of mountains, some, such as most of those in the Barisan Range, formed by uplift of sedimentary deposits and others, such as Mt. Kerinci, Mt. Sinabung, Mt. Merapi and Mt. Singgalang, formed by volcanic action. The physical environment changes up a mountain, thus providing a gradual change of conditions against which the fauna and flora can be examined. These differences between the montane ecosystems and the luxuriant, hot and humid lowland forest ecosystems are stimulating subjects for study.

The mountain about which most is known, Mt. Kerinci (3,800 m), is also the highest. It was the object of a large collecting expedition in 1914 and about 30 papers were published on different groups of animals and plants in the *Journal of the Federated Malay States Museum* between 1918 and 1931 (Pendelbury 1936). These papers are most useful for taxonomists, but for understanding the changes occurring with altitude, more informative papers are available on Mt. ni Telong (Frey-Wyssling 1931), the mountains in the Mt. Leuser National Park (van Steenis 1934a), Mt. Singgalang (Docters van Leeuwen) and Mt. Kerinci (Frey-Wyssling 1933; Jacobs 1958; Ohsawa 1979). A CRES team climbed Mt. Kemiri (3,340 m) in the Mt. Leuser National Park and Mt. Kerinci as part of the preparation for this book.

Climate

Temperature

There is a mountainward current of air blowing during the day which is caused by air warming in the lowlands and thus expanding and rising. As any gas rises and expands in response to lower pressure, so its temperature falls because the wider-spaced gas molecules, which move about in response to energy from the sun, collide less frequently and therefore create less heat. The montane environment does not warm up in the same manner as the lowlands because of the reduced air pressure (caused by the shorter and therefore lighter column of air above the mountain), the low density of water vapour, and the generally clearer air. These three factors result in less retention of infra-red (warming) radiation in the mountain air, even though the intensity of radiation is greater on mountains for the same reasons as above. Similarly, during the night, heat is lost quickly and the daily change of temperature at about 3,000 m can be as much as 15° to 20°C, but this depends on cloud cover.

The rate of temperature decrease is generally about 0.6°C per 100 m but this varies from place to place, with season, time of day, water vapour content of the air, etc. Readings of minimum temperature were taken of five consecutive nights during the ascent of Mt. Kemiri and the results are shown in figure 9.1. The four uncircled points, recorded when the sky was cloud-covered early in the morning, indicate a lapse rate of 0.63°C per 100 m and suggest that snow would settle at 4,200 m if such a mountain existed. The permanent snow line would be still higher than this and occurs at about 4,700 m in Irian Jaya. The circled reading was recorded after a totally cloudless night. The regression coefficient of the four uncircled points is 0.99 but this may be artificially high because other writers have found two lapse rates, the first up to the cloud or condensation level (about 2,500 m) of 0.6°C/100 m, and the second above that of 0.5°C (van Steenis 1962, 1972; Whitmore 1984). Interestingly, however, the regression coefficient calculated from the many temperature readings made by van Beek (1982) up a number of mountains in the Mt. Leuser National Park was much higher when all the points were taken together than when they were divided at the cloud level.

Of considerable importance to understanding some aspects of mountain ecology is the Massenerhebung effect which was first noticed in the European Alps. It describes the phenomenon that vegetation zones on a large mountain or the central parts of large mountain ranges (such as the Barisan Range) are higher than they are on a small mountain or outlying spurs (Grubb 1971). This must be remembered when reading about mountains in, for example, Peninsular Malaya (highest peak 2,200 m) or Irian Jaya (highest peak 5,300 m) because zones of vegetation and fauna will be lower in the former and higher in the latter compared with the majority of mountains in Sumatra.

Relative Humidity

The percentage saturation of a mass of air increases as its temperature falls. The dew point (the temperature at which condensation occurs and clouds or drops of dew form) at different altitudes depends therefore on the

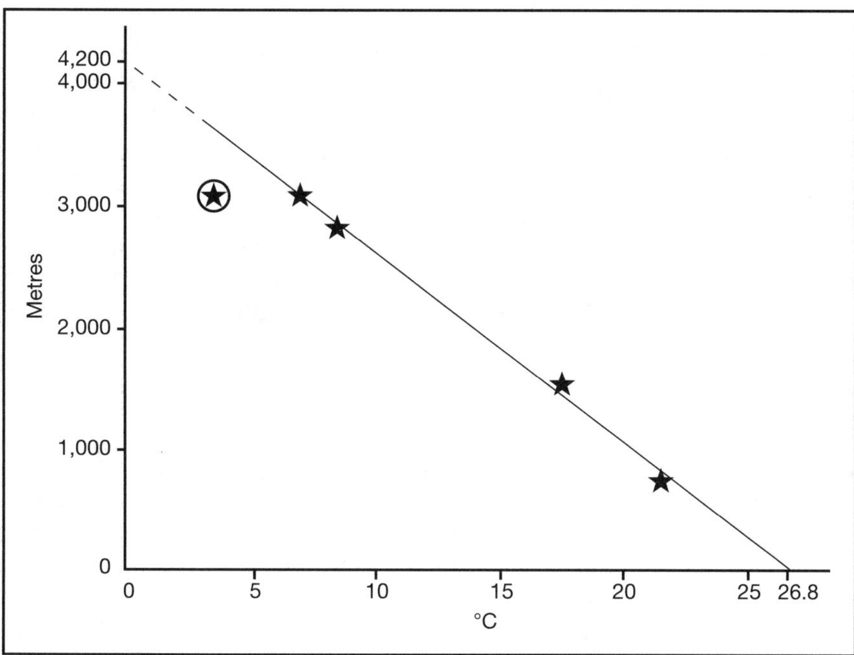

Figure 9.1. Minimum temperature readings on successive nights during the ascent of Mt. Kemiri by a CRES team.

temperature and the initial moisture content of the air. The forests of higher altitudes experience a very high relative humidity, particularly at night when the temperature falls, and the dew point is frequently passed so that water condenses on the leaves. During dry spells at altitudes above the main cloud layer, however, the relative humidity can be quite low during the day, leading to wide daily extremes.

Clouds

Clouds originate where ascending air reaches its dew point (see relative humidity). Once water droplets have been formed they tend to act as 'seeds', and more water condenses on them. During the wettest months, the slopes and peaks of mountains can be enveloped in clouds for days on end, but during the drier months, when the air is not saturated with water vapour, it is common for a belt of clouds to form around a mountain, often at about 2,000 m. As experienced mountain climbers know, a low, even continuous, layer of cloud at 2,000 m does not always mean that any cloud is present at 3,000 m or above.

Rainfall

Rainfall is generally greater on the windward (western) slopes of the Barisan Range and other mountains than on the lee (eastern) slopes (p. 14). There is, however, considerable local variation and there seems to be no guiding principle relating rainfall to elevation (McVean 1974). Rainfall on mountain slopes up to about 2,000 m will generally be higher than on the surrounding lowlands. At the cloud level, rainfall measurements are not particularly useful ecologically because water droplets in the air are utilised by the plants. Above the main cloud level, it seems reasonable to suggest that rainfall is less than below it because of the lower incidence of cloud cover.

Frost

Although the peaks of Sumatran mountains are not covered with permanent ice, they almost certainly occasionally experience frost in the hours before sunrise. Radiation of heat from the ground occurs both by day and by night but at night it is not compensated for by radiation from the sun. Thus as warm surfaces of plants, soil, rock, etc., cool so the thin layer of air immediately around them cools. Cold air is heavier than warm air and if there is not a slope for it to flow down or a wind to push it away, it will become progressively colder. Since heat loss from the earth is hampered by dust, fog and clouds the lowest temperatures will be reached on clear, dry nights. Maximum cooling will be from nonconductive surfaces such as dead twigs or grasses and dry sandy soil rather than from conductive surfaces such as rocks, water and living vegetation. Thus frost is most likely to occur on calm, dry, clear nights in flat hollows (van Steenis 1962). Such places are called frost pockets and can occur in silted-up small lakes or in the glaciated topography at the tops of the major mountains in the northern Barisan Range.

Ultraviolet Radiation

It has been suggested that ultraviolet radiation on tropical mountains is probably more intense than on mountains in any other region on Earth (Lee and Lowry 1980). This is due to the amount of ozone, which absorbs ultraviolet radiation in the stratosphere, being appreciably less near the equator, and to the atmosphere at low altitudes being more turbid and dense and thus more capable of absorbing or reflecting this radiation. Climbers must therefore protect their skins or rapidly suffer the effects of sunburn.

SOILS

The considerable changes in climate with altitude are reflected in the soils. Unfortunately the difficulty of identifying a consistent pattern parallels the confused pattern in rainfall which is influenced by local topography. Further variation obviously occurs due to the nature of the parent material. However, it is generally observed that the soils up to 1,000 m are more or less the same as the lowland soils (i.e., latosols and red-yellow podzolic soil). Between 1,000 m and the cloud zone at about 2,000 m (on the higher mountains), the increase in water supply leads to leaching, podzolization or water-logging. In addition, chemical weathering and biological activity are retarded by the lower temperatures. On the CRES expedition up Mt. Kemiri no termites were seen above about 1,500 m and earthworms apparently become scarcer as well (p. 303).

These features are accentuated in the cloud zone where the more or less continuously saturated air causes peat to be formed in blankets up to 1 m deep. The pH is often very low (pH 2.5-4.0). Above the cloud level, the peat layer remains but it is less thick and the soil is generally less wet (Burnham 1974, 1975; Whitmore and Burnham 1969). On Mt. Kemiri the percentage of carbon and the carbon:nitrogen ratio rose from 2.8% to 42.6% and from 10 to 24.6, respectively, from 880 m to 3,310 m. Mountain soils are frequently very poor in calcium and on Mt. Kemiri most of the soil samples collected by the CRES team had levels of only 0.11 m.eq/100 g or less. Similarly, many of the mountain soils examined by van Beek (1982) had no or very little calcium. These are indeed low compared with figures of 0.9 and 1.8 m.eq/100 g for the top 15 cm of lowland soils less than 100 m above sea level in North Sumatra. Soils on young volcanic debris are clearly very poor. A soil sample collected on the outer rim of the fuming Mt. Sibayak was, as expected, very low in bases (table 9.1). The important influence of soil nutrients on the growth of mountain plants is discussed on page 299 and the cycling of minerals is described on page 296. Volcano soils are discussed on page 301.

Table 9.1. Soil analysis of a sample from the outer unvegetated slope of the summit of Mt. Sibayak (Tanah Karo). Soil collected by a team from CRES.

pH H_2O	C (%)	N (%)	C/N	P-av (ppm)	Exchangeable (m.eq/100 g)				T.E.B. (m.eq/ 100 g)	C.E.B. (m.eq/ 100 g)	B.S. (%)
					K	Na	Ca	Mg			
3.4	0.22	0.02	11.0	22	0.13	0.08	0.23	0.51	0.95	10.82	9

VEGETATION

Introduction

A greater variety of plant communities is encountered on most tropical mountains than in any other area of comparable size in the world. Indeed, the vegetation changes observed when climbing a tropical mountain mimic the changes seen when travelling from the equator towards the poles. The tops of some Sumatran mountains are really very similar to, for example, Scottish moors in general appearance and some of the plants are even in the same genus. In fact, the plants of the high regions of tropical mountains have closer taxonomic links with plants of distant temperate regions than with the 'sea' of lowland forest plants that surrounds them.

The forest on the lower slopes of a Sumatran mountain seems indistinguishable from 'normal' lowland forest but then, with increasing altitude, the trees become shorter, and epiphytes, climbers and other features change in abundance. This zone is called lower montane forest. There then occurs a much more striking change when the tree canopy becomes even, the trees are even shorter, squat and gnarled, the leaves are thick and small, and mossy epiphytes are common. This zone is upper montane forest. Beyond this lies the subalpine forest, a complex of grass, heath and boggy areas. The grassy mountaintops of the subalpine zone have probably been caused by occasional fires, possibly lit by men for generations in their attempt to hunt animals such as deer and pig (van Steenis 1972). Analyses of soil samples taken in the subalpine zone of Mt. Kemiri by the CRES team revealed levels of potassium as high as 1.29 m.eq/100 g, levels which, in such generally nutrient-poor places, could have been caused only by fires. It has been suggested that the grassy slopes and valleys are caused by soil factors. In the wet grassy valleys near the summit this is more or less true because there is no tree species in the whole floral area of Malesia which can stand permanent inundation at high altitude (C.G.G.J. van Steenis, pers. comm.). The soil of the grassy slopes is, however, indistinguishable from the soil under the forest (unburnt) adjacent to it (C.G.G.J. van Steenis, pers. comm.). The question of regeneration of mountain vegetation after fire is discussed on page 379.

The general characteristics of the forest types on mountains is shown in table 9.2. During the ascent of Mt. Kemiri by the CRES team, the frequency of different forms of plant life were recorded at various altitudes and these are shown in table 9.3. In both these tables, the gradual decrease in vegetation height (illustrated in fig. 9.2) is recorded. This is not, however, accompanied by a proportional decrease in bole diameter which tends to increase with altitude for a given height of tree (fig. 9.3). The changes in vegetation structure up Mt. Kerinci have been described in general terms (Frey-Wyssling 1931; Jacobs 1958; Ohsawa 1979).

The division of montane vegetation into forest types is based on both floristics (van Steenis 1972) and on general structure. Many plant species cross the ecotones (transition areas between two types of forest or other ecosystems), but occur in different forms. For example, in the subalpine zone at 3,250 m on Mt. Kemiri, the CRES team found a flowering *Leptospermum flavescens* which was only 15 cm high, but 300 m lower down, the same species was a common tree about 4 m high. The dwarfed specimen had a massive root and this illustrates the fact that the measure of biomass called root-to-shoot ratio (ratio of dry weight of below-ground parts to above-ground parts) tends to increase with elevation. In the lower montane forest studied by Edwards the ratio was 0.1 (Edwards and Grubb 1982), but in harsh, cold habitats such as the subalpine zone the figure is often 3.0-4.0 (Ricklefs 1979).

Table 9.2. Characteristics of four types of forest found on mountains; the most useful characteristics in italics. * the leaf-size classes refer to a classification of leaves devised by Raunkier and modified by Webb (1959). The definitions are: mesophyll - 4,500-18,225 mm^2, notophyll - 2,025-4,500 mm^2, microphyll - 225-2,025 mm^2, nanophyll - less than 225 mm^2. An approximate measure of leaf area is $^2/_3$ (width x length).

	Lowland forest	Lower montane forest	Upper montane forest	Subalpine forest
Canopy height	25-45 m	15-33 m	1.5-18 m	1.5-9 m
Height of emergents	67 m	45 m	26 m	15 m
Leaf size class*	*mesophyll*	*notophyll/ mesophyll*	*microphyll*	*nanophyll*
Tree buttresses	*common and large*	*not common or small or both*	usually absent	absent
Trees with flower on trunk or main branches (cauliflory)	common	rare	absent	absent
Compound leaves	*abundant*	present	rare	absent
Leaf drip-tips	*abundant*	present	rare or absent	absent
Large climbers	*abundant*	usually absent	absent	absent
Creepers	usually abundant	common or abundant	very rare	absent
Epiphytes (orchids, etc.)	common	abundant	common	very rare
Epiphytes (moss, lichen, liverwort)	present	present or common	*usually abundant*	*abundant*

After Whitmore 1984 and Grubb 1977

Table 9.3. Frequency of plant life forms found in 10 m x 10 m plots at 10 locations on Mt. Kemiri. Location 1 is at the nearby Ketambe Research Station. Figures are means from three plots at each location.

Forest type	**Lowland forest**			**Lower montane forest**			**Upper montane forest**			**Subalpine forest**	
Forest subtype		**hill**	**hill**					**Moss forest**			
Number	**1**	**2**	**3**	**4**	**5**	**6**	**7**	**8**	**9**	**10**	**11**
Altitude	300	985	1,130	1,500	2,250	2,750	3,200	3,050	3,100	3,080	3,200
No. trees over 10 cm diameter	7	8	9	10	12	2	1	7	4	3	0
No. trees 2-10 cm diameter	18	10	14	12	16	6	15	12	6	6	0
Samplings	100s	400	100s	100s	100s	20	40	40	13	25	0
Creepers	3	0	3	6	6	0	0	1	0	0	0
No. climbers over 10 cm diameter	2	0	2	0	0	0	0	0	0	0	0
No. climbers 1-10 cm diameter	8	4	4	2	3	0	0	0	0	0	0
Herbs and grass	0	0	0	0	0	+	common	0	common	abundant	very abundant
Epiphytes (orchids)	+	+	+	many	many	0	0	+	0	0	0
Epiphytes (large ferns)	1	2	2	1	0	0	0	0	0	0	0
Epiphytes (small ferns)	+	+	+	+	100s	0	+	rare	+	0	0
Epiphytes (moss, lichen, liverwort)	0	0	0	0	+	abundant	abundant	very abundant	abundant	abundant	0
Tree ferns	0	0	0	0	3	0	0	0	0	0	0
Pinang palms	0	0	0	0	8	0	0	0	0	0	0
Canopy height (m)	30-45	25-35	20-35	20-30	10-15	4	3	9	4	2.5	0

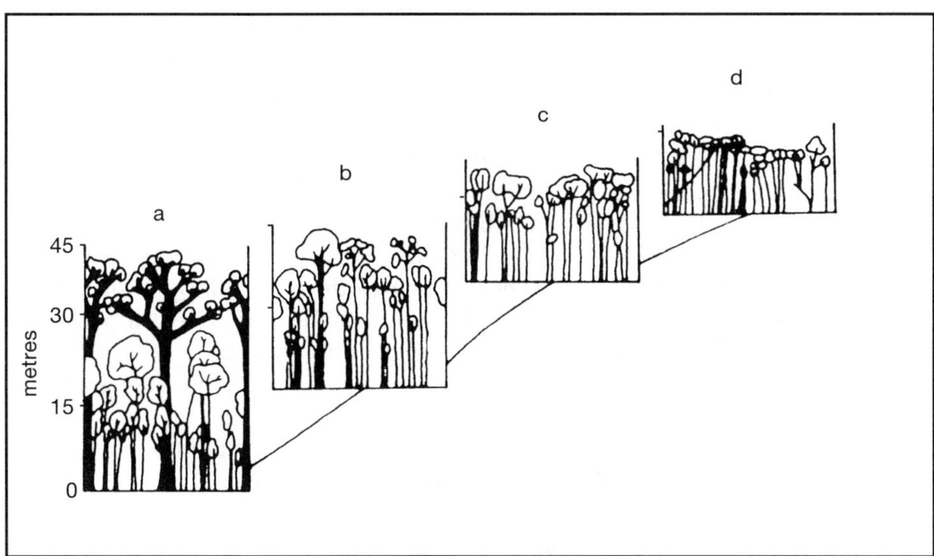

Figure 9.2. Cross-sections of: a - lowland forest; b - lower montane forest (low levels); c - lower montane forest (high levels); d - upper montane forest, to show decrease in tree height and simplification of structure.

Adapted from Robbins and Wyatt-Smith 1964

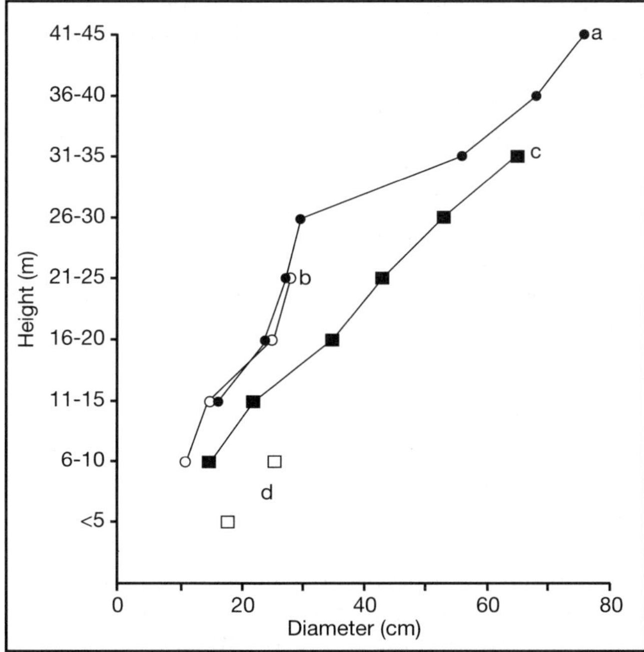

Figure 9.3. Relationship between tree height and trunk diameter at breast height in four montane forest types on Mt. Kemiri. a - lower montane, b and c - upper montane (two sites), d - subalpine.

Data collected by a CRES team

Leaf size and thickness change between lowland forest and subalpine forest. The CRES team measured the thickness of leaves from 10 species of trees/shrubs from the subalpine zone at 3,000 m and the mean thickness was 0.5 mm (range 0.3-0.85). The equivalent figure for a sample of leaves from lowland forest (Ketambe Research Station) was 0.2 mm (range 0.1-0.25). The thicker leaves of montane forests have been called 'pachyphylls' by Grubb (1974, 1977), which differ from other leaves chiefly in the thickness of their palisade layer (the long, photosynthetic cells of the leaf mesophyll) and the high frequency of a hypodermis (a layer immediately beneath the epidermis, often strengthened, forming an extra protective layer). See also the paper by Tanner and Kapos (1982).

The average heights of the various forest zones in Sumatra are as follows:

0 - 1,200 m	lowland forest
1,200 - 2,100 m	lower montane forest
2,100 - 3,000 m	upper montane forest
3,000 m +	subalpine forest

Temperature and cloud level are the major determinants of these forest zones (p. 299) but the vegetation zones are compressed on smaller mountains (the Massenerhebung effect, fig. 9.4). The presence of a montane plant does not necessarily correlate with altitude for reasons other than the Massenerhebung effect and this is represented by wavy lines in figure 9.4. From the locations where a particular species thrives, dispersion to other areas is achieved by upward and downward movement of its fruits, seeds or spores. Conditions of higher altitudes are mimicked by frost pockets (p. 280), beside mountain streams, on bare soil and near waterfalls. In the lowest parts of a species' range the individual plants are usually sterile. The population of a particular species in such an area therefore depends on seeds or spores being carried or blown down from above for its maintenance. A similar mechanism operates up the mountain. Thus it had been proposed by van Steenis (1962, 1972) that there is a zone of permanent occurrence which 'feeds' the populations of the upper and lower zones of temporary localities (fig. 9.5). Thus one mountain may lack a species of plant even though at the same height on a higher mountain (with a zone of permanent occurrence) that plant may be present.

Characteristic Plants

Whereas lowland forests are often characterised by trees of the family Dipterocarpaceae (p. 198), the lower montane forests are characterised by the occurrence of the families Fagaceae (oaks) and Lauraceae (laurels). Both these families are also present at lower levels and their apparent dominance is caused largely by dipterocarps becoming increasingly uncommon. Fagaceae are forest trees and are recognisable by their fruit which consists of a hard nut (containing the seed) sitting in an open or

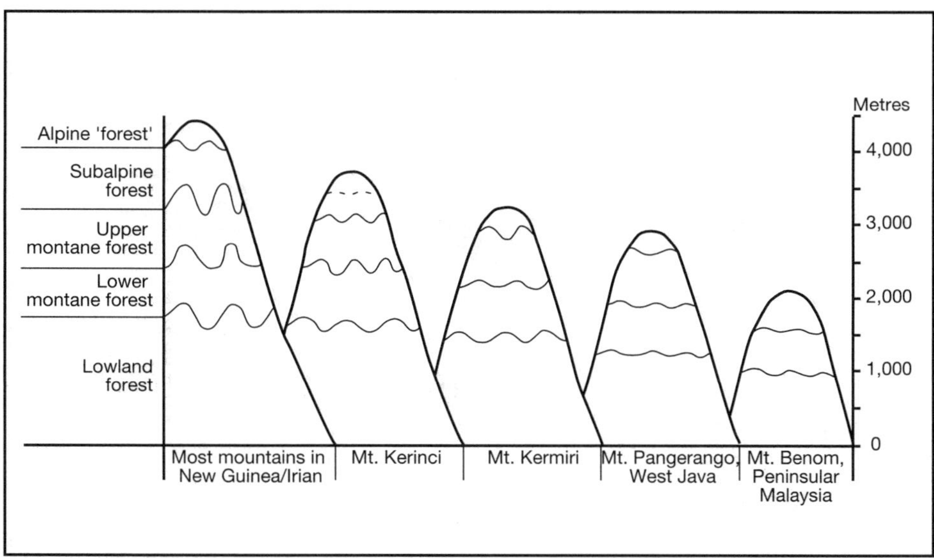

Figure 9.4. Vegetation zones on five mountains to show the compression found on smaller mountains.

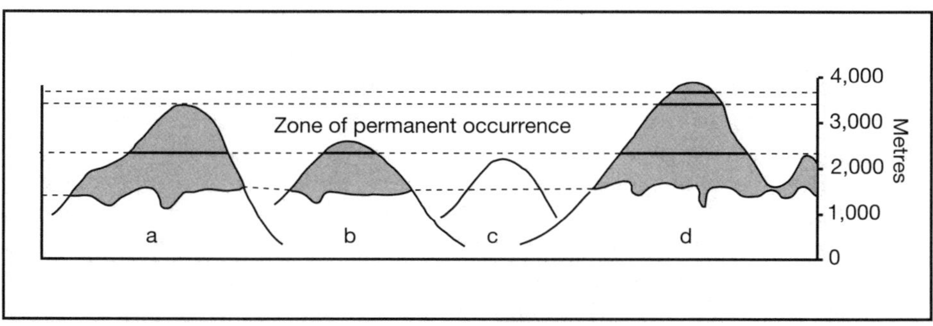

Figure 9.5. The zone of permanent occurrence on mountains and its significance for the distribution of species. Shaded areas represent localities where a hypothetical species might be found.

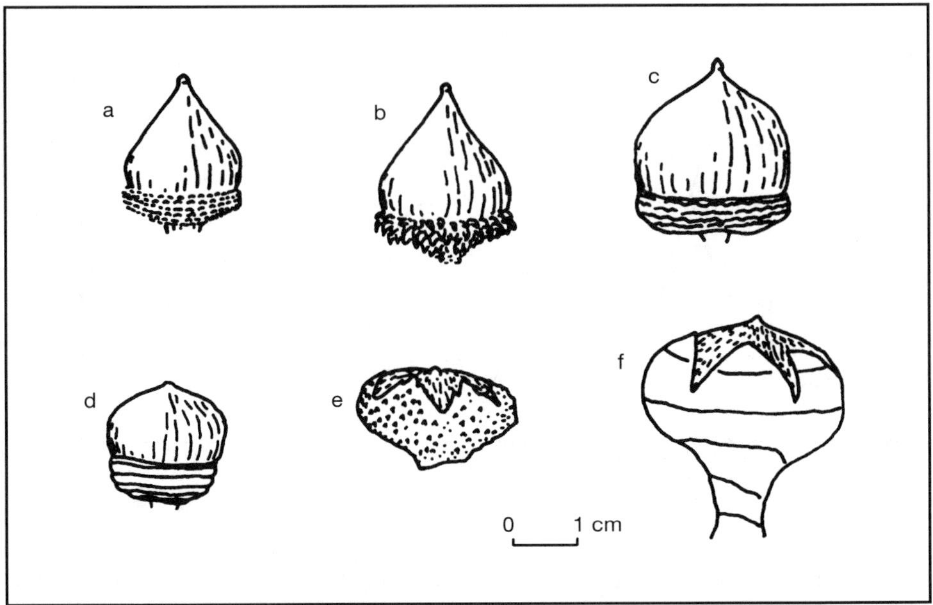

Figure 9.6. Fruits from trees of the genus *Lithocarpus* (Fagaceae) found in Sumatra: a-*L. conocarpus*, b-*L. hystrix*, c-*L. lamponga*, d-*L. rassa*, e-*L. blumeana,* f-*L. encleisacarpa*.

After Corner 1952

closed, usually spiky, cup. The fruit of various species of *Lithocarpus,* one of the commonest genera found in Sumatra, are shown in figure 9.6. Lauraceae (the family containing cinnamon *Cinnamomum burmansea* and avocado pear *Persea americana*) is represented in the mountain forest primarily by the genus *Litsea.* This genus has fruits which also sit in a cup, but the cup is not woody and the fruit is soft and pulpy.

Tree ferns are common in lower montane forests. All but one of the tree fern species in Sumatra belong to the genus *Cyathea* (the exception being a montane species of *Dicksonia*) and the identification of species is based on the form of reproductive parts and on the fine brown scales found covering young coiled leaves and the bases of older leaves (Holttum 1977). A simplified identification key to some of the species found in Sumatra is given by Holttum and Allen (1963). A recent demographic and growth study of *Cyathea* tree ferns in Jamaica has shown that a trunk can grow about 1 m per 15 years. A tree fern 1 m high will, however, be considerably more than 15 years old because many years elapse between the first growth of a young fern and the stage at which upward growth in the form of a trunk begins (C.G.G J. van Steenis, pers. comm).

The upper montane forest is characterized by the order Coniferae (the pines and related trees), particularly *Dacrycarpus imbricatus* (de Laubenfels 1969), and the families Ericaceae (e.g., bilberries *Vaccinium*, *Rhododendron*) and Myrtaceae (the family to which *Eucalyptus*, clove *Eugenia aromatica* and paperbark *Melaleuca* belong), particularly *Leptospermum flavescens*. These trees are often quite low, crooked and gnarled, and because of their mysterious shapes this part of the zone has earned the name of elfin forest. On ridge tops and in the upper regions of upper montane forest the conditions are quite dry and the commonest and most conspicuous epiphytes here are the beards of the bright yellow-green lichen *Usnea*. In the valleys and in the lower regions of this same forest more or less the same tree community is found but virtually every available space on the trees and on the ground is covered with thick carpets of liverwort and moss. This is commonly called moss forest and coincides with the most frequent level at which clouds form (p. 279).

The subalpine zone is characterized by dwarf specimens of upper montane forest species and by grasses (e.g., *Agrostis, Festuca*), rushes and sedges (*Juncus* of Juncaceae, *Carex, Scirpus* and *Cyperus* of Cyperaceae), and small herbs often with colourful flowers. Many subalpine plants form rosettes of leaves just above the ground. These rosettes are caused by the distance between successive leaves being extremely short, and only the flower stalk rises any distance above the ground, presumably to decrease water loss and to facilitate pollination or seed dispersal. The leaf under-surface of some species with rosette leaves and some other species are covered with dense, silky, white hairs (see p. 290).

It has been suggested that the high levels of ultraviolet radiation received by subalpine and alpine zones on high tropical mountains (p. 280), together with altitudinal shifts in vegetation zones during the Pleistocene (p. 17), may have led to accelerated rates of mutation and speciation (Lee and Lowry 1980). This may be a cause of high levels of endemism found in tropical subalpine and alpine plant communities. Figures have not been calculated for Sumatran mountains but 37% of the plant species on the higher parts of Irian Jaya mountains are not found in Papua New Guinea (Smith 1975), and 40% of the subalpine plants on Mt. Kinabalu, Sabah, grow only on that mountain (Smith 1970).

Repeated freezing of dead plant parts is thought by van Steenis (1972) to result in mummification. The dead material becomes an ashy-grey colour and when touched is easily reduced into fine grey dust. This was observed on Mt. Kerinci by Jacobs (1958) and by the CRES team on Mt. Kemiri.

Leaf Adaptations to Temperature and Radiation

In theory, structures lose heat by convection most rapidly at their edges where wind currents disrupt insulating boundary layers of still air; the more edges (or smaller), the cooler the leaf and the lesser the water loss (Ricklefs 1979). Although there are exceptions and effects opposing this theory, it explains the general pattern found in the shrubs and small trees of the subalpine and upper montane zones where midday temperatures can be high and where water availability is low. The same plants, however, need to cope with night-time temperatures approaching zero. The cooling effect of small leaves appears to be balanced by their close arrangement on the end of twigs, their dark colour and compact crown structure. Thus a compromise structure seems to have developed whereby the leaves will warm up quickly in the early morning sun so that efficient photosynthesis can begin as soon as possible before either clouds form or the solar radiation becomes too intense. The closeness and often vertical position of these small leaves have also been interpreted as a means of 'fog-stripping'. Water-saturated air or fog would be temporarily caught within the leaves on the end of a twig with a greater likelihood of water condensing on the plant and falling to the ground. Leaves in subalpine and dry areas of the upper montane zones probably experience longer periods of lower relative humidity than do leaves in the cloudy upper montane forest and so will run less risk of encouraging epiphyllae to grow on their leaf surfaces. Many leaves of the upper montane and subalpine zone plants have characteristics normally found in the areas subject to periodic drought: that is, xeromorphic characters. In fact, however, the mountain plants are not particularly efficient in restricting water loss (Buckley et al. 1980).

The woolly hairs of some subalpine plants have been variously attributed with the ability to protect against high temperatures (Lee and Lowry 1980), intense ultraviolet radiation (Mani 1980a) and frost/freezing (Smith 1979b; van Steenis 1972). Maybe they play a role in all three. For plants with leaf rosettes on the ground such as the silverweed *Potentilla borneensis,* however, the woolly hairs are on the lower surface and petioles only. This would make sense as frost protection but not against ultraviolet radiation or high temperatures. Other plants provide their growing buds and perennial parts with thermal insulation by retaining old leaves, and by having tufted branches and persistent scales. Such protection is necessary because the 'cost' to a subalpine plant of replacing a leaf is much greater than for plants growing in the more hospitable environments lower down the mountain or in the lowlands. Lee and Lowry (1980) list the adaptations of leaves to intense ultraviolet radiation as thick cuticles, wax deposits, and high concentrations of red, protective leaf pigments called anthocyanins in young leaves.

Origin and Dispersal of the Flora

The mountain flora of Sumatra is derived from two sources: those which originated locally (autochthonous) and those for which the centre of origin is outside the area concerned (allochthonous) (van Steenis 1972). The local source species are themselves divisible into groups: those which are characteristic of equatorial lowland forest, such as members of the families Dipterocarpaceae, Bombacaceae (e.g., durian) or the genus *Ficus* (figs), and those which have a wide latitudinal distribution throughout the globe and a wide altitudinal distribution in the tropics, such as Pinaceae (e.g., *Pinus*), Cruciferae (e.g., mustard), Theaceae (e.g., tea) and tree ferns. Of the first group very few are found over 2,000 m and most are found below 1,000 m. Genera and families of the second group, however, have wide tolerances of temperature and have both low- and high-temperature adapted (microtherm and megatherm) species. The low-temperature adapted species could possibly have originated outside the tropics and migrated towards Sumatra but they are equally likely to have evolved in or near Sumatra. None of the autochonous plants are therefore very useful in deciding how Sumatra gained its mountain flora.

The allochthonous flora, however, although a minority of the total mountain flora, do allow hypotheses regarding their origin to be formed. This part of the flora belongs to genera whose species are only found in cold climates (i.e., microtherm species), and in the tropics they are never found below 1,000 m and thus are generally found only in the subalpine forests on mountains 2,000-2,500 m high (see principle of the zone of permanent occurrence, p. 287). These genera, such as *Rhododendron,* the grass *Deschampsia* and the coloured herbs *Gentiana* and *Primula,* are found in many tropical and subtropical countries, yet none can tolerate a hot climate.

Wallace (1869) worked out more or less how the mountains of the Sunda Region had received their flora – that is, by dispersal at lower levels during cooler glacial periods when the Sunda Region was a single landmass. Van Steenis (1934a,b, 1962, 1972) made exhaustive studies of the region's microtherm allochthonous flora and revealed numerous interesting patterns of distribution. Plants must have extended their ranges, hopping by one means or another along mountain chains. Some of the 'chains' of passage are still quite complete. For example, the prostrate, branched herb with small white flower called *Haloragis micrantha* is found from New Zealand to Japan to the top of the Bay of Bengal, as well as on the top of Mt. Papandayan in West Java. It does not seem to have crossed over the relatively short gap to Sumatra (figure 9.7). Conversely, the stately herb with up to five layers of bright yellow flowers called *Primula prolifera* shows an enormous gap in its distribution (figure 9.8). Soils seem have little or no influence on the distributions since a single species will be found on soils originating from igneous, sedimentary or recent volcanic parent material. The age of the rocks seem to have an effect, though,

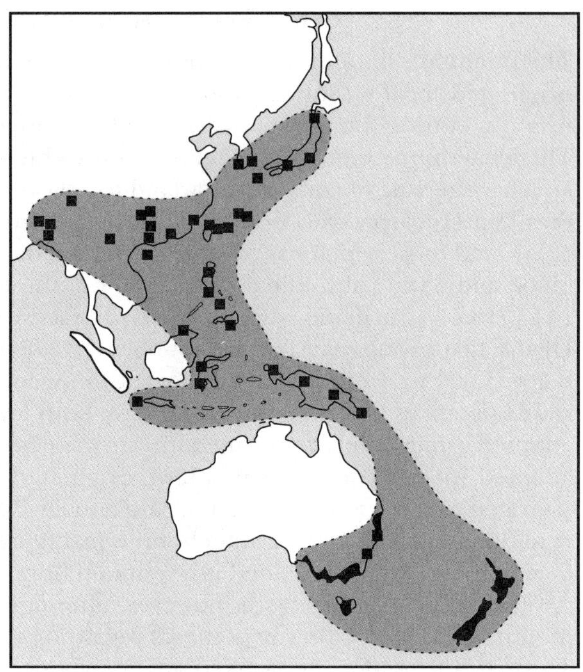

Figure 9.7. The distribution of *Haloragis micrantha.*
After van Steenis 1972

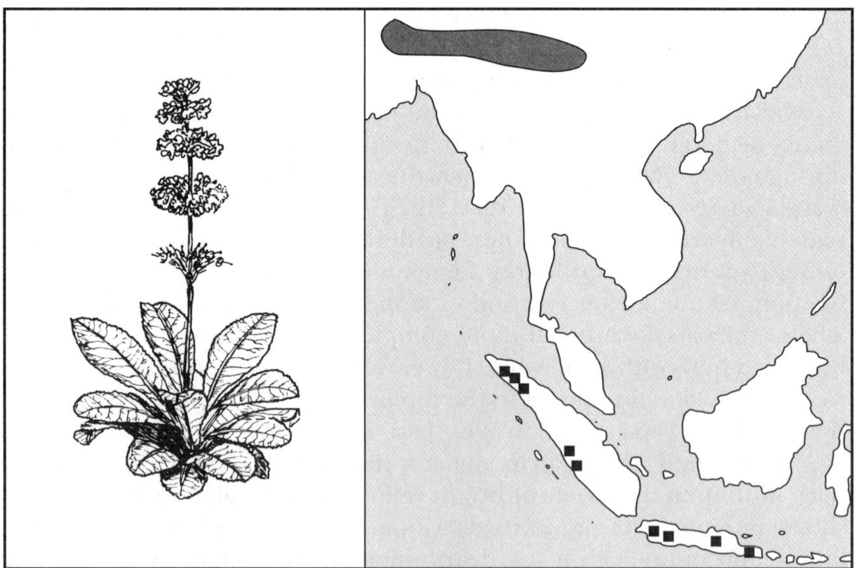

Figure 9.8. The distribution of *Primula prolifera.*
After van Steenis 1972

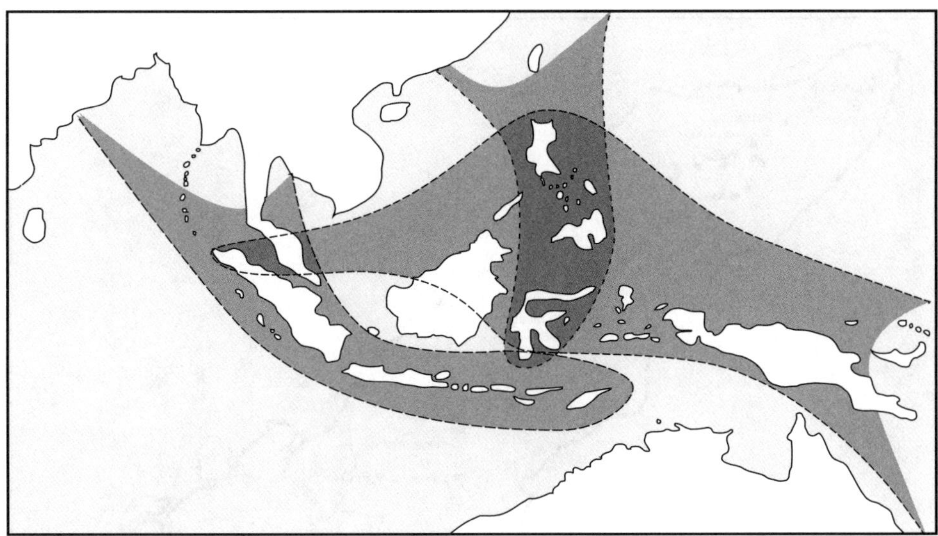

Figure 9.9. The three tracks by which microtherm allochthonous flora reached Sumatra.

After van Steenis 1972

with some species not being present on recent volcanic soils (van Steenis 1972) (p. 301).

From his analysis of the distribution of about 900 cold-adapted mountain species, van Steenis (1972) concluded the existence of three tracks by which plants arrived in Sumatra and other parts of the Sunda Region during some period or periods in the geological past. These three tracks are shown in figure 9.9. Continuous ranges of high mountains do not, of course, exist along the entire lengths of these tracks. During the coldest times of the Pleistocene the mean temperature dropped only about 2°C (p. 12) which, with rates of temperature change of 0.5-0.6°C/100 m (not necessarily applicable at that period), is equivalent to a drop in the levels of the forest zones of 350-400 m. The locations of mountains over 2,500 m high, and of mountains 2,000-2,500 m high (fig. 9.10), illustrates how the number of viable 'stepping stones' would have increased for microtherm plants during the coldest periods of the Pleistocene.

There seem to be no easy answers to the problem of how plants actually

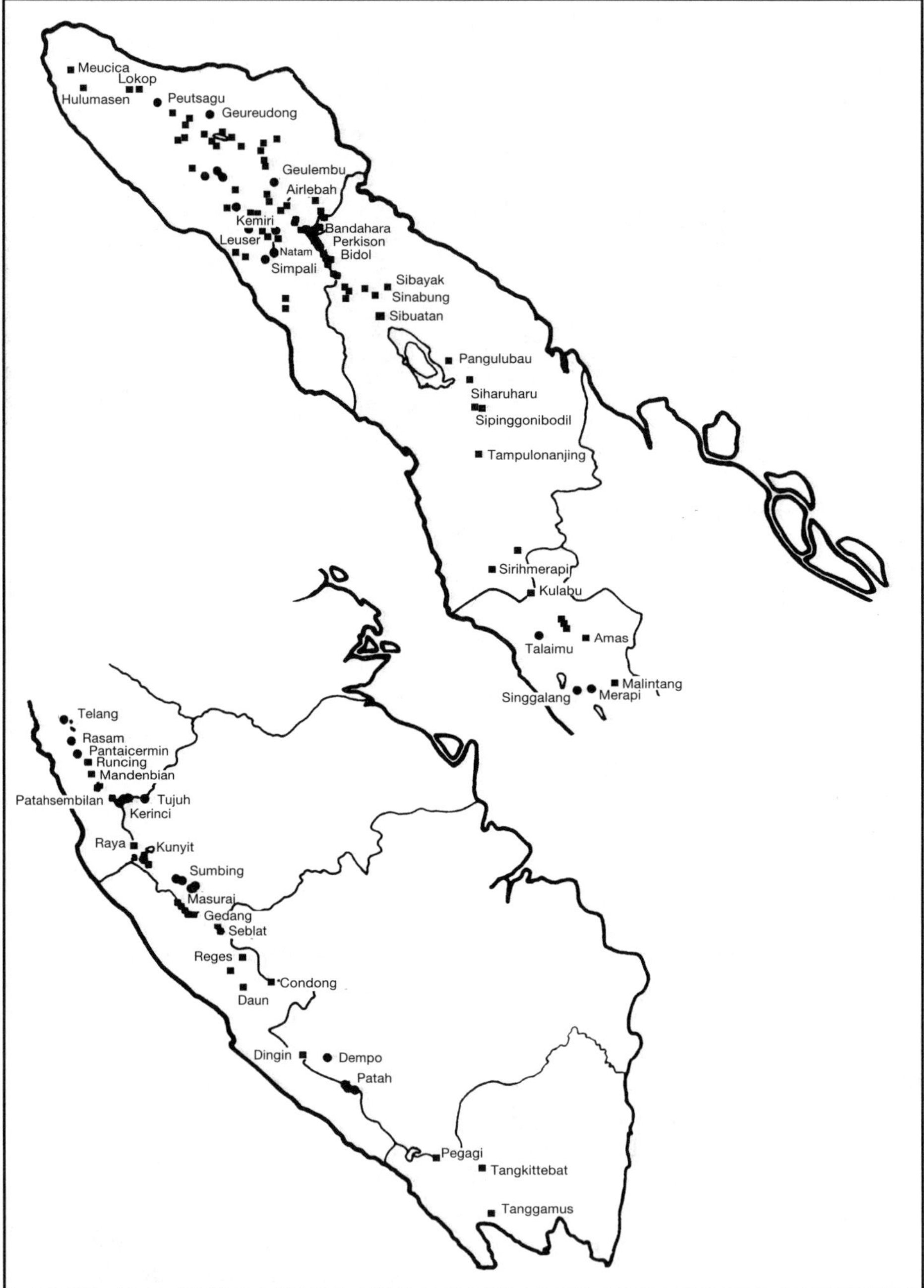

Figure 9.10. Location of mountains of 2,500 m and over (circles) and of 2,000-2,490 m (squares) on Sumatra.

dispersed (and currently disperse) themselves between their small islands of suitable habitat. There are four main ways in which plants disperse their fruit and seeds, or, in the case of ferns, mosses, fungi, etc., their spores:

a) by water;
b) by wind;
c) by sticking to, being eaten but not destroyed, or being removed for later ingestion but then forgotten by terrestrial mammals;
d) by sticking to or being eaten and not destroyed by birds or bats.

The first and third of these are clearly of no significance in explaining the distribution of microtherm plants. Water only flows downstream to unfavourable habitats, and the distances between high mountaintops are too great for terrestrial mammals to complete before either the seed is excreted or is rubbed off.

Wind seems a likely candidate for the dispersion of the minute, dust-like seeds of orchids, spores, or the light, parachute-like, plumed seeds (achenes) of the many montane species of Asteraceae (formerly Compositae). Wind must indeed transport such seeds over quite long distances but plumes only open in low humidity, and van Steenis (1972) points out that the expected wide distributions of such species is not always found.

Many of the shrubs and small trees in the upper montane and sub-alpine forests bear berries which, if not exactly tasty, are quite edible and form an important source of food for birds and mammals. Many of the seeds in these juicy berries pass through bird intestines without harm. However, since the time to pass through the intestine is little more than an hour, there is very little possibility of transferring seeds to mountaintops more than about 30 km away and, in addition, birds of mountaintops are mostly sedentary. Sticky or adherent seeds or fruits may attach themselves to feathers or legs, but before a long flight birds will generally preen themselves thoroughly so that their feathers give maximum flying efficiency.

If one assumes, however, that occasional, unusual dispersion of seeds occurs between neighbouring mountains by wind or by birds, it would be natural to expect that the percentage of plant species dispersed in these ways would increase up mountains in comparison with species whose seeds have no obvious dispersal device or are too heavy for wind dispersal. Although this has not been investigated in Sumatra, it has been investigated for Mt. Kinabalu in Sabah (van Steenis 1972) and Mt. Pangrango in West Java (Docters van Leeuwen 1920). The results show that the opposite is true.

The question of dispersal is further complicated by the observation that communities of plants having a variety of dispersal mechanisms have exactly overlapping ranges on different mountains as though means of dispersal was immaterial. In addition, limited experiments were conducted on Javan mountains to introduce a species not found on a particular mountain to grow there. They met with singular failure. Even if a species grew from a seed and survived, its own seeds did not survive (van Steenis 1972).

As stated above, the lowering of forest zones during the Pleistocene would have made chance dispersal easier if only because more mountain-tops would have become suitable habitat. But even then, large gaps would have existed and it is clear that the means by which plants or groups of plants effected their dispersal is not yet fully understood.

Biomass and Productivity

There appear to be no estimates of biomass or productivity in montane forests of the Sunda Region, but Grubb (1974) has summarised the trends he found from examining data for montane forests in all parts of the tropics. Thus:

a) biomass decreases proportionally less than height on passing from lowland forest to upper montane forest (see fig. 9.3, p. 285);
b) production of woody parts declines from lowland forest (3-6 t/ha/yr) to upper montane forest (± 1 t/ha/yr);
c) production of litter, particularly leaf litter, decreases proportionally much less than biomass or, it seems, production of woody parts;
d) the standing crop of leaves declines proportionally much less than the overall biomass;
e) the mean life span of leaves seems to lengthen only slightly from lowland forest to upper montane forest;
f) the leaf area index (m^2 of leaves per m^2 of ground) decreases proportionally much more than leaf standing crop because the leaves become thicker and harder. Thus the area of leaf per gram decreases from 90-130 cm^2/g in lowland forest, to ± 80 cm^2/g in upper montane forest.

It appears then, that as total production falls the montane plants invest relatively more of their production in making leaves, and so it is important to understand what advantages accrue to a plant from having thick, 'expensive' leaves which may last no longer than the thinner, 'cheap' leaves of the lowlands.

Mineral Cycling

In any type of forest there are two types of mineral turnover:

a) the rapid cycling in the small litter (leaves and twigs) and in throughfall (rain reaching the forest floor), and
b) the much slower cycling in the large woody parts of trees.

As stated above, moving from lowland to montane forests, the production of woody parts decreases proportionately more than does the production of leaves. The concentration of nutrients in leaves from upper montane forest are roughly half the concentration found in leaves in low-

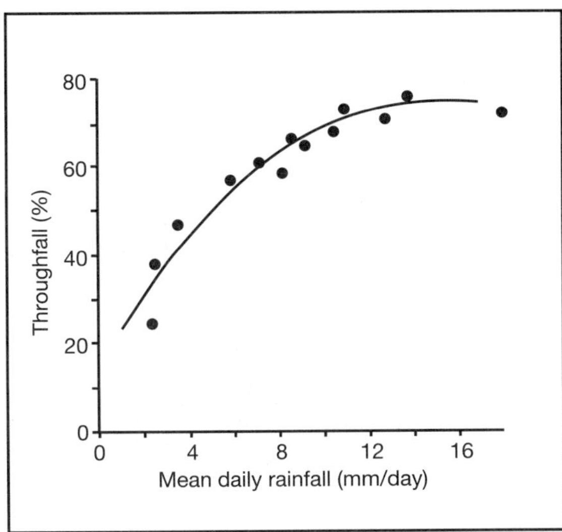

Figure 9.11. Percentage of rain falling over a lower montane forest that actually reaches the forest floor.

After Edwards 1982

land forests. This reduction is compounded because trees of upper montane forests seem to absorb half of the minerals from their leaves before shedding them, compared to about a quarter for trees of lower montane forests (Tanner and Kapos 1982). Thus, lower quantities of minerals cycle through leaves in montane forests than in lowland forest, and much lower quantities through woody parts.

The most detailed study of mineral cycling in montane forest close to the Sunda Region is by Edwards in Papua New Guinea (Edwards 1982; Edwards and Grubb 1977, 1982; Grubb and Edwards 1982). In the final paper of the series, Edwards (1982) summarised the major findings of the preceding papers. The major external input of minerals into the montane forest was from the rain, and these minerals may have been released into the atmosphere from fires (Ungemach 1969). Not all the rain reaches the forest floor as a certain percentage soaks into bark, evaporates, or gets trapped. The percentage which reaches the ground, the throughfall, varies with rain intensity, being greater when the rainfall is heavier (fig. 9.11). This throughfall contains minerals leached from the leaf and bark surface and from decomposition products so that the mineral concentration in rain reaching the forest floor is considerably higher than that reaching, say, a neighbouring cultivated field.

In addition to the throughfall, the other source of minerals falling to the forest floor is the litterfall. The proportions of minerals in the litterfall

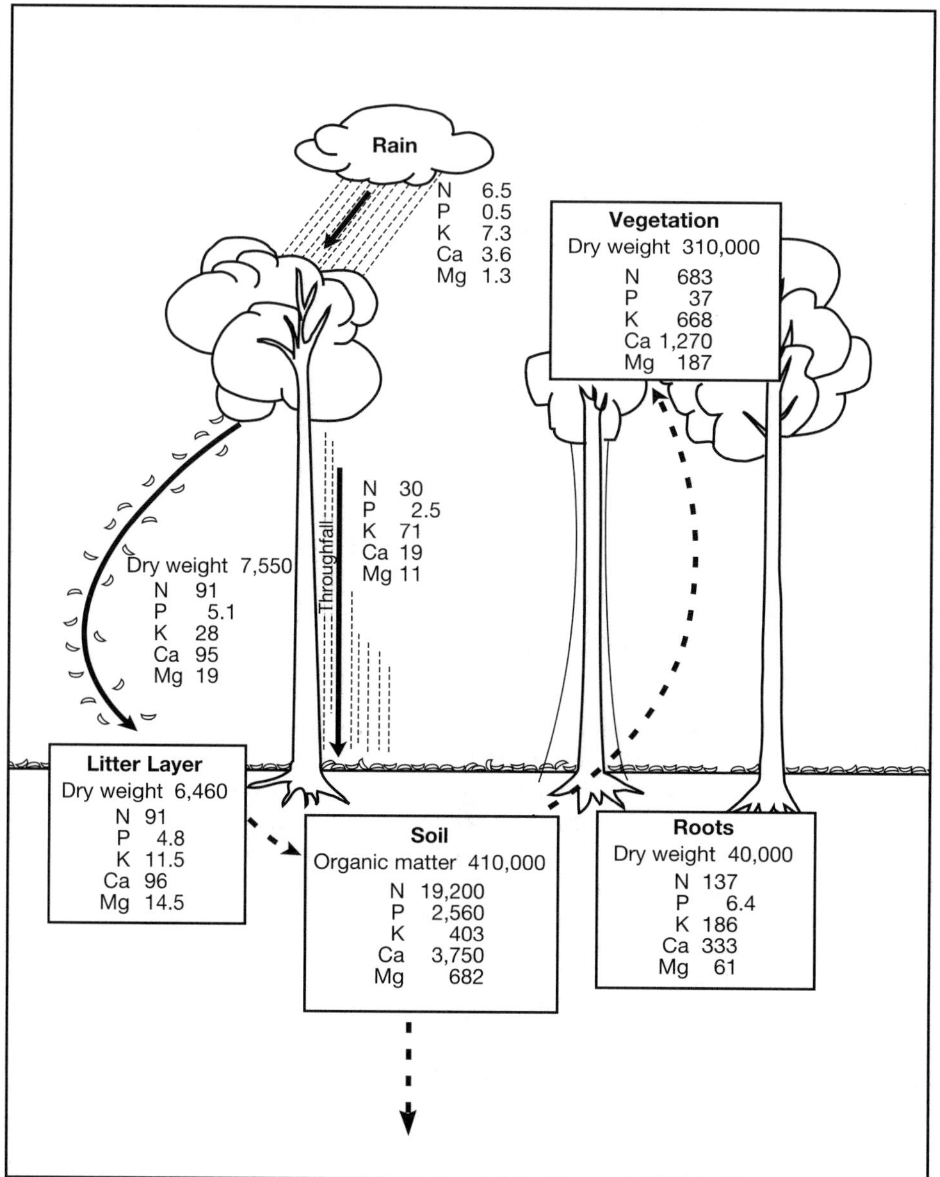

Figure 9.12. Summary of mineral cycling in a lower montane forest. Data in boxes are expressed in kg/ha. Complete arrows represent major pathways of mineral transfer, and data are expressed in kg/ha/yr. Dashed lines represent presumed pathways for which no data are available.

After Edwards 1982

is affected somewhat by the withdrawal of minerals from leaves before abscission (i.e., the parting of leaf and twig). Edwards compared mineral concentrations in healthy leaves with leaves in the litterfall, and found that calcium and magnesium did not seem to be absorbed, but that nitrogen and phosphorus were respectively 15% and 31% lower in the litterfall (Edwards 1982). The recycling of these two minerals may indicate that they were limiting in the soil (see p. 299). Potassium is very mobile in soil and quickly resorbed by plants so its retention is less important. The calcium levels in the top soil layers of the forest examined were extremely high (16-40 m.eq./100 g) (Edwards and Grubb 1982) compared to the levels in many mountain soils in Sumatra (p. 281) and elsewhere in the western Sunda Region. It would be interesting to determine whether calcium is absorbed before abscission in Sumatran mountain plants, and thus gain an indication of the limiting role of calcium in the mountain forest.

A summary of mineral cycling in the Papua New Guinea montane forest is shown in figure 9.12.

Limitations on Forest Distribution and Growth

The lower limit of a particular type of montane forest is probably set, not by temperature, but by the ability of its species to compete with species of the lower formation. The upper limit of a forest formation, however, is set by some factor related to temperature (Grubb 1977). The Massenerhebung effect (p. 278) appears to act on plants through the height at which cloud habitually settles and this is clearly related to temperature. The most important effect of cloud is that it prevents bright sunlight from raising the temperature of plant leaves. Those not adapted to living in the cloud zone may not reach their optimum temperature for photosynthesis. The amount of radiation available for photosynthesis is also reduced by cloud (Grubb 1974). The decrease with altitude of biomass, productivity, number of life forms and number of species can also be attributed to these factors.

Although temperature and radiation are the factors primarily responsible for limiting montane forest distribution and growth, the mineral supply is also highly important. Without conducting soil analyses or productivity studies, it can be seen that:

a) the crooked and gnarled growth of many montane trees is similar to the growth of trees on infertile lowland soils or to the miniature bonsai trees deliberately grown in nutrient-poor soils;
b) the slow rate of litter decay and the subsequent humus mineralisation may be assumed to lock up plant nutrients (especially nitrogen and phosphorus) in forms unavailable for plant growth;
c) forest types are found at lower levels and forest growth is less on sites most likely to be poor in nutrients (i.e., ridges);

d) the most dwarfed forests are found on peaty soils of very low pH which are likely to be too infertile for most plants;
e) some of the species found on peat soils in upper montane forest are found also in the similarly acid, infertile heath forests of the lowlands;
f) the insectivorous pitcher plants *Nepenthes* (p. 260) are often abundant in peaty upper montane forest (Grubb 1977).

Quantities of nutrients in leaves (see p. 258) reflect to some degree the nutrients available in the soil and so analysis of leaf nutrients is standard practice in investigative agriculture. Grubb examined both his own and other data on nutrient levels in leaves of montane species and found considerable variation within the forest types, but was able to conclude that the supply of phosphorus seemed particularly deficient in peaty upper montane forest. This supports the suggestion that phosphorus is the factor underlying the observations d), e), and f) above (Grubb 1977). Interestingly, Grubb (1974) also analysed the nutrients in leaves of pitcher plants *Nepenthes* to determine how similar they were to the surrounding vegetation that was not able to catch or digest insects. The results showed that nearly three times as much phosphorus and almost twice as much potassium was present in *Nepenthes* leaves compared with nearby trees and shrubs.

The major limitation on leaf life in the upper montane subalpine forests is probably not, as in most other forests, the destruction by insects, but the high humidity. This allows invasion by fungi and bacteria, and by epiphyllous lichens and mosses which are able to puncture the leaf's cuticle. The very thick outer walls of pachyphylls may be an adaptation to minimise such unwanted penetration (Grubb 1977).

On page 193 it was argued that drip-tips on leaves allow water to flow rapidly off the leaves and so prevent, or at least slow down, the growth of epiphylls. If that is so, then why do montane forest plants invest in much more expensive thick leaves? Grubb (1977) has suggested that whereas drip-tips are effective under conditions of alternating heavy rains and sun, they would not be so effective where the dampness of leaf surfaces is caused by very high humidity and very fine water droplets from clouds or fog.

Similarities between Upper Montane Forest and Heath Forest

The impoverished, acid soils of some higher mountain zones might lead to an expectation that upper montane and heath forests (p. 253) might be somewhat similar. The structure of the forests are indeed rather similar with both of their canopies being even and appearing rather pale (high albedo). The trees tend to have dense crowns, usually with microphyll leaves (p. 283) which are often held obliquely vertical and closely placed on the twigs. Large woody climbers are more or less absent and biomass and productivity also seem to be low (p. 296).

They also share some tree and shrub species which do not (or only very rarely) exist in other types of forests. Examples are sesapu *Baeckia frutescens* (p. 256), and species of bilberry *Vaccinium* and *Rhododendron.*

Volcanoes

Near volcanic craters the ground surface is generally rocky, dry, sterile, acid, exposed, lacking in organic matter, and often warm or hot from the gases beneath. The air contains such potentially toxic gases as sulphur dioxide (SO_2) and hydrogen sulphide (H_2S) (the smell of rotten eggs) as well as carbon monoxide (CO), nitrous oxide (NO), and chlorine (Cl_2) (van Steenis 1972). Water flowing out of craters is commonly little more than a stream of sulphuric acid. These are hardly conditions that favour plant growth, but van Steenis (1972) reports finding the tussock-forming sedge-like *Xyris capensis* (a species ranging from Central Africa to Australia) in a North Sumatran crater stream, the pH of which was 2.9. Blue-green algae are also commonly found in hot, sulphurous acid water.

Ash produced by eruptions occurs in screes on the tops of volcanoes, such as Mt. Kerinci, which have been active in the recent past. They are barren, pervious, sterile and unstable, moving downhill, particularly during heavy rains. The only plants able to colonise such ground must have long, deep roots for anchorage and to find moisture. Grasses (e.g., *Agrostis)*, sedges (e.g., *Carex*) and the composite *Senecio sumatrana* are among the first pioneers on Mt. Kerinci, but the highest-growing plant of all is a small fern (Frey-Wyssling 1933). *Melastoma* and pandans are common on the sulphorous screes of Mt. Sibayak and Mt. Sinabung, Tanah Karo, but they are absent from the non-sulphorous screes on Mt. ni Telong, Central Aceh (Frey-Wyssling 1931).

The magnificent edelweiss *Anaphalis javanica* is one of the very few plants confined to volcanoes and is known only from central and southern Sumatra. It is a long-lived pioneer of volcanic ash screes and crater soils and is frequently gregarious. It can grow to about 4 m (sometimes to 8 m) with its stem as thick as a man's wrist (van Steenis 1972), but 1 m specimens are now rare. The stem has fine, velvety white hair and the abundant flowers appear white, too, except for a central yellow disk. Most mountain plants have become dwarfed but the edelweiss is one of the giant, white-haired members of the Asteraceae which, in height at least, dominate the generally short subalpine vegetation on tropical mountains. Its equivalent on east African mountains are giant species of *Senecio* and in the Andes the 9 m tall *Espeletia* (Mani 1980a). The edelweiss has traditionally been regarded as a gift from heaven in some areas and climbers have taken small pieces with them when they descend. The habit has turned to vandalism on some mountains, such as Mt. Singgalang and Mt. Kerinci, where this strikingly beautiful plant does not simply have a portion removed, but is dug up

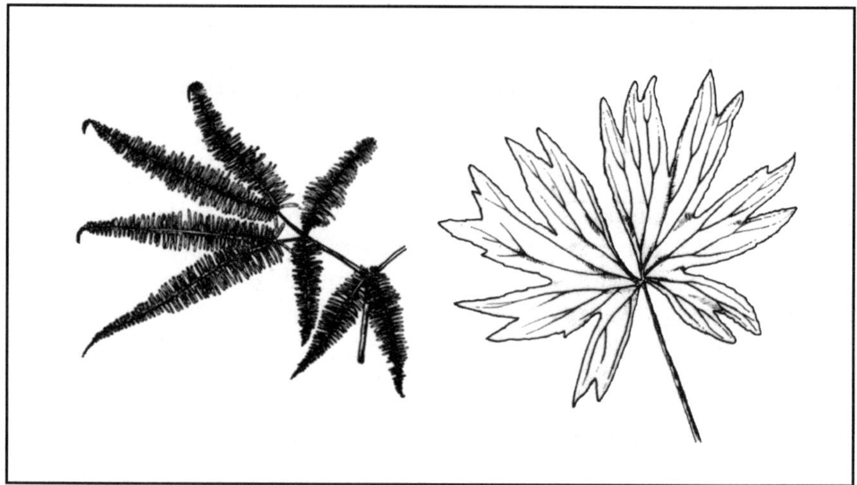

Figure 9.13. *Gleichenia* and *Dipteris* ferns of the higher mountain zones.
After Holttum 1977

completely in the naive hope that it might grow in the lowlands.

On volcanic scree such as on Mt. Kerinci (Jacobs 1958) the fern *Gleichenia arachnoidea* forms dense, prickly thickets. This and the common mountain fern *Dipteris* (fig. 9.13) are of interest because fossils of seemingly identical plants have been found in Greenland and Britain dating from the Cretaceous period, 65-135 million years ago (Holttum 1977).

Mountain Lakes

Mountain lakes in Sumatra are generally of two types: lakes in extinct volcanic craters such as Lake Maninjau (± 600 m a.s.l) and Lake Tujuh (1,996 m a.s.l) (de Wulf et al. 1981), and marshy lakes in small depressions formed by glacial action (see p.12).

After volcanic activity ceases, water begins to accumulate in the craters. At first the sulphurous content of the water prevents colonisation by most plants, but in time this is either leached away or incorporated into sediments, and marshy plants can begin to grow. The depth of a crater lake obviously depends on the height above the lake floor of an outlet, but is usually quite considerable. The flora and fauna of high, deep, cold, oligotropic (nutrient-poor) lakes is impoverished and, since the sides are generally steep, there is little or no characteristic fringe habitat.

Of more interest, however, are the marshy fringes of shallow lakes in glacial hollows such as the one at 3,000 m on Mt. Kemiri. The rushes, sedges and grass such as *Juncus*, *Carex*, and *Scirpus mucronatus* form clumped mats, while nearer the water's edge the white, round-flowered *Eriocaulon browniana* is commonly found in hummocks. The most characteristic plant, however, is the moss *Sphagnum* which is sometimes also found in moss forest (Chasen and Hoogerwerf 1941). This genus has a worldwide distribution and is generally found in shallow, cold, usually acid, nutrient-poor water or in similar waterlogged soil where few other plant species thrive. *Sphagnum* productivity varies between 100-600 g/m^2/year depending on environmental conditions (Clymo and Hayward 1982). Sumatran *Sphagnum* is very poorly known, but Johnson (1960) reports one species, *S. junghuhnianum* from mountaintops in Peninsular Malaysia and from China, Japan and Taiwan, which is also found on Sumatran mountains.

Sphagnum grows like other mosses but the dead parts decay very slowly (even though the leaves are only one cell thick) and form the basis of much of the peat in temperate regions. The reasons for this slow decay are perhaps twofold: the very low nitrogen content (± 1% of dry weight) makes it unattractive to decomposing organisms, and the acid conditions and the low oxygen levels in the water enclosed within the carpets of the moss, also create unfavourable conditions for breakdown. *Sphagnum* may be just a lowly moss but it is extremely successful: virtually nothing eats it, it covers at least 1% of the earth's surface, and the sum of its living and decomposing parts is probably greater than that of any other plant genus in the world (Clymo and Hayward 1982).

The fauna associated with *Sphagnum*-dominated lakes is poor and at the lake on Mt. Kemiri the only animals found were: a few bloodworms (larvae of chironomid midges) which feed on detritus; hunting spiders which can walk on the water surface between clumps of *Sphagnum* and probably prey on emerging midges; and predatory larvae of libellullid dragonflies.

Animals and Their Zonation

Invertebrates

The number of invertebrates declines with altitude. On Mt. Mulu, Sarawak, for example, the decrease in soil macroinvertebrates was accounted for largely by the gradual reduction in the abundance of ants and termites. Biomass, however, reached a maximum in the peaty soils of the taller parts of upper montane forest where beetle larvae took over from termites as the major detritivores. At higher altitudes centipedes and spiders took over from ants as the major predators (Collins 1979, 1980). A disjunction in the distribution of beetles was found up the mountain such that no beetle species found below 500 m (lowland forest) was found above 1,500 m (upper montane forest) (Hammond 1979; Hanski 1983). The highland beetles were thus a distinct association of species. A study of large moths in Papuan New Guinea showed that the total number of species dropped from 774 at 2,200 m (lower montane forest) to 379 at 2,800 m (upper montane forest) (Hebert 1980).

Whereas in the lowlands the most conspicuous bees are the carpenter bees (Xylopodidae), these are rare above 1,500 m. During sunny periods in the subalpine forest bumble bees *Bombus* sp. (Apidae) are a common sight as they visit flowers for pollen and for nectar. It must be remembered that a visit to a flower by a potential pollinator (bird, bat, moth, butterfly, wasp, bee, ant, fly or beetle) does not necessarily constitute actual pollination. Notes have been made on mountain plant pollination by van Steenis (1972), and for detailed, fascinating information on pollination ecology see Faegri and van der Pijl (1979).

It is well known to entomologists that moths and many other insects such as winged ants, beetles and flies exhibit the phenomenon of summit seeking, in which they fly up mountains and sometimes congregate in large numbers. The basic factors underlying this behaviour are not known and a bewildering variety of explanations have been proposed (Mani 1980b).

Reptiles and Amphibians

Nothing seems to be known about the altitudinal limits of reptiles and amphibians of Sumatran mountains but these were investigated up Mt. Benom, an isolated, granitic mountain in Peninsular Malaysia. No lizards were found above 1,000 m but five snakes were. Only one of these was found above 2,000 m. Similarly, seven species of frogs or toads were found above 1,000 m but only one was found above 2,000 m (Grandison 1972). None of the montane species found on Mt. Benom are known from Sumatra but a similar type of reduction in species would be expected on Sumatran mountains.

Birds

As part of their report on a collection of birds from Aceh, Chasen and Hoogerwerf (1941) list the species found in different habitats and at different altitudes. Thus, in the primary and dense secondary growth between 300 and 1,200 m, 134 species were found. In the upper montane forest they found only nine, but eight of these were also found lower down. In the subalpine zone, however, 11 species were found, but only two of these were found in either the lowlands or in the upper montane forest. This illustrates that the montane avifauna is also not just an impoverished selection of hardy species but is a distinct community of species.

Similarly, on Mt. Benom, only 3% of the lowland birds reached 1,200 m (the lower region of Fagaceae/Lauraceae forest in lower montane forest) and the species of the upper parts of lower montane forest and of upper montane forest were quite distinct from the lowland species (Medway 1972b). Similar results were found for the birds of Mt. Mulu (Croxall 1997; Wells et al. 1979). Characteristic and conspicuous birds of Sumatran mountaintops are the medium-size (± 30 cm) Sunda whistling thrush *Myiophoneus glaucinus*, shiny whistling thrush *Myiophoneus melanurus*, scaly thrush *Zoothera dauma* and island thrush *Turdus poliocephalus* (King et al. 1975; van Strein 1977). Perhaps because humans are so rarely seen on high mountains, many of these birds are very tame, and can easily be watched eating *Vaccinium* berries and similar foods. Above the summit of Mt. Kemiri, the CRES team observed a crested serpent eagle *Spilornis cheela* and a high-flying swift *Apus* sp. Large mountains clearly are no barriers to these master flyers.

In Papuan New Guinea, Kikkawa and Williams (1971) found that a distinct discontinuity of bird species occurred at about 1,500-2,200 m, which corresponded to the change from lower to upper montane forest. They continued their analysis to examine the differences in the niche occupancy of birds at high altitudes. Thus, with increasing altitude:

- the proportion of tree-nesting insectivorous species increased;
- the proportion of tree-nesting frugivorous species decreased;
- the proportion of tree-nesting omnivorous species stayed roughly the same;
- the proportion of predatory birds decreased; and
- the proportion of ground-living birds increased.

Mammals

At least 11 species of mammals found in Sumatra are more or less restricted to the mountains:

a) Grey shrew *Crocidura attenuata*, known from subalpine zones of Mt. Leuser, Mt. Kerinci and one caught by the CRES team on Mt. Kemiri. Also known from the highest mountain in Peninsular Malaya, Mt. Gede in West Java, and lowland areas of northern Indochina and southern China (Chasen 1940; Jenkins 1982; Lekagul and McNeely 1977).

b) Grey fruit bat *Aethalops alecto*, known only from mountain forests above 1,000 m in Sumatra, Peninsular Malaysia and Kalimantan (van der Zon 1979).

c) Sumatran rabbit *Nesolagus netscheri*, known only from Mt. Kerinci and the Padang Highlands (Jacobson 1921; Jacobson and Kloss 1919) and Mt. Leuser National Park where local people sometimes eat it (fig. 9.14).

d) Volcano mouse *Mus crociduroides*, known only from 3,000 m on

Figure 9.14. The Sumatran rabbit *Nesolagus netscherii* has not been seen by scientists since 1937.

Mt. Kerinci and 2,400 m on Mt. Gede in West Java (Chasen 1940; Marshall 1977b; Robinson and Kloss 1918). Other localities may yet be discovered.

e) Giant Sumatran rat *Sundamys infraluteus,* known from the length of the Barisan Mountains between 700 m and 2,500 m and also on Mt. Kinabalu, Sabah. This species is known on Sumatra from only six specimens (Musser and Newcomb 1983).

f) Edward's rat *Leopoldamys edwardsi,* which is widely distributed and rarely found below 1,000 m (Marshall 1977; Musser and Newcomb 1983).

g) Hoogerwerf's rat *Rattus hoogerwerfi,* known only from 800 m near Blangkejeren, Aceh (Chasen 1940; Musser and Newcomb 1983).

h) Kerinci rat *Maxomys hylomyoides,* known only from 2,200 m on Mt. Kerinci (van der Zon 1979; Musser et al. 1979; Musser and Newcomb 1983; Robinson and Kloss 1918).

i) Kerinci rat *Maxomys inflatus,* known only from 1,400 m on Mt. Kerinci (van der Zon 1979; Musser et al. 1979; Musser and Newcomb 1983; Robinson and Kloss 1918).

j) Kinabalu rat *Rattus baluensis,* known only from 2,200 m on Mt. Kerinci and from 2,500 m on Mt. Kinabalu, Sabah (van der Zon 1979; Musser et al. 1979; Musser and Newcomb 1983; Robinson and Kloss 1918).

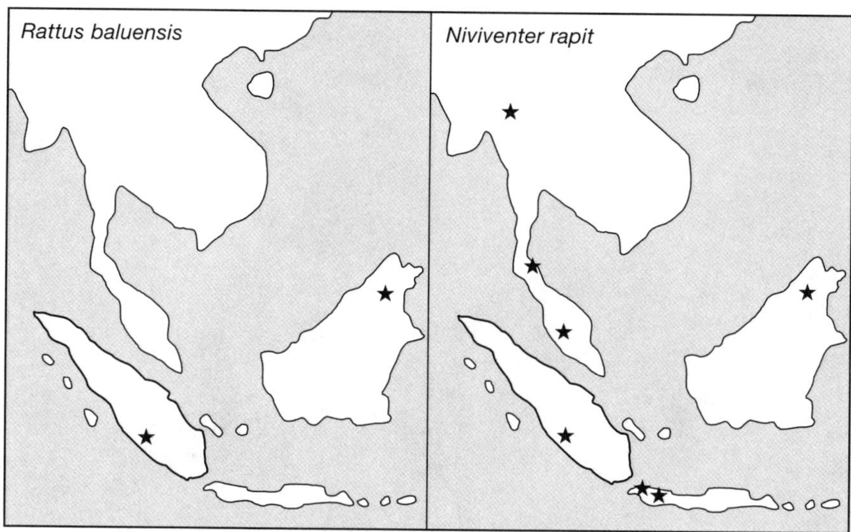

Figure 9.15. Distribution of two Sumatran rat species.
After Marshall 1977

k) Mountain spiny rat *Niviventer rapit,* known from 1,400 m on Mt. Kerinci and between 1,000 m and 2,600 m on widely separated peaks in northern and southern Thailand, Peninsular Malaysia and Sabah (Marshall 1977; Medway 1977b; Musser and Newcomb 1983). The distribution of these last two species (shown in figure 9.15) is particularly interesting. More localities in Sumatra may yet be discovered.

l) Serow, or Mountain goat *Capricornis sumatraensis,* known from 200 m to the vegetated summits of mountains in Sumatra, Peninsular Malaysia and north to the Himalayas of India. In Sumatra this agile animal also lives on forested limestone hills with nearly vertical sides, even though the hills may surrounded by cultivation. It leaves characteristic piles of faeces in small caves or the entrances to larger caves (Jacobson 1918; Lekagul and McNeely 1977).

The mammal fauna of Mt. Benom became markedly impoverished above 1,200 m (the lower level of Fagaceae/Lauraceae forest) (Medway 1972b). This same depauperisation has been recorded for Sumatran mountains by Robinson and Kloss (1918), and van Strien (1978), and by the CRES expedition to Mt. Kemiri. Those mammals with a wide

Figure 9.16. Grey shrew on Mt. Kemiri.

altitudinal range tend to be large – thus, above 2,500 m in the subalpine and upper montane zones on Mt. Kemiri, signs (tracks or faeces) were found of Sumatran rhino *Dicerorhinus sumatrensis,* tiger *Panthera tigris,* bear *Helarctos malayanus,* pig *Sus* sp., serow *Capricornis sumatraensis,* and barking deer *Muntiacus muntjak.*

The most surprising mountain mammal is probably the grey shrew. This is one of the smallest mammals on Sumatra, having a body length of only about 8 cm, and a cold mountaintop is the least likely habitat to find such an animal (fig. 9.16). Small mammals lose body heat very quickly (a problem exacerbated in cold regions) because the surface area : body weight ratio is very high and the heat loss must be compensated by a high energy intake in food. Shrews are almost entirely carnivorous and eat small insects, snails, earthworms, millipedes, centipedes and spiders, resources which are largely scattered, not storable and often unpredictable. Shrews therefore have to spend a great deal of time searching for food, which itself uses up energy. It is not surprising then that shrews eat about their own body weight in food each day and are active at most times of day although they rest briefly every few hours. If more than a few hours pass

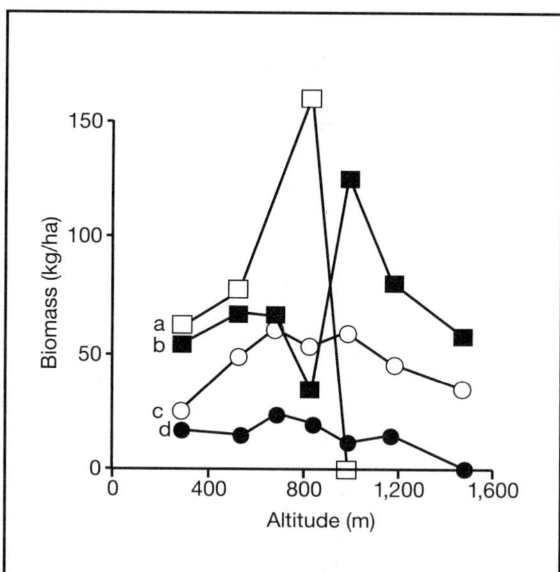

Figure 9.17. Changes in biomass of four primate species up Mt. Benom. a- white-handed gibbon, b- siamang, c- banded leaf monkey, d- dusky leaf monkey.

After Caldecott 1980

without a meal, a shrew will starve to death. Shrews make underground burrows in which they build a nest, and during the cold nights the temperature there is probably significantly higher than outside.

The volcano mouse experiences similar conditions but its diet is primarily vegetable, the sources of which are generally quite large (for a mouse) and storable. Mice will eat small invertebrates when they can catch them.

The only detailed study that seems to have been conducted on altitudinal effects on mammals in the Sunda Region is by Caldecott (1980) and concerns the altitudinal limits of monkeys and gibbons on Mt. Benom (2,108 m). From his repeated censuses conducted from 150 m to the summit he found that the changes in biomass of the six species present varied up the mountain (fig. 9.17). He asked two questions:

- why are the species limited by altitude?
- why are the species limited at different altitudes?

The answer to the first question lies in a combination of energetics and food supply. As described on page 282, with increasing altitude the forest becomes shorter and more crooked and gnarled. There are also few large boughs. Thus, at higher altitudes, the gibbons, which normally travel by swinging below boughs and branches, would experience increasing difficulty and thus greater energy costs in moving through the tree canopy. In addition, because productivity levels are lower at higher altitudes (p. 296), the young leaves and fruit eaten by leaf monkeys and gibbons would not be available in the same quantity as in the lowlands. Also, few tree species with leaves or fruit eaten by primates in the lowlands are found in the higher zones. Although no data exists it may be that leaves of the higher forest types (on acid, nutrient-poor soils) contain high concentrations of defence compounds (pp. 175, 230).

The answer to the second question (why are species limited at different altitudes?) is a combination of food preferences, body size and morphology. The altitudinal limit for white-handed gibbons, which are primarily frugivorous, is lower than for siamang which, although still mainly frugivorous, eat a larger proportion of leaves. This represents the differences in altitude at which these two species could reliably collect the different types of food they require to support the energy expended during food collection. The larger body size of the siamang (10-20 kg compared with 5-6 kg for the white-handed gibbon) and its thicker fur would also decrease the relative energetic costs of keeping warm at higher (colder) altitudes (Caldecott 1980). The dark-handed gibbon *Hylobates agilis*, found south of Lake Toba, is ecologically equivalent to the white-handed gibbon (Gittins 1979; Gittins and Raemaekers 1980), and so the ecological separation from the siamang described above would doubtless also occur in southern Sumatra.

The two leaf monkeys weigh about the same (about 7 kg) (Chivers and Davies 1978), but during a study of their comparative ecology, Curtin (1980) recorded 137 food species for the banded leaf monkeys and only 87 food species for the dusky leaf monkey. The apparently greater dietary versatility of the banded leaf monkey may be one reason it is able to live at higher altitudes than the dusky leaf monkey. Another possible reason is that the morphology and muscular anatomy of the banded leaf monkey are more adapted to the use of smaller supports and travelling on the ground (both of advantage in montane forest) than that of the dusky leaf monkey which seems to depend on larger branches in the forest canopy (Fleagle 1978, 1980). Similar reasoning can be applied to the two macaque species – the long-tailed macaque *Macaca fascicularis* and the pig-tailed macaque *Macaca nemestrina.* The former is an arboreal frugivore which is generally found beside rivers below 300 m a.s.l. (Aldrich-Blake 1980) whereas the pig-tailed macaque eats a varied diet including many terrestrial arthropods, and its morphology is dog-like and thus suitable for travel on the ground (Caldecott 1980).

Finally, there are many traditions which hold that mammals climb mountains to die; Mt. Leuser in the Gajo language means 'the mountain where animals go to die'. Jacobson (1921) and van Steenis (1938) report often finding dead or dying flying squirrels on barren mountain summits and Gisius (1930) and van der Bosch (1938) report similar observations for other animals. There are no data to negate the hypothesis but it should be remembered that:

- all animals (dead or alive) are far more conspicuous on bare mountains than in forest;
- scavenging beetles, etc., are rare at high altitudes; and
- rates of microbial decay are much slower at low temperatures and so corpses will remain for a longer time.

Chapter Ten

Caves

Introduction

Very few studies have ever been conducted in Sumatran caves. Musper (1934) explored a number of caves in the hills around Gumai (near Lahat), and Bronson and Asmar (1976) describe a little of the biology of Tiangko Panjang cave in Sarko, Jambi (p. 53). Some brief studies of a few caves in northern Sumatra have been conducted by teams from CRES but the most detailed collecting work was undertaken by van der Meer Mohr (1936), also in North Sumatra. He describes the fauna of two caves south of Medan (also visited by a CRES team) and one near Balige, south of Lake Toba.

Most caves are found in limestone areas (fig. 8.10) but 'caves' were also created during the Japanese occupation when tunnels were constructed. These have many of the same features as natural caves and are not without interest. For example, the moss-nest swiftlet *Aerodramus vanikorensis* was found for the first time in Sumatra in a tunnel in the famous Ngarai Simanuk Gorge near Bukittinggi (Wells 1975). Even road tunnels, such as that between Kota Panjang and Bankinang, Riau, support populations of bats and swiftlets.

Cave Structure

Rain water contains carbon dioxide from the atmosphere and is therefore slightly acid. This weak acid dissolves calcium carbonate (the main constituent of limestone) and forms channels which, in time, achieve the dimensions of caves, often with a stream running through them. Water dripping from the roof in the main chamber of the Quarry Cave, Lho'Nga, North Aceh, had a pH of 5.0. Dry caves exist where the stream flow has been diverted or where the limestone has been raised or tilted relative to the surrounding land. In Sumatra, most of the limestone emerges at intervals along the length of the Barisan Range and dates from the early and late Tertiary. The caves themselves can date from anytime between then and the present.

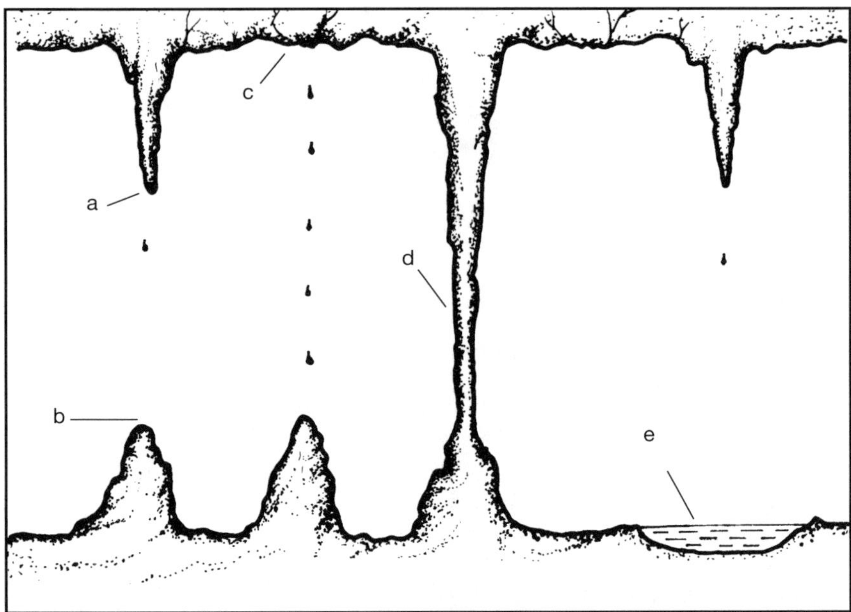

Figure 10.1. The formation of stalactites and stalagmites. a - water evaporates from drips, leaving deposits of calcium carbonate and impurities thereby forming a stalactite; b - water evaporates, also leaving a deposit but forming a squatter stalagmite; c - where water drips quickly no stalactite is formed; d - stalactite and stalagmite eventually join to form a single column; e - where the drips from a stalactite fall (or fell) into a river, no stalagmite is formed.

Two of the commonest features within a cave are stalactites and stalagmites which are columns of calcium carbonate with various impurities (which are the cause of the wide range of colours found). Their formation is described in figure 10.1. The surface of stalactites and stalagmites increases the surface area of a cave and therefore the living area available to its inhabitants.

The nature and depth of cave deposits and the manner in which they lie on one another can provide information on the past history of the cave and the surrounding area. Pollen grains may be found in organic and inorganic deposits and these too can help us to reconstruct the conditions of palaeo-environments.

The Cave as a Habitat

The salient features of a cave are its definable limits, its enclosed nature and the consequent reduction of light and comparative stability of climatic factors such as temperature, relative humidity and air flow (Bullock 1966). Variations in these features create a surprisingly wide range of effects which determine the type and number of animals which can inhabit a cave. Cave animals are divided into three ecological groups: (a) troglobites, obligate cave species unable to survive outside the cave environment; (b) troglophiles, facultative species that live and reproduce in caves but that are also found in similar dark, humid microhabitats outside the cave; and (c) trogloxenes, species that regularly inhabit caves for refuge but normally return to the outside environment to feed. Some other species wander into caves accidentally but cannot survive there (Howarth 1983).

The floor of most dry caves is composed largely of a layer of material formed from waste products and bodies of animals. A sample of 'soil' was taken from Kotabuluh cave, Tanah Karo, and the analysis (table 10.1) showed very low levels of carbon and nitrogen, but a high level of potassium and an extremely high level of phosphorus. Not surprisingly, local villagers are extracting this 'guano' to sell as fertiliser (see also the paper by Khobkhet [1980]).

Darkness

Exclusion of Photosynthesising Plants. Since light is essential for photosynthesis, it would be reasonable to expect that green plants are excluded from the dark parts of caves. This is essentially true for almost all caves but a form of 'lichen' consisting of a colony of gram-positive bacteria with a few blue-green algae between it and the substrate, has been reported from a cave wall in 'absolute' darkness in Nepal (Wilson 1977, 1981). The blue-green algae had high densities of thylakoids (the membranes on which photosynthesis occurs), and increased thylakoid density leads to greater

Table 10.1. Results from an analysis of guano from Kotabuluh cave.

pH H_2	C (%)	N (%)	C/N	P-av (ppm)	Exchangeable (m.eq/100 g) K	Na	Ca	Mg	T.E.B. (m.eq/100 g)	C.E.B. (m.eq/100 g)	B.S. (%)
3.7	0.21	0.03	7.0	160	0.96	0.12	0.76	0.50	2.34	14.20	16

Figure 10. 2. 'Lichen' from Kotabuluh cave, Tanah Karo.

photosynthetic efficiency. Interestingly, what appears to be a very similar 'lichen' was found at Kotabuluh cave by a CRES expedition. The most important effect of this virtually total exclusion of green plants is to make all cave-dwellers dependent on material brought in from the outside and to exclude all animals which feed directly on the above-ground parts of green plants. Plant roots can penetrate fissures above a cave and can commonly be seen attached to or hanging from the cave roof. Fungi and bacteria are not, of course, dependent on light and their role in caves is discussed below.

Vision. In near total darkness a cave-dweller becomes reliant on senses other than sight to detect food or enemies. This is not peculiar to cave animals – many nocturnal and cryptic animals depend almost exclusively on hearing, smell and touch (Bullock 1966). Clearly, however, diurnal animals

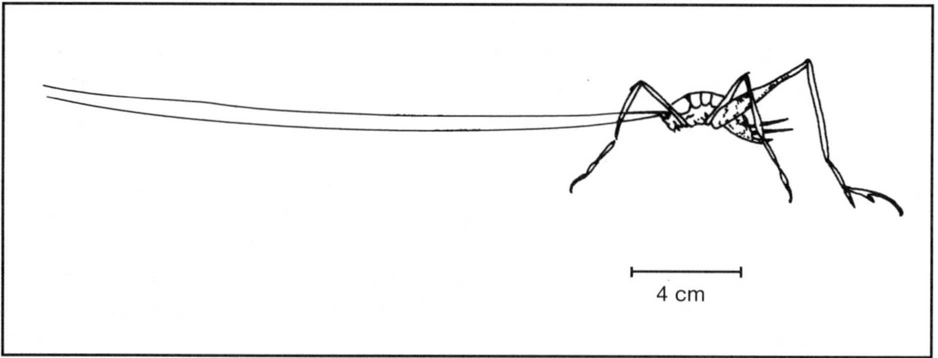

Figure 10.3. A Sumatran cave cricket showing its extremely long antennae.

can live in a cave provided their other senses are sufficiently acute, and sight may even be useful to animals ranging into the twilight areas, and also to those that range outside caves. For obligate cave-dwellers (troglobites) the presence or absence of eyes is immaterial, although the lack of evolutionary selection to maintain good sight permits deleterious variation to appear, resulting in the poor visual acuity characteristic of cave-dwellers in general and the total blindness of some (Holthuis 1979; Peck 1981; Roth 1980).

Adaptation and Modification to Life in Darkness. Adaptations to life in darkness are not particularly associated with cave-dwelling. Troglobites, together with nocturnal animals and troglophiles, tend to have their non-visual senses well developed so that they are able to reproduce, eat and move around in the darkness. Some, for instance, have very long appendages, such as the legs of scutigerid centipedes and the antennae of cave crickets (fig. 10.3). Antennae can also function as chemoreceptors and may be sensitive to relative humidity (Howarth 1983). Other animals such as many bats and most swiftlets *Aerodramus* spp. (Medway 1962) have developed the faculty of echo-location. A sound is produced and the echoes which reflect back from solid objects are interpreted to give a 'picture' of the surroundings. The swiftlets click their tongues and produce a low-frequency sound (1.5 to 5.5 kHz) which is audible to humans and sounds like a wooden rattle. This enables them to detect large objects, allowing them to navigate, nest and breed within a totally dark cave, but is not sufficiently accurate to enable them to catch insects at night (Medway 1969a). It might be thought that the different activity periods of bats and swiftlets would represent temporal partitioning of a common food resource. In fact, swiftlets feed mainly on small wasp-like insects (Hymenoptera) (Hails and Amiruddin 1981; Medway 1962), whereas insectivorous bats concentrate on various moths and beetles (Gaisler 1979; Yalden and Morris 1975). Bats do interact ecologically with nocturnal birds and this is described elsewhere (Fenton and Fleming 1976).

Echo-location in bats has evolved independently at least twice, but in each case it is characterised by high-frequency sounds (20-130 kHz), above the threshold of human hearing, originating from the larynx or speech-box, and by the reception of echoes in complex and often large ears.

Mouse-eared bats (Verspertilionidae) use a predominantly frequency-modulating (FM) system; that is, the frequency of the sound they emit through their mouth varies and is given in very short pulses. When cruising in the open a pulse is emitted and some time is spent listening for echoes. When closing in on the prey, however, pulses are emitted rapidly so that the exact locations for the flying insect can be determined. A flying mouse-eared bat can detect objects less than 1 mm across.

Horseshoe bats (Rhinolophidae and Hipposideridae) mainly use a single rather than a variable frequency and each species uses a characteristic frequency. Instead of emitting the sound through their mouth, these bats keep the mouth shut and emit the sound through their nostrils which are positioned half a wavelength apart to give a stereo impression. The peculiar 'horseshoe' around the nostrils has the function of a megaphone, causing the sound to be emitted in a concentrated beam.

It used to be thought that bats using echo-location had no difficulty catching their insect prey, but it now appears that some moths can detect bats from 40 m away and before they have been detected. These moths have developed the ability to utter clicks which confuse the bat. Other moths have a variety of responses which makes it difficult for the bats to predict their behaviour. Some bats in their turn do not keep their echo-locating system 'switched-on' continuously so as to give little warning as possible to the moths, while others emit frequencies above the moth threshold of hearing (Fenton and Fullard 1981).

Whereas the bats mentioned above catch flying insects, false vampires (Megadermatidae) feed by picking insects off leaves, or lizards, frogs and small rodents off the ground. They have also been known to eat bats caught in mist-nets without themselves getting caught in the net (Medway 1967). To avoid swamping the echoes from their prey they 'whisper' their sounds which are FM like those of mouse-eared bats. Sometimes false vampires locate their prey solely by homing in on sounds made by the prey itself. This is similar to a frog-eating bat which has been studied in Panama and which can differentiate between the calls of edible and poisonous frogs and between the calls of small frogs and frogs too big to capture. Its efficiency at catching frogs has probably led to adaptations in the frogs' calls so that the males still call to attract females but in such a way as to reduce their chances of being caught (Tuttle and Ryan 1981).

Finally, the only fruit bat to echo-locate, the cave-dwelling rousette bat *Rousettus*, uses a tongue-click like swiftlets (Fenton and Fullare 1981; Yalden and Tuttle 1975).

As well as the loss of eyes, many troglobite invertebrates have also lost cuticle pigments (van der Meer Mohr 1936) and wings. Some have developed a thinner cuticle, a larger, more slender body than their relatives outside the cave, adaptations to their feet allowing them to walk on wet surfaces, and also a lower metabolic rate (Barr 1968; Howarth 1983).

Diurnal Rhythms. Most animals have a clearly defined daily cycle of activity. Nocturnal species are active at night, diurnal species during the day, and crepuscular species around dawn and dusk. Such cycles are obviously associated with daylight and darkness and thus may not occur in a cave community. There are, however, certain events within a cave which may impose a daily rhythm on the inhabitants. The most important of these is the departure, and later return, of the bats. In their absence, food is not available for the predators and free-living ectoparasites in the roosts, and there is a halt to the rain of fresh faeces from the roof (Bullock 1966). Although data are lacking, it is probable therefore that a diurnal rhythm exists despite the absence of night and day.

Temperature, Humidity and Air Flow

The insulating role of the walls and roofs of caves effectively buffers the wide daily variations in temperature and humidity of the outside world. During the CRES expedition to the dry cave near Batu Katak, Langkat, it was found that the humidity in the cave did not fall below 97%, although the outside humidity at midday was about 75%. Similarly, the temperature in the cave varied between 24°-32°C. As would be expected the maxima and minima of humidity and temperature in the cave occurred some time after those outside. Conditions thus remain fairly stable from day to day, but there are still seasonal changes which can greatly alter conditions in caves. For example, during a rainy period the humidity and amount of free water within a cave tends to increase. Thus a chironomid midge which requires free water for breeding may only be able to reproduce when seasonal pools are full of water, while at other times the pools are dry and its numbers dwindle. Fluctuations also occur in other insect and bat numbers; some bats show distinct periodicity in breeding and mortality rates, but there is no actual evidence that this is caused by changes in the environment (McClure et al. 1967). These bat cycles affect other organisms: heavy mortality of bats permits a rapid increase in corpse-feeding organisms which would diminish later in the year due to lack of food.

Air movement is also buffered by the cave walls but still occurs as air is drawn out of the cave during the day when the air outside is warmer and lighter. This air movement follows a regular pattern but leaves pockets of stagnant air where spiders can weave delicate and complex webs, and preserves pockets of high humidity. Under such stable conditions the air disturbance caused by the approach of predators or prey may be discerned.

The constant high humidity in the deeper parts of caves appears to have led to many troglobite arthropods becoming morphologically similar to aquatic arthropods. In these deeper areas, the carbon dioxide levels may become relatively high if there is no inflow of air except from the cave mouth. Howarth (1983) has suggested that the lower metabolic rate of some troglobite arthropods may be a physiological response to this high carbon dioxide concentration. Many, if not most, of these troglobites can survive long periods of starvation, gorging themselves when food is available and storing a large amount of fat.

Food

As noted on page 315, all cave-dwellers are dependent for food on material brought into the cave from outside. Some animals feed on the plant roots attached to the cave roof, wood and other material washed in during floods (if the cave has a river running through it), or the organic matter percolating through from the surface. In Sumatra, however, the major providers of food are the bats and swiftlets which roost and breed in the cave but feed outside. Other cave shelterers such as porcupines may also contribute, but to a negligible extent in most caves.

Bats and swiftlets supply food in a number of ways:

- during their lives they produce faeces, collectively known as 'guano', which has nutritive value and is fed upon by various animals (coprophages) as well as providing a source of nourishment to fungi and bacteria;
- as live animals, they are hosts to many parasites, both internal and external, and provide food for predators;
- they moult hair and feathers and shed pieces of skin;
- they produce progeny which may be susceptible to different predators and parasites. For instance, eggs of swiftlets are attacked by a cave cricket *Rhaphidophora oophaga* which does not attack the birds themselves (Chopard 1929);
- when the bats and swiftlets die, their bodies form a source of food for various corpse-feeding organisms (necrophages).

Almost all animals provide food for others in these five ways, but within the cave ecosystem, in the absence of green plants, these are the only major sources of food.

The bats themselves roost on the roof and walls and form the basis of one community; their faeces and dead bodies fall to the cave floor and form the basis of another. Thus there is a distinct division of the animals into a roof community and a floor community (Bullock 1966).

Roof Community

The roof community includes all those animals which feed on live bats and swiftlets. These include only one major predator, the snake *Elaphe taeniura.* The snake waits, coiled in a crevice, until a bat alights nearby. Then, bracing itself with part of its body within the crevice, the snake lunges out and captures its victim, which is crushed in a quick coil and swallowed before the snake withdraws back into the crevice. The only other major predator on bats is the hawk *Macheiramphus alcinus* which roosts near large caves (Eccles et al. 1969).

There are many parasites, some internal, but mostly external. Some, such as the wingless nycteribiid flies (Marshall 1971), live almost their entire lives on bats, others, such as streblid flies and chigger mites (Trombiculidae), spend only part of their life cycle on bats, while some, such as the soft tick *Ornithodoros,* attach themselves to bats only when they are hungry (fig. 10.4). Most of these suck blood, but the large earwigs of the family Arixeniidae eat skin debris. These earwigs are quite large – equivalent to an insect 32 cm long on an average Sumatran man. A survey of parasites on bats in Thailand revealed no less than 116 species belonging to 13 families of insects and mites (Lekagul and McNeely 1977). It has been suggested that the great majority of insect species parasitic on tropical bats do in fact have only one species as a proper host (Marshall 1980). In caves containing nests of swiflets, the community is even more diverse, for these birds are also hosts to a variety of mites and insects.

The parasitic insects on bats show considerable convergence in many characteristics, although six families from four orders are represented. Many of them are flattened, some vertically and some horizontally, to ease their movement between the bat hairs. Most of the insects have tough but expandable 'skin' which allows for large meals of blood from the host. The skin often bears backward-pointing spines which lessen the chance of being dislodged by a scratching bat. They also have well-developed grasping claws. Wings have been lost in species belonging to all but one of the insect families concerned. The loss of wings and general lack of light have led, not surprisingly, to the loss or reduction of eyes in these parasitic insects (Marshall 1980).

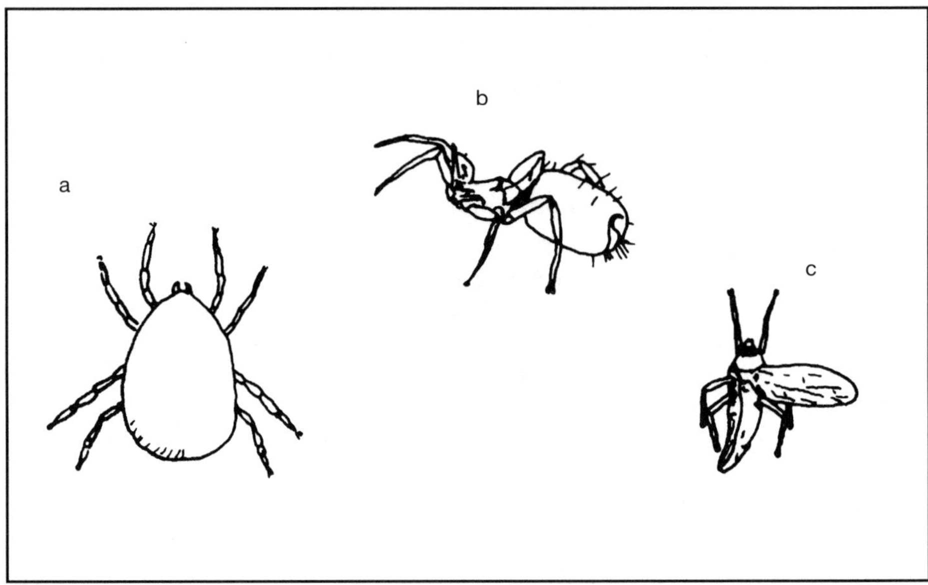

Figure 10.4. Common parasites of bats: a) *Ornithodoros* - a soft tick; b) a nycteribiid fly; and c) streblid fly.
After Lekagul and McNeely 1977

Floor Community

On the cave floor coprophages and necrophages predominate. It is difficult to distinguish between these two; although a few forms are exclusively necrophagous, many of the coprophages will include dead bats or swiftlets in their diet. The majority of cave-dwelling bats are insect-eating and the faeces they produce are hard and dry and readily exploited by coprophages such as wood lice, cocoon-encased caterpillars of *Tinea* moths, flies and beetles. The faeces produced by the few species of fruit-eating bats that roost in caves, however, are amorphous and wet and are not generally utilised by coprophages. In this case, cockroaches ingest the faeces and the general coprophages feed in turn on the faeces of the cockroaches (Doyle 1969). Fungi and bacteria develop on the faeces and some coprophages exploit this food resource too. The cave crickets and small Psocoptera flies may feed on fungi (McClure et al. 1967).

The floor community includes many predators such as the scutigerid centipede, assassin bug *Bagauda,* and medium-sized spiders which feed on the coprophages and the small *Tinea* moths. Some of these predators and others such as large spiders, live on the walls and wait for wandering coprophages to come to them, or only venture to the ground when hungry. Small predators may enter the diet of the larger predators such as shrews *Crocidura,* the toad *Bufo asper* and the spider *Liphistius. Liphistius* is the world's most primitive spider as shown by its abdomen which, unlike that of other spiders, is still segmented. One species lives on dry stalactites, and builds large web-cases 4-5 cm long with a door about 2 cm across. These are loosely constructed, decorated with debris from the surrounding area, and the inside is lined with smooth white silk. The hinged flap of the door is held partly open by the spider crouching behind it. Radiating from the entrance in a semicircle are six to ten strands of silk, 12 to 15 cm long. If a small insect touches one of these the spider races out, catches the prey and returns (McClure et al. 1967). No species of *Liphistius* has yet been collected in Sumatran caves but the larger *Liphistius sumatranus* can be found on road banks near Bukittinggi (Bristowe 1976a,b; Whitten, pers. obs.).

Food Webs and Pyramids

The various relationships described may be drawn as a food web (fig. 10.5). As the understanding of a cave ecosystem grows, so its food web becomes more and more complex. For instance, for simplicity's sake, figure 10.5 omits the role of certain wasps parasitic on the tinaeid moths which themselves must be preyed upon by one of the predators. As food webs become more complicated (and it should be remembered that caves probably represent Sumatra's simplest ecosystem), there is good reason to produce generalised representations or models of food webs. In 1927, Charles Elton, one of the 'fathers' of ecology, noted that the animals at the base of a food chain are relatively abundant while those at the top end are relatively few in number, with a progressive decrease between the two extremes. This 'pyramid of numbers' is found in ecosystems all over the world and provides a useful means of comparing communities. To construct the pyramid, species are grouped together according to their food habits. Thus all the autotrophs (plants) are called 'primary producers', herbivores are called 'primary consumers', predators on herbivores are called 'secondary consumers', predators on secondary consumers are called 'tertiary consumers', etc. Generally, in a defined area, a large number of primary producers support a smaller number of primary consumers supporting an even smaller number of secondary consumers supporting one or two tertiary consumers (fig. 10.6a). Variations of the pyramid shape occur when, for instance, a single tree is considered (fig. 10.6b). An inverted pyramid can be formed if one considers a single animal, such as a bat, or a plant, which carries a large number of parasites, which are themselves parasitised by an even larger number of hyperpara-

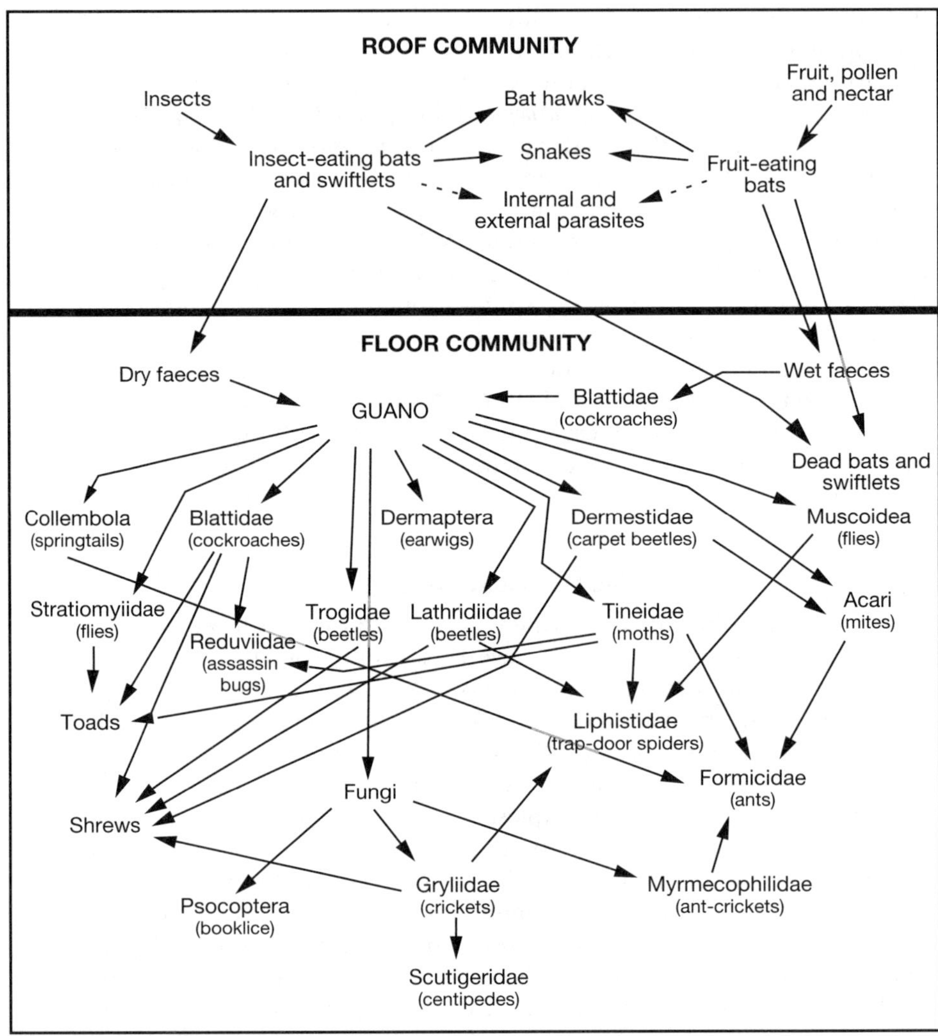

Figure 10.5. Simplified food web of a cave ecosystem.

sites (fig. 10.6c). The information contained in a pyramid of numbers permits us to state the number of herbivores supported by a certain number of plants and so on. But for comparisons between ecosystems, a better approach is to use the weight of organisms (biomass) rather than numbers, and to show this as a 'pyramid of biomass' which usually has a similar shape to the pyramid of numbers (Phillipson 1966).

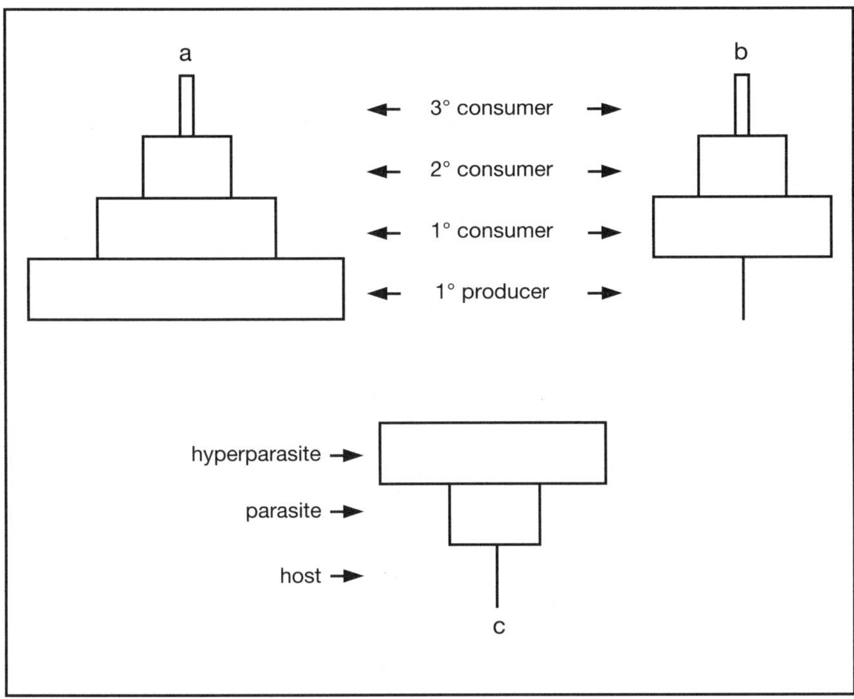

Figure 10.6. Pyramids of numbers: a) with a large number of primary producers, b) with a single primary producer, c) the case of parasites and hyperparasites.
After Phillipson 1966

In Liang Pengerukan near Bohorok, Langkat, a team from CRES counted all the animals in 25 m^2 of cave floor (or, in the case of wood lice, several 1 m^2 samples within that area), and samples of these animals were later weighed. The total biomass of the 25 m^2 was 63.38 g (2.5 g/m^2), 90% of which was contributed by a single species of wood louse which also accounted for over 99% of the animals found. The heaviest animal, a cave cricket, was 900 times heavier than the lightest, the wood louse. A pyramid of numbers would overemphasise the wood lice, but a pyramid of biomass (fig. 10.7) demonstrates clearly the relationship between the 'amount' of primary and secondary consumers. Note the absence of producers. Tertiary consumers such as large spiders were seen in the vicinity but not in the sample area.

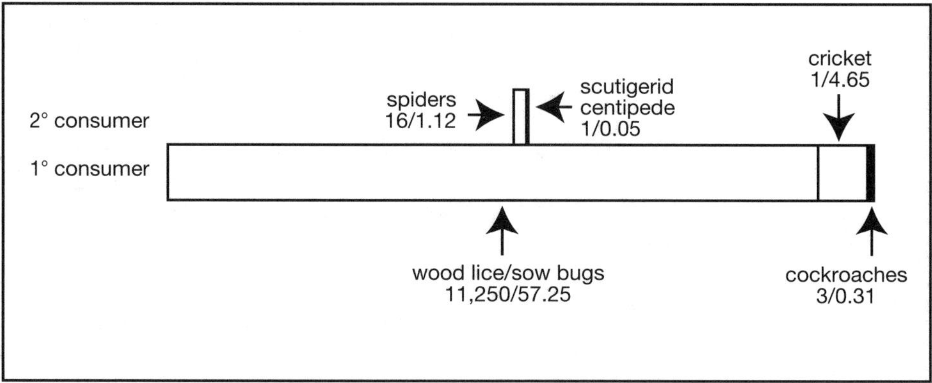

Figure 10.7. Pyramid of biomass using data from Liang Pengerukan.

Differences within Caves

Although a cave at first sight appears to be a fairly uniform habitat, this is not the case. One of the chief factors causing variation in the cave habitat is the distribution of bats, the producers of guano. Cave maps produced by CRES and by Doyle (1969) (fig. 10.8) show that the occupation of the roof by bats is very patchy. Physical factors also vary from place to place in the cave; thus, during a rainy period, standing water will accumulate in one part but not another. Similarly, some parts are subject to air movement. Others are not.

It is differences such as these that make different parts of the cave suitable for different species. Such factors probably explain the occurrence in Batu Caves, Selangor, Malaysia, of no less than 11 species of Psychodidae (McClure 1965), a family of moth-like flies (fig. 10.9). (Outside of caves the larvae of this family are important in sewage purification because they feed on microorganisms which would otherwise block the filter beds of the sewage plant.) The coexistence of this number of species in a cave requires that they each exploit a different set of limiting resources or, possibly, a similar set of limiting resources to different intensities. This is the principle of competitive exclusion, or Gause's theorem (named after the Russian scientist who first tried to make closely-related species coexist on the same nutritive medium in the laboratory). Many closely related species cannot live in the same place because their ecology and behaviour are too similar – that is, they compete for the same limiting resource to the exclusion of one or more of the species. This is illustrated by the exclusive distribution of three ecologically and behaviourally similar species of leaf monkey on Sumatra for

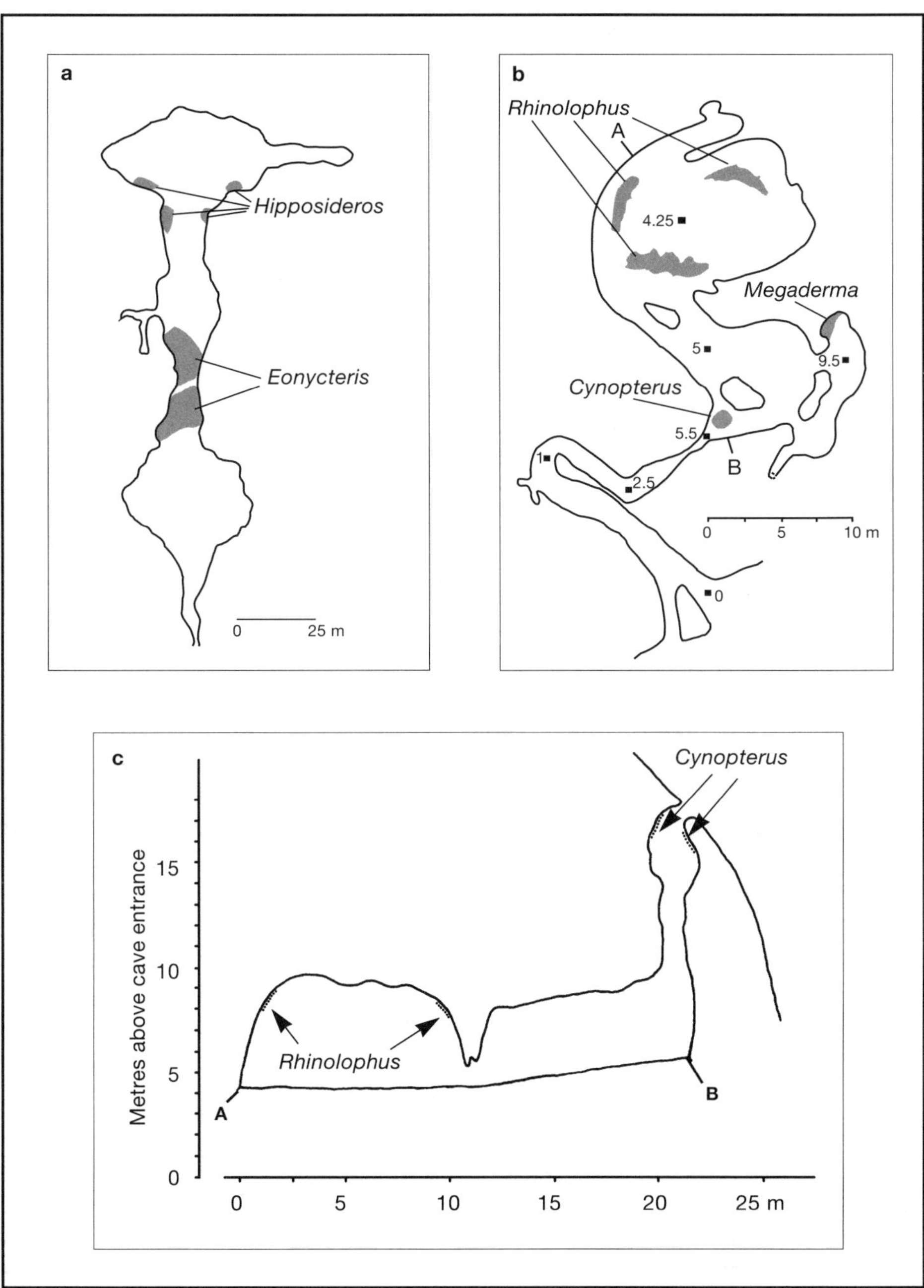

Figure 10.8. Maps of caves to show distribution of bats: a) a plan of Gua Pondok, Malaysia; b) a plan of Liang Pengerukan; c) an elevation of Liang Pengerukan taken between points A and B.

After Doyle 1969

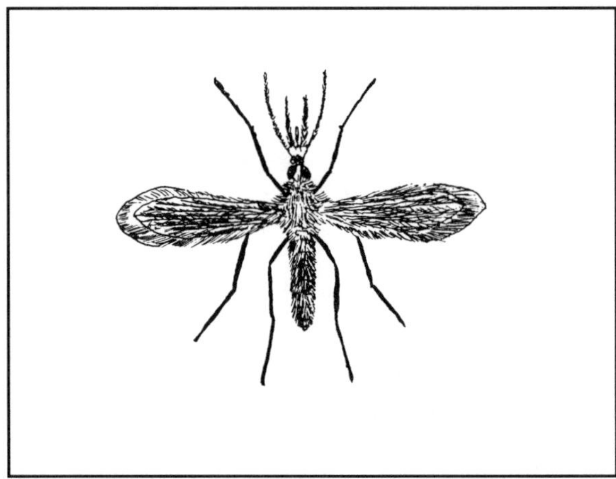

Figure 10.9. A psychodid fly.

which no area of overlap is known, and of the two species of freshwater crayfish described on page 155. The word 'limiting' is included in the statement of the competitive exclusion principle because only resources that limit population growth can provide the basis for competition. Non-limiting resources, like atmospheric oxygen, are superabundant compared to the needs of organisms, and their use by one organism does not preclude their use by another organism. So, although the 11 species of Psychodidae are all blood-suckers they must, somehow, be successfully dividing up the limiting resources between themselves.

Differences between Caves

If different caves are considered, conditions vary still further – one may be well drained while another may have a river running through it; one may contain swiftlets, another may not. There are almost no data by which the fauna of different Sumatran caves can be compared, but some data were collected by CRES for the caves around Batu Katak. Although not exhaustive, the lists (tables 10.2 and 10.3) serve to show considerable differences.

A knowledge of comparative cave fauna richness is of great value because it provides a useful biological input to environmental impact assessments for limestone mines (such as at Lho'Nga and Indarung). In most cases it would be possible to avoid disturbing a particularly rich cave and to exploit another area. It should be remembered that the disappearance of a cave fauna can affect an area far wider than the cave itself. For instance, people would no longer be able to exploit (in a controlled, ecologically sustained manner) the extremely rich guano for fertiliser; the loss of insectivorous bats would mean an increase in local insect populations; and the loss of the cave fruit bat *Eonycteris spelaea* would result in fewer durians *Durio zibethinus* which they pollinate.

Durian trees usually have two flowering and fruiting periods each year, although the timing of these varies between localities. Durian flowers are large, feathery and white with copious nectar, and give off a heavy, sour and buttery odour. These features are typical of flowers which are pollinated by certain species of bats while they eat nectar and pollen. From research conducted in Peninsular Malaysia it appears that durians are pollinated almost exclusively by cave fruit bats (Soepadmo and Eow 1977) and thus we owe the production of durians to these bats and thence to the limestone caves in which they live.

When durian trees are in flower, cave fruit bats feed enthusiastically upon them, but they also feed on and assist the pollination of other food trees such as banana, mango, petai bean, jackfruit and rose-apple (Marshall 1983, 1984; Nur 1976; Start and Marshall 1975). They also feed on the flowers of a common tree of mangrove forests *Sonneratia alba*, and studies of pollen remains in the faeces of cave fruit bats in Batu Caves near Kuala Lumpur have revealed that these bats regularly fly to mangrove areas nearly 40 km away (Marshall 1983; Start and Marshall 1975).

Only two roosts of cave fruit bats have been recorded in North Sumatra, both of them discovered by teams from CRES. The first roost was found in one of the caves, Liang Rampah, near Penen south of Medan, and contained an estimated 1,000-2,000 individuals; none of the other five caves in the area were inhabited by this species. Interestingly, this bat does not seem to have been present when van der Meer Mohr (1936) visited the cave in 1935. A colony of similar size was found at Kotabuluh cave. Assuming that these bats can also fly at least 40 km to a food source at night, the areas they cover (± 1,250 km^2 per roost) include the famous durian-growing area around Sembahe/Sibolangit. However, only the Penen bats are likely to visit the other famous durian farm around Binjai (fig. 10.10). Although there may well be other roosts of this bat in the area, it is reasonable to assume that Penen cave is the only source of cave fruit bats for at least some of the area shown. If the value of the intact, undisturbed cave is to be estimated we would need to assess the value of durians for durian orchard owners, small-scale owners and collectors, roadside vendors, wholesalers, distributors and town vendors. Owners of mature durian orchards can earn considerable sums of money.

Some people may believe that cave fruit bats are 'only bats' and do not deserve protection, but our choice is clear: leave the bat roosts undisturbed and enjoy durians, or destroy the roosts and be content to have only the

Table 10.2. List of invertebrates found in various caves in North Sumatra.

Class	Order	Family	Genus and species	Liang Pengurukan upper	Liang Pengurukan lower	Batu Katak	Liang Rampah	Liang Terusan	Sipegeh	Kotabuluh major	Kotabuluh minor
Mollusca	Stylommatophora (snails)	Subulinidae/ Stenogyridae	*Opeas gracile*	—	—	—	X	X	—	—	—
			Prosopeas achatinaceum	—	—	—	X	X	—	—	—
			Prosopeas paioense	—	—	—	—	—	X	—	—
Crustacea	Isopoda (wood lice, sowbugs)	?Oniscidae	sp. A	—	X	—	—	—	—	—	—
		Porcellionidae	*Porcellio pruinosus*	—	—	—	—	—	X	—	—
		Armadillidiidae	*Armadillo thienemanni*	—	—	—	—	—	X	—	—
			Cubaris meemohri	—	—	—	—	X	—	—	—
Arachnida	Scorpiones (scorpions)	?	*Mormurus australasiae*	—	—	—	X	—	—	—	—
	Pseudoscorpiones (pseudoscorpions)	?	*Megachernes grandis*	—	—	—	—	—	—	—	X
	Araneae (spiders)	?	sp. A	—	X	—	—	—	—	—	—
		?	sp. B	—	X	—	—	—	—	X	—
		?	sp. C	—	—	X	—	—	—	X	—
		?	sp. D	—	—	—	—	—	—	X	X
		Lyosidae	*Trochosa inops*	—	—	—	X	X	—	—	—
		Amaurobiidae	*Titanoeca fulmeki*	—	—	—	X	X	—	—	—
	Opilones (harvestmen)	?	*Stylocellus weberi*	—	—	—	X	X	—	—	—
		?	*Beloniscus quinquespinosus*	—	—	—	—	X	—	—	—

Table 10.2. List of invertebrates found in various caves in North Sumatra. (Continued.)

Class	Order	Family	Genus and species	Liang Pengurukan upper	Liang Pengurukan lower	Batu Katak	Liang Rampah	Liang Terusan	Sipegeh	Kotabuluh major	Kotabuluh minor
Chilopoda centipedes		?	sp. A	—	—	—	—	X	X	—	—
		?	sp. B	—	—	—	—	X	X	—	—
		?	sp. C	—	—	—	—	X	X	—	—
	Scutigero-morpha (scutigerids)	Scutigeridae	sp. A	—	—	—	X	X	X	X	—
			sp. B	—	X	—	—	—	—	—	—
Diplopoda (millipedes)			sp. A	—	—	—	—	X	—	—	—
			sp. B	—	X	X	—	—	—	—	—
Insecta	Thysannura (silverfish)		sp. A, etc.	—	—	—	X	X	—	—	—
	Collembola (thrips)		sp. A, etc.	—	—	—	X	X	—	—	—
	Orthoptera (crickets grass-hoppers, etc.)	Stenopelmatidae	*Raphidophora fulva*	—	X	X	—	—	—	—	—
		Gryllacidae	*Diastremmena vandermeermohri*	—	—	—	X	X	—	—	—
		Gryllidae	*Parendacusta cavicola*	—	—	—	—	—	X	—	—
			Parendacusta sp. A	—	—	—	—	—	—	X	X
	Dictyoptera (cockroaches)	Polyphagidae	sp. A								
		Pycnoscelliddae	*Symploce cavernicola*	—	—	—	X	X	—	—	—
			Pycnoscellus sp. A	—	—	—	—	—	—	X	—
			Pycnoscellus sp. B	—	X	—	—	—	—	—	—
			Pycnoscellus cf *suri-namensis*	—	X	—	—	—	—	—	—

Table 10.2. List of invertebrates found in various caves in North Sumatra. (Continued.)

Class	Order	Family	Genus and species	Liang Pengurukan upper	Liang Pengurukan lower	Batu Katak	Liang Rampah	Liang Terusan	Sipegeh	Kotabuluh major	Kotabuluh minor
	Hemiptera (bugs)	Reduviidae	*Bagauda* cf *lucifugus*	—	—	—	—	—	X	—	—
			Reduvius cf gua	—	—	—	—	—	X	—	—
			sp. A	—	X	—	—	—	—	—	—
	Diptera (flies)	Milichiidae	*Phyllomyza* sp.	—	—	—	X	—	—	X	—
	Coleoptera (beetles)	Aderidae	*Aderus kempi*	—	—	—	X	X	X	—	—
		Carabidae	*Anaulacus fasciatus*	—	—	—	X	X	—	—	—
	Lepidoptera (moths and butterflies)	Tinaeidae	*Tinea palaechrysis*	—	X	—	X	X	—	—	—
			Tinea sp. A	—	—	—	—	—	X	—	—

Data from CRES expeditions and from van der Meer Mohr 1936

Table 10.3. List of vertebrates found in various caves in North Sumatra.

Class	Order	Family	Genus and species		Pengurukan upper	Pengurukan lower	Batu Katak	Liang Rampah	Liang Terusan	Sipegeh	Kotabuluh major	Kotabuluh minor
Amphibians	Anura	Bufonidae	*Bufo asper*	black toad	—	—	X	—	—	—	—	—
Birds	Apodiformes	Apodidae	*Collocalia esculenta*	white-bellied swiftlet	X	—	—	—	—	—	X	—
			Aerodramus maxima	black-nest swiftlet	X	—	—	—	—	—	X	—
Mammals	Chiroptera	Pteropidae	*Cynopterus brachyotis*	common dog-faced fruit bat	—	—	X	—	—	—	—	—
			Eonycteris spelaea	cave fruit bat	—	—	—	X	—	—	X	X
		Rhinolophidae	*Rhinolophos stheno*		—	—	—	—	X	—	—	—
			Rhinolophus affinis	intermediate horseshoe bat	X	X	—	—	—	—	—	—
		Hipposideridae	*Hipposideros larvatus*		—	—	—	—	—	X	—	—
			Hipposideros galeritus	lesser roundleaf horseshoe bat	X	X	—	—	—	—	—	—
			Hipposideros diadema	diadem roundleaf horseshoe bat	X	—	—	—	—	—	—	—
		Megadermatidae	*Megaderma spasma*	false vampire bat	—	—	X	—	—	—	—	—
		Vespertilionidae	*Miniopterus schreibersi*	greater bent-winged bat	X	—	—	X	—	X	X	—
			Miniopterus pusillus	lesser bent-winged bat	X	—	—	—	—	—	—	—
	Rodentia	Histricidae	*Atherurus macrourus*	brush-tailed porcupine	—	—	X	—	—	—	—	—

Data from CRES expeditions and from van der Meer Mohr 1936

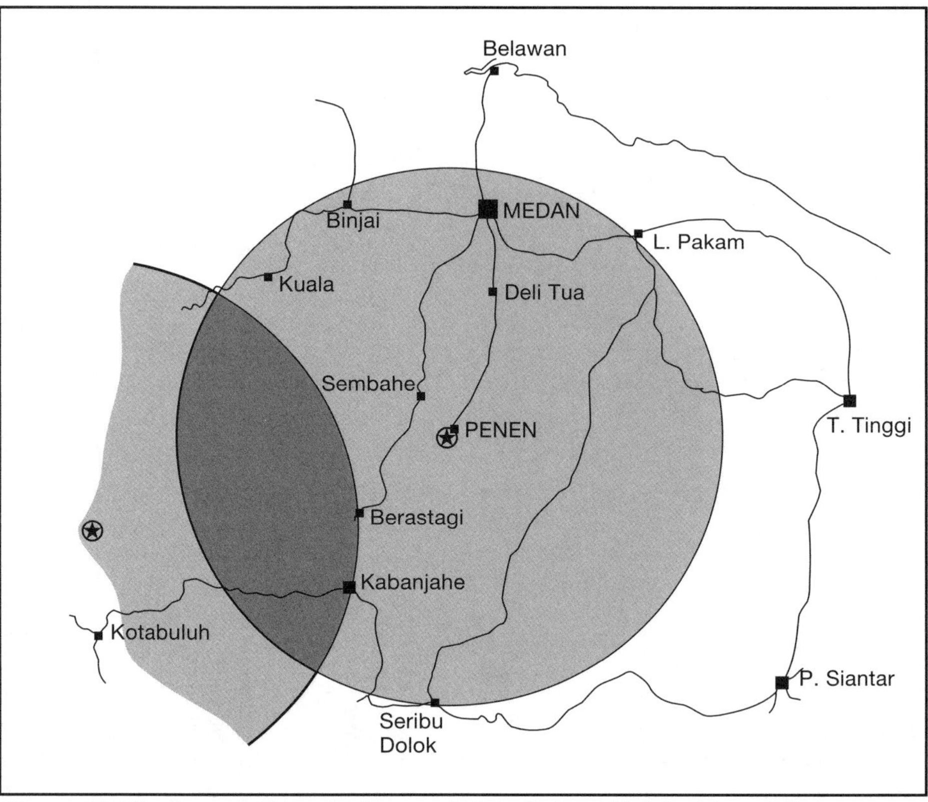

Figure 10.10. Areas within which cave fruit bats from Kotabuluh and Penen caves could pollinate durian flowers.

rarest taste of the fruit. There is almost no chance of successfully translocating bat colonies because, although we know that the cave fruit bat inhabits very few of the available caves, we do not know precisely what cave qualities it requires.

Caves as Islands

For some cave animals, such as some species of cockroach and scutigerid centipede, the outside world is an acceptable living place, and they merely move into a cave when the opportunity arises. But many cave animals, such as species of the primitive spider *Liphistius*, the cave cricket and the assassin bug *Bagauda*, are more or less confined to caves. Their caves, then, represent habitat islands set among inhospitable habitat seas or, perhaps, barely hospitable habitat seas occupied by inhospitable animals.

As with other form of islands, dispersal of animals between them represents a serious problem. Parasitic animals on bats can be transferred between caves if they move off their host when it is visiting a food source and then attach themselves to a new bat from another cave when it in turn visits the food source (Start 1974), or when the host changes roosts. Some species of non-parasitic cave animals are quite widely distributed but they are poor competitors in the outside world and their means of dispersal is not known. It may be that generations ago they had a means of dispersal (such as strong flight) but this has been lost through a lack evolutionary pressure to select for it. Thus each cave or group of caves may contain a population isolated from others of the same species which is gradually shifting or has shifted away from the species type to evolve distinct subspecies or even species. Nowhere in the Sunda region has sufficient work been conducted to show if certain caves have endemic fauna, but this extremely likely.

Another respect in which caves are like islands, is in the relationship between cave size and number of species. This was studied in Greenbrier Valley (West Virginia, U.S.A) where a series of seven limestone exposures are set amid other rocks or divided by rivers. Thus cave-dwelling animals could move within a limestone block with relative ease but not between the blocks. The number of terrestrial species was significantly correlated with 'island' area, but there was no correlation between 'island' size and species number for the aquatic animals. These species, unlike the terrestrial ones, were often found near cave mouths – for example, in pools formed by drips from the top of the entrance – and thus had more chance of dispersal. In addition, there were probably subterranean aquatic connections between the caves, meaning that the caves do not form truly isolated islands for these species (Gorman 1979).

Part C

Man-Made Ecosystems

Natural ecosystems do not hold exclusive rights to ecology, and ecological study is not the prerogative of those who gain pleasure from working in the grandeur of Sumatra's wild areas. Disturbed, agricultural and urban areas have biological components which interact, change in abundance, adapt to physical constraints and impinge upon human life according to the same principles described in the preceding chapters. There is no reason, for example, for people in Lampung, one of the most disturbed provinces in Indonesia, to regard ecology as an irrelevant science. Thus the Government of Indonesia now employs ecologists to work in transmigration areas.

As will be described in chapter 11, a characteristic of man-made ecosystems is their relative simplicity. This in itself should encourage those people who have been surprised at the complexity of the ecology of Sumatra's natural ecosystems to gain an understanding of how a simple ecosystem works. Valuable studies can be made of a pond, a street, a rice field or a cattle pasture. Although these may not appear at first sight to be very exciting, a short acquaintance with the areas and the first attempts to understand what eats what, when, how, in what quantities and why, are likely to produce enthusiasm to learn more and to apply the knowledge gained to increasingly complex systems.

Few of the present or planned industrial developments in Sumatra have major impacts on entirely natural ecosystems because they tend to be sited near road or rail networks and near centres of population which can provide the necessary manpower. Large-scale developments are now required by law to be preceded by an environmental impact assessment. This cannot be effectively conducted until a data base exists concerning the distribution and movements of species, their yearly (or longer) population cycles, species food and reproductive requirements, competition, predators, prey, etc. Possible impacts can then be judged against the status quo. If monitoring starts when an industrial project begins to be built, many ecological 'impacts' might be detected. However, these changes may be nothing more than normal fluctuations. The time to start some limited form of ecological monitoring is now.

Consider the efforts of the Meteorological Department. The actual recording of one day's weather details is not in itself particularly useful – no one needs to be told that his town has just experienced floods or extremely high temperatures. The value of the information derives from the data being recorded regularly. Yearly patterns can be established, and frequency of extreme records and trends can be calculated. Agriculture benefits most from such information because planting times can be geared to maximum productivity. Ecological monitoring should be seen in the same light and should be undertaken as normal practice by departments concerned with different aspects of the biological and physical environment.

Chapter Eleven

Effects of Disturbance

Introduction

Natural ecosystems are not static. The plants and animals within them die and others are born to replace them. Energy and nutrients pass through the organisms and are removed from the system in water, by emigration or by 'visitors'. If part of an ecosystem is destroyed in some way, the biotic community and the physical features of the soil will generally be rebuilt. This can take decades or even a thousand years depending on the ecosystem and the extent of the disturbance. The disturbed area is colonised by pioneer species which are gradually replaced by others until an ecosystem closely resembling the original is formed which is itself subject to disturbance. This sequence of changes is called succession. In areas where repeated disturbance occurs or where the process of succession is halted, an ecosystem may never have a chance to recover.

Where plants colonise a newly formed substrate, primary succession is said to occur. This has already been described for coastal mud (p. 84) and sandy beaches (p. 115), and the primary secession of plants on the remains of Mt. Krakatau is described on page 343.

The return of an area to its natural state following major disturbances, such as tree falls and landslides on steep, unstable slopes, forms part of the natural pattern of ecosystem functioning. These disturbances are the major cause of the mosaic of gap, building and mature phases in forests described on page 193. The disturbances envisaged in this chapter are of a large scale, probably caused by man, possibly repeated and relatively unselective in their effects.

Four basic ecological principles can be formulated from the study of succession (Ricklefs 1979).

1) Succession proceeds in only one direction. That is, fast-growing, tolerant colonisers are replaced by slower-growing species with more specific requirements but with great competitive ability where those requirements are met.
2) As new species colonise an area, they will inevitably alter the environment by their presence. The new conditions are generally less suitable for their own seedlings but more suitable for those of other

species; hence the succession continues.

3) The understanding of a climax community should not be too rigid. A particular 'climax' vegetation is probably just one of a continuous range of possible 'climaxes' for that area. The actual climax which is formed is determined by local climate, soil, topography, the nature and duration of the preceding disturbance and the completeness of the community of herbivores and seed-dispersing animals.
4) A climax is itself a mosaic of successional stages (p. 193).

In the following sections disturbance is discussed with reference to all the natural ecosystems discussed in previous chapters. For some, it has been difficult to find any relevant information; for others, such as lowland forest, there is a great deal known, albeit not enough. No study of the effects of disturbance to a tropical lowland forest can even be compared to the detailed work at Hubbard Brook in the U.S.A. (Bormann and Likens 1981; Likens et al. 1977) but many of the principles are common to both tropical and non-tropical forests (Bazzaz and Picket 1980; Whitmore 1984). An equivalent study in the Sunda Region is not imminent, and because of this, research must be aimed at understanding particular problems. There is no shortage of problems.

General Effects of Disturbance on Forests

Introduction

In brief, the effects of large-scale disturbance on forested ecosystems are:

1) the creation of open, hot, simple habitats containing relatively few, small, widespread species with broad niches and great reproductive potential. These species are rarely, if ever, found in mature ecosystems;
2) a huge decrease in biomass (± 30 kg dry weight/m^2 in lowland forests to 0.2 kg dry weight/m^2 in alang-alang plains);
3) the temporary or permanent impoverishment of the soil;
4) the creation of increasingly small and isolated patches of natural vegetation whose animal and plant diversity also progressively declines;
5) less rain water enters the ground water and more flows in the surface runoff because virtually all the rainfall reaches the soil surface, and far more rapidly than in a forest (p. 296). This can lead to a loss of soil, a decrease in ground water supplies and a increased propensity to flooding.

These ecological effects can be observed whether the disturbed area becomes wasteland or valuable agricultural land.

The major effect of less intense disturbance (where the forest retains at least some of its former structure) is a simplification of the ecosystem caused by deleterious changes to the soil, hydrology and microclimate as well as by the actual removal of plant material such as logs. As the taller trees are removed, so the volume of living space available to the forest biota is considerably reduced (Ng 1983). There is also less substrate available for use as nest sites, aerial pathways or growing sites for epiphytes and climbing plants. In addition, there is obviously also the loss of resources, particularly food.

The Relevance of Island Biogeographic Theory

As the disturbance of natural ecosystems proceeds, so smaller and smaller pockets of pristine vegetation are left, and these contain reduced populations of animals and plants. The study of island biogeography (p. 46) is not yet 20 years old and academic arguments continue over what predictions the mathematical models can actually make about the extinction and colonisation of species in different-sized areas of land, and therefore what shapes are theoretically best for nature reserves (Cole 1981; McCoy 1982; Wright 1981; Eright and Biehl 1982). Despite this, a number of important principles concerning the remaining 'islands' of natural habitats can be agreed.

1) The smaller the area of a particular habitat type, the higher will be the rate of extinction.
2) Even the largest nature reserves (± 10,000 km^2) will lose about half their larger mammal species and many of their bird species over the next 1,000 years or so, through natural extinction.
3) 'Corridors' of suitable habitat linking reserves or the close proximity of two or more reserves will increase the effective population for only a proportion of species. Others are distinctly averse to travelling through or over open or disturbed areas.

Corridors may also permit the transmission of diseases, and the risk of epidemics should not be underestimated. If a fatal disease were to be introduced to the wild orangutan population through orangutans released from a rehabilitation station, it could reduce the population to such a low level that extinction would soon follow. Virtually all the areas of forest occupied by Sumatran orangutans are connected and any areas unaffected by the disease would not contain enough animals for a viable breeding population. It is conceivable that the road to Blangkejeren which is splitting the Mt. Leuser National Park in two might one day be seen as a blessing if an epidemic were to occur – but this is no argument for dividing up Sumatra's reserves! As domestic plants and animals – both major reservoirs of contagious diseases – press ever more heavily against natural ecosystems, the risk of epidemics becomes ever greater. The extinction of certain species which play key roles – such as dipterocarp trees, fig wasps, and animals which are the major dispersers of particular fruit – would have disproportionate effects on the remaining biota (Frankel and Soulé 1981).

As far as Sumatra is concerned, the subject of nature reserve design is largely one for academic debate. The question of what percentage of species will become extinct in different-shaped reserves in 50, 500 or 5,000 years is not of great relevance when it is by no means certain how much land will still be covered by natural vegetation in even 25 years. The shapes of Sumatran reserves do not generally conform to the neat models of the theories (see Gorman 1979) because the rationale behind the establishment of reserve boundaries has been set primarily by human settlement patterns. In Sumatra today the major priority is simply maintaining the integrity of reserves against the pressure of legal and illegal forms of habitat disturbance. Every day the natural ecosystems are being destroyed at a vastly faster rate than they are being added to by succession. Anyone acquainted with the major Sumatran reserves knows that not only non-reserved areas are being destroyed, but that incursions into reserved areas occur also, in the form of logging, settlement, and wood cutting.

Genetic Erosion and Conservation

Destruction of tropical forests, and particularly lowland forest, represents an extremely serious global loss of plant and animal genetic resources. These play a vital role in the improvement of crop species, and in the development of industrial and medicinal products, and as such play an essential role in world economic productivity. If their availability and diversity are reduced or lost, the effects on man and his many and growing needs will be severely felt. Anyone who has the slightest doubt that conservation of Sumatra's and other tropical genetic resources is one of the highest environmental priorities should consult some of the many papers and books on the subject (Adisoemarto and Sastrapradja 1977; Anon. 1980b, 1981; Djajasasmita and Sastraatmadja 1981; Ehrlich and Ehrlich 1981; Frankel and Soulé 1981; Jacobs 1982; Myers 1979, 1980; Oldfield 1981; Sastrapradja 1977, 1978a,b; Sastrapradja and Rifai 1975).

As stated in the previous section, reduced forest area leads to a reduced number of species as well as a reduced number of individuals. The resultant inbreeding of organisms often reduces their 'fitness' (adaptability, genetic stability and variation) and can fix deleterious traits in a population which can lead to extinction. Many social mechanisms exist in man and other animals to prevent inbreeding but for animals 'trapped' in a small population within a patch of forest there may be no choice.

How large should a population of animals be to prevent disastrous inbreeding? There is no hard and fast rule but it has been suggested that an 'effective population' of 50 may be the absolute minimum possible before deleterious characters begin to become fixed and that a minimum of 500 individuals should be the goal. An 'effective population' of 50 is not the same as 50 individuals nor necessarily the same as 25 breeding animals of each sex. As explained on page 237, many animals live in social groups which are not simply pairs of adults. The effective population number, N_e, can be calculated as follows:

$$N_e = \frac{4\,N_m\,N_f}{N_m + N_f}$$

where N_m and N_f are the number of breeding males and breeding females, respectively (Frankel and Soulé 1981). Thus for tigers, in which one breeding male may reproduce with three females, 17 males and 51 females are required for an effective population of 50. These 68 adult animals will require 3,400 km^2 or 340,000 ha. Thus each of the three largest reserve areas in Sumatra – the Kerinci-Seblat National Park (± 1,000,000 ha), Mt. Leuser National Park (± 800,000 ha) and South Sumatra I (356,800 ha, much destroyed) – are large enough to retain a fully functioning population of tigers. At present the areas of habitat used by tigers extends outside the reserves, and those at Kerinci could conceivably walk to South Sumatra I. How much longer it will be before the tigers are confined to reserves, and disturbed reserves at that, is a matter for those in positions of authority.

Atmospheric and Climatic Changes

The destruction of forests and changes in land use contribute to the rise in atmospheric carbon dioxide levels since forests fix a great deal more carbon dioxide than grasslands or other secondary growth. Levels of carbon dioxide have been increasing over the last 100 years or so and if present trends persist the concentration will have doubled from 0.03% in the middle of last century to 0.06% in the middle of next century. Although still a small proportion of the atmosphere, the increase is likely to cause major climatic changes.

PRIMARY SUCCESSION—THE CASE OF KRAKATAU

Krakatau is the name of a group of four islands in the Sunda Straits halfway between Tanjung Karang and Labuan (fig. 11.1). These islands represent the visible remains of Mt. Krakatau, the whole cone of which is thought to have once been above the sea surface. Periods of volcanic activity have altered the shape of the islands and it was only in 1930 that Anak Krakatau

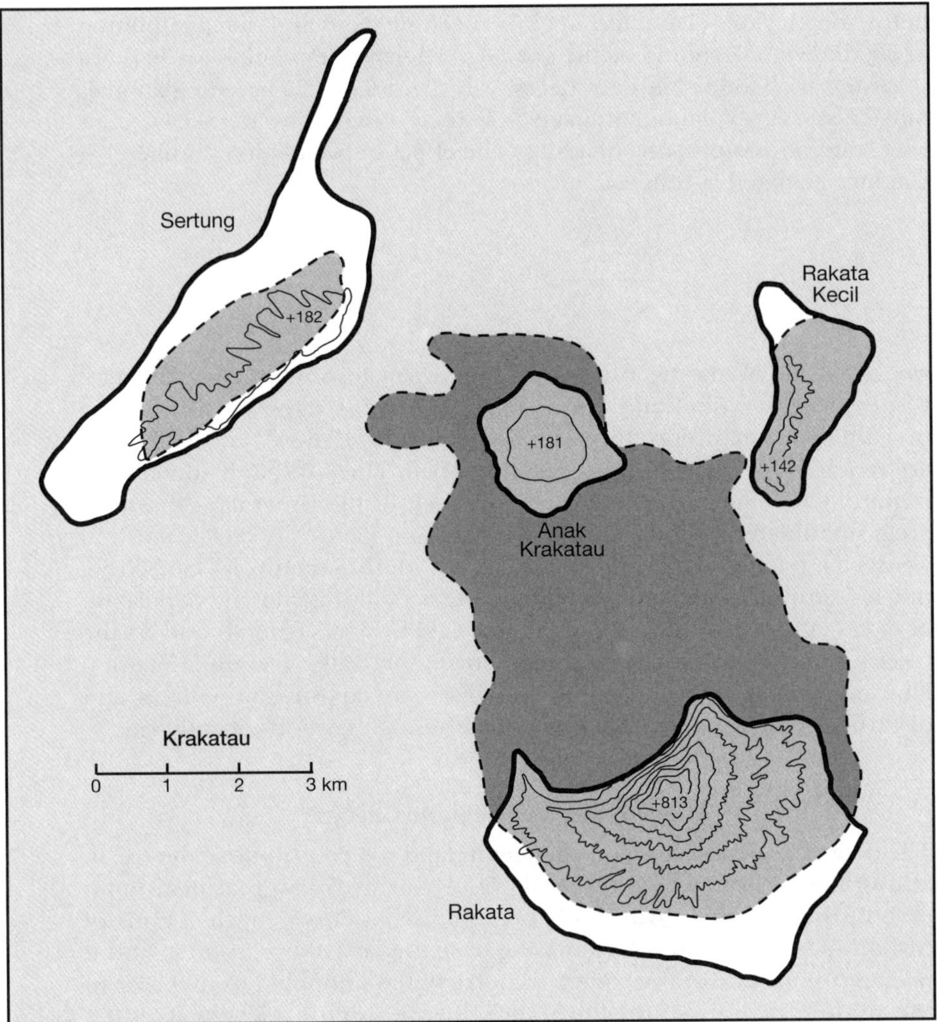

Figure 11.1. The Krakatau Islands showing contours every 100 m and (dotted line) the coastlines in 1883.

After Richards 1982

Island emerged from beneath the sea after three years of activity (fig. 11.2) (Hehuwat 1982; Partomihardjo 1982; Richards 1982).

It was the massive eruption of 1883 that made Krakatau famous. The force of the explosion was equivalent to 2,000 Hiroshima bombs, and it was heard as far away as Sri Lanka, the Philippines and Australia. Pumice and ash shot up 80 km above the earth and this dust darkened the sky so much

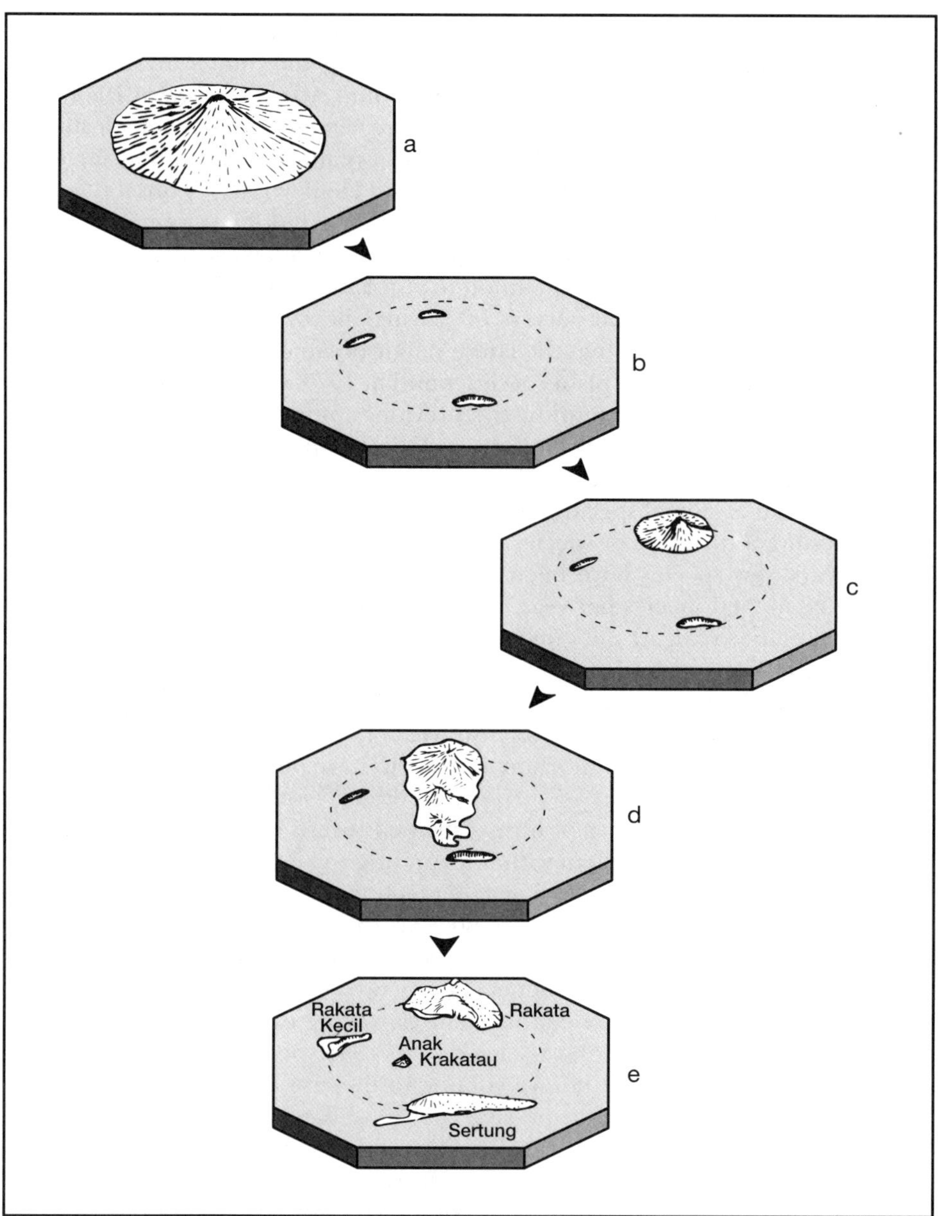

Figure 11.2. The geological history of the Krakatau Islands. a - the original volcanic cone of Mt. Krakatau in prehistoric times; b - the three small islands left after a prehistoric eruption; c - the growth of Rakata as a result of prehistoric volcanic activity; d - Krakatau before 1883; e - Krakatau in 1930 (it was not until this year that Anak Krakatau emerged).

After Richards 1982

that lamps had to be used during the day in Jakarta and Bandung. Enormous tidal waves claimed 36,000 lives along the Sunda Strait and were detected in Alaska, San Francisco and South Africa (Hehuwat 1982).

The explosion of 1883 made Krakatau very important ecologically. Not only was all the vegetation burned, but it is almost certain that the entire land surface was sterilised (Whittacker and Flenley 1982). Thus it is possible to observe the ways in which plants and animals colonise virgin land – a natural laboratory was formed in which to study primary succession.

The first botanical expedition to visit Krakatau after the explosion was in 1886 and irregular surveys of varying thoroughness continued up to 1951. It was in 1979 that the latest, major botanical survey was conducted. The total number of plant species found in 1979 was about 200 and this was similar to the total found by the previous comprehensive survey in 1934. This does not, however, indicate a static equilibrium because the species are changing (fig. 11.3).

That is, not all species found in the past appear to be present now, suggesting a turnover of species. It has been calculated that over the last 50 years new species have been appearing on the islands at an average of 2.28 to 2.60 species per year, representing an immigration rate of 1.14%-1.30% per year. At the same time, however, extinctions have been occurring at 1% per year and so the net increase in plant species actually present on the islands is between 0.14% and 0.30% per year. Thus, very roughly, a new plant species arrives every two years (Whittacker and Flenley 1982).

So, a dynamic equilibrium appears to have been reached of about 200 species. This is, perhaps, surprising because within about 150 km of the islands there are areas of little-disturbed forests containing thousands of plant species. A problem with interpreting the data may be the time scale we are imposing. There is a nagging feeling that more species should have arrived in the second period of 50 years since the major eruption. In fact, as was stated above, the number of species is not at a perfect equilibrium because one new species appears every two years or so. That is equivalent to about 50 species per century. The flattening of the curve in figure 11.3 may just be because the r-selected (p. 190) species of the pioneer and building phases (p. 193) are coming to the end of their domination. We may now be witnessing the slow arrival of K-selected species which will contribute to the mature phase of forest development. Only one tree typical of primary forest has so far been recorded on Krakatau (Whittacker and Flenley 1982). It may well take a thousand years or more before the forest on Krakatau becomes properly mature but, at the present rates of destruction, there is no evidence to suggest that a pool of 'parent' primary lowland forest trees will exist anywhere in Lampung, South Sumatra or West Java in even 100 years time.

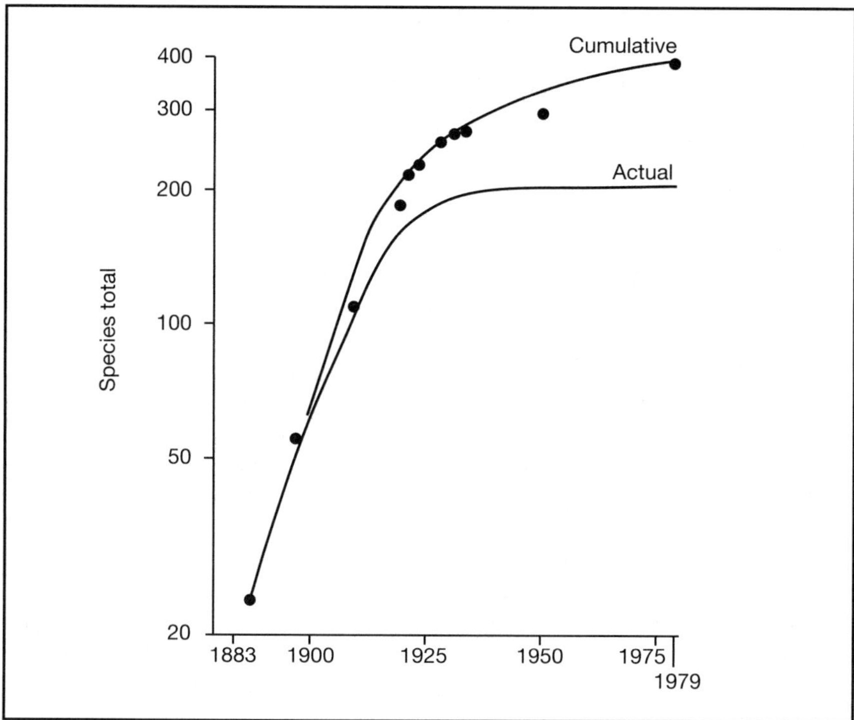

Figure 11.3. The cumulative total number of plant species recorded on Krakatau since the eruption of 1883 and the total of plant species actually present on the islands.

After Whittacker and Flenley 1982

MANGROVES

The management of mangrove forest for sustained yield forestry is possible but there is very little known about natural regeneration rates or processes (Aksornkoae 1981). Unless great care is taken, the result of large-scale disturbance is not the regeneration of productive forest but rather the appearance of a degraded type of vegetation. This is typically dominated by nipah (DKDPI Riau 1979) or *Avicennia,* or the fern *Acrostichum aureum* and sea holly *Acanthus ilicifolius.* Mangrove soils cannot be used for successful agriculture except when careful land management practices have been followed (Soegiarto and Polunin 1980; van Beers 1962; Hanson and Koesoebiono 1977; Koesoebiono et al. 1982).

Large-scale disturbance of mangroves can lead to coastal erosion because the shoreline is no longer protected by trees. The shore may be reduced to a narrow sandy beach or to inhospitable salt pans. The coastal population centres are then more susceptible to the effects of storms, such as flooding.

The replacement of mangrove with tambak ponds (for prawn and fish production) is increasing along the northeastern coast of Sumatra. As described on page 77, mangroves are vital for the continuance of coastal fisheries (Koesoebiono and Chairul 1980; Martosubroto 1979), but the establishment of tambak necessitates the removal of mangrove. In moderation this can be successful, but in some areas a situation is developing which has been called the 'Tragedy of the Commons' (Hardin 1968). The 'commons', or unowned useful resource, in this case is a mangrove forest. One person establishes an area of tambak in order to provide an income. Other people are attracted to the idea of owning tambak and more mangrove is felled. Yields stay more or less the same for a while but then they begin to drop. Tambak owners increase the area of their tambak to compensate and to increase their income, and this is achieved by clearing more mangrove. The first tambak is now some way from the source of the fish fry and young prawns (i.e., the mangroves) and the first owner may try to move towards the fringes of the tambak area (nearer the mangroves) where yields are higher. The first tambak is taken over by ferns and other scrub. Other owners experience the same drop in yields and so the process continues. Avoidance of this situation requires either a strong environmental ethic among the tambak owners, or strong enforcement of land use regulations. In fact, some tambak owners have learned, and others are discovering, that the area suitable for tambak is limited by certain conditions of tidal range, shore elevation and soil characteristics (Soegiarto and Polunin 1980).

The *Showa Maru* oil tanker accident of 1975 showed just how sensitive mangroves are to oil pollution (Soegiarto and Polunin 1980; Wisaksono, N.D.; Baker 1982). The oil probably acts by clogging the trees' pneumatophores which act as organs of gas exchange, and by raising the water temperature and lowering the dissolved oxygen levels; leaves may also be easily damaged by oil (Lugo et al. 1978; Mathias 1977). Mangroves that survive exhibit signs of chronic stress such as reduced productivity and gradual leaf loss (Lugo et al. 1978; Saenger et al. 1981). A covering of oil on the mangrove mud would obviously have extremely adverse effects on the mangrove trees. The susceptibility of mangroves to other industrial and domestic pollutants requires more study (Bunt 1980), and although many mangrove species may be resistant to particular types of pollution, the equilibrium of the system will probably be upset (Saenger et al. 1981).

Some palm oil and rubber factories discharge their effluents into rivers near the coastal zone. One study of this was across the Malacca Straits along the Puloh River near the estuary of the Kelang River, the river that flows through Kuala Lumpur. The head of the Puloh River was the site of a plantation factory outfall and Seow (1976) monitored the macrofauna composition and various water parameters at different locations along the river through the mangrove. The water quality near the outfall was obviously the least conducive to aquatic life because it had hardly been diluted by the river water. The effluent had caused several species of fish to disappear from the estuary and others had moved downstream from the outfall. Unfortunately, no observations were made of the effects of the effluent on the surrounding vegetation.

Other Coastal Ecosystems

Beach Vegetation

Destruction of fringing coral reefs (see p. 350) and the commercial exploitation of sand (such as near Banda Aceh) can almost totally destroy beach vegetation. By their pioneer nature, however, the smaller plants can recover quickly, particularly in the pes-caprae association (p. 115). The *Barringtonia* association (p. 119) has been replaced in most areas with coconut plantations and other types of agricultural areas. Where she-oak forest has been felled near Singkil, South Aceh, the vegetation has become dominated by the sensitive plant *Mimosa pudica, Melastoma* and various rushes, sedges and grasses.

The removal of beach vegetation may not be regarded as a particularly serious loss in itself, but its ability to hold together a loose, sandy substrate means that in its absence more or less continuous coastal erosion occurs. The erosion and the impact on human settlements are particularly serious during bad storms, because the power of the wind and waves is no longer countered by deep-rooted vegetation.

Oil pollution of sandy beaches in Sumatra is not generally a serious problem but it does occur. Even on the remote western coast of Siberut Island, patches of tar have been found (Whitten 1982b). Organic pollution by, for example, domestic waste can reduce the depth of the aerated layer of sand and so affect the zonation of beach biota. Serious pollution by oil or other organic wastes effectively removes the fauna, and the processes of soil formation in which they play an important role are greatly slowed.

The overexploitation of turtle eggs and turtles for food makes survival for the rest of the turtle population progressively less likely because the mass nesting behaviour of turtles is ecologically akin to the gregarious fruiting behaviour of dipterocarps (p. 221). Both act to satiate the appetites of their predators with the result that at least some of their eggs or seeds will get the chance to develop. The smaller the population of turtles, however, the larger the proportion of eggs destroyed.

Brackishwater Forest

The disturbance of this forest generally results in a great simplification and effective domination by nipah palms. No account of other effects specifically concerning brackishwater forest has been found.

Rocky Shores

Rocky shores are not usually subject to disturbance because the waves, strong currents and ruggedness and steepness of terrain are not compatible with agriculture or other forms of development.

Coral Reefs

Coral reefs experience different scales of disturbance. One of the most common forms of serious disturbance is caused by the mining of reefs for limestone either for lime or for road fill (Soegiarto and Polunin 1980). Mining wipes out the entire ecosystem and the fisheries that depend on it because all the components in this ecosystem depend on the biological and physical features of the coral for their existence (p. 127). Natural disturbances may be equally devastating; corals were more or less wiped out around Krakatau following the eruption of 1883.

Destruction of reefs also occurs when fishermen 'cheat' and catch fish using explosives. Explosions are totally unselective in the species and size of fish they kill and can reduce live coral to mere rubble (Odum 1976). Recovery of the coral, if it occurs at all, is very slow (Parrish 1980). The rubble is, of course, a 'habitat' of sorts but the biomass of plankton in such areas is little greater than over plain sand (Porter et al. 1977). When considering different impacts on coral it is worth remembering that most coral has a growth rate of between only 1 and 10 mm/yr (Soegiarto and Polunin 1980).

Waves generally break over a fringing reef, thereby dissipating the wave energy and leaving the sea between the reef and the shore relatively calm. Destruction of the reefs allows waves to break with their full force against the shore (fig. 11.4) and the land erosion this causes has been described for the north coast of Java (Praseno and Sukarno 1977).

Land use which increases the sediment load of rivers and thus coastal waters can have a negative impact on corals because the polyps are suffocated by a rain of silt. This distribution of dead and live coral around a log-loading pier on Siberut Island is shown in figure 11.5. Many hectares of

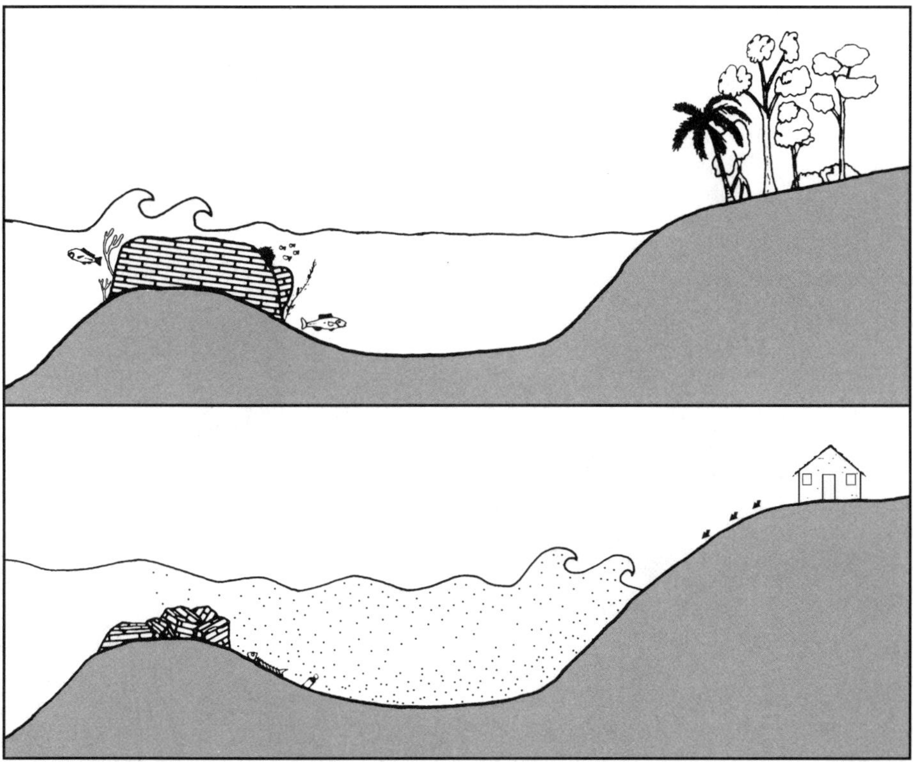

Figure 11.4. The consequences of clearing coastal forest and of destroying living coral reefs which protect the coast are that the land suffers from erosion, the coral no longer supports fish and other animals, the water is turbid, and fishing becomes less successful.

After Odum 1976

muddy logging roads and log-ponds caused extremely muddy water to flow into the sea. The disturbance of reef communities by changes in patterns of sediment deposition caused by the construction of a pier near Padang is described by Umbgrove (1947). The land uses which increase sediment load in rivers also increase the seasonality of water flow (p. 132). Thus, at times of flood, large quantities of fresh water are discharged into the sea and the salinity level may fall, causing the death of certain corals.

Changes in the community structure of coral reef fishes caused by different types of fishing methods can affect the composition and growth of corals. This subject is reviewed by Parrish (1980).

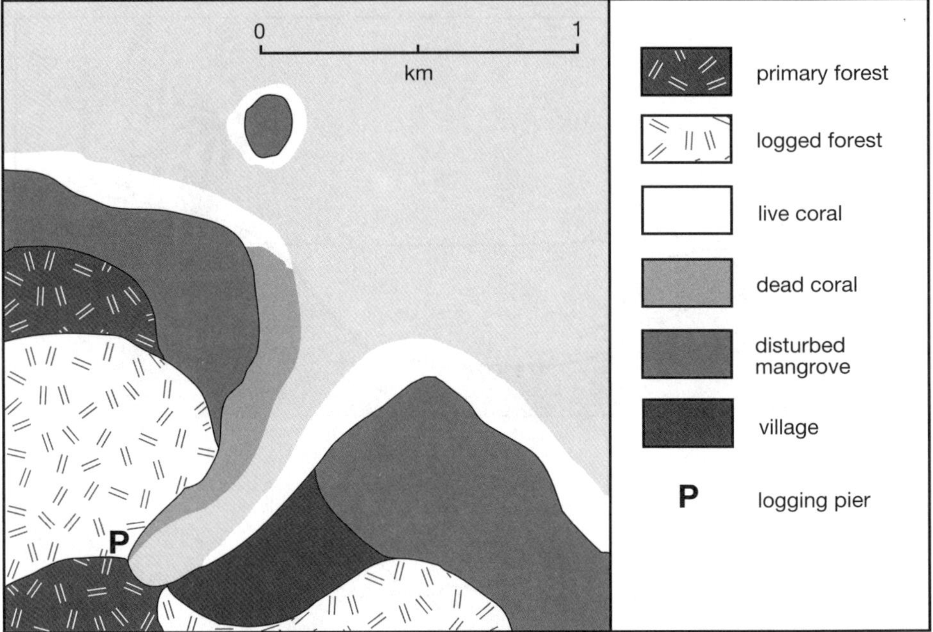

Figure 11.5. The distribution of living and dead coral around a logging company pier on Siberut Island.

After Anon. 1980

The detailed effects of domestic and industrial pollution on corals have not been investigated but they are clearly deleterious (Soegiarto and Polunin 1980). Corals have been observed to exhibit a 'shutdown reaction'. When they are stressed in some way – for example, by the effects of reduced water quality or explosions – they are weakened to such an extent that they may die if another stress is imposed, even though the second stress alone would not have caused their deaths. This means that the total cumulative disturbances on coral should be assessed in a study of potential environmental impact, not just the likely stresses caused by one particular development (Antonius 1977).

RIVERS AND LAKES

Introduction

Disturbance of aquatic ecosystems can, as with clearance of forest, be deleterious to some species but also extend the habitat of other more opportunistic species. Disturbance can occur in a variety of different ways. Industrial and domestic pollution reduce the water quality with consequent effects on the whole biota: fish poisons and bombs cause indiscriminate fish death; overexploitation of fish stocks to supply the aquarium trade can have devastating effects; introduced fish can cause the demise of indigenous species; and forest clearance causes changes in temperature, turbidity, insolation, stream flow, water input, etc., thereby altering the habitat of many species.

Industrial and Domestic Pollution

Pollutants of water have been divided into four categories—pathogens, toxins, deoxygenators and nutrient enrichers (Prowse 1968). Pathogens include a wide range of bacteria, protozoa and parasitic worms harmful to man and other organisms. These are all associated with untreated sewage. Toxins derive from industrial waste and from agricultural chemicals. Their effects can be both dramatic and cumulative and affect not only aquatic animals but also plants and, of course, man. Deoxygenation is caused by bacterial and fungal decay of organic matter and by animal and plant respiration. It can also be related to certain weather conditions (Johnson 1961a). Where large quantities of organic wastes are disposed of, deoxygenation and the subsequent death of animals by suffocation can be expected.

Nutrient enrichment is known as eutrophication. Eutrophic habitats are excellent for fisheries because of their high productivity. Indeed, lakes and ponds are often artificially fertilised to increase fish production. In the Philippines, for example, yields of over 1,000 kg/ha/yr can be expected from such habitats, whereas a normal, medium-sized oligotrophic lake might produce only one-hundredth of this yield (Ricklefs 1979). Primary production of the phytoplankton is similarly enhanced.

Eutrophication is not, of itself, a major problem for aquatic ecosystems. Naturally eutrophic systems are usually well balanced, but the addition of artificial nutrients can upset this balance and cause devastating results. Algal 'blooms' are the most spectacular of these effects where the combination of high nutrient levels, and favourable conditions of temperature and light, stimulate rapid algal growth. These blooms are a natural response of algae to the environment, but when the environment changes and can no longer support such high algae populations, the algae that accumulate during the bloom then die and decay. The ensuing rapid decomposition of organic debris by bacteria robs the water of its oxygen, sometimes to the extent that fish and other aquatic organisms suffocate.

Various studies (e.g., Schindler 1974) have shown that phosphorus is usually the limiting factor in eutrophication (rather than nitrogen or potassium which might have been suspected) and 0.01 mg phosphorus per litre of water has been suggested as the figure above which algal blooms are likely to occur in North America. Research is needed to determine the appropriate level for Sumatran lakes, so that it can be used for the establishment of rational effluent control standards.

In spite of the disturbing effects of artificially enriching lakes to such high levels, these lakes can recover their original condition if the artificial inputs cease. Sediments at the bottom of most lakes have a high affinity for phosphates which are 'locked' into the sediment even if there is an overturn (p. 145). The classic example of this is Lake Washington which in 1963 was receiving 75,600 m^3 of phosphate-rich human waste from the city of Seattle each day; it stank, algal blooms occurred and the fish died. Six years after the human waste was diverted, the phosphorus levels had fallen to those measured before Seattle became a city (Edmondson 1970).

Sumatra has yet to experience serious, large-scale, artificial eutrophication of lakes but it has been reported from a Malaysian reservoir (Arumugan and Furtado 1981). It is important to know and understand the processes involved in order to prevent any such occurrence in Sumatra. Lake Kerinci is perhaps the most likely place to experience such eutrophication given its area, and relatively dense surrounding human population. Saravanamuthu and Lim (1982) describe how algae can be used as indicator plants to determine the trophic status of fresh water.

Chemical, physical and microbiological parameters were studied at points along the Kelang River (which passes through Kuala Lumpur) and the results were examined against fish catches at the same locations (Law and Mohsin 1980; Mohsin and Law 1980). The species diversity, richness and the number of fish per m^2 were much higher in stations above Kuala Lumpur than in the city itself or downstream. In the worst polluted places only the guppy *Poecilia reticulata* (p. 412) was found. Downstream of Kuala Lumpur the water was unsuitable for most freshwater fishes but a few air-breathing fish species were found. The primary causes of the change in fish abundance and variety were heavy siltation, low pH and high biological oxygen demand. Heavy silt load kills certain fish by clogging their gills, thereby causing suffocation.

Lakes and rivers in North America and Europe are experiencing the serious environmental effects of acid rain. Under completely natural conditions rain is in fact slightly acid, about pH 5.7, because atmospheric carbon dioxide dissolves to form carbonic acid. Acid rain is understood to be rain which is more acid than this. The increased acidity of rain has been caused primarily by gaseous nitrogen and sulphur oxides which form nitric and sulphuric acids. These gases are waste products from the combustion of oil and coal, from metal smelting and from other industrial processes. Most water bodies have some buffering capacity but these are limited and so eventually their acidity will increase. At pHs below 5.0 most fish will die, and almost all other aquatic organisms are affected in some way. The problem is worsened because when the acid rain percolates through soil it releases certain metals, particularly aluminium, which are toxic to most organisms, even at very low concentrations. Acid rain is not currently an environmental problem in Sumatra and is unlikely to become one in the near future. However, acid rain is a transnational problem – that is, industrial processes in one country can produce waste gases which fall as acid rain in another. Sumatran environmental scientists should be aware of the problems of acid rain and it would be useful to begin a low-key, long-term monitoring program of water bodies.

Poisons and Bombs

The most common traditional fish poison is 'tuba', made from the roots of the climber *Derris* (various species) in which the active, identifiable chemical is rotenone (Sinnappa and Chang 1972). Other poisons, typically commercial insecticides, are also commonly used. Poisons and bombs are very effective and kill or debilitate large numbers of fish (the very reasons they are used), and because of this, many urban rivers such as the Deli and Babura in Medan, have virtually no fish left in them. Bombs are even used in rivers within reserves such as the rivers Alas and Bohorok and their tributaries around the Mt. Leuser National Park. Apart from any other arguments against the use of poisons and bombs, these forms of fishing are selfish, needlessly destructive and, in the case of poisonous insecticides, can cause sickness in humans.

Aquarium and Food Trade

The keeping of freshwater fish in aquaria is a popular, low-cost hobby the world over. Most of these fish originate from the tropics, and Southeast Asia is a source of many popular species. Most of the indigenous species exported from Sumatra are caught in the wild rather than bred in captivity. The scale of the trade is hard to gauge but in Peninsular Malaysia and in Singapore uncontrolled collecting has completely robbed small rivers of some of their species (Johnson 1961a; Johnson et al. 1969), with unknown effects on the river ecosystem.

Similarly, the effects of catching certain species of fish for the food trade has not been investigated. In Medan, large featherbacks *Chitala lopis*

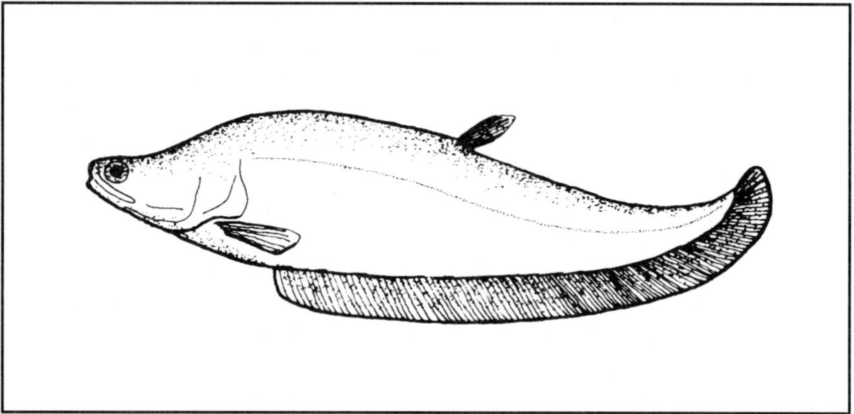

Figure 11.6. Featherback *Chitala lopis*.
After Tweedie and Harrison 1970

(fig. 11.6) can be seen being sold by roadside traders. They are especially esteemed for making fish krupuk (Adisoemarto and Sastrapradja 1977). Featherback females lay their eggs on submerged timber in slow-moving water and these are then jealously guarded by the male. During this time he is extremely easy to catch and without his protective presence the eggs will be eaten by smaller fish and other animals.

Introduced Species

Introduced fish can have significant and detrimental effects on the natural ecosystems of rivers and lakes, but this does not seem to be documented for Sumatra. Tilapia, *Oreochromis mossambica*, are extremely productive pond fish but where stock is uncontrolled they can enter waterways. They are aggressive fish and compete very successfully (both by exploitation and interference competition – p. 159) to the detriment of the indigenous fish fauna (Prowse 1968).

Forest Clearance

Forest clearance is by far the most serious threat to natural river and lake ecosystems. Many aquatic animals depend on allochthonous material (material falling into the river – p. 160) for their existence, for they feed directly or indirectly on dead leaves and other vegetable matter. If forest around the river is cleared, plant matter fails to accumulate regularly in sufficient quantities and many organisms such as loaches, catfish, prawns, crabs and dragonfly nymphs die even if other conditions are still suitable

(Johnson 1973).

The species that are least resistant to such changes tend to be the specifically Sumatran or Sunda species with restricted ranges, and they are ousted by species with wider ranges more typical of northern Southeast Asia where open habitats occur naturally.

No systematic explanation of the effects of forest clearance on the aquatic ecosystems of Sumatra appears to have been published but it is certain that the experience in Singapore (Alfred 1966) and Peninsular Malaysia (Crowther 1982b; Johnson 1973) is extremely relevant. Unforested streams may indeed have considerable numbers of fish but few species are represented and these tend to be widely-distributed or introduced species. With loss of forest those species which are exclusive to Sumatra will generally be lost.

Peatswamp Forest

In areas where this forest is disturbed on a small scale, such as where careful controlled logging occurs, regeneration of timber trees is successful and weed trees (list in Whitmore 1984) dominate the regrowth along the heavily disturbed roads.

The patterns of secondary growth on peat soils after large-scale disturbance in Sumatra and Kalimantan are described by Kostermans (1958). Fire often prevents the natural succession and instead paperbark *Melaleuca cajuputih* develops extensive and virtually single-species stands in areas where its roots can reach the mineral soil beneath the peat. Fire occurs most years in southern Sumatra where peat has been drained for transmigration projects. Paperbark occurs naturally in peatswamp forest but generally in the drier, better-drained parts. In areas where the peat is too thick for paperbark, *Macaranga maingayi* (p. 376) forms extensive, almost pure stands (Wyatt-Smith 1963) such as observed on the CRES expedition to peatswamp areas of Labuhan Batu. In the areas where the peat itself has burned, small, shallow lakes have formed. Some of these are almost completely covered with floating or semi-floating islands of grasses and herbs. This has been paralleled by farmers who grow young rice on floating rafts (Vaas et al. 1953).

In the areas of deeper peat, plants of the early stages of succession commonly include sedges and grasses and occasional trees of *Combretocarpus rotundatus* and *Campnosperma macrophylla* amongst others. In the inner regions of the swamp where nutrients are poor and the tree growth stunted, the predominant tree is *Tristania obovata* (p. 173) and this persists in the secondary growth as smaller, thinner trees. The secondary growth is also characterized by pitcher plants *Nepenthes* being even more abundant.

Freshwater Swamp Forest

The only source of information concerning disturbance in freshwater swamp is the management plan for Way Kambas National Park (Wind et al. 1979). Apart from large-scale logging (especially from 1968 to 1974) there have been several large fires (1972, 1974, 1976, 1982). These disturbances and the wide yearly fluctuations in water level hinder regeneration of the original vegetation types, and areas of swamp grass and bush grow in their place. If these were left undisturbed, it would still take several centuries for a vegetation similar to the primary forest to form. The alang-alang grass and bush would probably first give way to a forest dominated by legumes and oaks with *Commersonia bartramia* (Sterculiaceae), a common 10 m tree of secondary forest. Dipterocarp seedlings are not light-tolerant and so can only grow once a secondary forest has grown up. The seeds cannot last many years in or on the soil, so dipterocarps may have to spread from parent trees. Also, since dipterocarp trees may be 100 years old before they fruit, and seeds rarely fall more than 100 m from the parent, the replacement of these economically important trees would be painfully slow.

The opening up of the swamp forests has allowed weeds such as the aggressive, shade-intolerant creeper *Mikania micrantha* to form dense mats over grass and bush. In addition, the small fern *Salvinia molesta* and water hyacinth *Eichhornia crassipes* form thick mats on some of the waterways.

In Peninsular Malaysia, and possibly in parts of Sumatra, paperbark is a common secondary growth tree in freshwater swamps (Whitmore 1984; Wyatt-Smith 1963). The papery bark peels off and, with the resin-rich leaf litter, forms a highly inflammable soil cover. When burning occurs the paperbark tree itself is relatively resistant, but not so the other trees of the secondary growth. Grass will burn down to the soil, but the rhizomes will remain alive.

Lowland Forest

Introduction

It is often stated that after forest clearance the loss of soil fertility and the loss of nutrients in plant biomass pose a serious threat to the integrity of most natural ecosystems in the Sunda Region. While this is certainly true for heath forest, mountain vegetation and certain peat swamps, there are no descriptions of disturbance to other types of forest resulting in succession which is not directed towards the re-establishment of 'primary' forest where continued disturbance is absent. Unfortunately, once disturbed, most forests in Sumatra experience a long series of destructive disturbances which never allow the succession to progress. Effects which are sometimes attributed to the loss of nutrients from soils, may possibly be due to competitive interaction between plants. Studies have shown that the fertility of soil does indeed decline following forest removal, but the evidence of short-term loss in soil fertility is usually accompanied by evidence that the successional vegetation is remarkably good at regenerating soil fertility (von Baren 1975; Harcombe 1980; Soerianegara 1970). This is not an argument against all possible precautions to reduce nutrient loss, but it indicates that further studies are required to elucidate the situation.

A detailed comparison of soil chemistry, soil respiration, seed storage and plant growth between soils of primary forest, cut-over forest, and cut-over and burned forest both before and after rain, has produced many interesting results (Ewel et al. 1981). Amongst these it was noted that after the disturbed vegetation had been burned, 57% of the initial amount of nitrogen and 39% of the initial amount of carbon remained in the soil.

The number of viable seeds in the forest soil (the seed bank) was 8,000 seeds per m^2 (67 species) but 3,000 seeds per m^2 (37 species) after the burn. Thus vigorous and relatively diverse growth followed the burn because only a proportion of the seeds were killed. Mycorrhizal fungi (p. 206) survived, and nutrients were released from the burnt material. The 5,000 seeds that did not survive the fire may have either been killed in the heat of the fire or, as might be expected for primary forest trees, be dependent on high humidity for germination (Ng 1983). If humidity remains low for a certain length of time, the seeds will die.

A 1 ha plot of Queensland lowland forest (which contains many genera found in Sumatran lowland forest) was felled and then burned shortly afterwards. Two years afterwards the regeneration of the tree species which had reappeared were studied (Stocker 1981). Many of the 82 tree species present had regenerated in more than one way: 74 had formed shoots from the base of the old trunk, 10 had formed shoots from roots, and 34 had germinated from seed. This last category appeared to have the greatest growth rate, but the frequency of shoots from old trunks shows that this is an extremely important mode of regeneration. The ability of these trees to form shoots is probably limited, however, and repeated cutting or repeated burning may give smothering, light-demanding creepers the advantage and prevent regeneration in this form.

When forest is cut selectively for timber, recovery is relatively rapid because the seed bank has many types of viable seeds and many tree species have the ability to sprout from stumps. The silvigenetic cycle has been described on page 193. However, when forest is cut, dried and burned and then abandoned, the succession proceeds more slowly because part of the seed bank has been destroyed and because some of the sprouting species are not resistant to fire (Uhl et al. 1981).

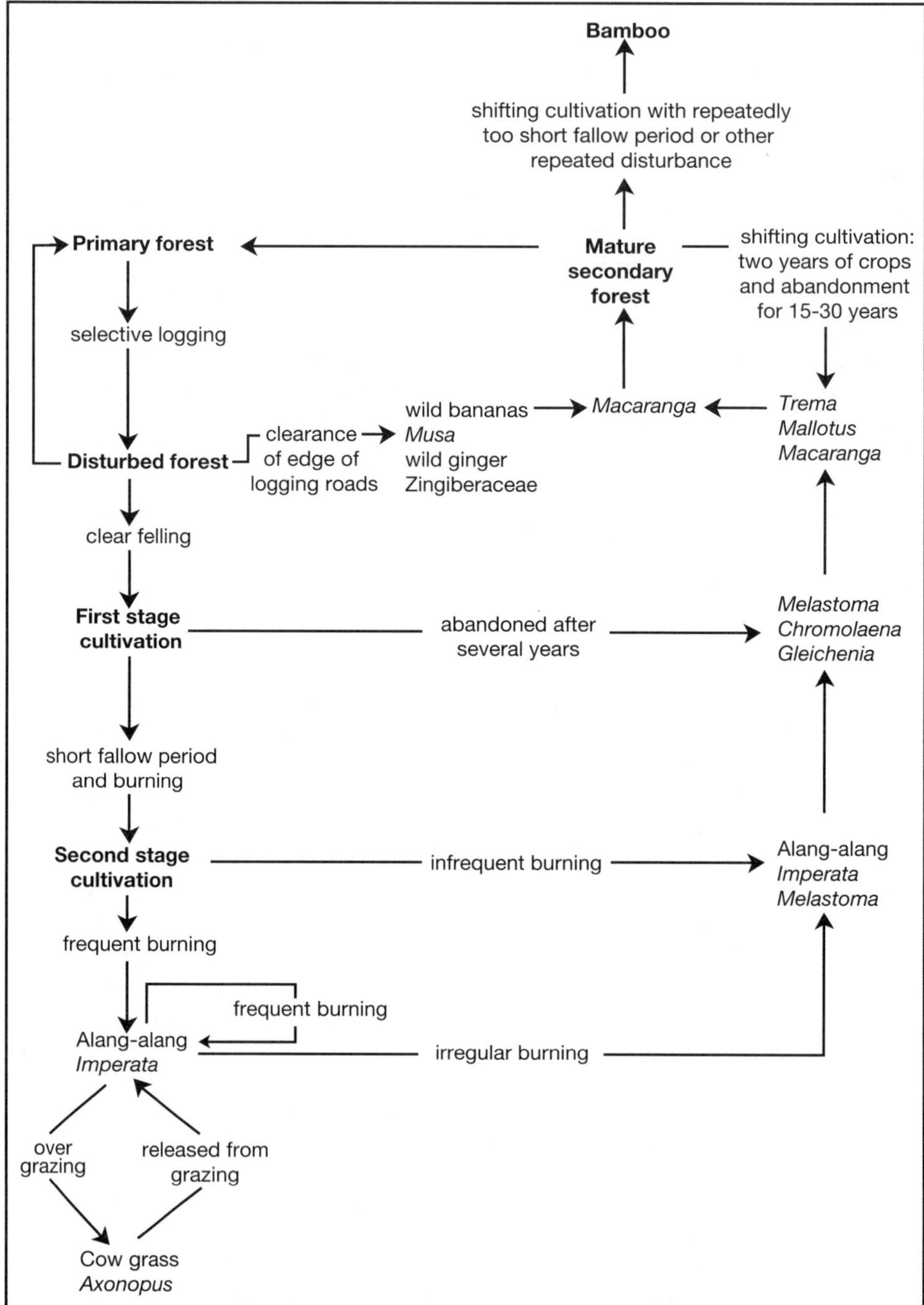

Figure 11.7. Land-use and forest succession.

After Tanimoto 1981; A.J. Whitten, pers obs., and Wyatt-Smith 1963

Thus, when forests are cut, burned, farmed, weeded or burned and farmed again, recovery of the forest (if allowed) is extremely slow. In places where the only remaining stands of primary forest (source of seeds) are some kilometres away, full regeneration would take hundreds if not thousands of years (Uhl et al. 1981). Even if a few mature trees have been left to stand there is no guarantee that these will regenerate successfully because some species have seeds which have to pass through an animal intestine before they will germinate (Ng 1983). Repeatedly burned and farmed areas should not, however, be regarded as ecological deserts or as being without interest, particularly since their rehabilitation to useful land requires an ecological approach.

The vegetation that grew up during the seven-year fallow period between tobacco crops in the Deli area has been described in detail by Jochems (1928). It contained those species which were able to fruit within seven years and whose seeds could remain viable for at least one year. The shorter fallow period allowed nowadays has presumably reduced the diversity of the secondary vegetation but it still avoids invasion by the pernicious alang-alang grass (BIOTROP 1976).

A common strategy in reforestation programs for wastelands is to plant pines or other plantation trees. Where there is a local human population these programs do not always succeed. It has been suggested that assisted natural regeneration might, in the long term, produce a larger yearly biomass increment than plantations, particularly in areas where financial resources were limited (Jordan and Farnworth 1982). Assisted natural regeneration would require strips or other shaped areas to be planted with a range of good, light-tolerant timber trees and other species. One likely candidate is *Anthocephalus chinensis.* Regeneration of neighbouring areas would then occur outwards from these centres.

The typical (but not inevitable) vegetation types resulting from different forms of disturbance are summarised in figure 11.7.

Shifting Cultivation

Shifting cultivation, in its purest sense, is the repeated use of a patch of forested land for agriculture, characterised by long fallow periods between short periods of intensive production. An area of forest is cleared, the remains are left to dry and are burned. Crops are subsequently sown in the ashes between the tree trunks which were too large to burn.[1] Nutrients are lost in the smoke, and rain washes ash into rivers before it has had a chance to become incorporated into the soil (Geertz 1963). After one or two crops, yields decrease, weeds become a serious problem (Chapman 1975) and the area is abandoned for 15-30 years to allow the soil to recover (von Baren 1975; Soerianegara 1970). The farmer then moves to another area of forest. This is usually one of mature secondary forest on a former

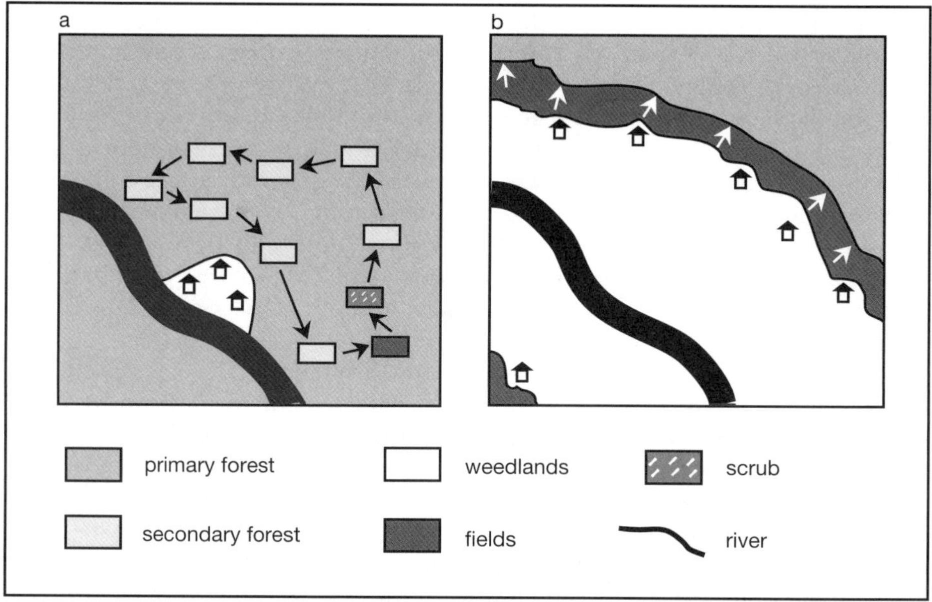

Figure 11.8. The distinction between traditional shifting agriculture (a) and the now prevalent slash-and-burn agriculture (b) which is endangering so many of Sumatra's natural ecosystems.

After Rijksen 1978

cultivated area because it is easier to clear than primary forest. This secondary forest may contain mature fruit trees such as durian which would have been visited in the intervening years. This form of shifting cultivation can support 30-40 individuals per km^2 and depends on subsistence agriculture being the predominant use of the cleared areas, and on cultural restraints which had evolved as part of the traditions bound up in animistic religions. The environmental care implicit in Christianity and Islam (Achmad 1979; Schaeffer 1970) does not seem to find its way into the lives of new generations of converts. True shifting cultivation has virtually vanished from Sumatra and only a minuscule percentage of the population is now involved. Some publications still quote figures for areas under shifting cultivation which were determined some decades ago. It should be remembered that Sumatra's population has more than doubled since the 1940s.

Use of cleared and burned land for shifting cultivation alters the early stages of succession (p. 340). Weeding removes shoots growing from tree stumps and although herbaceous plants are removed, their relative and absolute abundance increases. These herbs dominate the area as soon as the land is abandoned, but after a year, fast-growing pioneer trees dominate these cultivated areas because they have many viable seeds in the soil and/or the seeds are easily dispersed.

It is clear that when the shifting cultivation cycle is long enough (e.g., 30 years), species of herbaceous weeds are not present in the secondary forest felled prior to cultivation. When the cycle is reduced to, say, just five years and succession has not been allowed to progress very far, domination by herbaceous weeds such as *Chromolaena* (formerly *Eupatorium*) *odoratum* can become a serious problem (Kushwaha et al. 1981).

Shifting cultivation is not, despite what one might be led to believe, the relentless slash-and-burn cultivation practised in many areas. The distinction is illustrated in figure 11.8. Slash-and-burn is practised with almost no cultural or social restraint and the cultivation of cash rather than subsistence crops is the major use of cleared land. The change from shifting cultivation to slash-and-burn is caused largely by population increase (Geertz 1963), whereby rested land is virtually nonexistent and primary forest is cleared when the former cultivated area is exhausted or when alang-alang grass can be kept at bay no longer by simple (cheap) weeding. Regeneration of forest in such areas would take many centuries.

Effects of Logging on the Soil

The damage that logging causes to soil has barely been studied in Indonesia but the general subject has been reviewed by Kartawinata (1980). The major problems are soil removal by bulldozers when making roads, soil compaction caused by heavy vehicles, and soil loss due to rain striking the ground with its full force and the greater quantity of rain reaching the soil. The removal or destruction of trees also reduces litterfall and hence the organic inputs to the soil, but in areas where the soil has not been too greatly damaged, soil nutrient levels and microorganisms can recover within a few years. Since only the larger portions of the tree trunks are removed, and most of a tree's nutrients are in its leaves and roots, the loss of biomass after logging is proportionally greater than the loss of nutrients.

Effects of Logging on Hydrology

The study of the effects of logging and other forms of forest clearance on hydrology is also beset by inadequate data and differences in interpretation (Nordin and Meade 1982). It is obvious, however, from a brief visit to a logging camp or a flight over areas where logging occurs, that a great deal of water runs off the roads, carrying soil with it. This has been quantified by Hamzah (1978) who measured the silt content:

a) of a river near, but not influenced by, a particular logging area,
b) from a river in the logging area, and
c) from a ditch by a logging road.

Figure 11.9. Traditional swidden plot on Siberut Island, West Sumatra.
A.J. Whitten

The silt contents were 0.01%, 0.05% and 0.1%, respectively – a tenfold increase. The increased silting of Sumatra's main rivers could probably be traced to logging as well as other forms of unprotective land use.

In the lower stretches of major rivers the actual volume of water is probably not greatly affected by forest clearance, but it is in the higher river regions, before high water levels can be dampened, that dramatic effects are noticed. For example, the consequence of clearing just 10 ha of forest for fields on slopes above the Mengkudu River near the edge of the Mt. Leuser National Park in May 1981 was that 13 people died (Robertson and Soetrisno 1981, 1982). The Mengkudu River is normally only 1 m wide and 8-15 cm deep but landslides on the steep, deforested slopes and the increased surface runoff, both caused by the 10 ha ladang, turned the Mengkudu into a river capable of moving 1-3 m^3 boulders. A nearby river also flooded and destroyed new bridges and roads (which took months to repair), rice fields and most of a town, and by luck claimed only one more

Figure 11.10. Even selective logging causes damage beyond simply the felling of trees, but if done carefully and if given time, the forest will recover.

life. Few floods in Sumatra have been so well documented as these but they are commonly reported in the newspapers. The floods are generally attributed to illegal or irresponsible logging or other forest clearance.

Effects of Logging on the Forest

The felling of trees and extraction of logs are clearly the primary causes of disturbance during a logging operation. Felling damages tree crowns, boles and saplings, exposes wood, making it susceptible to fungus damage, and also covers seedlings. Extraction of the logs exposes bare soil and damages large areas of the forest floor (Kartawinata 1980b).

Selective logging sounds a very mild and benign activity to many people who have not visited or worked in a logging concession and it comes as a surprise to learn how much damage is caused to the forest as a whole. One estimate is that five times more timber is destroyed or badly damaged than is extracted (Burgess 1971). It has also been reported that for every large tree felled, 17 similar- or smaller-sized ones are destroyed (Abdulhadi et al. 1981).

Most of the destruction is caused by the access roads which remove all cover from the soil and form channels which are further scoured by water during storms. It has been estimated that 20%-30% of a logged area is completely bare, being composed of roads and log yards (Kartawinata 1980; Meijer 1970). Some of the roads cut across small rivers, acting as dams, and the subsequent flooding kills most of the inundated trees and other plants (Anon. 1980a; Kartawinata 1980b). The lorries used to drag logs out of the forest cause compaction of the soil surface, and this disturbance is traceable in some forests by the occurrence of the pioneer tree *Anthocephalus chinensis* even 40 years after the logging. By this time other pioneer trees such as *Trema, Macaranga* and *Homolanthus* have died or become rare as different genera take over in the succession. The soil compaction caused by lorries is, however, only a fraction of that caused by the bulldozers used to make the roads. Additional disturbance is caused when trees are felled on both sides of the main logging roads to hasten the drying of the road surface after rain. These 'daylighting' areas have been estimated as occupying 8 ha/km of road (Hamzah 1978).

A very thorough study of the effects of logging on the forest was conducted on South Pagai Island in the Mentawai Islands, West Sumatra (Alrasjid and Effendi 1979). A total of 2,416 trees (20 cm diameter and over) originally stood in the 15 1-ha plots, half of which were commercial species and half non-commercial species. A total of 194 trees (13/ha) were felled and extracted. On average, about half of the remaining trees had been noticeably damaged, broken or knocked down, and the other half had escaped damage (fig. 11.11). This proportion seems to be relatively consistent between studies (Abdulhadi et al. 1981; Burgess 1971; Tinal and Palinewan 1978). In areas of South Pagai Island where many trees were extracted, over 70% of the remaining trees had been damaged or killed. Damage to saplings and seedlings was also extremely high.

Damage to a forest during logging can be reduced if all climbing plant stems are cut. When a tree is felled the tangle of these stems often pulls other trees over and the climbers also compete with the trees for light, water and nutrients. This cutting may be sound forestry but the death of climbers must have a significant effect on frugivorous animals. For example, 25% of the food plants of orangutans are climbers (Rijksen 1978), as are 42% of the food plants of Mentawai gibbons (Whitten 1982e).

The cutting of climbers is obviously disastrous for the famous *Rafflesia* and other members of its family (p. 217), all of which are rare enough as it is, because they live only as non-deleterious parasites on the stems and roots of *Tetrastigma* climbers. In addition, the climbers of the genus *Aristolochia* have spectacular flowers and their leaves are the sole food of some of the larger (and saleable) butterflies and moths. It is difficult to resolve the conundrum of whether to cut or not to cut climbers. The answer will probably have to be total and enforced protection of such plants in reserves since they are unlikely to survive otherwise.

Selective logging is, in theory, a repeatable exercise in any given area.

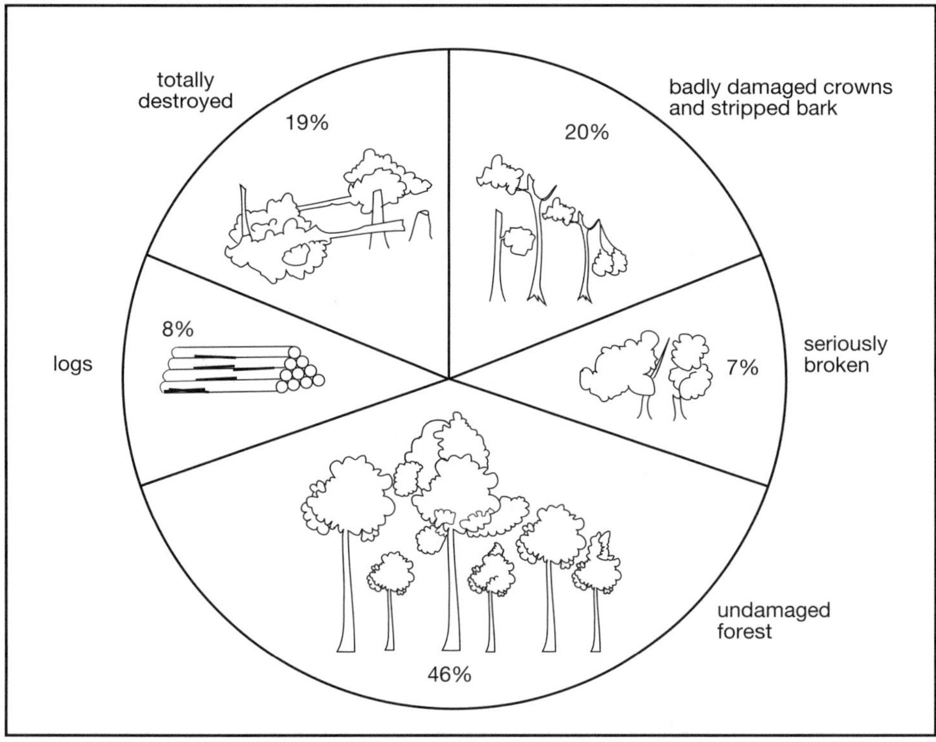

Figure 11.11. The effects of extracting 8% of the trees by selective logging on an area of forest on South Pagai Island (Mentawai). The other categories are (clockwise) totally destroyed, badly damaged crowns and stripped bark, seriously broken, and undamaged forest.

From data in Alrasjid and Effendi 1979

After the largest commercial trees have been extracted the area should be left for about 70 years before the next timber crop is harvested. The average volume of dipterocarp timber in a forest logged forty years earlier is only half the volume found in adjacent unlogged forest (Meijer 1970). Logging of a forest at intervals of less than about 70 years is unsound forestry practice but as more species become economically worth exploiting and when timber prices increase, and when smaller and smaller trees are permitted to be cut, repeated logging occurs. Over a hundred species of Sumatran trees are probably commercially exploited but a tree species which is non-commercial today can easily become commercial tomorrow if a particular use is found for its timber properties or if supplies of better species are insufficient to meet the demand. Each time logging occurs plant succession is set back a step, the forest becomes progressively poorer in desirable species and progressively richer in 'weed' species (Whitmore 1984; Wyatt-Smith 1963).

The microclimatic changes caused by logging are obvious – it is hotter, lighter and drier in logged-over forest. These changes result in the dieback of crowns, scalding of intolerant trunks and branches, water stress and even an increased likelihood of insect attack, any one of which might lead to the death of a tree (Kartawinata 1980b).

While selective logging clearly has many advantages, its also removes the best individual trees, leaving only inferior ones to produce seeds for future 'crops'. This genetic erosion is potentially extremely serious for future forestry (Ashton 1980; Kartawinata 1980b; Sastrapradja et al. 1980; Whitmore 1984) but in many logged areas it is doubtful whether in fact a full cycle of regeneration will ever be allowed to occur.

The regeneration of lowland forests in natural gaps has been described on page 193. Regeneration after selective logging takes a similar course and has been studied in parts of West Sumatra, Jambi, South Sumatra and Lampung (Geollegue 1979; Geollegue et al. 1981; Huc 1981; Huc and Rosalina 1981a). Plots studied one, three, four, and ten years after logging each had distinctly different structure and species composition (table 11.1). After 10 years it was generally found that the secondary forest was forming a layered canopy and entering the building phase (Geollegue 1979; Geollegue and Huc 1981). It seems unlikely that regenerated forests will grow to their original heights and Ng (1983) has suggested that a height decrease of 25%-50% would not be surprising.

The daylighting areas at the sides of the roads are quickly colonised by wild bananas (although in some areas these appear to be quite rare), smaller weed trees and various members of the ginger family (Zingiberaceae). This early community plays an important role in stabilising otherwise vulnerable soil.

Effects of Disturbance on the Fauna

About 70% of the mammal and bird species found in the Sunda Region are dependent on more or less intact primary lowland forest (Medway 1971; Wells 1971). Considering the very large number of insects which are restricted to a single species of plant (p. 192), the percentage of the region's insects dependent on undisturbed forest is probably even greater. In view of this, there is clearly a need to know what effects different types of disturbance have on Sumatra's fauna. The trends and general patterns, at least for mammals and birds, however, are now quite well understood. The forest fauna is affected by disturbance in at least three ways:

(1) the noise and shock of disturbance may cause immediate changes in behaviour;

Table 11.1. Structure and composition of trees in four study plots in selectively logged forest in South Sumatra.

	One-year-old plot	**Three-year-old plot**	**Four-year-old plot**	**Ten-year-old plot**
Sylvigenetic phase	Pioneer	Pioneer	Pioneer	Building
Characteristics	Presence of decaying logs Seedlings of pioneers Dense crown cover	Logs in advanced decay Presence of vines and saplings of pioneers Heterogeneous but closed crown cover	Decaying logs still present Vines on top of canopy Clear understorey	Decaying logs absent Vines in lower canopy only Crown stratified into three layers
Mean height of dominant trees (m)	2	4	8	22
Mean trunk diameter of dominants (cm)	2	5	10	35
Dominant species	*Macaranga gigantea* (Euphorbiaceae) *Alchornea villosa* (Euphorbiaceae)	*Macaranga trichocarpa* *Croton laevifolius* (Euphorbiaceae) *Anthocephalus villosa* (Rubiaceae) *Shorea ovalis* *Shorea leprosula* *Shorea escimia* *Quercus gemelliflora* (Dipterocarpaceae) (Fagaceae)	*Anthocephalus chinensis* *Macaranga gigantea* *Macaranga trichocarpa* *Macaranga hypoleuca* *Glochidion sericeum* (Euphorbiaceae)	**Upper Storey:** *Anthocephalus chinensis* *Sapium discolor* (Euphorbiaceae) **Middle Storey:** *Vitex vestita* (Verbenaceae) *Glochidion* sp., *Euodia alba* (Rutaceae) *Tetrameles nudiflora* (Dattiscaceae) **Lower Storey:** *Buchania sessifolia* (Anacardiaceae) *Polyalthia rumphii* (Annonaceae) *Macaranga trichocarpa* *Anthocephalus villosa*

After Geollegue and Huc 1981

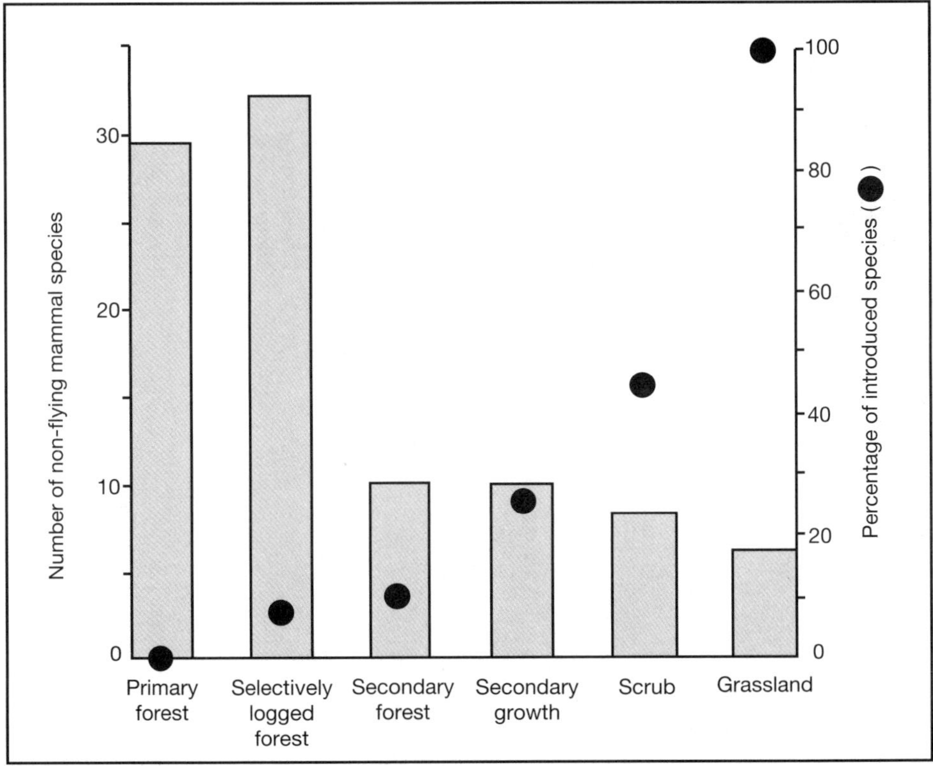

Figure 11.12. The total number of non-flying mammal species in six types of vegetation and the percentage of introduced species.

From data in Harrison 1968

(2) the actual removal of parts of the forest canopy will alter ranging patterns and diet which may in turn affect social behaviour and population dynamics;

(3) the slow regeneration rate may cause permanent changes in population density (Johns 1983).

Different species show different degrees of tolerance to disturbance and some animals will usually be found even in the most disturbed areas. A long-term study of small mammals in different habitats ranging from primary forest to an area of alang-alang grass showed, not surprisingly, that the total number of species decreased. It also demonstrated that the proportion of introduced species (rats) increased from 0%-100% (Harrisson 1968) (fig. 11.12). Similar results were reported by Yong (1978). Surveys for

Figure 11.13. Logging roads cause serious damage to the soil, the forest and hydrology.

noticeable animals along transects in forest with different levels of disturbance in the Sekundur Reserve (Langkat) also showed a dramatic decrease in species present with increasing disturbance (Rijksen 1978) (table 11.2).

Properly executed selective logging is not disastrous for much of the forest fauna, although some squirrels and birds fare badly (Johns 1981, 1983; Marsh and Wilson 1981; Marsh et al. 1984; Wilson and Wilson 1975; Wilson and Johns 1982). 'Properly executed' in the ecological sense used here means an average of 8-10 and an absolute maximum of 15 trunks removed per ha, no relogging for at least 50 and preferably 70 years (p. 365), no replanting with foreign tree species, and an adjacent area of primary forest from which fruits can be dispersed into the logged area. If such practices are adhered to, there is no reason why nature conservation and forestry should conflict. Logged forest supports a lower species diversity but is able to maintain viable populations of many species.

The initial effects of selective logging are probably the most serious,

Table 11.2. Direct observations (observations/km) (A), and indirect observations (signs/km) (B), of animals encountered along transects of different lengths in four forests in the Sekundur Reserve, Langkat.

	Secondary vegetation (5 km)	**Forest being logged (4 km)**	**Slightly disturbed forest (14 km)**	**Undisturbed primary forest (7 km)**
A) Siamang *Hylobates syndactylus*	-	-	0.07	0.14
White-handed gibbon *Hylobates lar*	-	-	0.35	0.57
Long-tailed macaque *Macaca fascicularis*	-	-	0.14	0.14
Pig-tailed macaque *Macaca nemestrina*	-	-	0.07	0.14
Thomas' leaf monkey *Presbytis thomasi*	-	0.25	0.07	0.29
Silvered leaf monkey *Presbytis cristata*	0.6	-	-	-
Elegant flying squirrel *Hylopetes elegans*	-	0.25	-	-
Horse-tailed ground squirrel *Sundasciurus hippurus*	-	-	0.21	0.57
Slender squirrel *Sundasciurus tenuis*	0.2	0.25	0.28	0.57
Common treeshrew *Tupaia glis*	-	-	0.21	0.28
Flying lemur *Cynocephalus brachyotis*	-	-	-	0.14
Pig *Sus scrofa*	-	-	0.14	0.14
Argus pheasant *Argusianus argus*	-	0.25	0.64	0.71
Helmeted hornbill *Rhinoplax vigil*	-	-	0.14	0.86
Rhinoceros hornbill *Buceros rhinoceros*	-	-	-	0.42
Bushy-crested hornbill *Anorrhinus galeritus*	-	-	0.14	0.14
Total	**0.8**	**1.00**	**2.46**	**5.11**
B) Orangutan nests	-	-	0.28	1.57
Elephant tracks	0.4	-	0.21	0.14
Deer tracks	0.4	0.25	0.28	0.28
Pig tracks	1.0	0.50	0.86	0.57
Total number of species	**5**	**6**	**15**	**17**

After Rijksen 1978

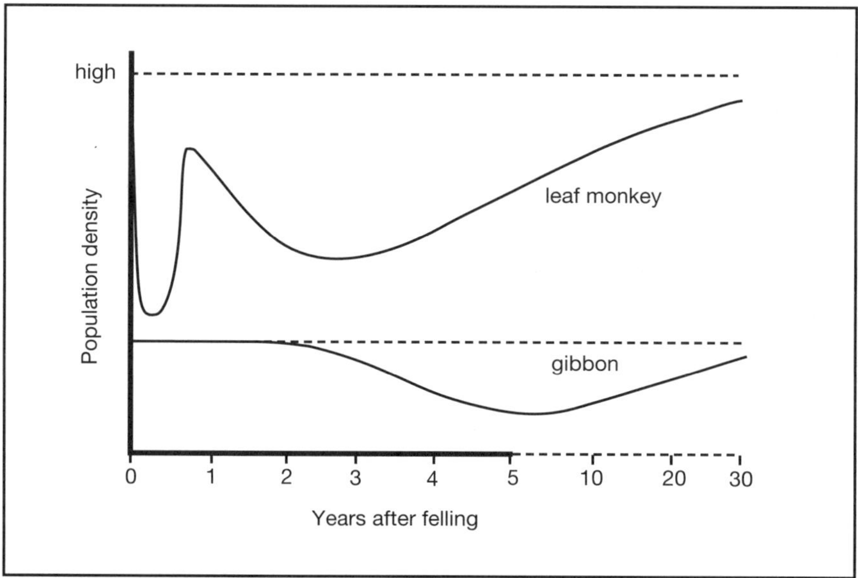

Figure 11.14. Changes in population density of leaf monkeys (non-territorial) and gibbons (territorial) after selective logging.
After Marsh and Wilson 1981

and Johns (1981) found a lower birth rate and a greater infant mortality amongst primates in such areas. These effects were temporary and the populations eventually began to return to normal. It is likely, however, that the effects of disturbance will last several decades because it takes that long for the lost resources such as food sources and nest sites to be replaced and for microclimatic conditions to recover.

The effects of disturbance depend in part on the social system and diet of the species concerned. Territorial species are worst affected because they are unable, even temporarily, to move out of the disturbed area because they will probably be surrounded by defended territories of the same species. Non-territorial species may be able to withstand some temporary crowding before moving back into the disturbed forest (fig. 11.14). Species with relatively specialised diets (e.g., p. 240) would be expected to fare rather worse in disturbed forest than unspecialised species.

Some indigenous animals, such as bulbuls *Pycnonotus,* tapirs and, to some extent, elephants, actually seem to benefit from selective logging because they favour the r-selected foods (p. 190) which are abundant in disturbed areas.

When forest is completely cleared for large-scale agriculture or other purposes the animals that once lived in it will eventually die. The territorial species are very unlikely to be able to find unoccupied areas in the neighbouring undisturbed areas, and the non-territorial species may be able to move into adjacent forests (assuming there are any) but sooner or later the food resources will limit the population to more or less its original level.

It is worthwhile to quantify roughly the losses in order to make the effects more understandable. If 10,000 ha of lowland forest were felled for, say, an oil palm plantation, the following would be among the subsequent deaths:

30,000	squirrels
5,000	monkeys
15,000	hornbills
900	siamang
600	gibbons
20	tigers
10	elephants

More than these would be affected by the disturbance, but some may later be able to exploit the plantation in some way.

Those people involved with logging or forest clearance and those who remain behind after the initial disturbance occasionally see the animals whose habitat has been disturbed. Some of these are shot by hunters and others are caught (often illegally) and sold. The trade in young orangutans has decreased enormously because of the efforts of the Directorate-General of Nature Conservation but the problem of what to do with these displaced animals is not easily solved.

Rehabilitation of animals displaced by forest clearance or caught and sold as pets is not really an answer because most of them found themselves homeless because of forest destruction. Rehabilitation centres for orangutans, such as at Bohorok (Langkat), are wonderfully valuable as sites for conservation education (Aveling and Mitchell 1982) but in terms of long-term conservation of orangutans their role is doubtful for two main reasons[2]:

(a) The surrounding forests have their own resident population of orangutans, and their birth-rates in areas close to logging operations (that is, near the forest edge) are lower than average, indicating that they are suffering from stress (MacKinnon 1974). Introducing rehabilitant orangutans into this already disturbed population could have any number of consequences.

(b) There is the danger that diseases contracted from humans during the time the orangutans were in captivity could spread to the wild population with possibly disastrous effects. The orangutans are kept in quarantine for a while but this will not necessarily reveal those diseases with a long incubation period or animals which are 'carriers' of the disease but do not show any symptoms.

Orangutans are particularly amenable to rehabilitation because they do not live in social groups. Other Sumatran primates are social, however, and attempts to introduce a monkey into an established group are quite likely to lead to the new animal being killed, probably by the adult male. There is no reason why the adult male should bother to defend his group or territory for any individual except his mate(s) and offspring, because the introduced animal will not increase the adult male's genetic legacy.

Bamboo

Repeated disturbance of an area of lowland forest allows various species of bamboo to gain a toehold and they can form a more or less permanent vegetation cover. Many species are shade-tolerant and so persist even when forest has grown around and above them. Patches of bamboo in the forest almost invariably indicate sites disturbed intensively by man at some time in the past. Bamboos are giant members of the large grass family, Poaceae (formerly Graminae), which includes rice, sugar cane, alang-alang and maize. Like rice, bamboo dies after flowering and fruiting, but it can take over a century for flowering to occur (Janzen 1976b; Soderstrom and Calderon 1979). Stranger than this, however, is that under natural conditions most species flower simultaneously over wide areas. This is similar to the phenomenon of gregarious flowering found in dipterocarp trees (p. 221) and pigeon orchids (p. 403) but in these cases flowering is mainly governed by environmental factors. The flowering of bamboo is not. Even when individual plants of some bamboo species are grown in different parts of the world with different climates, the species will still flower simultaneously (Janzen 1976b).

Why should bamboo behave in this way and how are the cycles maintained? The reasons are in essence the same as for the dipterocarps. That is, long gaps between fruiting periods allow reserves to be accumulated so that very heavy fruiting occurs, and this satiates the available seed predators so that at least some seeds are left to form the next generation. If the plants did not synchronise the fruiting, heavy fruiting would simply result in large populations of seed predators. Bamboos provide very attractive food:

1) Bamboo seeds are extremely nutritious, having a nutrient quality slightly greater than either rice or wheat (Janzen 1976b; Rao et al. 1969).
2) There is no evidence to suggest that bamboo seeds are defended chemically against predation.
3) The fallen seeds are easily found beneath the clumps of bamboo because the dense shade produced by the bamboo stems and leaves precludes most other plants from growing beneath them.

Bamboo seeds therefore form an easily harvested, nutritious but periodic food source. The seeds are eaten by a wide range of opportunistic vertebrates such as rats, pigs, porcupines, peasants, jungle fowl, doves and parrots. None of these animals form defended territories and so can range widely around forests in search of food. They also have r-selected reproductive strategies; that is, they can increase dramatically in numbers in response to an abundant source of food which, in the case of bamboos, may last for two years. It has been hypothesised by Janzen (1976b) that the above animals would have been effective at devouring the majority of seeds produced by clumps of bamboo fruiting out of phase of, at the very beginning or very end of, a fruiting period. Conversely, when bamboo fruit was most abundant, the animals would eat proportionately less of the crop. Since the determination of the inter-fruiting period appears to be largely genetic rather than environmental, such 'weeding out' of those plants behaving independently would favour increasingly synchronous behaviour. When the fruiting period is over, the swollen population of certain animals, such as rats, often turn their attention to growing or stored crops.

Man disrupts this synchrony in three ways:

1) He removes (by hunting or habitat change) the animals which are the agents of selection.
2) He moves bamboo around deliberately, mixing different species or cohorts of the same species (Soderstrom and Calderon 1979). He does this because of the many and varied uses of bamboo (Sharma 1980; Soderstrom and Calderon 1979; Widjaya 1980).
3) He harvests (or harvested) the larger bamboo seeds when they are most abundant and so his impact is most when the other animals' impact is least. This is largely because it is not worth his while putting effort into collecting the seeds unless there are considerable quantities available. Conversely, a foraging animal would be able to occupy a bamboo area and gradually increase its intake of bamboo seeds.

Bamboo deserves more attention, particularly now that it is being grown as a minor plantation crop in Sumatra for use in the oil-palm industry (Widjaya 1980).

Macaranga

One of the most common genera of pioneer trees found in disturbed lowland forest (and peat swamp – p. 357) is *Macaranga* (Euphorbiaceae). This genus contains over 20 species found in Sumatra and the vast majority are shade intolerant and are only found in open conditions. Keys to the species have been prepared by Corner (1952) and Whitmore (1967), of which the second is the more useful.

About one-third of the species of *Macaranga* are peculiar in having hollow twigs in which live colonies of ants. Winged females bite their way into the twigs of young seedlings and subsequent holes are made by other members of the growing colony. It is usually the case that every branch of a tree will be occupied and only rarely is an 'ant-species' of *Macaranga* found without ants. Only one species of ant, *Cremastogaster borneensis macarangae,* occupies *Macaranga* in Peninsular Malaysia (Whitmore 1967) and it is likely that the same species is found on Sumatra.

The ants bring young scale insects (generally a species of *Coccus* [Khoo 1974]) into the twigs where the scale insects suck sap from the tree tissue. The scale insects excrete sugary fluid (honey-dew) on which the ants feed. *Macaranga* does not only provide the ants with a safe place to keep their 'cattle' but it also forms small, white, oily food bodies on the stipules (small leaf-like extension at the base of the petiole) which are utilised by the ants (Corner 1952).

While it is clear how the ants benefit from the *Macaranga,* no thorough investigation has been made of the other side of the relationship. It is presumed, however, that the ants protect the tree from caterpillar and other pest damage. It would be easy enough to test this hypothesis by comparing insect damage on two sets of trees of the same species in which one set had the ants systematically killed and the other set was left alone as a control.

Uncommon Lowland Forests

Heath Forest/Padang Vegetation

Disturbance of heath forest results in impoverished vegetation. Repeated disturbance results in vegetation which, because of the extremely low mineral status of the soil, appears unable to develop back into heath forest (Mitchell 1963) (p. 257). Disturbance to padang vegetation, which may possibly be a natural ecosystem in its own right (albeit impoverished), results in even poorer padang, easily burned and increasingly slow to recover. It has been suggested that the large sandy area of impoverished vegetation near Gunung Tua (Labuhan Batu/South Tapanuli) used to be heath forest (van der Voort 1939).

The only study on the effects of disturbance on heath forest in Indonesia was conducted by Riswan (1981) in East Kalimantan. He clear-felled two 0.5 ha plots in primary heath forest, burned the vegetation in one and removed all the surface vegetation from the other without burning. Transects 100 x 1 m were established and divided into 100 1 x 1 m quadrats and the regrowth was monitored. After 18 weeks all the quadrats in the unburnt plot were occupied by some living vegetation. There were only two seedlings (two species) but 29 species had sprouted from the base of the felled trunks. In contrast, 13 of the quadrats in the burnt plot remained bare after the same time period. In the other 87 quadrats there were five seedlings of one species but only 19 species were sprouting from felled tree bases. In a nearby 'mature' disturbed heath forest there were only eight tree species present compared with 27 species in an adjacent primary heath forest. Riswan (1981) joined virtually all other authors who have experience of heath forest in strongly recommending that:

1) All heath forests should be retained in their natural state and classified as conservation areas; and that
2) Heath forests should be used for education, research and recreation.

Reclamation of the bare, sandy, glaring-white tin mine tailings, which are in areas probably once covered with heath forest, presents a considerable challenge to a plant ecologist. Soepraptohardjo and Barus (1974) write, "Because of the low potential of tin tailings, they are generally forgotten about, becoming mounds of white in an otherwise green landscape. It is the responsibility of P.T. Timah (the major tin-mining company) to seek ways of rehabilitating the tailings." Similar views were expressed in the environmental impact assessment of the P.T. Koba Tin operation (LAPI-ITB 1980). Reclamation experiments with various foreign shrubs and trees have been started (Siagian and Harahap 1981) but tree species from heath forest may represent an important element in any such reclamation. Brunig (1973) estimated that a total of 849 species of tree from 428 genera lived in the heath forests of Sarawak and Brunei. The area of heath forest on Bangka and Belitung Islands is much less, so a correspondingly lower number of species would be expected. However, it is these trees, which have adapted to atrociously poor soils, that should be used in trials aimed at reclamation. Van Steenis (pers. comm.) has suggested *Ploiarium, Rhodamia* and *Rhodomyrtus* as possible trial genera. There is little hope of recreating heath forest (Mitchell 1963) but the use of local species is likely to meet with much more success than the use of foreign species adapted to quite different conditions. Enormously successful reclamation of kaolin soils in southwest England has been achieved through soil conditioning and the careful choice of appropriate indigenous plants (Allaby 1983).

Considering the size and influence of P.T. Timah on Bangka and Belitung Islands, it is recommended that they make strenuous efforts to protect heath forest in and around their concession areas, and to begin a program of identifying tree species suitable for use in reclamation projects.

Ironwood Forest

There appears to be no documentation on the effects of disturbance on ironwood forest. One of the characteristics of ironwood trees is that after being felled, large numbers of suckers grow freely from the base of the trunk. It is not clear how many, if any, of these survive to become mature trees. These suckers grow vigorously even in full light, such as at the side of logging roads. This, plus the initial predominance of ironwood (seeds and saplings) in the forest, probably means that ironwood would regenerate quite successfully in the long term. For notes on germination requirements for ironwood seed see Koopman and Verhoef (1938) and page 265.

Forest on Limestone

Limestone hills are generally drier than hills consisting of other rock, and fires, either natural or deliberate, are the chief causes of the destruction of forest on limestone. The thin soil is highly organic and can itself burn in the fires and so regeneration may be very slow. Bare rock is colonised first by bryophytes and ferns and then eventually clumps of shrubs and small trees grow where litter has accumulated (Anderson 1965). Forest on limestone has few commercially important tree species but is destroyed by mining.

Mountains

During the ascent of Mt. Kemiri by a CRES team it was repeatedly remarked that signs of prior expeditions were extremely obvious, particularly in the upper montane and subalpine zones. A ditch and a soil pit dug by a geomorphological expedition two years before still had sharp edges and the piles of earth at their sides were still quite loose. There can have been hardly any heavy rain to smooth the edges or to wash the piles of earth back into the ditch. Campsites used by army expeditions in the 1940's were still largely clear of vegetation. Van Steenis (1938) remarked on the flattened empty biscuit tins at the summit left by the expedition which set up the triangulation pillar six years before his 1937 expedition. In 1982 these tins were still there although they were slightly rusty. Lastly, in 1982 the burnt slopes were essentially unchanged in character from the photographs at the end of his 1938 paper.

Most people in Sumatra live in climates where vegetation can grow vertically as much as 8 m in three years. It is important, therefore, that those people who climb Sumatra's mountains for recreation or study should remember just how long-lasting the effects of disturbance can be on mountains. The campsite at about 3,200 m on Mt. Kerinci was, at the time that it was visited by a CRES team, littered with tins, paper and plastic wrappings and there were signs of fires which had got out of control. An untended fire on Mt. Sinabung destroyed most of the upper zone vegetation there in 1981.

The only known detailed studies of regeneration of tropical vegetation at high altitudes were both conducted in Costa Rica. The regeneration of subalpine vegetation studied by Janzen (1973b) followed a fire three years previously and he describes fire ecologically as a generalist herbivore. The two shrubs he studied in detail were a *Vaccinium* and a *Hypericum*, both of which have species occurring on Sumatran mountains. The fire killed the above-ground parts of the shrubs but suckers had shooted from the base of the plants. On average, the suckers had grown less than 50 cm in the three years. The soil surface exposed by the fire did not, as one would expect in lowland areas, become covered with fast-growing pioneer species. Instead, much of the bare soil remained unvegetated. In other places liverworts and mosses had started to grow. Very few seedlings of the surrounding plants were found.

Three years after the fire, dead stems of plants were still standing and even those that had fallen to the ground showed no signs of decay. This extremely slow rate of decomposition is caused partly by the low temperatures but also by the absence of (or extreme lack of) most of the lowland decomposers such as ants, termites, and earthworms. This in turn may be due to the extremely moist condition of the soil, branches, and logs which never warm up to any great extent (Janzen 1973b).

Regeneration of Costa Rican oak forest, which is similar to that found in Sumatra from 1,500-2,000 m, was also found to be extremely slow (Ewel 1980). Repeated cutting would deplete the seed stocks irrevocably and it was concluded that montane forests are "truly the tropics' most fragile ecosystems." Disturbance of oak/laurel forest in Peninsular Malaysia leads to a vegetation dominated by tree ferns and the scrambling fern *Gleichenia* (Wyatt-Smith 1963).

The Sumatran pine *Pinus merkusii* is an indigenous tree occasionally found at about 1,000-1,600 m in lower montane forest in northern and western Sumatra. Trunks of large specimens can reach a metre in diameter, considerably larger than most of the trees of the same species seen in plantations, such as around Lake Toba, the first of which were planted in 1927 (van Alphen de Veer 1953). Young pines in primary forest are generally found growing as pioneers on, for example, landslides. Large areas (possibly 150,000 ha) of unplanted pine forests occur in parts of central Aceh and these are sometimes thought to be natural. In fact, they have been caused by man. Felling and repeated burning of lower montane forest has increased the area of suitable habitat for the pioneer pines. Once over 3 m tall the trees are quite resistant to fires (Whitmore 1984), unlike most of the broadleaf trees with which they had once shared the land. These pine forests are maintained by the cattle owners who burn the undergrowth to encourage the growth of tender young grass for cattle to graze.

Figure 11.15. Limestone quarrying for roads, lime and cement can have major impacts on the range-restricted species living in caves.

Caves

The study of the effects of disturbance on caves is an open book. At one extreme, it is clear that opening up a cave by mining the limestone and allowing the sunlight in will utterly destroy the specialised cave communities. There is no such thing as regeneration of cave life. Instead, a succession of plants from the limestone flora (p. 273) will colonise the rocks where light newly penetrates.

The resilience of the cave fauna to disturbance is unknown. The extraction of insectivorous-bat guano should not be complete, to avoid endangering the cave floor community. This sort of exploitation would obviously be better undertaken after an ecological survey and assessment of the possible impacts of different collection techniques. Similarly, the harvesting of swiftlet nests for food should be conducted so as to cause as little disturbance as possible to the birds. A nest harvesting plan has been worked out for Niah cave in Sarawak where some 4,000,000 swiftlets nest (Medway 1957).

It is interesting that Ngalau Indah near Payakumbuh, the only Sumatran cave where tourists are catered for, still seems to have sizeable populations of bats and swiftlets despite the almost daily disturbance. It is not known, however, if some less-resilient species have abandoned the cave.

Bats are sensitive to disturbance and when caves are visited during daylight hours, strong lights and loud noises should be avoided in the darker chambers. Catching bats requires skill and practice and should not be attempted except with good reason. Scientific collecting of bats should be conducted in moderation and other studies should cause as little disturbance as possible. The consequences of bats abandoning caves are described on pages 329 and 334.

Limestone mining can cause disturbance in a variety of ways, even if it is not immediately adjacent to a cave. For example, explosions used to break rock free can cause shock waves that break stalactites and stalagmites or cause thin cave roofs to cave in (Sani 1976; Sardar 1980). Even so, there are still quite a number of bat species roosting in the Quarry cave at Lho'Nga but, again, it is not known what species were present before quarrying began.

Chapter Twelve

Agricultural Ecosystems

Introduction

This chapter does not attempt an exhaustive review of all the different agricultural ecosystems found on Sumatra, nor an analysis of ecological constraints to tropical agriculture (for which see Janzen 1983d). Nor does this chapter discuss the fascinating home garden ecosystems which cover 20% of the available arable land in some areas of Indonesia, particularly Central Java (Penny and Ginting 1980; Sastrapradja 1979; Widagda 1981). Elsewhere, where Javanese have been settled for 20 years or so, home gardens can also be found but they do not yet cover a significant area of the arable land of Sumatra.

Green Revolution

Some decades ago widespread famine was seen as the inevitable fate of the increasing population of Asia. The famines have not yet come, rice fields are still productive, and the real price of rice has halved. The first high-yielding variety (HYV) of rice from the International Rice Research Institute (IRRI) was distributed in 1966 and to this can be traced the turn-around in fortunes that was dubbed the Green Revolution. The first variety distributed widely was a cross between an Indonesian variety and a dwarf Chinese species, and it had shorter and stronger stems that were able to support the heavier seed heads produced with applications of nitrogen fertilizer and 'protected' by pesticides.

The Green Revolution brought few if any benefits to upland populations, largely because its fundamental features of improved seed and larger quantities of fertilizers cannot be applied to upland areas because of the greater range of topographies, soil types, and a lack of dependable water supplies.

The overall success of the rice intensification programmes known as BIMAS (or 'mass guidance') during the 1970s can be counted as one of the most impressive achievements of the post-1965 government, although the first programmes began in 1963. It involved major investments in new or rehabilitated irrigation systems, fertilizer plants, transport and storage networks, and the establishment of effective research facilities, extension services, and administrative bureaucracies, all with trained staff, as well as a rural banking network, and local cooperatives. The crowning achievement has been a remarkable increase in rice production, increasing from about 11 million to 30 million tons. As a result, Indonesia has moved from being the world's largest rice purchaser to the status (albeit precarious) of being self-sufficient. Even so, the programme has not been without its problems.

For example, after the HYV were introduced it was recognized that valuable local varieties were being lost and so the International Rice Research Institute in Los Baños, Philippines, set up a gene bank. The government's encouragement for the growing of the relatively few high-yielding varieties (by 1975 half of the sawah were planted to just four varieties) led to increasing genetic uniformity which, predictably, opened the crop to disease and insect pests. This vulnerability was aggravated by increasing uses of pesticides and fertilisers, closer spacing of plants, and double- or triple-cropping each year without pause. The first signs of trouble appeared in 1974 with attacks by the brown planthopper *Nilaparvata lugens* which damages the rice directly by sucking sap from the phloem of the leaf sheaths, and by introducing the 'grassy stunt' and 'ragged stunt' viruses. The viruses are a serious problem only when the pest populations are high, but the planthopper numbers proved difficult to control because planthoppers have:

- only about four weeks between generations;
- females which lay 100-300 eggs in the two-week laying period;
- males which can mate on the day after emergence and live for nearly a month;
- eggs which are lodged deep in the leaf sheath out of the reach of pesticides;
- tolerance to great crowding (up to 6,000 hoppers per rice hill);
- tremendous mobility, being able to fly for 10 hours;
- a complex life cycle that produces winged adults only every other generation; and
- voracious appetites which can turn a healthy green plant to a withered brown shadow in just two days. This condition is known as 'hopperburn'.

Four new IRRI varieties with the BPH-1 gene for resistance to the planthopper were introduced in 1975, just one year after the first outbreak, but these became susceptible after just 4-5 cropping seasons. This happened in part because the 'new' varieties were quite similar genetically to their predecessors, and because of the enormous reproductive potential of the female planthopper. In the first four years of infestation, an estimated three million tons of rice (worth some US$500 million) was lost as a result of the planthopper. One of the newer varieties, IR36, remains resistant.

Meanwhile, Indonesia's own rice breeding programme had produced some excellent varieties such as Cisadane, with good taste, even higher yields than IR36, and which fared well in wetter conditions. These served to increase the number of varieties planted, and helped Indonesia to achieve rice self-sufficiency and surplus by 1984, although this was founded upon a genetic base even more uniform than that of the 1970s. In 1985 the planthoppers were again causing serious damage to the rice crops, having evolved the ability to overcome the genetic resistance of almost all the varieties, and these increases in damage were generally preceded by increases in pesticide use. A new IRRI variety, IR64, appeared to be resistant to the insects and was very rapidly adopted throughout, causing Indonesia's rice production to be pivoted on a yet more narrow base, supported by still heavier use of government-promoted, highly-subsidized pesticides.

It was at this time that ecologists' views began to be heard. They had been intrigued that an insect which had been of such minor importance should so quickly become of such significance that the Cabinet itself was holding meetings to discuss its control. It was explained that, under 'natural' conditions, over 100 predators, parasites and diseases kept the planthopper numbers under control, neutralizing its tremendous reproductive potential, and that the broad-spectrum insecticides that had been used against the pests were even more damaging against the predatory spiders, beetles, dragonflies, and other insects. It was also explained that sublethal doses of insecticides caused the development of resistance and resurgence of the planthoppers, and that this was probably due to the stimulation of egg production, and to the destruction of natural enemies.

And the planthopper is just one of over 400 species of insects, mites and ticks which have developed resistance to one or more pesticides. The resistance spreads through populations because some individuals survive pesticide applications because of behavioural, biochemical or physiological adaptations, and their genes are therefore passed on to the next generation. Repeated applications thus lead to the entire population becoming resistant.

Integrated Pest Management

In late 1986 President Soeharto took a bold step and instituted a number of landmark ecological measures, the most radical of which was the ban on the use of 57 varieties of organophosphate chemicals on rice. These had been implicated in the resurgence or explosions of pest numbers and the unintentioned demise of their predators. Never before had any country adopted ecological solutions to the problems of their major crop in such a sweeping manner. The Presidential Instruction placed emphasis on promoting the use of a hormone which prevents the larvae from developing into adults, and a restricted range of conventional pesticides were permitted only when severe pest outbreaks occurred. These had been supposed to affect only the pest insects attacking the plants, but it is now known that other insects, including parasitoids, drink from the plants, and that the chemicals drip from the leaf tips into the water surface where many important insect predators live.

The Presidential Decree was effectively the start of Indonesia's world-leading position in IPM, replacing regular calendar spraying with a variety of biological and cultivation controls, and spraying only when defined levels of infestation were exceeded. In addition, within two years of the decree, Indonesia had removed all the pesticide subsidies. The benefits of the decree have been highly visible and this demonstrates how development which works with the natural ecosystem rather than against it can achieve dramatic results:

- yields have continued to increase and have become more stable;
- farmers are saving money previously spent on pesticides;
- the Government is saving $150 million annually on subsidies;
- the Government has made direct savings of $1 billion;
- no serious outbreaks of brown planthopper have occurred;
- water quality has almost certainly improved;
- over 300,000 farmers have been trained for one season, and 2,000 extension agents trained intensively for 15 months; and
- trained farmers pass on their knowledge and experience.

Thus the evidence supports the idea that insecticides have had a major destabilizing effect on overall yields, causing losses of millions of tons of rice, wasting billions of dollars in hard currency, and degrading the health and well-being of farmers and their environment. All this for the lack of understanding of how a rice field functions. The importance of ecology has perhaps never been so clear.

The problems of pesticide and fertilizer misuse will not be solved, however, until even more farmers have been trained and become confident to manage their fields, and until the banned pesticides stop being used on secondary crops on the same land where rice is grown, thereby eliminating useful predators and parasites. The continued availability of these pesticides also means that clandestine spraying on rice is very easy. It is clearly urgent to extend IPM to other crops.

Non-Insect Pests

Rats and birds are ubiquitous pests which also cause considerable damage. Soerjani (1980) considers rats to be the most important group of pests hindering agricultural production in Southeast Asia. Rats are known to occur in almost all agricultural crops and to damage many stored products (BIOTROP 1980; COPR 1978; Estioko 1980; Myllymaki 1979; Soekarna et al. 1980). They present the additional hazard of being a health risk to humans; the organisms causing scrub typhus, meningo-encephalitis, plague and leptospirosis are all known to be carried by parasites of agricultural or urban rats (Lim et al. 1980; Sustriayu 1980).

The number of bird species inhabiting rice fields, plantations and other areas of permanent agriculture is quite considerable. For example, 72 species were observed in these areas in Aceh by Chasen and Hoogerwerf (1941). This represents about half the number of bird species observed in primary and mature secondary lowland forest, but in fact only two species, or 1% of the combined species list, were seen in both areas. Seventy-five species were seen in clearings, pastures and young secondary growth, but when compared with those seen in agricultural areas only 23 (19%) were common to both lists. This serves to demonstrate how few of the forest birds can survive in disturbed areas. The birds of agricultural areas tend to have large geographic distributions and many can become pests. Among the more serious pests are:

White-rumped munia	*Lonchura striata*
White-headed munia	*Lonchura maja*
Baya weaver	*Ploceus philippensis*
Red avadavat	*Amandava amandava*
Pin-tailed parrotfinch	*Erythrura prasina*

all of which are members of the sparrow family (Ploceidae).

Bird control poses considerable problems since birds generally range far wider and into more habitats than insect pests. Poisoning does not seem to be a viable control method because effective poisons are usually toxic to other animals, including man. Alternative methods, such as destroying communal roosts of weaver birds, can have temporary, local effects. In his account of bird pests, Adisoemarto (1980) repeatedly states that more ecological information is required for effective control programs.

A locally problematic group of pests on a wide range of fruit trees is fruit bats. Village-level control programs are rarely sufficiently selective and do not confront the problem that whereas some bats cause economic losses to fruit harvests (Adisoemarto 1980), others are essential to the fruit harvest in their role as pollinators (Khobkhet 1980) (p. 329). The flying fox *Pteropus vampyrus* is easy to shoot because of its size, but although it does eat fruit, observations reveal that it takes only fruit of a ripeness beyond what could be harvested and transported to markets.

Rice-Field Ecology

Anyone who cares to sit by a rice field for a short while will see that the rice field is quite a complex ecosystem. The water surface is disturbed by the movements of mosquito larvae, the dragonflies hawk just above the rice and particularly around the bunds, and other creeping, crawling animals can be seen among the rice stems themselves. When it is considered that rice is the world's most important food crop, that it grows in probably the most complex ecosystem of any crop, it is remarkable that only now is the plant being studied from an ecological perspective.

The conventional wisdom is that the pests' natural enemies (predators and parasitoids) attack either herbivores or other natural enemies. In this simplistic system, changes in the size of the enemies' populations would necessarily follow behind those of the pests. There are in fact two additional pathways of energy flow, both of which are independent of the herbivores. In the first, predators and parasitoids feed on organisms such as the larvae of beetles, flies, and springtails, and bacteria, which decompose organic material or detritus in the soil, most of which is derived from weeds, rice straw, and algae. In the second, the bacteria are fed upon by zooplankton which also feed on phytoplankton. The large zooplankton are fed upon by filter-feeding organisms such as chironomid midges and mosquitoes which themselves become prey. In fields where insecticides are not used, the populations of filter-feeders and detritivores build up early in the season, causing the populations of natural enemies to increase accordingly. These in turn act as a buffer against damaging increases in pest populations. If pesticides are used early in the season, then the population of natural enemies is low, pest populations are released from natural controls, and the phenomenon of 'pesticide resurgence' occurs (W. Settle, pers. comm.). Linkages between these systems can be extended to form a simplified rice-field food web (fig. 12.1).

As a result of the Indonesian IPM programme, hundreds of thousands of farmers have learned that spiders are among the most voracious predators on rice hoppers and other small rice pests. Of the various species, the wolf spiders *Lycosa pseudoannulata* are the best performers, and are easily recognizable, having a fork-shaped mark on their back. They are highly mobile animals and colonize rice fields early in the development of the crop and can control the prey before damaging levels are reached. The female lives for three to four months and can lay 200-400 eggs during that time. After the spiderlings hatch, as many as 80 of them can be seen riding on her back. Wolf spiders do not spin webs but hunt their prey directly among the base of the plants, jumping over the water when disturbed. The spiderlings eat planthoppers and leafhopper nymphs, and the adults also eat stem borer moths. Each individual spider will eat 5-15 prey each day. IPM farmers know these statistics, not because they have

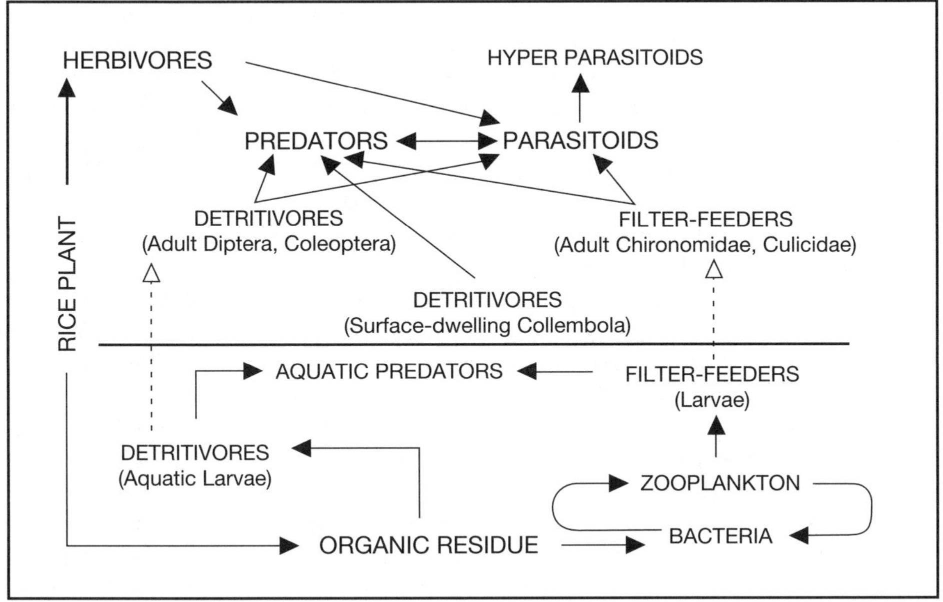

Figure 12.1. Simplified food web of a rice field.
After Whitten et al. 1997

been told them by an extension agent, but because they have all made these observations by housing planthoppers in cages – with and without spiders. The foundation of the Indonesian IPM programme is that farmers learn every important concept by doing experiments and making observations themselves.

Wolf spiders are just one of a wide range of spiders and insects, including beetles, damselflies, earwigs, pond skaters, ripple bugs, ants and wasps, that prey directly on the pests of rice. Of the approximately 650 species of arthropods collected in rice fields by IPM researchers, some 65% are predators and parasites, 20% are detritivores and filter-feeding species, and 15% are herbivores (W. Settle, pers. comm.). In addition to all these there are also numerous pathogenic fungi and viruses.

Certain birds probably play a role in controlling insect numbers, and the more complex (layered) the vegetation is around the crops, the more predatory birds can be supported. Also, even simple measures, like placing bamboo poles around the fields to act as perches and lookout posts, enable birds such as drongos, treeswifts, wood swallows and falconets to hunt more efficiently and thereby to help the farmers. Exactly how beneficial they are has yet to be assessed. The same applies to two of the most conspicuous birds in some rice-field areas: the Javan pond heron *Ardeola speciosa* and cattle egret *Bubulcus ibis.* The herons forage in rice fields mainly just after they have been ploughed or planted and eat mainly dragonfly and waterbeetle larvae, molecrickets and spiders. The egrets are generally seen in wet rice fields while they are being ploughed, and on dry harvested rice fields, where they feed mainly on grasshoppers, crickets and spiders. The egret is generally considered to be beneficial since their main prey damage crops and compete with cattle (Kalshoven 1981). These rice pests include the short-horned grasshoppers *Oxya japonica* and *Stenocatantops splendens,* and the locust *Locusta migratoria.* The long-horned grasshoppers *Conocephalus* are also eaten but this situation is complicated by the fact that some members of the genus are beneficial by the fact of their predilection for the eggs of the pestilential stinkbug *Leptocorisa oratorius.* They also eat spiders but it is not known how this affects pest populations. The mole cricket *Gryllotalpa* sp. is a pest because it lives among the rice roots, feeding on the lower stem and loosening the soil which causes the plants to wilt (Kalshoven 1981).

Rice-field frogs are also beneficial biological control agents. Some animals eaten by the frogs are predators themselves, but the vast majority are potential pests. Rice also suffers serious damage from rats and introduced golden snails, and these are discussed elsewhere.

The Ecology of Rice-Field Rats

Rats cause the loss of least 12 million tons of rice each year worldwide, and it is possible that rats and birds combined may have more serious impacts on rice crops than all other types of pests (COPR 1976). In the Sunda Region rats are probably responsible for a 5%-6% overall loss of rice yield, but losses in individual fields can be far greater.

Rice fields in Sumatra are inhabited to different extents by at least eight species of rodent. The most serious pest among them is the rice-field rat *Rattus argentiventer,* which damages rice crops throughout Southeast Asia, although the wood rat *R. tiomanicus* and Whitehead's rat *Maxomys whiteheadi* are locally the dominant species (Adisoemarto 1980; Lim 1974; Soekarna et al. 1980). Two species of mice, the ryukyu mouse *Mus caroli* and the fawn-cauled mouse *Mus cervicolor,* have been introduced into the northern half of Sumatra where they are confined to rice fields (Marshall 1977). It is possible that the large bandicoot *Bandicota indica,* which can measure 50 cm (head to tail) and weigh half a kilogram, has increased its range through introduction by man, because its flesh is commonly eaten. Two further species, the little rat *Rattus exulans* and the house rat *Rattus rattus diardi,* are generally found in peripheral areas but will enter rice fields to feed. These species are usually only found close to man's buildings or farms.

Analyses of stomach contents of these rodents has shown that they all eat insects and snails, and that the amount of plant material they ingest depends on the type of habitat and varies between species (Lim 1974). Rats eat rice at all stages of its development and although the theft of ripening rice grains is a direct loss from the harvest, the shredding of growing stems to eat the tender growing shoot does more absolute damage. A single, ripe seed head may represent a rat's daily food requirement, but the same rat may eat 100 growing shoots in a day before it is satisfied.

The reproduction of these rats clearly follows an r-strategy. The gestation period of the rice-field rat, for example, is 21 days, a litter generally numbers seven, and about eight litters are produced per female per year. Females could theoretically have more litters but reproduction tends to occur only when an abundant food supply is available, that is, towards the end of the rice-growing cycle. Male rats are capable of breeding at two months of age and female rats at 1½ months (Lam 1983). The mean life span of the rat is only about 4-7 months (Harrison 1956), but even by seven months a single female could have raised about 20 young.

The very high intrinsic rate of growth among rats is the major reason why attempts to exterminate them using traps or poisons are effective only locally or for short periods. If any rats are left, their litter size is likely to be larger than normal (up to 11) because competition for food is less and so the population will increase quickly. If, as rarely happens, all the rats in an area are killed, the unoccupied rice fields represent a wonderful opportunity for rats from other areas to colonise. If one thinks of an animal's niche as its 'profession', then when rats have been exterminated from a field, hundreds of 'jobs' become vacant. It should be remembered that rats do not produce large numbers of offspring for the good of the species or because that many are needed to fill all vacant 'jobs'. They are produced because rats are locked into their r-strategy and each female produces as many fast-growing, healthy young as she can; in this way her hereditary line is not smothered by others. Many of the rats die young (as is shown by the life-span figures above) but this is a consequence of the r-strategy. If one mother rat produced just a few offspring to whom she devoted her time in fetching food and defending them, the available 'jobs' would be filled by the larger number of other rats from other mothers.

Thus, control of rats in the long term has to be centred on land management. We can take an ecological lesson from the dipterocarps, the family of relatively common trees in lowland forest (p. 221). These trees avoid undue loss of their seeds by all fruiting together with long intervals between fruiting periods. This prevents any particular seed predators maintaining a population of size that would pose a serious problem to the dipterocarps. Rats can live for only a few days on a diet of just rice stems (COPR 1976) and this may be the only food at certain times, particularly if insect populations are also controlled in some way. Thus, if rice is planted and therefore harvested at the same time over large areas, and if scrub and other neighbouring habitats – where rats could find alternative foods when rice grain was not available – were removed or utilised, rat populations could control be kept within bounds. These are thus two of the main methods of rat control encouraged by the Department of Agriculture, the others being the digging out of rat holes and poisoning (Soekarna et al. 1980).

The niches of the different rodent species in rice fields clearly overlap – they share a major food item and many of them make their burrows in the banks between the fields. Each species must compete in some way with the others and this must be important in determining the relative abundances of the species and therefore the amount and type of rice damage. Very little is known about the actual ecology of these animals, and detailed studies of their movements (Harrison 1958; Taylor 1978), food (Lim 1974) and inter-specific relations could be of considerable economic benefit.

Plantations

Introduction

It is sometimes said, without a great deal of evidence, that plantations (monocultures of trees) mimic the functions of natural forest ecosystem. This analogy contains a certain amount of truth but should not be taken too literally. Mature plantations certainly protect the soil, water and a few of the indigenous biota more than, say, rice fields, but they cannot approach the efficiency of mature natural forest. There is tremendous soil loss, for example, when forest or old plantations are laid bare prior to planting. Unfortunately, the myth that plantations are as good as forest in many aspects is perpetuated and is sometimes used as an excuse for extensive forest clearance.

The two main plantation crops in Sumatra, rubber *Hevea brasiliensis* and oil palm *Elaeis guineensis,* have rather different ecological characteristics.

Rubber is intolerant of swampy conditions and, in the past, poorly drained areas within a plantation were left as forest or secondary growth from which various forms of wildlife could make sorties into the plantations. Conversely, under natural conditions, oil palm is a tree of riverine forest, but those now in cultivation are tolerant of drier conditions. Almost no residual areas of forest are left in an oil palm plantation because virtually all the land can be utilised.

The branches of rubber form a stable and long-lasting substrate on which nests of squirrels and birds can be built. Large birds-nest ferns

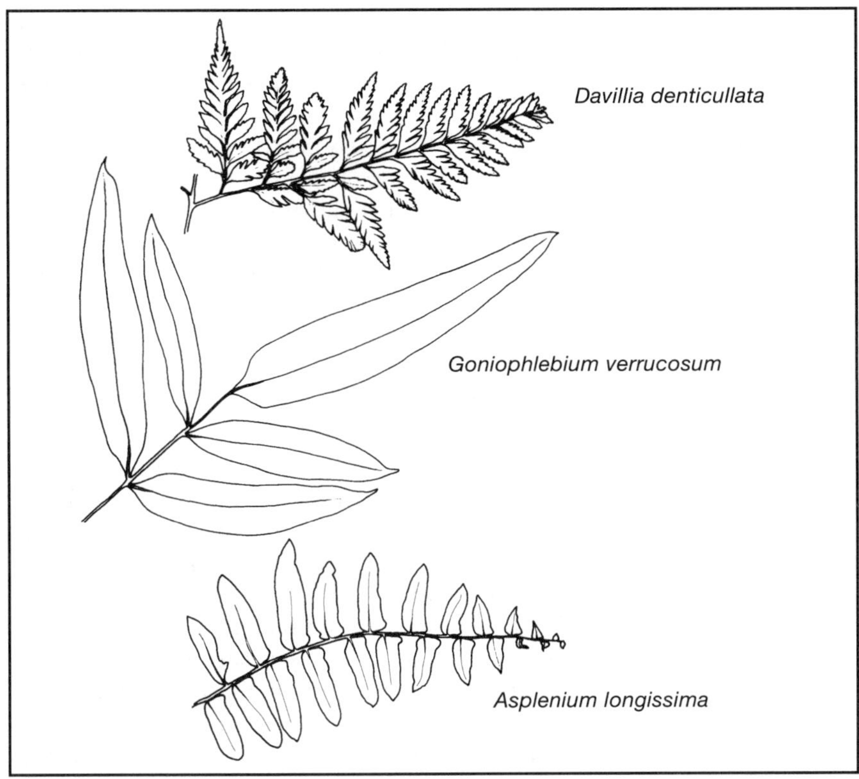

Figure 12.2. Three species of epiphytic fern commonly found in plantations.

Asplenium nidus and other epiphytic ferns are common on old trees. The leaves of oil palm, however, continually change their position – from the vertical young spear to the dangling dead leaf. Only a few animals, such as weaver birds, are able to make nests on these leaves. The dead leaf bases remaining on the tree are colonised by various epiphytic ferns which, because of the generally very moist air, are extremely abundant (fig. 12.2). A useful guide to epiphytic ferns has been written by Piggott (1979).

Harvesting the rubber latex usually disturbs nothing above head height, whereas the cutting of the infructescence away from the crown of the oil palm disturbs the only place where animals could set up a permanent base. The only animals that seem to be tolerant of this more or less weekly disturbance are squirrels (Duckett 1982) and rats, and the control of the latter is discussed below.

Rubber has one major fruiting season each year. The fruit are hard and have limited appeal to wildlife. Oil palm produces oil-rich, brightly coloured fruit which are obviously far more attractive as a food source (Duckett 1976).

The plant species covering the ground between the oil palm or rubber trees also influence the general abundance of animals (Ahmad 1980). For example, ferns and long grass between short oil palms provide an excellent environment for rats. Poorly kept smallholder estates where rubber trees are densely planted and many other plants are allowed to grow have a more diverse biota but do not make economic sense.

Large Pests

The major reason that plantations sustain a very low diversity of wildlife is the highly restricted floral diversity. A walk of 500 m through a well-maintained estate is unlikely to reveal more than 15 plant species (including ferns). This leads to an impoverished fauna of pollinators, leaf-eaters, parasites on other insects, etc. The impression gained in certain areas is that monkeys and other animals live in plantations. These have almost always come from an area of neighbouring forest or an isolated forest block (Bennett and Caldecott 1981), or are immature animals seeking unoccupied forest to avoid personal extinction. Visits by large animals to plantations, particularly young oil palm plantations, can cause severe economic loss. The main offenders are elephants and forest pigs. In these cases, which are reported almost every month and are a continual threat in certain places, it is pointless to blame the animals. The situation is almost always caused by careless land-use planning. If a favoured elephant food (such as sugar cane or young oil palm) is planted next to elephant habitat or across a traditional elephant pathway (Groeneveldt 1936, 1938; van Heurn 1937), of course the elephants will eat the food, and there are few effective ways of dissuading elephants from their avowed intention.

The answer to the elephant (bear/pig/tiger) problem lies in the formation of buffer zones around forested natural ecosystems. These zones aim to:

- contain plant species which are low-grade food items for potential agricultural pests so that the animals are not attracted to the area, and this in turn will make their predators search elsewhere for food; and
- provide local inhabitants with products they had traditionally taken from the forest (timber, rotan, medicine, firewood), or provide a cash income to enable the people to buy these products.

Buffer zones and the merits of different potential crops are discussed by MacKinnon (1981).

Rats and Their Predators

As described on page 368, one of the consequences of forested land being converted to other forms of land use is that some species of birds have increased in number or have entered and colonised new areas. One such is the barn owl *Tyto alba,* one of the world's most widely distributed birds, which has spread from Java into southern Sumatra (Holmes 1977) and has thence spread quickly into northern Sumatra.

In other parts of its range, the barn owl eats a range of small animals but in plantations it appears to feed almost entirely on rats, particularly the wood rat. The diet of owls is easy to determine because they regurgitate pellets containing the fur and bones of their prey about 8 to 12 hours after a meal. These pellets may be collected from beneath the roosts. Of 2,839 pellets examined in Peninsular Malaysia, 90% contained remains of rats and the rest were composed of insect, shrews and frogs (Lenton 1980). More than half (55%) of the rats were rice-field rats, a third (32%) were wood rats and the remainder (13%) were little rats. Whether this reflects the actual relative abundance of the rats is unknown. The rats have different activity patterns and some may successfully avoid owls in this way. For example, house rats and rice-field rats have finished most of their activity by midnight, but wood rats maintain moderate activity all night long. House rats start becoming active before dusk, rice-field rats at about dusk and wood rats after dusk (Ahmad 1980).

The owls reproduce rapidly; the clutch size averages 6.6 eggs per nest and two or three broods can be raised per year. At this level of production an adult pair accounts for about 1,300 rats per year in the feeding of themselves and their young (Lenton 1983).

Rats in oil palm plantations live at considerable densities – an estimate for the wood rat is 250/ha (Wood 1969) and so barn owls can be an extremely important control agent and some plantations are breeding owls in captivity and then releasing the young at a nest box site. Nest sites for barn owls in oil palm plantations are limited and so nest boxes (0.5 x 0.5 x 1 m) on the top of telegraph poles have been provided. There is a limit to how close adult pairs can live to each other and 20 ha has been suggested as the minimum area required. This area could contain 5,000 rats so that the owls could be responsible for the removal of about a quarter of the standing crop of rats per year. (It should be remembered that this is less than a quarter of all rats that live in an area during any period because rats are continually reproducing and dying.) In a study of the effects of owl predation on small mammal populations in England, it was found that 20%-30% of the standing crop of mice was taken every two months (Southern and Lowe 1982).

Snakes also appear to be significant predators on rats in oil palm plantations and for this reason plantation managers make little effort to control them. The stomach contents of five snake species from an oil palm plantation in Peninsular Malaysia were examined. Rats predominated, but other vertebrates included frogs, lizards and birds. The results are shown in table 12.1. Five live individuals of each of the five species were kept in cages and for one year five wood rats were placed in each cage at the start of each week. On the seventh day any remaining rats were removed prior to a new set being provided.

To calculate how many rats a snake of each species would devour in nature, where the whole range of prey would be available, the figures have to be adjusted. Thus the experimental feeding rate is multiplied by the percentage of dead snakes examined which had rats in their stomachs (table 12.2). So, if an average rat-eating snake eats 1.5 rats per week, it would consume 78 rats per year. The density at which these snakes live is not known but 2/ha would not be unreasonable. With a standing crop of 250 rats per ha, it can be seen that the snakes do take a considerable proportion (Lim 1976).

Pest Control by Predators

Caution is needed before one concludes that a predator, particularly a vertebrate predator, is actually controlling the number of its prey (Erlinge et al. 1984). Tigers and other large forest predators certainly do not control the numbers of deer, pigs, mouse deer or other common prey species. Detailed studies of large predators have dispelled the myths that these are master killers from which a prey animal, once detected, has no chance

Table 12.1. Results of stomach analyses of five snake species in a plantation.

	Total number examined	% containing only rat remains	% containing rat and other remains	% containing only other remains	% empty
Common cobra *Naja naja*	62	32	11	14	43
Indo-Chinese rat snake *Ptyas korros*	42	24	2	4	70
Brown snake *Xenalaphis hexagonotus*	35	40	3	3	54
White-bellied rat snake *Zaocys fuscus*	20	20	—	5	75
Keeled rat snake *Zaocys carinatus*	5	20	—	—	80

After Lim 1976

of escape. Instead, these large predators tend to take the young, the old and the sick. The remainder can successfully outrun, outwit or intimidate the predators such that a relatively small percentage of hunting attempts end with a kill.

Many insect predators, however, certainly do control the numbers of their prey species. This was clearly demonstrated in the 1960s and early 1970s when DDT and other long-lasting poisons were used to combat pests such as bagworms (a common pest on perennial crops) in oil palm plantations. In many cases the pests escaped the spraying because of behavioural characteristics. In the case of bagworms, the worst of the pests, for example, the larvae, pupae and adult females are protected by the bags they construct around themselves (fig. 12.3). In contrast, the 20 or so species of wasps that laid eggs in the young bagworms, and the larvae of which slowly ate the bagworm from the inside, had no such protection and were killed by the chemicals. The result was an increase in the bagworm population. A cessation of spraying resulted in an increase in predators and a decrease in bagworms (Conway 1972, 1982). Thus the predator wasps were truly controlling the bagworm population.

The major distinction between tigers and parasitic wasps in the way they predate is that only the wasp will certainly kill its prey every time it attacks. In addition, the tiger's prey can try to evade the tiger by running away but the wasp's prey cannot. Even with this efficiency, the wasps are highly unlikely to reduce bagworms to extinction because the more they predate upon the bagworms the harder it is for the wasps to find the remaining individuals or populations. If prey are hard to find, the population of wasps will decrease, which therefore reduces the pressure on the bagworms, thereby allowing them to increase, and so on. Thus, after the wasps have reduced the bagworm population to a certain level, it is the numbers

Table 12.2. Estimates of the number of rats eaten by snakes.

	Average number of rats eaten per snake per week while caged	Estimated number of rats eaten per snake per week if free
Common cobra	3.5	1.1
Indo-Chinese rat snake	5.1	1.2
Brown snake	4.6	1.9
White-bellied rat snake	5.6	1.1
Keeled rat snake	6.5	1.3

After Lim 1976

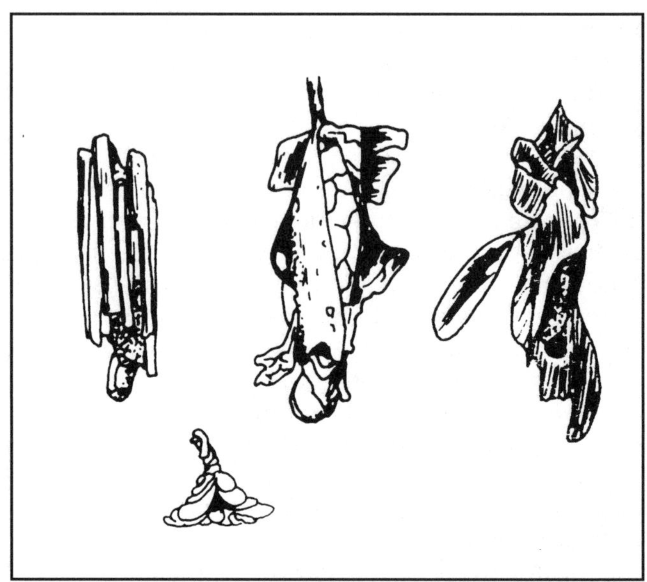

Figure 12.3. Bagworms (Psychidae) are serious potential pests on a range of perennial crops. The caterpillars make bags of silk which they cover with twigs, etc. (species differ in the material used). When moving, only the head, thorax and legs are exposed but if disturbed all these are pulled inside the bag which is then closed. The caterpillars pupate inside the bag and the male emerges as a typical grey/brown moth. The female emerges as a degenerate, wingless adult, capable of little more than building her own bag and producing eggs. The male copulates with her through the open end of her bag and the newly hatched caterpillars leave the parental bag and start making their own bags, often using material from the parental bag. The adult female dies inside her bag.

of prey that control the numbers of predators. The relationship between owls and rats, however, is similar to that between tigers and deer.

In the section on rats and their predators (p. 395) it was stated that pairs of barn owls are unlikely to live in areas of less than 20 ha, even if the prey density increases. So while territory size for animals is related to the abundance of available resources, there are strict upper and lower limits. Detailed studies of owls and their prey in temperate and arctic regions, where the size of rodent populations exhibit considerable natural fluctuations, have shown that the resident population of owls remains more or less the same, but the number of young birds raised varies considerably depending on the availability of prey (Pitelka et al. 1955; Sastrapradja 1979). This is a reflection of the striving to maximise reproductive success. Insufficient studies have been conducted in plantations and other relatively constant ecosystems in the Sunda Region to know whether similar natural fluctuations occur in their rodent populations (Flemming 1975). Fluctuations are created in plantations, however, by the rat poisoning programs which usually severely reduce the populations. The number of eggs laid by the owls, and the number of young raised, would be expected to reflect these population changes.

Owls are not going to defeat the rat problem but they may be seen as a means by which the period between rat poisoning programs can be lengthened because owls can slow the rats' population growth.

Chapter Thirteen

Urban Ecology

Introduction

Sumatran towns are not nearly as interesting ecologically as they could be because their vegetation is largely foreign and therefore supports few birds, squirrels and other animals. Many of the larger animals that are able to live in Sumatran towns are shot as soon as they are in the sights of an air rifle and inappropriate pesticide use probably also takes its toll. Towns in India are often home to large numbers of animals, including monkeys and squirrels, because the human inhabitants have a religious respect for these forms of life. India is a useful comparison, because the high human population density and the effects of development there cause more serious environmental problems than currently exist in Sumatra.

Urban ecology is an underexploited field of study. For example, some trees are better at exploiting certain urban conditions than others, some urban animals and plants can be used as indicators of pollution, some urban rats carry transmittable diseases, and some trees support a wide range of other harmless plants and animals, thereby making the environment more interesting. This type of knowledge should be sought by urban planners for they are charged with the responsibility of creating a healthy and fulfilling human environment.

A great deal of ecology can be taught in towns and cities. Studies of common toads or house geckos (pp. 405 and 411) are unlikely to change any national policy, but they are useful for learning about population structure and dynamics, and thus increasing the awareness of the way animal populations function. Urban lakes can be used for studying phytoplankton production. Bird watching is a recreation, but records kept over months or years can provide information on seasonality, new invaders, appropriateness of certain city trees, control of birds at airports and adaptations to a man-made environment. Some studies may be economically useful, such as the prevention of algae growing on white-washed and painted walls (p. 404).

As the proportion of Sumatra's population living in town increases, so urban ecologists will face great challenges to provide living space which is functional yet attractive, dynamic, and interesting.

VEGETATION

It may be felt that the vegetation of parts of Sumatran cities is really quite diverse. This is largely because we spend so much time in cities and thereby become familiar with the plants. If we spent an equivalent length of time in any of Sumatra's forested ecosystems it would soon become clear how poor city vegetation is in comparison.

How many tree species are there in Sumatran towns and where do they come from?[1] The ubiquitous acacia *Acacia auriculiformis* comes from Thursday Island in the Torres Strait. The tall mahogany *Swietenia macrophylla* comes from tropical America. The teak *Tectona grandis* comes from Burma and Thailand. The origins of the tamarind *Tamarindus indica* are uncertain but are possibly east Africa and western Asia. Of the common urban trees probably only the she-oak *Casuarina equisitifolia* is indigenous to Sumatra.

The two most common urban trees, acacia and she-oak, were chosen for good reasons. They have strong roots, grow quickly, give shade, require little attention, are evergreen and quite attractive, can grow in dry exposed habitats and both possess the capability of fixing atmospheric nitrogen in root nodules. This last property means that the trees can grow on young soils and help to improve the soil for other trees. But these advantages are not confined to the two species above. All the leguminous trees fix nitrogen and drought-resisting properties are only necessary in the most open places. If just some of the thousands of indigenous Sumatran trees, including those with fruit eaten by birds, were brought into towns, the fauna might be rather more interesting. Try identifying roadside and garden trees – the total will probably surprise you – to acquaint yourself with plant characters and the classifications of plants. Identifying trees in natural ecosystems will then be much easier. Most of the epiphytic ferns are also quite easily identifiable (Piggott 1979).

The number of species of insect herbivores associated with a particular plant species depends on various factors, such as the plant's geographical range, its taxonomic isolation, growth form, palatability, structural complexity, successional status and also on its relative abundance in the prehistoric past and the length of time it has been available for colonisation (Wratten et al. 1981). The acacia used so much in recent re-greening programs originated on a small Australian island and has only recently been introduced to Sumatra, and so it would not be expected to have many herbivores. This may be thought of as a point in its favour, but native trees survive with perhaps hundreds of associated species of insects. It should be remembered that a herbivore does not necessarily devastate its host. In fact, acacia leaves[2] in Medan are fed upon by larvae of bagworm moths (Psychidae) (p. 398), some of which build their bags from near-entire leaves and others (Eumeta) which build small conical bags from circles of leaf epidermis removed from the lower leaf surface.

Quite a wide range of urban and rural plants have some form of compound leaves which 'sleep' at night. That is, at night they hang down and/or fold up so that the leaf surfaces are held close or even touching. Common examples are the sensitive plant *Mimosa pudica* and young leaves of cassava *Manihot utillissima.* In 1880, Darwin suggested that this behaviour protected the plants from chilling, but this view did not gain wide acceptance. On page 290 it was stated that leaves with long edges relative to their length (such as compound leaves) would lose heat faster than simple leaves, and the chilling hypothesis has recently been shown to be valid. By experimentally preventing leaves from 'sleeping' it was found that the leaves became colder at night than those leaves which were allowed to sleep. The difference was only about 1°C but this could affect growth rates, and plants whose leaves sleep may have a competitive advantage over those plants with simple leaves (Enright 1982).

One of the most common orchids on big city trees is the pigeon orchid *Dendrobium crumenatum.* It is not frequently seen with flowers but if one plant is in flower, then all the plants in an area will be flowering. How is this coordination achieved? Similar gregarious flowering in bamboo was found to be genetic (p. 375) but the gregarious flowering of dipterocarp trees is probably a response to some environmental cue, possibly water stress (p. 221). The major trigger for pigeon orchids is a rapid fall in temperature, such as occurs during a rainstorm. Nine days after that temperature drop all the pigeon orchids will bear the white, sweet-smelling flowers. This was demonstrated in Medan during July 1983 and on the nights the stimulus occurred, the temperature fell below 20°C (fig. 13.1). Cold night temperatures alone (without a rapid drop) are not the primary stimulus, however, and analyses of meteorological data have been published (Burkill 1917; Coster 1926). The pigeon orchids presumably flower simultaneously to increase the likelihood of cross-pollination by the pollinating insects.

Plots of land awaiting development, roadsides, the gardens of unoccupied houses, and neglected corners of towns represent opportunities to study plant succession (Gilbert 1983). Which plants colonise open ground? How quickly does humus form? Assuming the first plants are grasses and herbs, how long until the first woody plant appears? Measure samples of the above-ground biomass at different times. Is the rate of increase constant? Does it show changes related to the seasons? What animals are associated with different stages of the succession?

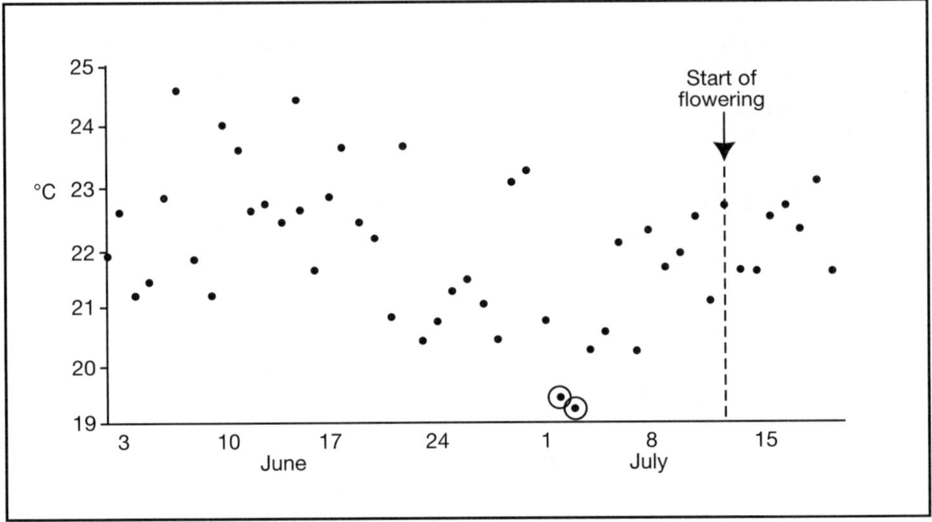

Figure 13.1. Minimum night temperatures in Medan during part of 1983 and the start of flowering in the pigeon orchid.

Life on Walls

Algae

White-washed or emulsion-painted walls quickly discolour. Areas of green, black and orange appear in patches or streaks, some in shade, some in the open. The 'stains' are caused by green and blue-green algae and by diatoms (p. 134). In Singapore the distribution of these microscopic plants on buildings has been investigated by Chua et al. (1972) and by Lee and Wee (1982), who also provide an identification key. The black stains are usually blue-green algae which have caught airborne dust particles in their mucilaginous sheaths. Some of the algae are able to fix atmospheric nitrogen and as more dust and soil particles become attached, so a favourable habitat is formed for higher plants to colonise. This obviously takes time, and regular maintenance will prevent succession from proceeding.

Try to establish the ecological requirements for different species of algae. Note the type and age of the substrate, the compass direction it faces, the times it is in the shade, etc. Clear small circles from the middle of a patch of algae and 'transplant' another species. Follow the success of the two species. Reverse the experiment by transplanting some of the first algae into a cleared patch of the second algae. Do certain paints or colours remain free of algae for longer than others? Paint manufacturers would be interested in the results.

Geckos

There are few houses or other buildings in towns without a resident population of geckos. Their ecology is not very well known and among the earliest and most complete appear to be those of Church (1962), Church and Lim (1961), and Chou (1975, 1978). Geckos are unusual amongst the animals which live alongside man in that they are not, for the most part, regarded as dangerous or undesirable.

It is a frequent matter of debate how geckos manage to climb on vertical surfaces or even walk upside down on a ceiling. Gecko feet do not have suckers but instead have small, overlapping flaps of skin (fig. 13.2). These flaps are covered with minute, closely set hairs which make contact with the slight irregularities of a surface and enable geckos to cling where other animals would fall.

Geckos are also known for their ability to shed their tails when caught – a response designed either to enable them to break free if caught by the tail, or to distract the 'predator' with a wriggling tail, or perhaps both. Tail shedding is also known in some snakes and lizards and involves muscular contractions which cause a fracture to occur across a vertebra, not between two vertebrae. The tails which regenerate are not usually as long or as symmetrical as the original; for a discussion of the adaptive strategies and energetic costs involved in tail loss, see Vitt et al. (1977).

There are three common species of gecko in houses over most of Sumatra – the common house gecko *Hemidactylus frenatus*, the flat-tailed gecko *H. platurus* and the four-clawed gecko *Gehyra mutilata* which has no claw on any of its inner digits. The much larger (35 cm) tokay *Gekko gecko*, although quite common in Java, does not seem to be found in the northern half of Sumatra but would be expected to occur in the south as an introduction from West Java. The spotted gecko *Gekko monarchus* is intermediate in size (about 20 cm) but is not commonly seen. The tokay has a proper voice produced in a larynx but the common gecko call is produced by tongue-clicking. The tongue-clicking serves to space males to prevent overcrowding but has little or no effect on females (Marcellini 1977).

Geckos generally feed on insects but the tokay feeds on smaller geckos and even mice and small birds. Geckos in Medan have also been observed

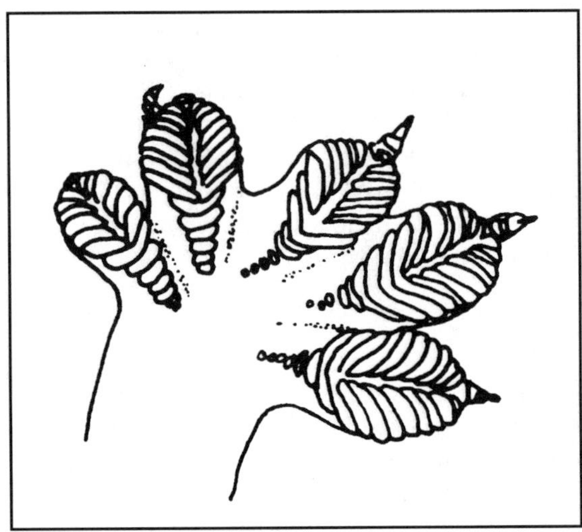

Figure 13.2. The foot of a gecko to show the hair-covered folds of skin.

licking the juice from mangoes which have been part-eaten by bats.

The three common house gecko species are more or less the same size ± 11 cm – what forms of competition and niche separation allow them to coexist? Kill several of each species at intervals of several weeks over a period and analyse the stomach contents. Without necessarily identifying the animal food remains to species or even family, it is possible to detect consistent differences in composition of size of prey between the gecko species. Watch the geckos – does one species sit and wait for prey and another species search? Does one species become inactive and another active in the middle of the night? How do species space themselves around light bulbs? Is aggression shown within or between species? Are resting places (in cracks, behind pictures and mirrors, under stones, in atap roofing, in corners of ceilings or behind cupboards) different between species? For clues and guides see Church and Lim (1961), Pianka and Huey (1978) and Pianka et al. (1979).

Lichens — Monitors of Pollution

Lichens are not strictly a group of plants but rather the results of a mutually beneficial relationship, or symbiosis, between a fungus and an alga. The dominant partner is the fungus and it derives its nutritional needs from the alga trapped within its strands. Despite their dual nature they behave like

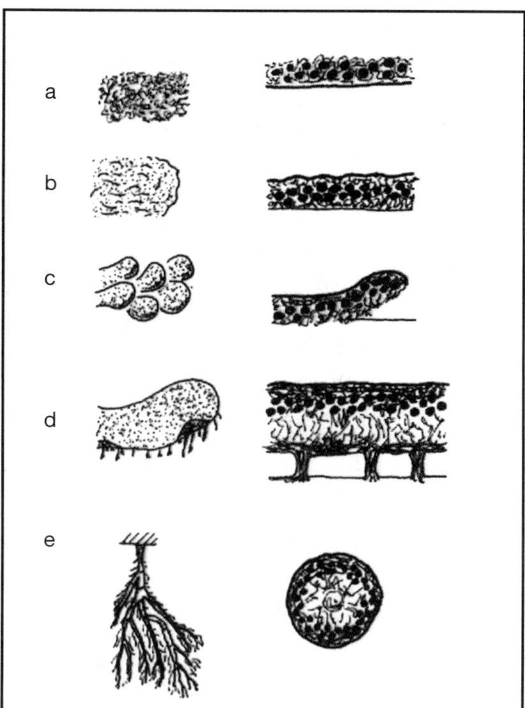

Figure 13.3. Five forms of lichen. a. leprose; b. crustose; c. squamulose; d. foliose; e. fruticose.
After Hawksworth and Rose 1976

a single organism and for convenience are treated as such (see description of coral polyps on p. 126).

Lichens of the Sunda Region are poorly known but without knowing their specific names they can be divided into various growth forms:

leprose - powdery collection of fungal hyphae (root strands) and algal cells with no organisation;

crustose - crust-like with algae situated below a distinct layer of fungal material;

squamulose - crustose but with the margins raised above the substrate surface;

foliose - like a leaf with distinct upper and lower layers, often attached by hair-like 'roots' but easily peeled from the substrate;

fruticose - erect or beard-like, attached to the substrate only at the base (fig. 13.3) (Hawksworth and Rose 1976).

Lichens are probably the longest-living organisms on earth. In the cold wastes of the Arctic, lichens have been found which are over 8,000 years old, and in a part of Antarctica where it is warm enough for growth for only

300 hours per year, a lichen of about 10,000 years old has been found. For comparison, the oldest known trees are not quite 5,000 years old.

Even in the tropics, lichens grow slowly – 5 mm of radial growth per year would be a normal rate, although some grow faster. Since many species occur on exposed surfaces they have to withstand considerable extremes in environmental conditions. When conditions are moist, photosynthesis and respiration can occur rapidly but when lichens are dried out in hot sun, such processes halt because, unlike leaves, lichens have no protective cuticle. Mineral nutrients are either received from the rain or from the substrate (Hawksworth and Rose 1976).

Lichens occur on tree bark, rock and other substrates in most natural ecosystems, and different species or groups of species are confined to particular types of substrate. Tree bark, for example, varies in its acidity and so lichens of different tolerance will be found on different trees. In natural and urban ecosystems lichens are often the only organisms occupying a particular substrate. In towns, for example, lichens are often the sole inhabitants of certain types of wall, exposed tree bark, gravestones, building timber, etc.

The environmental interest in lichens stems from their sensitivity to pollutants: no organisms are more sensitive to sulphur dioxide than lichens. They can thus be used as indicators in programs of environmental monitoring.

Sulphur dioxide is a by-product of combustion of coal and some types of oil, and in its various forms (solutions of sulphate, sulphite and bisulphite ions, sulphorous acid, gaseous sulphur dioxide and sulphur dioxide and sulphur trioxide) affects many plants but particularly lichens. The effect of sulphur dioxide is primarily through disrupting photosynthesis. This can cause a reduction in the reproduction and growth rates or in death. This can be recognised by chlorosis, a bleaching of the lichen, and a tendency for it to peel away from the substrate. The centres of lichens (the oldest parts) generally die first. Whitewash, cement containing lime and certain tree barks tend to neutralise the acidic effects on the lichens they bear, whereas substrates with acidic properties such as some other tree barks and sandstones, will quickly lose much of their lichen flora. Some lichens are more resistant than others to the damaging effects of sulphur dioxide, but the mechanism of this resistance is complex (Hawksworth and Rose 1976). Thus, if a survey transect is established outwards from a source of sulphur dioxide pollution, the following effects are likely to be found:

- the quantity of sulphur in the lichen tissue will decrease;
- the size of a given species of lichen will increase (fig. 13.4);
- the species assemblage found on a particular type of substrate will change and probably become less diverse.

Lichens show similar accumulation of, and sensitivity to, other pollutants such as fluorides, radio nuclides and heavy metals. Many species

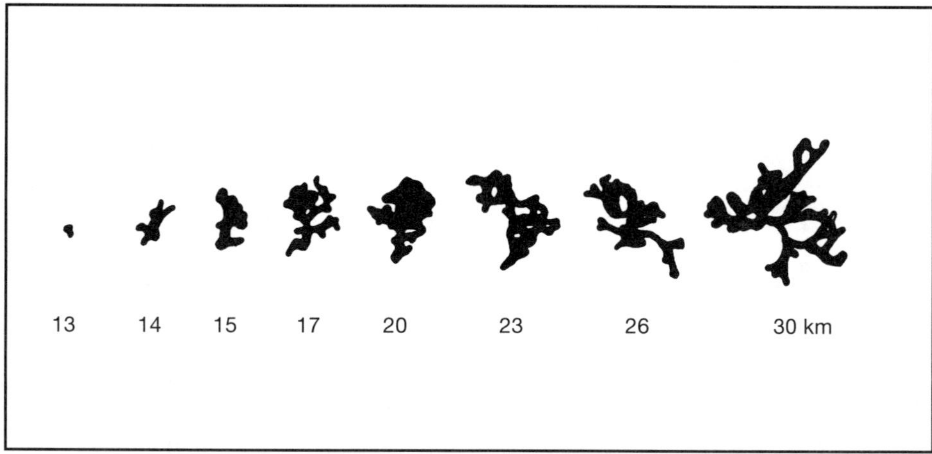

Figure 13.4. Changes in size for a species of lichen at different distances from a source of pollution.
After O.L Gilbert in Hawksworth and Rose 1976

accumulate heavy metals without overt effects and thus metal content in lichens will generally show a gradual reduction with increasing distance from, for example, a metal smelter.

The study of lichens, let alone their use as pollution monitors, has barely begun in Sumatra or even elsewhere in Southeast Asia. It is obviously preferable to know the identity of different lichen species before beginning environmentally oriented studies. Even if scientific names have not been determined, temporary names referring to colour, shape or location, or simple codes, can be given.

It is important that a reference collection be made so that, at least within one group of researchers, names can be standardised. Whole specimens, not fragments, should be collected, dried gently, and placed in individual air-tight containers containing silica gel or some other desiccant.[3] The containers and specimens should be handled carefully because lichens can be very brittle when dry.

The following preliminary studies are suggested:

1) Count the number of lichen species on a single type of substrate. Repeat in other areas or along a transect suspected of being a pollution gradient. Plot the number of species on a graph with distance from a possible pollution source along the horizontal axis. Suitable substrates for study include tombstones, trunks of a single tree species, walls of buildings, roof tiles, and milestones. Care

should be taken that the samples of substrates chosen are of more or less similar age and aspect to the sun. If it is possible to conduct a number of transects around a possible pollution source, a 'contour' map may be drawn of the number of lichen species.

2) Decide on one particular and relatively abundant species of lichen and measure the concentration of sulphur or metal in this species along a transect from a possible pollution source. Display results graphically as in 1).
3) Measure the size of one particular and relatively abundant species of lichen on one substrate of similar age along a transect as above (Hawksworth and Rose 1976).

Sulphur dioxide levels in Sumatran towns are not particularly high (± 1 ppm was measured in Medan) but lichens on trees are clearly killed by some agent near busy and confined roads. The death of lichens can have unexpected effects, as is shown by the following story of the peppered moth. Soot originating from uncontrolled gaseous factory effluent killed many lichens on tree trunks around Britain's industrial centres up to the middle of this century. The death of one particular light-coloured lichen species resulted in interesting changes in the peppered moth *Biston betularia.* This is generally a pale-coloured moth with flecks of black on its wings which afford good camouflage when it rests on certain lichens. Black forms of this species, however, were found to be more common than the normal pale form in the industrial areas where pollution had killed the lichens. Experiments by Kettlewell (1955, 1959) demonstrated that the variation in the abundance of the two colour forms was caused by natural selection, one of the main mechanisms by which evolution occurs. He placed equal numbers of light and dark forms onto tree trunks in two forests – one in a polluted area and the other in an unpolluted area. He observed the frequency with which predatory birds caught the moths and found that light forms in the polluted area and dark forms in the unpolluted area were more susceptible to being eaten (fig. 13.5). His results are shown in table 13.1.

Thus, over a long period, pollution caused changes in the genetic com-

Table 13.1. Number of light and dark forms of the peppered moth eaten by birds in unpolluted and polluted forests.

	Total of moths eaten by birds	
	Pale-coloured moths	**Dark-coloured moths**
Unpolluted forest	26	164
Polluted forest	43	15

From Kettlewell 1959

Figure 13.5. Both forms of peppered moth on unpolluted (left) and polluted (right) bark.

position of a moth population, and insectivorous birds acted as the agents for the change. Controls of factory emissions in Britain began to come into effect in the early 1950s and, as one might predict, the light form of the moth and the lichen are becoming common again (Cook et al. 1970).

If Sumatra's lichens have been barely studied, studies of their associated fauna have not even begun. The example above of the peppered moth comes from Europe but it serves to illustrate the perhaps unexpected dependence of one component of an ecosystem on another. Soot is not a serious pollutant in Sumatra but the death of lichens caused by other pollutants might well be expected to have more widespread effects than simply a change in the distribution of that lichen.

DITCHES

To an engineer, urban roadside drains are simply means of preventing floods and of removing household water to large water courses. To an ecologist, a ditch is a simple, small river, the life in which can give indications of water quality.

Perhaps the most obvious animal in many ditches is the guppy *Poecilia reticulata* (fig. 13.6). This small fish is a native of South America but was probably first introduced to Indonesia as an unwanted aquarium fish. It is mainly an algae eater but its growth rate is greater if the diet includes animal material such as insect larvae (Dussault and Krammer 1981). Guppies do not lay eggs like most fish but give birth to small fry.

Guppies are most obvious in urban ditches when they mouth at the surface, looking as though they are breathing air. Some fish can use atmospheric oxygen but the guppies, and many other fish which exhibit similar behaviour, are in fact taking in oxygen-rich water from the air-water interface (Kramer and Mehegan 1981). This aquatic surface respiration is generally only used where oxygen levels are low (such as slow-moving ditch water with a high organic content), and when oxygen levels are raised the fish respire normally. Aquatic surface respiration does not allow guppies to stay alive for long periods in highly deoxygenated water (Kramer and Mehegan 1981) but it confers an advantage such that guppies can survive where other fishes cannot. Thus guppies can be used as an indicator species.

Investigate different urban ditches, and try to catch one of each of the fish species present. It is more informative if the fishes caught can be given a scientific name (Kottelat et al. 1993), but simply the number of different species provides useful data. Plot a graph of the number of species against the biological oxygen demand or the oxygen concentration of the water. Repeat in a different section of town and in irrigation canals. You now have a means of determining approximate oxygen levels in the field using biological indicators instead of resorting to expensive laboratory tests. Johnson (1968) suggests some other organisms which could be used as pollution indicators.

Another ubiquitous inhabitant of ditches and their surroundings is the common toad *Bufo melanostictus*. An average-sized garden in Medan can harbour nearly 30 toads and these can be individually recognised by clipping toes and noting colour (variable in this species), weight or length. These data can be used to estimate the population size (Caughley 1978). Do the toads grow at a constant rate throughout the year? Is there a clear breeding season (as indicated by large numbers of small toads appearing at about the same time) or not? Kill a few and examine the stomach contents– does the diet change through the year? Take a stretch of road and remove toad corpses. Make daily counts of the toads freshly killed. What proportion of the toad population is killed by cars each year (Gittins 1983)? What other forms of mortality are important?

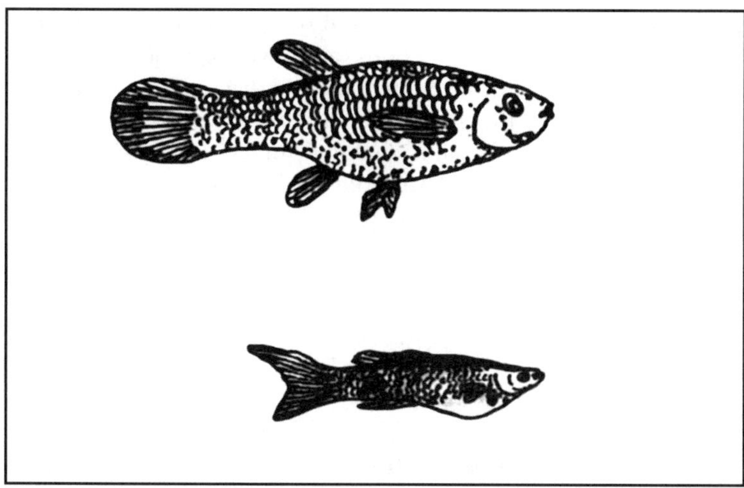

Figure 13.6. The guppy *Poecilia reticulata*; female above (about 4 cm), male below (about 2.5 cm). The males found in ditches are very variable but not as fancy as those bred for aquaria.

Bats

Some cities, such as Medan, have huge populations of resident bats, but others, such as Palembang, have rather fewer. Some bats, usually insectivorous ones, often roost in roofs of houses, churches and mosques. Others, both insectivorous and frugivorous species, live amongst palm fronds, particularly the hanging, dead leaves of oil palms, and in tree holes. A Sumatran city is home to probably 15-30 bat species.

At first sight, bats in flight all look the same but with a little patience different groups can be distinguished (Gould 1978b). The medium-sized, roof-dwelling, long-winged tomb bat *Taphozuos longimanus* is usually the first to start flying. Before it leaves its roost it becomes quite vocal and this is easily heard in the house below. It and other smaller insectivorous bats fly rather erratically as they swoop to catch insects. Frugivorous bats generally fly in a more direct manner.

Stand outside at dusk and watch for bats and note the time the different types appear. Where do they seem to come from? Finding roosts is sometimes easier at dawn when bats tend to fly round and round their roost.

Over a series of days try to pin down the actual roosts. Do the bats emerge at the same time each day? What effect does rain have? Are roosts available all over the city or are they concentrated in one area? How many bats emerge from each roost? Is it the same number day after day? Take a series of 1 ha plots in a town and collect data to estimate the number of resident bats. From data in Lekagul and McNeely (1977) and Medway (1969), estimate the biomass.

Birds

Introduction

In general the bird fauna of towns tends to have a lower species richness and diversity than nearby forests, but a higher biomass and density, and a very few dominant species. In addition, the major feeding niche shifts from bark- and canopy-insect eaters to ground-feeders (Ward 1968).

No comprehensive list of birds appears to have been compiled for Sumatran towns despite the ease of observation and data collecting, but lists for Kuala Lumpur and Singapore can be used as indications of what might eventually be recorded. A two-year study of birds in Kuala Lumpur revealed the presence of 24 species of common resident birds, nine species from rural areas which visited towns irregularly, one species which remained only between April and September, and eight species which were resident only between September and April. In addition to these, about 20 other species were seen only occasionally (McClure 1961). A similar number and composition of species was recorded for Singapore and it was noted that the total number of species was lower than for towns in other tropical areas, for instance west Africa (Ward 1968). An examination of the natural habitats of urban and suburban birds in Peninsular Malaysia and Singapore suggests that over half originate in coastal and riparian vegetation, only about 5% from lowland forest and a similar percentage are normally cliff or cave-mouth nesters. About 25% have been introduced or are recent immigrants. The similarities between cliffs and buildings are obvious and swifts have taken advantage of that. The similarity between coastal and riparian vegetation and towns is less clear, but a common factor is their simple plant communities with few species (p. 402) (Ward 1968). Because of this, generalised foragers are the major town invaders. The most common invader from the above habitat is the yellow-vented bulbul *Pycnonotus goiaveri*. Unlike Africa, the Sunda Region has no large areas of natural open country or savannah which might be expected to form a source of urban birds, and so the number of urban birds originating from indigenous natural habitats is limited.

For certain birds from the dryer land north of the Sunda Region, roads, railways, disturbed vegetation and towns present opportunities for colonising areas which had previously been closed to them because of the intervening large areas of species-rich forests (Ward 1968). This has yet to be documented for Sumatra but birds from the dryer, more open land in Java are starting to spread into Sumatran towns (Harver and Holmes 1976; Holmes 1977).

Sumatra and Peninsular Malaysia have witnessed a spectacular invasion by the tree sparrow *Passer montanus*. This bird is a native of the Palaearctic Realm (fig. 1.23, p. 41), that is Europe, Russia and China, and probably arrived at ports aboard ships in the sixteenth and seventeenth centuries. It has now spread to every urban area on Sumatra. A similar spread of a new species may occur if the introduced house crow *Corvus splendens* (not to be confused with the shy forest crow *Corvus enca*) or the common mynah *Acridotheres tristis* spreads from the island of Penang across the Malacca Straits to northern Sumatra (Charles 1978). The crow fills the scavenging niche which in most other tropical regions is also filled by various crows and birds of prey.

Those concerned with environmental affairs in Sumatra should keep an eye open for species new to a region. They are likely to occur first in the coastal urban areas. Even if they are unlikely to compete severely with indigenous birds, they may be considered undesirable for other reasons and the time to undertake a management program is at the beginning of an 'invasion', not when the 'invaders' have established a firm foothold. House crows are somewhat unpopular because of their raucous calls, gregarious nesting habits, fouling of public places, and stealing of food or young chicks (Charles 1978), but they do process urban waste, which might otherwise be utilised by rats which, unlike crows, harbour and transmit several important diseases (Hadi et al. 1976) (p. 390). Charles (1978) has concluded that one of the best ways to control these birds is by controlling the disposal of household waste, which would clearly have other benefits.

It is sometimes remarked that the total number of birds, not just the number of species, is low in Sumatran towns. One reason, of course, is the thoughtless shooting by the air-rifle toting, motor-bike riding, urban cowboys on Sunday afternoons. Additional ecological reasons are that few of the urban trees produce fruit suitable for birds and there are few insects able to utilise the 'foreign' trees and therefore there is less food for insectivorous or partially-insectivorous birds (see p. 223).

Birds represent excellent subjects for a study of urban ecology. Observation conditions are as near ideal as one could wish for and the number of food species and competing bird species are relatively few. The study of an urban bird community in Sumatra is unlikely to be of island-wide environmental significance, but it will furnish those involved in such a study with an awareness of ecological complexity and principles (see, for example, Ward and Poh 1968). These will be of immediate practical use when those involved turn to a more complex ecosystem, an environmental impact assessment or similar study.

Flowers that are pollinated by birds are generally recognisable by the following features:

1) open during daylight;
2) vivid colours, often scarlet or striking contrasting colours;
3) lip or margin absent or curved back;
4) hard flower wall, filaments stiff or united, nectar retained at rear of flower;
5) no odour;
6) abundant nectar;
7) no nectar guides (lines running along the petals indicating nectar source);
8) a relatively large distance between nectar source and sexual parts (Faegri and van der Pijl 1979);
9) relatively low sugar concentration (± 25%) in the nectar in comparison with nectar in flowers pollinated by bees (± 75%) (Baker 1975).

A common urban flower which fulfils all these criteria is the hibiscus *Hibiscus rosa-sinensis.* It is visited by sunbirds (Nectariniidae) but, because most hibiscus plants are sterile, the plants do not seed (van der Pijl 1937; Prendergast 1982). What other plants do sunbirds drink nectar from? Do all the plants conform to the characters listed above? Are sunbirds the only birds to visit the flowers? Design experiments to ensure that only sunbirds gain access to certain flowers. Repeat the experiment but exclude all possible pollinators. Are the birds actually pollinating the flowers? Where on the bird's body is pollen carried and what features of the flower design ensure that pollen is transferred?

Swifts

Swifts and swift-like birds belong to three families: the swifts and swiftlets (Apodidae), treeswifts (Hemiprocnidae), and swallows (Hirundinidae). Only the first and last of these are particularly common in towns and the two most common species can be distinguished from each other as shown in table 13.2 and figure 13.7. Other species occur in towns, however, and for proper identification reference should be made to King et al. (1975).

Swifts are masters of the air. A swift in Africa has been recorded as flying at 170 km/hr and is easily the fastest-moving vertebrate. In normal flight, however, swifts usually fly at about 50 km/hr. Swifts feed, drink, mate and even sleep while flying. As dusk approaches, non-breeding birds rise steadily in the sky and probably reach a few thousand metres above the

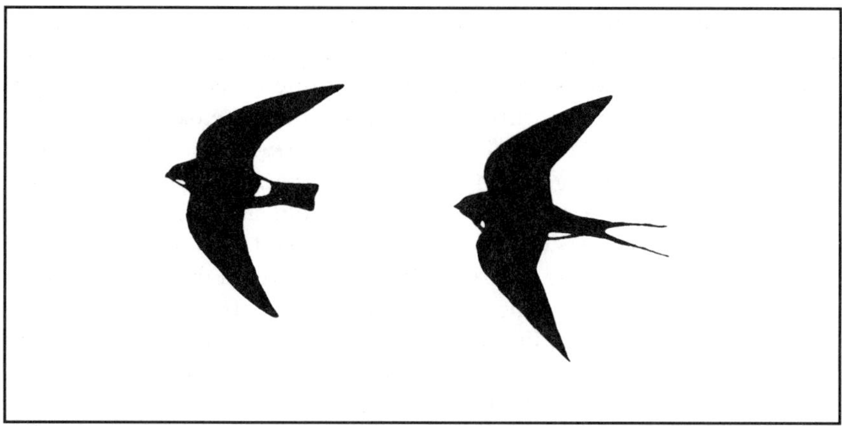

Figure 13.7. House swift (left) and barn swallow (right) in flight.
After King et al. 1975

ground where they sleep and beat their wings irregularly. When a young swift leaves its nest it may be at least two years before it lands again. Some swifts migrate to and from the cooler northern climates. The oldest known wild swift was ringed as an adult 15 years before it was found dying in England. It was calculated that it had flown over 7 million km during its life. Swifts normally live about six years (Bromhall 1980).

Swifts feed by chasing or filtering insects and airborne spiders from the air. They are, in effect, feeders on 'aerial plankton'. The food that adult European swifts give to their chicks has been examined and produced

Table 13.2. Characteristics of the house swift and barn swallow.

House swift	Barn swallow
Black except for white rump and pale throat	Black above with orange throat and white belly
Wings curved smoothly in flight	Wings somewhat bent in flight
Tail only slightly forked	Tail strongly forked
Nest round, and made of mud, built beneath overhanging roofs and frequently in groups	Nest like a cup, built in inconspicuous places; does not usually breed in Sumatra
Roosts on vertical walls or on nest, rarely if ever on telephone wires	Roosts, often in large groups, along telephone wires
Permanent resident	Migrant, appearing in about September and leaving in about March

some surprising figures (Bromhall 1980). A single mouthful collected over half an hour or so might contain over 500 animals of many species. During a fine day a pair of swifts with two or three young to feed may catch up to 20,000 insects and spiders (Bromhall 1980). Under the eaves of the grand Central Post Office in Medan, there are some 100 swift nests which are occupied during the breeding season (approximately April to June). Swift chicks generally leave their nest when about six weeks old and a simple calculation shows that the breeding swifts and their chicks account for about 40,000,000 insects and spiders during that six-week period. It is not difficult to estimate how many insects would be consumed in one year by this one colony of swifts alone and thus how important the birds are to insect control.[4]

If the nest sites chosen in Medan are any indication, it is clear that modern buildings are not favoured. Most of the nest sites were built at least 30 years ago and are used repeatedly. When the Medan Central Post Office was repaired in 1982, all the nests were scraped away from the eaves. It was not many months, however, before nests were rebuilt in the same places.

Examine the locations used by swifts for nesting. What features do they have in common? How many nests are there in each location? Is there any correlation between number of nests and age of buildings? Is there any connection between direction faced by the wall and the number of nests? Can you suggest ways in which modern buildings could be made more attractive to swifts? Do the numbers of nests occupied vary from year to year? How many nests are occupied in which months? How does the number of occupied nests relate to rainfall or other patterns? Do the patterns tie in with the cycles discussed on page 222?

It is hoped that these few examples will instill some enthusiasm in readers to venture into the urban environment with newly opened eyes, a questioning mind and a notebook. Sumatra desperately needs a wider and deeper data base of ecological information and students, school teachers, and university lecturers should take up the challenge now.

Notes

Chapter One: Background

1. (p. 3) The Sunda Region comprises the present-day land areas of Sumatra, Peninsular Malaysia, Borneo, Palawan, Java, Bali and their neighbouring islands.

2. (p. 18) A 'species' is a group of actually or potentially interbreeding populations that are reproductively isolated from all other types of organisms. Within a species, populations can sometimes be recognized as being distinct on the basis of colour, shape or behaviour; these are called 'subspecies'. Similiar species are grouped within a 'genus' and every know organism is named using its genus and species, and these are correctly written either in italics or underlined. The genus begins with a capital letter, the species (and subspecies if necessary) with a small letter. Thus most cats are in the genus *Felis,* which has various species such as the domestic cat *Felis domesticus,* the jungle cat *Felis bengalensis,* and the flatheaded cat *Felis planiceps.* The tiger *Panthera tigris* and the clouded leopard *Neofelis nebulosa* belong to different genera (plural of genus) but are in the same 'family', Felidae, as the *Felis* cats. Family names are written in normal script, starting with a capital, and usually ending in '-dae' for animals and '-aceae' for plants; Felidae is just one family in the 'order' Carnivora which includes dogs, civets and bears. Order names are written in normal script starting with a capital letter and generally end in '-a'. Thus, to summarize, the levels of classification used in this book are:

Order → Family → Genus → Species → Subspecies

The use of these names may seem complicated, but it is a very elegant system conforming to many rules and is recognized and used by scientists throughout the world.

3. (p. 39) Bearded pigs *Sus barbatus* are also know to migrate, probably in response to changes in food availability, but records of this seem to be confined to Borneo and Peninsular Malaysia.

Chapter Two: Mangrove Forests

1. (p. 69) In this chapter the term 'mangrove forests' includes not only the vegetated area but also the exposed land or silt area between it and the sea.

2. (p. 86) The measurement of dry weight biomass of a forest requires access to drying ovens, time and patience. If one or more of these is not available it is possible to use the allometric relationships determined by Ong et al. (1980). In these the above-ground dry weight (wa.g.)for *Bruguiera* trees is:

$$\text{Wa.g.} = 0.0033\pi \text{ (diameter at chest height)}^{2.167}$$

and for *Rhizophora* trees is:

$$\text{Wa.g.}=0.0277\pi \text{ (diameter at chest height)}^{2.167}$$

Relationships for other species can be determined from first principles using a number of samples.

3. (p. 110) The binding of tannins onto protein is what is felt when one bites through unripe banana or banana skin. Bitter tastes, such as coffee and papaya leaves, are caused by compounds called alkaloids.

Chapter Four: Rivers and Lakes

1. (p. 134) The entire *Wolffia* plant is less than 0.5 x 1 mm.

2. (p. 146) Saprophytes are organisms that derive their energy from dead organic matter and are the chief agents in the process of decay.

3. (p. 158) The goldfish *Carassius auratus* is the popular aquarium fish; the 'goldfish' of fishponds and lakes is in fact the golden carp *Cyprinus carpio flavipennis.*

Chapter Five: Peatswamp Forests

1. (p. 167) Soil which contains 35%-65% organic material is called 'muck.' (Anderson 1964).

2. (p. 171) Secondary compounds are compounds which do not play a role in metabolism.

3. (p. 173) Anderson (1976) used the term 'padang' for peatswamp forest which contained many low trees (about 800/ha). 'Padang vegetation' (p. 265) is poor in species, dwarfed and relatively open, so in this chapter Anderson's padang forest is referred to as 'pole forest'. Anderson (1959) had used this term earlier in descriptions of Sarawak and Brunei peatswamp forests.

4. (p. 175) 'Basal area' is a forestry term for the cross-sectional area of a trunk at breast height. The total basal area of trees in a plot is a rough measure of the amount of forest present.

5. (p. 176) The term 'predator' is used because the seed represents a new individual which can be killed outright. A herbivore grazing on leaves or eating fruit is clearly not a predator.

Chapter Seven: Lowland Forests

1. (p. 189) The Simpson Index of Diversity is given by: $D = 1 - \Sigma\ (p_i)^2$ where p_i is the proportion of species 'i' in a community. Zero is the lowest possible index (when only one species is present) and the highest index is

$$1 - \frac{1}{\text{number of species}}$$

2. (p. 234) Pheromones are volatile chemicals produced by animals. The ingestion or smell of a pheromone produced by one individual of a particular species can determine or influence the behaviour of another individual of the same species.

3. (p. 238) The remarkable nesting behaviour of hornbills has been described by various authors (Bartels 1956; Bartels and Bartels 1937; Johns 1982).

4. (p. 240) There are persistent reports of a ninth species, 'orang pendek' or 'short man', in the Kerinci area. Since 1995 small teams have been trying to collect solid support for its existence. A very impressive cast of a large foot print has been obtained which baffles mammal specialists, and the field workers, even those who started as sceptics, have reported positive sightings. Photo traps have been set in the forest, but despite a fascinating range of exceptional large animal photographs, the orang pendek has remained elusive.

5. (p. 240) The distribution of tarsiers and orangutan used to overlap in southern Sumatra before the orangutan became extinct there. It is doubtful whether the range of the eastern leaf monkey has ever overlapped with that of the orangutan.

Chapter Eight: Uncommon Lowland Forests

1. (p. 264) Maps (1:50,000) redrawn in 1945 from older Dutch maps were made available to the CRES team by the Head, Bureau of Ecology and Land Affairs, Bangka Tin Mining Unit.

2. (p. 266) 'Kulim' is the local name for *Scorodocarpus borneensis* (Afriastini 1982). This also grows naturally in species-poor stands but because of its value, there are only a few damaged skeletons of it left north of Muara Tembesi.

Chapter Eleven: Effects of Disturbance

1. (p. 361) On Siberut Island, the shifting agriculture traditionally practised does not include a burn; instead, fruit trees are planted amongst the fallen vegetation which remains to protect the soil until it rots down, slowly releasing nutrients to the soil. This 'wise' land practice is probably a result of the very high rainfall which makes a large burn difficult, and of the fact that the people do not traditionally grow hill rice (Anon. 1980a; Whitten 1982b).

2. (p. 374) A 1997 decree has stopped orangutan rehabilitation in Indonesia.

Chapter Thirteen: Urban Ecology

1. (p. 402) Corner (1952) and van Steenis (1981) can be used to identify virtually all the trees planted in towns and villages.

2. (p. 402) *Acacia* 'leaves' are in fact modified petioles or leaf stems and are correctly called phyllodes.

3. (p. 409) Tea dried gently in an oven is an effective and cheap desiccant.

4. (p. 418) A study of urban swiftlets has been published by Langham (1980).

Bibliography

(Note: numbers following references are the catalogue numbers given in the library of the Centre for Resource and Environmental Studies, University of North Sumatra, Medan)

Abdulhadi, R., Kartawinata, K. and Sukardjo, S. (1981). Effects of mechanised logging in the lowland dipterocarp forest at Lempake, East Kalimantan. *Malay. Forester* 44: 407-418.

Abe, T. (1978). The role of termites in the breakdown of dead wood in the floor of Pasoh study area. *Malay. Nat. J.* 30: 391-404. (12.4.31)

Abe, T. (1979). Food and feeding habits of termites in Pasoh Forest Reserve, *Jap. J. Ecol.* 29: 121-136.

Abe, T. and Matsumoto, T. (1978). Distribution of termites in Pasoh Forest Reserve. *Malay. Nat. J.* 30: 325-335. (12.4.32)

Abe, T. and Matsumoto, T. (1979). Studies on the distribution and ecological role of termites in a lowland rain forest of West Malaysia III. Distribution and abundance of termites at Pasoh Forest Reserve. *Jap. J. Ecol.* 29: 337-353.

Achmad, A.G. (1979). Menyadarkan masyarakat untuk keberhasilan dalam lingkungan hidup ditinjau dari sudut agama. In *Lokakarya Membudayakan Kesadaran Lingkungan hidup di Masyarakat,* pp. 168-178. Pusat Kajian Lingkungan Hidup, Universitas Sumatera Utara, Medan. (20.2.26)

Addicott, J.F. (1974). Predation and prey community structure: an experimental study of mosquito larvae on the protozoans of pitcher plants. *Ecology* 55: 475-492. (12.4.20)

Adisoemarto, S. (1980). *Binatang Hama.* Lembaga Biologi Nasional, Balai Pustaka. (11.1.4)

Adisoemarto, S. and Sastrapradja, S. (1977). *Sumber Protein Hewani.* Lembaga Biology National, Bogor. (9.2.2)

Afriastini, J. (1982). Kayu bawang. Suara *Alam* No. 15: 20-21. (12.1.95)

Ahmad, A. (1978). Klasifikasi rawa-rawa pesisir (coastal swamps) di Pasaman Barat. *Terubuk* 10: 30-36 (12.10.11)

Ahmad, I. (1980). *Tinjauan umum tentang beberapa aspek ekologi dan lakukan tikus liar dari ladang kelapa sawit.* Tesis Sarjana Muda, Universiti Kebangsaan Malaysia, Kuala Lumpur. (12.8.80)

Aksornkoae, S. (1981). Distribution, growth and survival of seedlings of mangrove forest in Thailand. *BIOTROP Spec. Publ.* 13: 23-27. (10.4.18)

Aldrich-Blake, F.P.G. (1980). Long-tailed macaques. In *Malayan Forest Primates: Ten Year's Study in Tropical Rain Forest* (ed. D.J. Chivers), pp. 147-165. Plenum, New York. (12.8.25)

Alexander, M. (1964). Biochemical ecology of soil micro-organisms. *Ann. Rev. Microbiol.* 18: 217-252. (12.4.59)

Alfred, E.R. (1961). Singapore fresh-water fishes. *Malay. Nat. J.* 15: 1-19. (12.5.20)

Alfred, E.R. (1966). The fresh-water fishes of Singapore. *Zool. Verh. Leiden* 78: 1-68. (12.5.28)

Allaby, M. (1983). The greening of the Cornish Alps. *New Scient.* 98: 138-141. (10.1.2)

Allen, B. (1954/5). Some common ferns of the open country. *Malay Nat. J.* 8: 95-110, 133-151. (12.1.90)

Alrasjid, H. and Effendi, R. (1979). *Pengaruh eksploitasi dengan traktor terhadap kerusakan tegakan sisa di kelompok hutan hujan tropis Pulau Pagai Selatan, Sumatera Barat.* Laporan No. 293, Lembaga Penelitian Hutan, Bogor. (9.1.13)

Altieri, M.A., Martin, P.B and Lewis, W.J. (1983). A quest for ecologically based pest management systems. *Env. Manag.* 7: 80).

Altieri, M.A. Martin, P.B. and Lewis, W.J. (1983) A quest for ecologically based pest management sustems. *Env. Manag.* 7: 91-100. (11.1.16)

Anderson, J.A.R. (1959). Observations on the ecology of the peatswamp forests of Sarawak and Brunei. In *Proc. Symp. Humid Tropics Vegetation, Tjiawi* 1958, pp. 141-148. UNESCO, Paris. (12.10.14)

Anderson, J.A.R. (1961). *The Ecology and Forest Types of the Peat Swamp Forest of Sarawak and Brunei in Relation to Their Silviculture.* Ph.D. thesis, Edinburgh University.

Anderson, J.A.R. (1963). The flora of the peat swamp forests of Sarawak and Brunei, including a catalogue of all recorded species of flowering plants, ferns, and fern allies. *Gdns' Bull. S'pore* 29: 131-228. (12.10.18)

Anderson, J.A.R. (1964). The structure and development of the peatswamps of Sarawak and Brunei. *J. Trop. Geong.* 18: 7-16. (12.10.12)

Anderson, J.A.R. (1965). Limestone habitat in Sarawak. In *Proc. Symp. Ecol. Res. Humid Trop. Veg.*, pp. 49-57. UNESCO, Paris. (12.10.69)

Anderson, J.A.R. (1976). Observations on the ecology of five peat swamps in Sumatra and Kalimantan. In *Proc. ATA 106 Midterm Seminar, Peat and Podzolic Soils and Their Potential for Agriculture in Indonesia.* pp. 45-55. Soil Research Institute, Bogor. (12.10.13)

Anderson, J.M. and Swift, M.J. (1983). Decomposition in tropical forests. In *Tropical Rain Forest: Ecology and Management* (ed. S.L. Sutton, T.C. Whitmore and A.C. Chadwick), pp. 287-309. Blackwell, Oxford.

Anderson, J.M., Proctor, J. and Vallack, H.W. (1983). Ecological studies in four contrasting lowland rain forests in Gunung Mulu National Park, Sarawak III. Decomposition processes and nutrient losses from leaf litter. *J. Ecol.* 71: 505-527.

Andriesse, J.P. (1974). *Tropical Lowland Peats in South-east Asia.* Koninklijk Instituut voor de Tropen, Amsterdam. (23.4.12)

Anon. (1979a). *Fisheries Statistics of Indonesia 1977.* Direktorat Jenderal Perikanan, Jakarta.

Anon. (1979b). Jenis penyu laut di Indonesia. *Suara Alam* 5: 34-36. (12.6.34)

Anon. (1980a). *Saving Siberut: A Conservation Masterplan.* World Wildlife Fund, Bogor. (10.2.1)

Anon. (1980b). World Conservation Strategy: Living Resource Conservation for Sustainable Development. International Union for the Conservation of Nature and Natural Resources, *Gland.* (20.3.11)

Anon. (1981). Conclusions and recommendations from the BIOTROP symposium on forest regeneration in Southeast Asia. *BIOTROP Spec. Publ.* 13: 224-226. (10.4.18)

Antonius, A. (1977). Coral mortality in reefs; a problem for science and management. Proc. Third Intnl. *Coral Reef Symp.* 2: 617-623. Referred to in Parrish (1980).

Anwar, S., Mukhtar, A., Dasni, S., and Hamdan, A (1980). Plankton air tawar Pekanbaru. *Terubuk* 6:2-17. (12.2.12)

Arumugan, P.T. and Furtado, J.I. (1981). Eutrophication of a Malaysian reservoir: effects of agroindustrial effluents. *Trop. Ecol.* 22: 271-275. (25.1.18)

Ashton, P.S. (1964). Ecological studies in the mixed dipterocarp forest of Brunei State. *Oxf. For. Mem.* No. 25. (Referred to in Whitmore, 1975).

Ashton, P.S. (1971). The plants and vegetation of Bako National Park. *Malay Nat. J.* 24: 151-162. (12.10.58)

Ashton, P.S. (1976). Mixed dipterocarp forest and its variation with habitat in Malayan lowlands: a reevaluation at Pasoh. *Malay. Forester* 39: 56-72. (12.10.86).

Ashton, P.S. (1980). The biological and ecological basis for the utilisation of dipterocarps. *BioIndonesia* 7: 43-53. (9.1.25)

Ashton, P.S. (1982). Dipterocarpaceae. *Flora Malesiana I* 9: 237-552.

Aveling, R.J. and Mitchell, A.. (1982). Is rehabilitating orang utans worthwhile? *Oryx* 16: 263-271. (12.8.65)

Ayal, Y. and Safriel, V.N. (1982). Species diversity of the coral reef - a note on the role of predation and of adjacent habitats. *Bull. Mar. Sci.* 32: 787-790. (12.3.21)

Baillie, I.C. and Ashton P.S. (1983). Some aspects of the nutrient cycle in mixed dipterocarp forests in Sarawak. In *Tropical Rain Forest: Ecology and Management* (ed. S.L. Sutton, T.C. Whitmore and A.C. Chadwick), pp. 347-356, Blackwell, Oxford

Baker, H.G. (1975). Sugar concentrations in nectars from hummingbird flowers. *Biotropica* 7: 37-41. (12.1.117)

Baker, J.M. (1982). Mangrove swamps and the oil industry. *Oil Petrochem. Poll.* 1: 5-22.

Baldwin, I.T. and Schultz, J.C. (1983). Rapid changes in tree leaf chemistry induced by damage; evidence for communication between plants. *Science* 221: 277-279.

Barlocher, F. and Kendrick, B. (1975). Leaf conditioning by microorganisms. *Oecologia* 20: 359-362. (3.2.22)

Barr, T.C. Jr. (1968). Cave ecology and the evolution of troglobites. In *Evolutionary Biology* (ed. T. Dobzhansky, M. Hecht and W Steere), vol. 2, pp. 35-102. Appleton-Century-Crofts, New York. (12.13.4)

Bartels, H. (1956). Waarnemingen bij een

Broedhol van der Jaarvogel *Aceros u. undulatus* (Shaw) op Sumatra. *Limosa* 39: 1-18.

Bartels, M. and Bartels, H. (1937). Uit het leven der neushoornvogels (I), (II), (III). *Trop. Natuur* 26: 117-127, 140-147. 166-172. (12.7.49)

Bauchop, T. (1978). Digestion of leaves in vertebrate arboreal folivores. In *The Ecology of Arboreal Folivores* (ed. G.G. Montgomery), pp. 193-204. Smithsonian Institution, Washington, D.C. (12.8.47)

Bauchop, T. and Marrucci, R.W. (1968). Ruminant digestion of the langur monkey. *Science* 161: 698-700. (12.8.8)

Bazzaz, F.A. and Pickett, S.T.A. (1980). Physiological ecology of tropical succession: a review. *Ann. Rev. Ecol. Syst.* 11: 287-310. (12.10.92)

Beadle, N.C.W. (1966). Soil phosphate and its role in molding segments of the Australian flora and vegetation with special reference to xeromorphy and sclerophylly. *Ecology* 47: 992-1007. (12.1.47)

Beaver, R.A. (1979a). Leafstalks as a habitat for bark beetles. (Col: Scolytidae). *Z. Entomol.* 88: 296-306. (12.4.40)

Beaver, R.A. (1979b). Non-equilibrium 'island' communities. A guild of tropical bark beetles. *J. Anim. Ecol.* 48: 987-1002. (12.4.39)

Beaver, R.A. (1979c). Biological studies of the fauna of pitcher plants (*Nepenthes*) in West Malaysia. *Annals soc. ent. Fr. (N.S.)* 15: 3-17. (12.4.18)

Beaver, R.A. (1979d). Fauna and foodwebs of pitcher plants in West Malaysia. *Malay. Nat. J.* 33: 1-10. (12.4.19)

Beekman, H.A.J.M. (1949). *Houttelt in Indonesia*. H. Veenan and Zonnen, Wageningen. Referred to in Suselo (1981).

Bennett, E.L. and Caldecott, J.O. (1981). Unexpected abudance: the trees and wildlife of the Lima Belas Estate Forest Reserve, near Slim River, Perak. *The Planter* 57: 516-519. (10.2.14)

Bennett, E.W. (1984). *The Banded Langur : Ecology of a Colobine in West Malaysian rainforest.* PhD thesis, University of Cambridge, Cambridge.

Benzing, D.H. (1981). Mineral nutrition of epiphytes: an appraisal of adaptive features. *Selbyana* 5: 219-223.(12.1.85)

Benzing, D.H. (1983). Vascular epiphytes: a survey with special reference to their interactions with other organisms. In *Tropical Rain Forest: Ecology and Management* (ed. S.L. Sutton, T.C. Whitmore and A.C. Chadwick), pp. 11-24. Blackwell, Oxford.

Benzing, D.H. and Ott, D.W. (1981). Vegetative reduction in epiphytic Bromeliaceae and Orchidaceae: Its origin and significance. *Biotropica* 13: 131-140. (12.1.65)

Bergmans, W. and Hill, J.E. (1980). On a new species of *Rousettus*, Gray, 1921, from Sumatra and Borneo (Mammalia: Megachiroptera). *Bull. Br. Mus. nat. Hist.* 38: 95-104. (12.8.66)

Berkhout, A.H. (1895). Boschbouwkundige beschrijring v.h. eiland Banka. *Tijd. v. Nijverh. en Landb. in Ned. Indie.* 50: 11-66. (12.10.23)

Bernays, E.A. (1978). Tannins: an alternative viewpoint. *Ent. exp. appl.* 24: 44-53. (12.13.31)

Bernays, E.A. and Chamberlain, D.J. (1980). A study of tolerance of ingested tannin in Schistocerca gregaria. *J. Insect Physiol.* 26: 415-420. (12.4.55)

Berry, A.J. (1963a). An introduction to the non-marine molluscs of Malaya. *Malay. Nat. J.* 17:1-17. (12.3.27)

Berry, A.J. (1963b). Faunal zonation in mangrove swamps. *Bull. natl. Mus. Singapore* 32: 90-98. (12.9.22)

Berry, A.J. (1972). The natural history of West Malaysian mangrove faunas. *Malay. Nat. J.* 25: 135-162. (12.9.11)

Berry, A.J. (1974). Freshwater bivalves of Peninsular Malaysia with special reference to sex and breeding. *Malay. Nat. J.* 27: 99-110. (12.3.26)

Berry, P.Y. (1965). The diet of some Singapore Anura (Amphibia). *Proc. Zool. Soc. Lond.* 144: 163-174.

Berry, P.Y. and Bullock, J.A. (1962). The food of the common Malayan toad *Bufo melanostictus* Schneider. *Copeia* 1962: 736-741.

Bertness, M.D. (1981a). Competitive dynamics of a tropical hermit crab assemblage. *Ecology* 62: 751-761. (12.3.37)

Bertness, M.D. (1981b). Pattern and plasticity in tropical hermit crab growth and reproduction. *Am. Nat.* 117: 754-773. (12.3.36)

Biere, J.M. and Uetz, G.W. (1981). Web orientation in the spider *Micrathena gracilis* (Araneae: Araneidae). *Ecology* 62: 336-344. (12.4.38)

BIOTROP (1976). *Proceedings of BIOTROP Workshop on Alang-alang, Bogor, 27-29 July 1976.* BIOTROP, Bogor.

BIOTROP (1980). *Symposium on Small Mammals: Problems and Control.* BIOTROP Special Publication No. 12, BIOTROP, Bogor. (11.1.6)

Bird, E.C.F. (1982). Coastal landforms of the Asian humid tropics. In *Man, Land and Sea* (ed. C. Soysa, L.S. China, W.L. Collier), pp. 3-14. Agricultural Development Council, Bangkok. (2.4.25)

Bird, E.C.F. and Barson, M.M. (1977). Measurement of physiographic changes on mangrove-fringed estuaries and coastlines. *Mar. Res. Indonesia* 18: 73-80. (2.4.7)

Birkeland, C. and Neudecker, S. (1981). Foraging behaviour of two Caribbean Chaetodontids. *Copeia* 1981: 169-1787. (12.5.17)

Biro Pusat Statistik (1980). *Statistik Indonesia.* Biro Pusat Statistik, Jakarta. (19.6.1)

Bishop, J.E. (1973). *Limnology of a Small a Malayan River, Sungai Gombak. Junk,* The Hague. (3.2.15)

Bishop, J.E. and Hynes, H.B.N. (1969). Downstream drift of the invertebrate fauna in a stream ecosystem. *Arch. Hydrobiol.* 66: 59-90. (3.2.25)

Black, H.C. and Harper, K.T. (1979). The adaptive value of buttresses to tropical trees: additional hypotheses. *Biotropica* 11: 240. (12.1.64)

Boon, D.A.(1936). De inrichting van de voor exploitatie in aanmerking komende bosschen in de afdeeling Bengkalis, benevens eenige ipmerkingen omtrent de samenstelling der terplaatse voorkomenden moerabosschen. *Tectona* 29: 344-373. (12.10.113)

Bormann, F.H. and Likens, G.E. (1981). *Pattern and Process in a Forested Ecosystem.* Springer-Verlag, New York. (12.10.94)

Borner, M. (1979). *A Field Study of the Sumatran Rhinoceros* Dicerorhinus sumatrensis *Fischer 1914: Ecology, Behaviour and Conservation Situation in Sumatra.* Juris Druck, Zurich. (12.8.48)

Borowitzka, M.A. (1981). Algae and grazing in coral reef ecosystems. *Endeavour* 5: 99-106. (12.3.22)

Bovbjerg, R.V. (1970). Ecological isolaion and competitive exclusion in two crayfish (*Orconectes virilis* and *Orconectes immunis*). *Ecology* 51: 225-236

BPPP/Soepraptohardjo, M., Soekardi, M., Kurnia, H.D. and Suharjo, S. (1979). *Sumber daya lahan/tanah dan potensinya di Pulau Sumatera.* Konsultasi Penelitian Pertanian Menunjang Pembangunan Sumatera 28 Nopember 1979. Lembaga Penelitian Tanah, Bogor. (23.4.14)

Brafield, A.E. (1978). *Life in Sandy Shores.* Studies in Biology No. 89. Edward Arnold, London.

Brehaut, R.N. (1982). *Ecology of Rocky Shores.* Studies in Biology No. 139. Edward Arnold, London.

Bridges, E.M. (1970) *World Soils.* Cambridge University Press, Cambridge. (23.4.1)

Bristowe, W.S. (1976a). A contribution to the knowledge of liphistiid spiders. *J. Zool., Lond.* 178: 1-6. (12.4.11)

Bristowe, W.S. (1976b). Rare arachnids from Malaysia and Sumatra. *J. Zool.* 178: 7-14. (12.4.37)

Brock, T.D. (1981). Calculating solar radiation for ecological studies. *Ecol. Modelling* 14: 1-19. (19.5.18)

Bromhall, D. (1980). *Devil Birds: The Life of the Swift.* Hutchinson, London. (12.7.33)

Bronson, B. and Asmar, T. (1976). Prehistoric Investigation at Tiangko Panjang cave, Sumatra: An Interim Report. *Asian Perpectives* 18: 128-145. (23.7.14)

Bronson, B. and Wisseman, J. (1978). Palembang as Srivijaya. *Asian Perpectives* 19: 220-239. (23.7.6)

Brotonegoro, S. and Abdulkadir, S. (1979). Penelitian pendahuluan tentang kecepatan gugur daun dan penguraiannya dalam hutan bakau Pulau Rambut. In *Prosiding Seminar 'Ekosistem Hutan Mangrove'* (ed. S. Soemodihardjo, A. Nontji, A. Djamali), pp. 81-85. Lembaga Oseanologi National, Jakarta. (12.9.2)

Brown, R. (1822). An account of a new genus of plants named Rafflesia. *Trans. Linn. Soc. Lond.* 13: 201-234. (12.1.80)

Brown, R. (1845). Description of the female flower and fruit of Rafflesia arnoldi with remarks on its affinities. *Trans. Linn. Soc. Lond.* 19: 221-247. (12.1.81)

Brown, S. and Lugo, A.E. (1982). The storage and production of organic matter in tropical forests and their role in the global carbon cycle. *Biotropica* 14: 161-187. (12.10.19)

Brunig, E.F. (1973). Species richness and stand diversity in relation to site and succession of forests in Sarawak and Brunei (Borneo). *Amazoniana* 4: 293-320. Referred to in Whitmore (1975).

Bruning, E.F. (1965). A guide and introduction to the vegetation of the kerangas forests and the padangs of the Bako National Park. In. *Proc. Symp. Ecol. Res. Humid. Tropics Veg.,* Kuching, UNESCO, Kuching. (12.10.60)

Bruning, E.F. (1969). Forestry on tropical podzols and related soils. *Trop. Ecol.* 10: 45-58. (9.1.20)

Bruning, E.F. (1970). Stand structure, physiognomy and environmental factors in some lowland forests in Sarawak. *Trop. Ecol.* 11: 26-43. (12.10.52)

Bruning, E.F. (1973). Species richness and

stand diversity in relation to site and succession of forests in Sarawak and Brunei (Borneo). *Amazoniana* 4: 293-320.

Buckley, R.C., Corlett, R.T. and Grubb, P.J. (1980). Are the xeromorphic trees of upper montane rain forests drought-resistant? *Biotropica* 12: 124-136. (12.1.39)

Budowski, G. (1963). Forest succession in tropical lowlands. *Turrialba* 13: 42-44. (12.10.109)

Bull, G.D. (1982). Scleractinian coral communities of two inshore high island fringing reefs at Magnetic Island, N. Queensland. *Mar. Ecol. Prog. Ser.* 7: 267-277. (12.3.18)

Bullock, J.A. (1966). The ecology of Malaysian caves. *Malay. Nat. J.* 19: 57- 63.

Bunning, E. (1944). Botanische Beobachtungen in Sumatra. *Flora N.S* 37: 334-344. (12.1.12)

Bunt, J.S. (1980). Degradation of mangroves. In *Marine and Coastal Processes in the Pacific: Ecological Aspects of Coastal Zone Management.* UNESCO, Jakarta. (2.4.33)

Bunt, J.S. and Williams, W.T. (1981). Vegetational relationships in the mangroves of tropical Australia. *Mar. Ecol. Prog. Ser.* 4: 349-359. (12.9.48).

Burbridge, P.R. (1982a). Problems and issues of coastal zone management. In *Man, Land and Sea* (ed. C. Soysa, L.S. Chia and W.L. Collier) pp. 309-320. Agricultural Development Council, Bangkok. (2.4.26)

Burbridge, P.R. (1982b). Valuation of tidal wetlands. In *Man, Land and Sea* (ed. C. Soysa, L.S. Chia and W.L. Collier) pp. 43-64. Agricultural Development Council, Bangkok. (2.4.29).

Burges, N.A. (1965). Biological processes in the decomposition of organic matter. In *Experimental Pedology.* (ed. E.G. Hallsworth and D.V. Crawford), pp. 189-199. Butterworth, London.

Burgess, P.F. (1971). The effect of logging on hill dipterocarp forests. *Malay. Nat. J.* 24: 231-237. (9.1.29)

Burham, C.P. (1975). Soils. In *Tropical Rain Forests of the Far East* (by T.C. Whitmore), pp. 103-120. Clarendon, Oxford. (12.10.8)

Burhanuddin and Martosewojo, S. (1979). Pengamatan terhadap ikan gelodok, Periophthalmus koelreuteri (Pallas) di Pulau Pari. In *Prosiding Seminar 'Ekosistem Hutan Mangrove'* (ed. S. Soemodihardjo, A. Nontji, A. Djamali), pp. 86-92. Lembaga Oseanologi Nasional, Jakarta. (12.9.2).

Burkill, I.H. (1966). *Dictionary of the Economic Products of the Malay Peninsula.* Department of Agriculture, Kuala Lumpur.

Burkill, I.N. (1917). The flowering of the pigeon orchid, *Dendrobium crumenatum* Lindl. Grdns'. *Bull. Straits Settlement* 1:400-105.

Burnham, C.P. (1974). The role of the soil forming factors in controlling altitudinal zonation on granite in Malaysia. In *Altitudinal Zonation in Malesia* (ed.J.R. Flenley), pp. 59-74. University of Hull, Hull. (12.10.44)

Burnham, C.P. (1975). Soils. In *Tropical Rain Forests of the East* (by T.C. Whitmore), pp. 103-120. Clarendon, Oxford. (12.10.8)

Burton, R. and Ward, N. (1827). Report on a journey into Batak country in the interior of Sumatra in the year 1824. *Trans. R. As Soc.* 1: 485-513. (23.7.3)

Caldecott, J.O. (1980). Habitat quality and populations of two sympatric gibbons (Hylobatidae) on a mountain in Malaya. *Folia primatol.* 33: 291-309. (12.8.12)

Caughley, G. (1978). *Analysis of Vertebrate Populations.* Wiley, Chicester. (19.4.3)

Chaiba, S. (1978). Numbers, biomass and metabolism of soil animals in Pasoh Forest Reserve. *Malay. Nat. J.* 30: 313-324. (12.10.87)

Chambers, M.J (1980). The enviroment and geomorphology of deltanic sedimentation. (Some examples from Indonesia). In *Tropical Ecology and Development* (ed. J. Furtado), pp. 1091-1095. University of Malaya, Kuala Lumpur. (2.4.27).

Chambers, M.J.G. and Sobur, A. (1977). *Problems of Assessing the Rates and Processes of Coastal Changes in the Province of South Sumatra.* Research Report 3, Centre for Natural Resource, Management and Environmental Studies, IPB, Bogor. (2.4.28).

Chambers, R.E. and Manan, M.E. (1978). *Agroclimatology of the Musi-Banyuasin Coastal Zone, South Sumatra.* Research Report 005, PSPSL, Bogor. (1.1.3)

Chaphekar, S.B. (1978). Urban ecosystems- a challenge for plant ecologists. *Int. J. Ecol. Environ. Sci.* 4: 19-31. (12.1.79)

Chapman, E.C. (1975). Shifting agriculture in tropical forest areas of south-east Asia. In *The Use of Ecological Guidelines for Development in Tropical Forest Areas of South-east Asia,* pp. 120-135. IUCN, Gland. (20.1.14)

Chapman, V.J. (1976). *Coastal Vegetation,* 2nd Edition. Pergamon, Oxford. (2.4.32).

Chapman, V.J. (1977a). Introduction. In *Wet Coastal Ecosystems* (ed. V.J.Chapman), pp. 1-29. Elsevier, Amsterdam. (2.4.28)

Chapman, V.J. (1977b). Wet coastal formations of Indo-Malesia and Papua New Guinea. In *Wet Coastal Ecosystems* (ed. V.J. Chapman), pp.

261-270. Elsevier, Amsterdam.

Charles, J.K. (1978). Management of the house crow *Corvus splendens*. *BIOTROP Spec. Publ.* 8: 191-197. (12.11.2)

Chasen, F.N. (1937). The birds of Billiton Island. *Treubia* 16: 205-238. (12.).11)

Chasen, F.N. (1940). The mammals of the Netherlands Indies Mt Leuser Expedition 1937 to North Sumatra. *Treubia* 17: 479-502. (12.8.13)

Chasen, F.N. (1940). A handlist of Malayan mammals. *Bull. Raff. Mus. S'pore* 15. (12.8.32)

Chasen, F.N. and Boden Kloss, C. (1927). Spolia Mentawiensia - Mammals. *Proc. zool. Soc. Lond.* 1927: 797-840. (12.8.63)

Chasen, F.N. and Hoogerwerf, A. (1941). The birds of the Netherlands Indies Mt Leuser Expedition 1937 to North Sumatra. *Treubia* 18 (suppl.): 1-125. (12.7.12)

Chaston, I. (1968). Endogenous activity as a factor in invertebrate drift. *Arch. Hydrobiol.* 64: 324-334. (3.2.37)

Chaston, I. (1969). The light threshold controlling the periodicity of invertebrate drift. *J. Anim. Ecol.* 38: 171-180. (3.2.36)

Cheke, A.S., Nanakorn, W. and Yankoses, C. (1979). Dormancy and dispersal of seeds of secondary forest species under the canopy of a primary tropical rain forest in northern Thailand. *Biotropica* 11: 88-95. (12.1.66)

Cheng, L. (1965). The Malayan pond skaters. *Malay. Nat. J.* 19: 115-123. (12.4.4)

Chin, S.C. (1977). The limestone hill flora of Malya. Part I. *Gdns' Bull. S'pore* 30: 165-220. (12.1.115)

Chiu, S.C. (1979). Biological control of the brown planthopper. In *Brown Plant Hopper: Threat to rice production in Asia*, pp. 335-355. International Rice Research Institute, Los Banos.

Chivers, D.J. and Davies, A.G. (1978). Abundance of primates in the Krau game Reserve, Peninsular Malaysia. In *The Abundance of Animals in Malesian Rain Forests* (ed. A.G. Marshall), pp. 9-36. University of Hull, Hull. (12.8.22)

Chopard, L. (1929). Fauna of Batu Caves, Selangor. xii Orthoptera and Dermaptera. *J. Fed. Malay. States Mus.* 14: 366-371.

Chou, L.M. (1975). Systematic account of the Singapore house geckos. *J. Natl. Acad. Sci. Singapore* 4: 130-138.

Chou, L.M. (1978). Soma bionomic data on the house geckos of Singapore. *Malay. Nat. J.* 31: 231-235. (12.6.20)

Chua, N.H., Kwok, S.W., Tan, K.K., Teo, S.P. and Wong, H.A. (1972). Growths on concrete and other similar surfaces in Singapore. *J. Singapore Inst. Arch.* 51: 13-15.

Church, G. (1962). The reproductive cycles of the Javanese house geckos, *Cosymbotus platyurus, Hemidactylus frenatus* and *Peropus mutilatus*. *Copeia* 2: 262-269. (12.6.30)

Church, G. and Lim, C.S. (1961). The distribution of three species of house geckos in Bandung (Java). *Herpetologica* 17: 119-201.

Climap (1976). The surface of the ice-age earth. *Science* 191: 1131-1137.

Clutton-Brock, J. (1959). Niah's Neolithic dog. *Sarawak Mus. J.* 9: 142-145. (12.13.36)

Clymo, R.S. and Hayward, P.M. (1982). The ecology of *Sphagnum*. In *Bryophyte Ecology* (ed. A.J.E. Smith), pp. 229-289. Chapman and Hall, London. (12.1.19)

Cockcroft, V.G. and Forbes, A.T. (1981). Tidal activity rhythms in the mangrove snail Cerithidea decollata (Linn.) *S. Afr. J. Zool.* 16: 5-9. (12.3.16).

Cole, B. (1981). Colonising abilities, island size and the number of species on archipelagos. *Amer. Nat.* 117: 629-638. (12.11.4)

Colinvaux, P.A. (1973). *Introduction to Ecology.* Wiley, New York. (23.2.3)

Collins, N.M. (1979). Soil invertebrates from Mulu's forests. Paper presented at Gunung Mulu Symposium, 1979, Royal Geographical Society, London. (12.4.17)

Collins, N.M. (1980). The distribution of soil macrofauna on the west ridge of Gunung (Mount) Mulu, Sarawak. *Oecologia* 44: 263-275.

Collins, N.M. (1982). The importance of being a bugga-bug. *New Scient.* 94: 834-837. (12.4.41)

Collins, N.M. (1983). Termite populations and their role in litter removal in Malaysian rain forests. In *Tropical Rain Forest: Ecology and Management* (ed. S.L. Sutton, T.C. Whitmore, A.C. Chadwick), pp. 311-325. Blackwell, Oxford.

Conway, G.R. (1972). Ecological aspects of pest control in Malaysia. In *The Careless Technology* (ed. M.T. Farvar and J.P. Milton), pp. 467-488, Natural History Press, New York.

Conway, G.R. (1982). (ed.) *Pesticide Resistance and World Food Production.* Imperial College, University of London, London. (11.4.27)

Cook, L.M., Askew, R.R. and Bishop, J.A. (1970). Increasing frequency of the typical form of the peppered moth in Manchester. *Nature* 227: 1155. (12.4.56)

COPR (1976). *Pest Control in Rice.* Centre for Overseas Pest Research, London. (11.1.2)

COPR (1978). *Pest Control in Tropical Root Crops*. Centre for Overseas Pest Research, London. (11.1.3)

Corner, E.J.H. (1952). *Wayside Trees of Malaya*. 2 vols., 2nd Ed., Government Printer, Singapore. (12.1.15)

Corner, E.J.H. (1962). An introduction to the distribution of figs. *Reinwardtia* 4: 325-355.

Corner, E.J.H. (1964) *The Life of Plants*. University of Chicago Press, Chicago. (12.1.13)

Corner, E.J.H. (1978). The freshwater swamp-forest of South Johore and Singapore. *Gdns' Bull. S'pore Suppl*. No. 1. (12.10.10)

Coster, C. (1926). Periodische Blütterscheinungen in den Tropen. Ann. *Jard. Bot. Buitenzorg* 35: 125-162.

Coulter, J.K. (1957). Development of the peat soils of Malaya. *Malay. Agric. J.* 40: 161-175. (23.4.9)

Cowling, E.B. and Linthurst, R.A. (1981). The acid precipitation phenomenon and its ecological consequences. *BioScience* 31: 649-654. (25.7.11)

Cremer, M.C. and Duncan, B.L. (1979). *Brackish Water Aquaculture Development in Northern Sumatera, Indonesia*. Auburn University, Auburn. (2.2.14)

Critchlow, R.E. and Stearns, S.C. (1982). The structure of food webs. *Am. Nat.* 120: 478-499. (12.13.8)

Crockett, C.M. and Wilson, W.C. (1980). The ecological separation of *Macaca nemestrina* and *Macaca fascicularis* in Sumatra. In *The Macaques: Studies in Ecology, Behaviour and Evolution* (ed. D.G. Lindburg), pp. 148-181. van Nostrand Reinhold, London.

Crowther, J. (1982a). Ecological observations in a tropical karst terrain, West Malaysia. I. Variations in topograpy, soils and vegetation. *J. Biogeogr.* 11: 65-78. (12.10.112)

Crowther, J. (1982b). The thermal characteristics of some West Malaysian rivers. *Malay Nat. J.* 35: 99-109. (3.2.18)

Croxall, J.P. (1976). The composion and behaviour of some mixed species bird flocks in Sarawak. *Ibis* 118: 333-346. (12.7.26)

Croxall, J.P. (1979). The montane birds of Gunung Mulu. Paper presented at Gunung Mulu Symposium, 1979, Royal Geographical Society, London. (12.7.15)

Cummins, K.W. (1974). Structure and function of stream ecosystems. *Bioscience* 24: 631-641. (3.2.21)

Curtin, S.H. (1980). Dusky and banded leaf-monkeys. In *Malayan Forest Primates: Ten Years' Study in Tropical Rainforest* (ed. D.J. Chivers), pp. 107-145. Plenum, New York. (12.8.25)

Dahl, A.L. (1981). Monitoring coral reefs for urban impact. *Bull. Mar. Sci.* 31: 544-551.

Dammerman, K.W. (1926). The fauna of Durian and Rhio-Lingga Archipelago. *Treubia* 8: 281-326. (12.13.9)

Dammerman, K.W. (1930). The orang pendek or ape-man of Sumatra. *Proc. 4th Pac. Sci. Cong., Java 1929*, 3: 1-6. (12.8.103)

Davies, A.G. (1984). *The Ecology and Behaviour of the Rufous Leaf-monkey Presbytis rubicunda in Sabah*. PhD thesis, University of Cambridge, Cambridge.

Davis, C.C. (1955). *The Marine and Freshwater Plankton*. Michigan University Press, Michigan. (12.2.2)

Davison, G.W.H. (1980). Territorial fighting by lesser mouse deer. *Malay. Nat. J.* 34: 1-6. (12.8.75)

Davison, G.W.H. (1981a). Sexual selection and the mating system of *Argusianus argus* (Aves: Phasianidae). *Biol. J. Linn. Soc.* 15: 91-104. (12.7.27)

Davison, G.W.H. (1981b). Diet and dispersion of the Great Argus *Argusianus argus*. *Ibis* 123: 485-494. (12.7.28)

De Jong, J.K. (1938). Een an ander over Enggano. *Natuur Tijd. ned.-Indie* 98: 3-46. (12.13.39)

De Silva, M.W.R.N., Betterton, C. and Smith, R.A. (1980). Coral Reef Resources of the East coast of Peninsular Malaysia. In: *Coastal Resources of East Coast Peninsular Malaysia: An Assessment in Relation to Potential Oil Spills* (ed. Chua T.E. and J.k. Charles), pp. 95-158. Universiti Sains Malaysia, Penang. (2.4.3)

De Wit, (1949). A note on Eusideroxylon T. and B. (Laur.) *Bull. Bot. Gard. Buitenzorg* 18: 200-208. (12.1.42)

de Beaufort, L.F. (1939). On collection of freshwater fishes of the island of Billiton. *Treubia* 17: 198-203. (12.5.27)

de Hass, C.P.J. (1950). Checklist of the snakes of the Indo-Australian Archipelago. *Treubia* 20: 575-653.

de Laubenfels, D.J. (1969). A revision of the Malesia and Pacific rainforest conifers, I. Podocarpaceae, in part. *J. Arn. Arb.* 50: 274-369. (12.1.11)

de Leeuw, H.A.L. (1936). Het boschbedrijf, meer in het bijzonder de boschverjonging op Bangka en Billiton. *Tectona* 29: 915-928. (9.1.33)

dela Cruz, A.A. and Banaag, J.F. (1967). The ecology of a small mangrove patch in

Matabungkay Beach, Batangas Province. *Natur. Appl. Sci. Bull.* 20: 486-494. (12.9.33).

Delsman, H.C. (1929). The distribution of frehwater eels in Sumatra and Borneo. *Treubia* 11: 287-292. (12.5.21)

den Hartog, C.(1958). Hydrocharitaceae. In *Flora Malesiana, I.* 5: 381-413.

Dickinson, R.E. (1981). Effects of tropical deforestation on climate. In *Blowing in the Wind: Deforestation and Long-range Implications,* pp. 411-441. College of William and Mary, Williamsburg. (10.4.16)

Diehl, E.W. (1980). *Heterocera Sumatrana, Band 1: Sphingidae.* Distributor: E.W. Classey Ltd., London. (12.4.27)

Dieterlen, F. (1982). Fruiting seasons in the rain forest of Eastern Zaire and their effect upon reproduction. Poster presentation at Tropical Rain Forest Symposium, 1982. Leeds University, Leeds. (12.8.49)

Ding Hou (1958). Rhizophoraceae. *Flora Malesiana I* 5:429-493. Introductory section on ecology (pp. 431-441) by C.G.G.J. van Steenis.

Djajasasmita, M. and Sastraatmadja, D.D. (1981). *Penelitian Peningkatan Pengdayagunaan Sumber Daya Hayati.* Lembaga Biologi Nasional, Bogor. (9.2.6)

Djuhanda, T. (1981). *Dunia Ikan.* Armico, Bandung. (12.5.30)

DKDPI Riau (1979). Hutan mangrove di Propinsi Riau. In *Prosiding Seminar Ekosistem Hutan Mangrove* (ed. S. Soemodihardjo, A. Nontji, A. Djamali) pp. 176-185. Lembaga Oseanologi National, Jakarta. (12.9.2)

Docters van Leeuwen, W. (1920). Naar de top van de Singgalang bij Fort de Kock. *Trop. Natuur* 9: 97-104.

Docters van Leeuwen, W. (1933). Biology of plants and animals occuring in the higher parts of Mount Pangrango-Gedeh in West Java. Ver. Kon. *Akad Wet. Amst.* 31: 1-1278. (12.13.7)

Doyle, M.E. (1969). Factors affecting the distribution of fauna in Gua Pondok. *Malay. Nat. J.* 23: 21-26. (12.4.9)

Dransfield, J. (1972). The genus *Johannesteijsmannia* H.E. Moore Jr. *Gdns' Bull. S'pore* 26: 63-84. (12.1.84)

Dransfield, J. (1974). Notes on the palm flora of Central Sumatra. *Reinwartia* 8: 519-531. (12.10.61)

Driessen, P.M. (1977). Peat soils. Soils and Rice Symposium, Manilla. (23.4.8)

Driessen, P.M. and Soepraptohardjo, M. (1974). *Soils for Agricultural Expansion in Indonesia.* Soil Research Institute, Bogor. (23.4.3)

Dubois, E. (1891). Voorlooping bericht omtrent het onderzoek naar de Pleistocene en tertiarire vertebraten-fauna van Sumatra en Java, gedurende heb jaar 1890. *Natuur Tijd. Ned.-Indie* 51: 93-100. (23.6.11).

Duckett, J.E. (1976). Plantations as a habitat for wild-life in Peninsular Malaysia with particular reference to the oil palm (*Elaeis guineensis*). *Malay. Nat. J.* 29: 176-182. (11.3.6)

Duckett, J.E. (1982). The plantain squirrel in oil palm plantations. *Malay. Nat. J.* 36: 87-98.

Dudgeon, D. (1982a). Aspects of the hydrology of Tai Po Kau Forest Stream, New Territories, Hong Kong. Arch. *Hydrobiol. Suppl.* 64: 1-35. (3.2.19)

Dudgeon, D.(1982b). Spatial and seasonal variations in the standing crop of periphyton and allochthonous detritus in a forest stream in Hong Kong, with notes on the magnitude and fate of riparian leaf fall. *Arch. Hydrobiol. Suppl.* 64: 189-220. (3.2.17)

Duellman, W.E. (1979). The numbers of amphibians and reptiles. *Herp. Review* 10: 83-84. (12.6.27).

Dunn, D.F. (1974). Zoogeography of the Irenidae (Aves: Passeres). *Biotropica* 6: 165-174. (12.7.18)

Dunn, F.L. (1965). Gua Anak Takun: Ecological observations. *Malay. Nat. J.* 19: 75-87. (12.4.8)

Dussault, G.V. and Krammer, D.L. (1981). Food and feeding behaviour of the guppy, *Poecilia reticulata* (Pisces: Poeciliidae). *Can. J. Zool.* 59: 684-701. (12.5.29)

Dyck, V.A. and Thomas, B. (1979). *The brown planthopper problem. In Brown Planthopper: Threat to rice production in Asia.* 3-17. International Rice Research Institute, Los Banos.

Eccles, D.H., Jensen, R.A.C. and Jensen, M.K. (1969). Feeding behaviour of the bat hawk. *Ostrich* 40: 26-27.

Edmondson, W.T. (1970). Phosphorus, nitrogen and algae in Lake Washington after diversion of sewage. *Science* 169: 690-691. (3.1.10)

Edwards, P.J. (1982). Studies of mineral cycling in a montane rain forest in New Guinea. V. Rates of cycling in throughfall and litter fall. *J. Ecol.* 70: 507-828. (12.20.26)

Edwards, P.J. and Grubb, P.J. (1977). Studies of mineral cycling in a montane rain forest in New Guinea. *J. Ecol.* 65: 243-969. (12.10.24)

Edwards, P.J. and Grubb, P.J. (1982). Studies of mineral cycling in a montane rain forest. IV. Soil characteristics and the division of mineral elements between the vegetation and soil. *J. Ecol.* 70: 649-666. (12.10.25)

Edwards, P.J. and Wratten, S.D. (1980). *Ecology of Insect-Plant Interactions.* Edward Arnold, London. (23.2.11)

Edwards, P.J. and Wratten, S.D. (1982). Wound induced changes in palatability in birch (*Betula pubescens* Ehrh. ssp *pubescens*). *Amer. Nat.* 120: 816-818. (12.1.87)

Edwards, W.E. (1967). The late-pleistocene extinction and diminution in size of many mammalian species. In *Pleistocene Extinctions: A search for a cause.* (ed. P.S. Martin and H.E. Wright), pp. 141-154. Yale University Press, New Haven. (23.6.9)

Ehrlich, P.R. and Ehrlich, A.H. (1981). *Extinction: The Causes and Consequences of the Disappearance of Species.* Victor Gollancz, London. (9.2.13)

Endert, F.H. (1920). De Woudboomflora van Palembang. *Tectona.* 13: 113-160. (12.10.15 / 12.9.34).

Endler, J.A (1982). Pleistocene forest refuges: fact or fancy? In *Biological Diversification in the Tropics* (ed. G. Prance), pp. 641-657. Columbia University Press, New York. (23.6.17)

Enright, J.T. (1982). Sleep movements of leaves: In defence of Darwin's interpretation. *Oecologia* 54: 253-259. (12.1.88)

Erlenge. S., Göransson, G., Hogstedt, G., Jansson, G. Liberg, O., Loman, J., Nilsson, I.N., von Schantz. T. and Sylven, M. (1984). Can vertebrate predators regulate their prey? *Am. Nat.* 123: 125-133.

Ernst, A. and Schmidt, E. (1983). Uber Blüte und Frucht von *Rafflesia.* Ann. *Jard. bot. Buitenz.* 27: 1-58. (12.1.82)

Erwin, T.L. (1982). Tropical forests: their richness in Coleoptera and other arthropod species. *Coleopterists' Bull.* 36: 74-75 (12.4.46)

Erwin, T.L. (1983). Tropical tree canopies: the last biotic frontier. *Bull. ESA* 1983: 14-19. (12.4.47)

Estioko, B.R. (1980). Rat damage and control in Philippine sugar cropping. *BIOTROP Spec. Publ.* 12: 159-168. (11.1.6)

Evans, J.H.N. (1918). Preliminary report on cave exploration, near Lenggong, Upper Perak. *J. Fed. Malay States Mus.* 7: 227-234.

Ewel, J. (1980). Tropical succession : manifold routes to maturity. *Biotropica 12 (Tropical Succession Suppl.)* pp. 2-7. (12.1.89)

Ewel, J., Brish, C., Brown, B., Price, N., and Reach, J. (1981). Slash and burn impacts on a Costa Rican wet forest site. *Ecology* 62: 816-829. (10.4.13)

Faegri, K. and van der Pijl, L. (1979). *The Principles of Pollination Ecology.* 3rd rev. Ed., Pergamon, Oxford (12.1.27)

Fairchild, D. (1928). A jungle botanic garden. *J. Hered.* 19: 145-158. (12.13.17)

FAO/ de Wulf, R, Djoko, S., and Kurnia, R. (1981). *Usulan Taman Nasional Kerinci-Seblat: Rencana Pengelolaan Pendahuluan 1982-1987.* UNDP/FAO National Parks Development Project, Bogor. (10.2.11)

FAO/ MacKinnon, J. (1981). *National Conservation Plan for Indonesia, Vol. VIII: National Park Development and General Topics.* FAO, Bogor. (10.2.15)

FAO/ MacKinnon, J. (1982a). *National Conservation Plan for Indonesia, Vol. II. Sumatra.* FAO, Bogor. (10.2.16)

FAO/ MacKinnon, J. (1982b). *National Conservation Plan for Indonesia. Vol. 1. Introduction: Evaluation methods and Overview of National Nature Richness.* FAO, Bogor. (10.2.10)

FAO/ MacKinnon, J.R. and Wind, J. (1979). *Birds of Indonesia* (draft). UNDP/FAO, Bogor. (12.7.41)

FAO/ van der Zon, A.P.M. (1979). *Mammals of Indonesia.* FAO, Bogor. (12.8.36)

FAO/ Wind, J., Soepomo, D. and Isnan, W. (1979). *Way Kambas Game Reserve Management Plan 1980/81-1984/5.* Field Report No. 5, INS/78/061, FAO, Bogor. (10.2.9)

Farnworth, E.G., Tidrick, T.H. and Smathers W.M. (1981). The value of natural ecosystems: an economic and ecological framework. *Environ. Conserv.* 8: 275-282. (9.2.10).

Fenton, M.B. and Fleming, T.H. (1976). Ecological interactions between bats and nocturnal birds. *Biotropica* 8: 104-110. (12,7.8)

Fenton, M.B. and Fullard, J.H. (1981). Moth hearing and the feeding strategies of bats. *Amer. Scient.* 266-275. (12.8.11)

Ferguson, M.W.J. and Joanen, T. (1982). Temperature of egg incubation determines sex in *Alligator mississippiensis. Nature* 296: 850-853. (12.6.12)

Fernando, C.H. (1963). Notes on aquatic insects colonising an isolated pond in Mawai, Johore. *Bull. Nat. Mus. Singapore* 32: 80-89.

Fisher, J.B. (1976). Adaptive value of rotten tree cores. *Bitropica* 8: 261. (12.1.123)

Fleagle, J.G. (1978). Locomotion, Posture, and habitat utilization in two sympatric, Malaysian leaf-monkeys (*Presbytis obscura* and *Presbytis melalophos*). In *The Ecology of Arboreal Folivores* (ed. G.G. Montgomery), pp. 243-251. Smithsonian Institution, Washington. (12.8.19)

Fleagle, J.G. (1980). Locomotion and posture. In *Malayan Forest Primates: Ten Years' Study in Tropical Rain Forest.* (ed. D.J. Chivers), pp. 191-207. Plenum, New York. (12.8.25)

Fleming, T.H. and Heithaus, E.R. (1981). Frugivorous bats, seed shadows and the structure of tropical forests. *Biotropica (Reprod. Bot. Suppl.)* pp. 45-53. (12.10.111)

Flemming, T.H. (1975). The role of small mammals in tropical ecosystems. In *Small Mammals: Their Productivity and Population Dynamics* (ed. F. Golley, K. Petrusewicz and L. Ryszkowski), pp. 269-298. Cambridge University Press, Cambrige. (12.8.81).

Flenley, J.R. (1980a). *The Equatorial Rain Forest: a geological history*. Butterworths, London. (23.6.13)

Flenley, J.R. (1980b). The Quaternary history of the tropical rain forest and other vegetation of tropical mountains. *IV Int. Palyn. Conf. Lucknow (1976-7)* 3: 21-27. (12.10.69)

Flenley, J.R. and Morley, P.J. (1978). *Isoetes* in Sumatra. *J. Biogeog.* 5: 57-58. (12.1.102)

Flynn, R. and Abdullah, M. (1983). Distribution and numbers of Sumatran rhinoceros in the Endau-Rompin region of Peninsular Malaysia. *Malay. Nat. J.* 36: 219-247.

Fogden, M.P.L. (1972). The seasonality and population dynamics of equatorial forest birds in Sarawak. *Ibis* 114: 307-343. (12.7.48)

Forbes, H.O. (1885). *A Naturalist's Wandering in the Eastern Archipelago. A Narrative of Travel and Exploration from 1878 to 1883.* Sampson Low, London.

Francis, P. (1983). Giant volcanic calderas. *Scient. Amer.* 248: 60-70. (23.3.10)

Frankel, O.H. and Soulé, M.E. (1981) *Conservation and Evolution.* Cambridge University Press, Cambridge. (12.12.7)

Franken, N.A.P. and Roos, M.C. (1981). Studies in lowland equatorial forest in Jambi Province, Central Sumatra. Unpublished report, BIOTROP, Bogor. (12.10.103)

Freeland, W.J. and Janzen, D.H. (1974). Strategies in herbivory by mammals: the role of plant secondary compounds. *Amer. Natur.* 108: 269-289. (12.8.46)

Frey-Wyssling, A. (1931a). Over de struikwildernis van Habinsaran. *Trop. Natuur* 20: 194-198. (12.1.103)

Frey-Wyssling, A. (1931b). Over de vegetatie van den Boer Ni Telong en omstreken in Gajolanden (Noord Sumatra). *Trop. Natuur* 20: 37-49.

Frey-Wyssling, A. (1933a). Over de flora van den piek van Kerintji (3800m). *Trop. Natuur* 22: 1-10.

Frey-Wyssling, A. (1933b). Over de zandsteppen van Kota Pinang ter Oostkust van Sumatra. *Trop. Natuur* 22: 69-72. (12.10.55)

Frith, D.W., Tantanasiriwong, R. and Bhatia, O. (1976). Zonation of macrofauna on a mangrove shore, Phuket Island. Phuket *Mar. Biological Center, Res. Bull.* No. 10. (12.9.49)

Furtado, J.I and Mori, S. (ed.) (1982). *Tasek Bera: The Ecology of a Freshwater Swamp,* Junk, The Hague.

Gaisler, J. (1979). The ecology of bats. In *Ecology of Small Mammals* (ed. D.R. Stoddart), pp. 281-342. Chapman and Hall, London. (12.8.20)

Galil, J. (1977). Fig biology. *Endeavour N.S.* 1: 52-56. (12.1.109)

Geertz, C. (1963). *Agricultural Involution: The Processes of Ecological Change in Indonesia.* University of California Press, Berkeley. (23.7.10)

Geollegue, R.T. (1979). Notes on sylvigenesis in the tropical lowland primary rainforest and loggedover areas in Semangus, South Sumatra. Unpubl. report, BIOTROP, Bogor. (9.1.31)

Geollegue, R.T. and Huc, R. (1981). Early stages of forest regeneration in South Sumatra. *BIOTROP Spec. Publ.* 13: 153-161. (10.4.18)

Geyh, M.A., Kudrass, H.R. and Streif, H. (1979). Sea level changes during the late Pleistocene and Holocene in the Strait of Malacca. *Nature, Lond.* 278: 441-443.

Gilbert, O. (1983). The wildlife of Britain's wasteland. *New Scient.* 97: 824-829. (12.13.21)

Gisius, A. (1930). Het sterven der dieren in de wildernis. *Trop. Natuur* 19: 197-198. (12.8.18)

Gittins, S.P. (1978). The species range of the gibbon *Hylobates agilis.* In *Recent Advances in Primatology* vol.3. Evolution (ed. D.J. Chivers and K.A. Joysey), pp. 319-321. Academic Press, London.

Gittins, S.P. (1979). *The Behavoriour and Ecology of the Agile Gibbon (Hylobates agilis).* Unpubl. thesis, University of Cambridge, Cambridge. (12.8.26)

Gittins, S.P. (1980). Territorial behaviour in the Agile gibbon. *Int. J. Primatol.* 1: 381-399. (12.8.41)

Gittins, S.P. (1983). Road casualties solve toad mysteries. *New Scient.* 97: 530-531. (12.6.19)

Gittins, S.P. and Raemaekers, J.J. (1980). Siamang, lar and agile gibbons. In *Malayan Forest Primates: Ten Years' Study in Tropical Rain Forest* (ed. D.J. Chivers), pp. 63-105. Plenum, New York. (12.8.25)

Glass, A.D.M. (1973). Influence of phenolic acids on ion uptake. I. Inhibition of potassium absorption. *J. Exp. Bot.* 25: 1104-1113.

(12.1.36)

Glass, A.D.M. (1974). Influence of phenolic acids on ion uptake. II. Inhibition of phosphate uptake. *Plant Physiol.* 51: 1037-1041. (9.1.23)

Glocek, G.S. and Voris, H.K. (1982). Marine snake diets: prey composition diversity and overlap. *Copeia.* 1982: 661-666. (12.6.4)

Glover, I.C. (1977). The Hoabinhian: hunter-geatheres or early agriculturalists in SE Asia? In *Hunters, Gatherers and First Farmes Beyond Europe* (ed. J.V.S. Megaw), pp.145-166. Leicester Univ. Press, Leicester. (23.6.2)

Glover, I.C. (1978). Report on a visit to archaeological sites near Medan, Sumatra Utara, July 1975. *Bull. Indo-Pacific Prehistory Ass.* 1:56-60. (23.6.19)

Glover, I.C. (1979a). Prehistoric plant remains from Southeast Asia, with special reference fo rice. *South Asian Archaeology* 77: 7-37. (23.6.7)

Glover, I.C. (1979b). The Late Prehistoric Period in Indonesia. In *Early South-East Asia* (ed. R.B. Smith and W. Watson), pp. 167-184. Oxford Univ. Press, Oxford. (23.6.3)

Goh, A.H. and Sasekumar, A. (1980). The community structure of a fringing coral reef, Cape Rachado. *Malay. Nat. J.* 34: 25-37. (12.3.19)

Gomez. E.D. (1980). *The Present State of Mangrove Ecosystems in southeast Asia and the Impact of Aquatic Pollution.* South China Sea Fisheries Development and Coordinating Programme, Manilla. (12.9.24)

Gorman, M. (1970). *Island Ecology.* Chapman and Hall, London. (23.2.14)

Gould, E. (1977). Foraging behaviur of *Pteropus vampyrus* on the flowers of *Durio zibethinus. Malay. Nat. J.* 30: 53-57. (12,8.87)

Gould, E. (1978a). Opportunistic feeding by tropical bats. *Biotropica* 10: 75-76. (12.8.18)

Gould, E. (1978b). Foraging behaviour of Malaysian nectar feeding bats. *Biotropica* 10: 184-193. (12.8.43)

Gould, E. (1978c). Rediscovery of *Hipposideros ridleyi* and seasonal reproduction in Malaysian bats. *Biotropica* 10: 30-32. (12.8.57)

Grandison, A.G.C. (1972). The Gunung Benom Expedition 1967. Reptiles and amphibians of Gunung Benom with a description of a new species of Macroca. *Bull. Brit. Mus. Nat. Hist.* (Zool.) 23: 43-107. (12.6.6)

Green, J. Corbet, S.A., Watts, E. and Oey, B.L. (1976). Ecological studies on Indonesian lakes. Overturn and restratification of Ranu Lamongan. *J. Zool., Lond.* 180:315-354. (3.1.8)

Green, J. Corbet, S.A., Watts, E. and Oey, B.L. (1978). Ecological studies on Indonesia lakes. The montane lakes of Bali. *J. Zool. Lond.* 186: 15-38. (3.1.9)

Greer, A.E. (1971). Crocodilian nesting habits and evolution. *Fauna* 3: 20-28.

Greeser, E. (1919). Bijdragen Resumeerend Repport over het voorkomen van ijzerhout op de olieterreinen Djambi I. *Tectona.* 12: 283-304. (9.1.23)

Grigg, G.C. (1981). Plasma homeostasis and cloacal urine composition in *Crocodylus porosus* caught along a salinity gradient. *J. Comp. Physiol. B.* 144: 261-270. (12.6.13)

Groeneveldt, W (1938). Een overzicht van de vaste trekwegen in Zuid Sumatra alsmede enige gegevens over rhinocerossen. *Ned. Comm. Intern. Natturbesch. Med.* 12: 73-109.

Groeneveldt, W. (1936). Beschouwingen over wild en landbouw. *Ned. Ind. Ver. Natuurbesch.* 1935: 143-151.

Groeneveldt, W. (1938). Een oversicht van de vaste trekwegan in Zuid Sumatra lsmede a enige gegevens over rhinocerossen. *Ned. Comm. Intern. Natuurbesch. Med.* 12: 73-109.

Grubb, P.J. (1971). Interpretation of the 'Massenerhebung' effect on tropical mountains. *Nature* 229: 44-46. (12.10.45)

Grubb, P.J. (1974). Factors controlling the distribution of forest types on tropical mountains: new factors and a new perspective In *Altitudinal Zonation in Malesia.* (ed. J.R. Flenley), pp. 13-46. University of Hull, Hull. (12.10.27)

Grubb, P.J. (1977). Control of forest growth and distribution on wet tropical mountains with special reference to mineral nutrition. *Ann. Rev. Ecol. Syst.* 8: 83-107. (12.10.28)

Grubb, P.J. and Edwards, P.J. (1982). Studies of mineral cycling in a montane rain forest in New Guinea. III. The distribution of mineral elements in the above-ground material. *J. Ecol.* 70: 623-648. (12.10.29)

Hadi, T.R., Atmosoedjono, S., van Peenen, P.F. and Sukaeri (1976). Acarine ecto-parasites of small mammals in Jakarta, Indonesia. *S.E. Asian J. Trop. Med. Public Health* 7: 144-148.

Haffer, J. (1982). General aspects of the Refuge Theory. In *Biological Diversification in the Tropics* (ed. G. Prance), pp. 6-24. Columbia University Press, New York. (12.13.11)

Haile, N.S. (1975). Postulated late Cainozoic high sea level in the Malay Peninsula. *J. Malay. Br. R. Asiatic Soc.* 48: 78-88. (23.6.20)

Hails, A.J. and Yaziz, S. (1982). Abundance, breeding and growth of the ocypodid crab

Dotilla myctiroides (Milne-Edwards) on a West Malaysian beach. *Estuar. Coast. Shelf Sci.* 15: 229-239.

Hails, C.J. (1982). A comparison of tropical and temperate aerial insect abundance. *Biotropica* 14: 310-313.

Hails, C.J. and Amiruddin, A. (1981). Food samples and selectivity of white-bellied swiftlets *Collocalia esculenta. Ibis* 123: 328-333. (12.7.40)

Haines, T.A. (1981). Acidic precipitation and its consequences for aquatic ecosystems: A review. *Trans. Amer. Fish. Soc.* 110: 669-707. (25.1.17)

Hamdan, A. and Rasoel, H. (1980). Beberapa jenis ikan hias di S. Siak, Pekanbaru. *Terubuk* 6: 30-43 (12.5.19)

Hammond, P.M. (1979). Distribution and diversity of Coleoptera in relation to forest type and altitude. Paper presented at Gunung Mulu Symposium, 1979, Royal Geographical Society, London. (12.4.18)

Hamzah, Z. (1978). Some observations on the effects of mechanical logging on regeneration, soil and hydrological conditions in East Kalimantan. *BIOTROP Spec. Publ.* 3: 73-78. (9.1.32)

Hansell, J.R.F. (ed.) (1981). *Transmigration settlements in central Sumatra: identification, evaluation, and physical planning.* Land Resource Study No. 33. Land Resources Development Centre, Surbiton. (15.4.1)

Hanski, I. (1983). Distributional ecology and abundance of dung and carrion eating beetles (Scarabaeidae) in tropical rain forests in Sarawak, Borneo. *Acta Zool. Fenn.* 167: 1-4. (12.4.51)

Hanson, A.J. and Koesoebiono (1977). *Settling coastal swamplands in Sumatra: a case study for integrated resource management.* PSPSL Research Report 004, Institut Pertanian Bogor. (10.4.8)

Harahap, Z. (1979). Breeding for resistance to brown planthopper and grassy stunt virus in Indonesia. In *Brown Planthopper: Threat to rice production in Asia,* pp. 201-208. International Rice Reasearch Institute, Los Banos.

Harcombe, P.A. (1980). Soil nutrient loss as a factor in early tropical secondary succession. *Biotropica* 12 (Tropical Succession Suppl.) : 8-15. (12.10.97)

Hardin, G. (1968). Tragedy of the Commons. *Science* 162: 1243-1248. (10.4.15)

Hardjosuwarno, S., Shalihuddin, D.T., Slukahar, A., Pudjoarinto, A., and Purwoto (1978). *Studi mengenai ekosistem dari mangrove community Cilacap (Jawa Tengah).* Study Group Pencemaran, LEMIGAS, Jakarta. (12.9.4)

Hardon, H.J. (1937). Padang soil, an example of podzol in the tropical lowlands. *Proc. Kon. Akad. Wet.* 40: 530-538. (23.4.4)

Harris, W.V. (1957). An introduction to Malayan termites. *Malay. Nat. J.* 12: 20-32. (12.4.33)

Harrison, J.L. (1951). A Kuala Lumpur garden by night. *Malay. Nat. J.* 5: 193-201. (12.8.102)

Harrison, J.L. (1955). Data on the reproduction of some Malayan mammals. *Proc. zool. Soc. Lond.* 125: 445-460. (12.8.58)

Harrison, J.L. (1956). Survival rates of Malayan rats. *Bull. Raffles Mus.* 27: 5-26.

Harrison, J.L. (1958). Range of movement of some Malayan rats. *J. Mammal.* 39: 190-206. (12.8.44)

Harrison, J.L. (1962). The distribution of feeding habits among animals in a tropical rain forest. *J. Ecol.* 31: 53-63. (12.13.30)

Harrison, J.L. (1968). The effects of forest clearance on small mammals. In *Conservation in Tropical Southeast Asia.* IUCN, Morges. (12.8.71)

Harrison, J.L. (1969). The abundance and population density of mammals in Malayan lowland forests. *Malay. Nat. J.* 22: 174-178.

Harrison, T. (1958). The Caves of Niah: A History of Prehistory. *Sarawak. Mus. J.* 8: 549-595. (23.6.21)

Harrison, T. (1965). Some quantitative effects of vertebrates on the Borneo flora. In *Symp. Ecol. Res. Humid Trop. Vegetation, Kuching, Sarawak, 1963,* pp. 164-169. UNESCO, Paris. (12.13.7)

Harrison, T. (1967). Niah caves progress Report to 1967. *Sarawak Mus. J.* 15: 95-96. (23.6.22)

Harrison, T., Hooijer, D.A. and Medway, Lord (1961). An extinct giant pangolin and associated mammals from Niah Cave, Sarawak. *Nature, Lond.* 189: 166. (12.13.35)

Hartono, W., Soeroso, S. and Bambang, S. (1979). Pengelolaan hutan payau Cilacap. In *Prosiding Seminar Ekosistem Hutan Mangrove* (ed. Subagjo, S., Nontji, A. and Asikin, D.J.), pp. 72-80. Lembaga Oseanologi National, Jakarta. (12.9.2)

Hartshorn, G.S. (1978). Tree falls and tropical forest dynamics. In *Tropical Trees as Living Systems* (ed. P.B. Tomlinson and M.H. Zimmermann), pp. 617-638. Cambridge University Press, Cambridge. (12.10.108)

Harvey, W.G. and Holmes, D.A. (1976). Additions to the avifaunas of Sumatra and Kalimantan, Indonesia. *Bull. Brit. Orn. Club* 96: 90-92. (12.7.51)

Hawksworth, D.L. and Rose, F. (1976). *Lichens as Pollution Monitors.* Edward

Arnold, London. (19.5.19)

Hazewinkel, J.C. (1933). *Rhinoceros sundaicus* in Zuid Sumatra. *Trop. Natuur* 22: 101-109. (12.8.17)

Hazlett, B.A. (1981). The behavioural ecology of hermit crabs. *Ann. Rev. Ecol. Syst.* 12: 1-22. (12.3.38)

Heaney, L.R. (1978). Island area and body size on insular mammals: Evidence from the tri-coloured squirrel (*Callosciurus prevosti*) of Southeast Asia. *Evolution* 32: 29-44. (12.8.31)

Heaney, L.R. (1984). Mammalian species richness on islands on the Sunda Shelf, Southeast Asia. *Oecologia* 61: 11-17.

Hebert, P.D.N. (1980). Moth communities in montane Papuan New Guinea. *J. Anim. Ecol.* 49: 593-602. (12.4.14)

Heckman, C.W. (1979). *Rice Field Ecology in Northeastern Thailand. The Effect of Wet and Dry Seasons on a Cultivated Aquatic Ecosystem.* Junk, The Hague.

Hehuwat, F. (1982). Krakatau: Punah dan lahirnya sebuah Gunung Api. *Suara Alam* No. 15: 29-31 and 34. (23.3.9)

Hehuwat, F. (1982). Tao Toba: Tumor Batak, Tufa Toba, Depresi Vulcanotektonik; Sejarah Terjadinya Danau Toba. *Suara Alam* 16: 18-20. (3.1.6)

Heine-Gelde R. (1972). The archaeology and art of Sumatra. In *Sumatra: Its History and People* (by E.M. Löeb), pp. 305-331. Oxford Unip. Press, London. (23.7.13)

Henderson, M.R. (1939). The flora of the limestone hills of the Malay peninsula. *J. Malay. Br. R. Asiat. Soc.* 17: 13-87. (12.10.50)

Henry, L.E. (1980). Coral Reefs of Malaysia. Longman, Kuala Lumpur.

Henwood, K. (1973). A structural model of forces in buttressed tropical rain forest trees. *Biotropica* 5: 83-93.

Henwood, K. and Fabrick, A. (1979). A quantitative analysis of the dawn chorus: temporal selection for communicatory optimalization. *Amer. Nat.* 114: 260-274

Herrera, C.M. (1982). Defense of ripe fruit from pests: its significance in relation to plant-disperser interactions. *Am. Nat.* 12: 218-241. (12.1.55)

Hill, J.E. (1983). Bats (Mammalia: Chiroptera) from Indo-Australia. *Bull. Brit. Mus. Nat. Hist. (Zool.)* 45: 103-208

Hingkley, A.D. (1973). Ecology of the coconut rhinoceros beetle *Oryctes rhinoceros* (L.) (Coleoptera: Dynastidae). *Biotropica* 5: 111-116. (11.1.12)

Hoffman, R.L. (1979). A siphoniulid millipede from Central America. *Rev. Suisse Zool.* 86: 535-540. (12.4.54)

Holmes, D.A. (1976). A record of White-winged Wood Duck *Cairina scutulata* in Sumatra. *Bull. Brit. Orn. Club* 96: 88-89. (12.7.50)

Holmes, D.A. (1977a). A report on the White-winged Wood Duck in southern Sumatra. *Wildfowl* 28: 61-64. (12.7.6)

Holmes, D.A. (1977b). Faunistic notes and further additions to the Sumatran avifauna. *Bull. Brit. Orn. Club* 97: 68-71. (12.7.50)

Holthuis, L.B. (1979). Cavernicolous and terrestrial decapod durstaceans from northern Sarawak, Borneo. *Zool Verh. Leiden* No. 171: 1-47. (12.3.34)

Holttum, R.E. (1963). Cyathaceae. *Flora Malesiana II* I: 65-176.

Holttum, R.E. (1977). *Plant Life in Malaya*, 2nd Ed. Longman, Kuala Lumpur. (12.1.14)

Holttum, R.E. (1979). *Plant Life in Malaya*, 2nd Ed. Longman, Kuala Lumpur. (12.1.14)

Holttum, R.E. and Allen, B.M. (1963). The tree-ferns of Malaya. *Gdns' Bull. S'pore* 22: 41-51. (12.1.31)

Hooijer, D.A. (1962). Prehistoric bone: The gibbons and monkeys of Niah Great Cave. *Sarawak Mus. J.* 9: 428-449

Hooijer, D.A.(1975). Quaternary mammals west and east of Wallace's Line. *Neth. J. Zool.* 25: 46-56. (23.6.8).

Hook, D.D., Brown, C.L. and Kormanik, P.P. (1971). Inductive flood-tolerance in swamp tupelo (*Nyssa sylvatica* var. biflora Walt., Sarg.). *J. Exp. Bot.* 22: 78-89. (12.1.128)

Howarth, F.G.(1983). Ecology of cave arthropods. *Ann. Rev. Entomol.* 28: 365-389. (12.4.6)

Howe, H.F. (1980), Monkey dispersal and waste of a neotropical fruit. *Ecology* 61: 944-959. (12.1.72)

Howe, H.F. (1981). Dispersal of a neotropical nutmeg (*Virola sebifera*) by birds. *Auk* 98: 88-98. (12.1.71)

Howe, H.F. and Smallwood, J. (1982). Ecology of seed dispersal. *Ann. Rev. Ecol. Syst.* 13: 201-228. (12.1.54)

Howe, H.F. and Vande Kerckhove, G.A. (1980). Nutmeg dispersal by tropical birds. *Science* 210: 925-927. (12.1.51)

Howe, H.F. and Vande Kerckhove, G.A. (1981). Removal of wild nutmeg (*Virola surinamensis*) crops by birds. *Ecology* 62: 1093-1106. (12.1.52)

Hubback, T.R. (1939). The two-horned Asiatic rhinoceros (*Dicerorhinus sumatrensis*). *J. Bombay. nat. Hist. Soc.* 40: 594-617. (12.8.96)

Hubback, T.R. (1941). The Malay Elephant. *J. Bombay nat. Hist. Soc.* 42: 483-509. (12.8.97)

Huc, R. (1981). Preliminary studies on pioneer trees in the dipterocarp forest of Sumatra. Unpubl. report, BIOTROP, Bogor. (12.10.104)

Huc, R. and Rosalina U. (1981a). Aspects of secondary forest succession in logged-over lowland forest areas in East Kalimantan. Unpubl. report, BIOTROP, Bogor. (12.10.96)

Huc, R. and Rosalina, U. (1981b). Chablis and primary forest dynamics in Sumatra. BIOTROP report, Bogor. (12.10.19)

Humner, P.J. (1980). Determining the population size of pond phytoplankton. *Amer. Biol. Teach.* 42: 545-548. (12.2.10)

Hutchings, P.A. and Reicher, H.F. (1982). The fauna of Australian mangroves. *Proc. Linn. Soc. N.S.W.* 106: 83-121. (12.9.50)

Huxley, C.R. (1978). The ant-plants *Myrmecodia* and *Hydnophytum* (Rubiaceae), and the relationship between their morphology, and occupants, physiology, and ecology. *New Phytol.* 67: 231-268. (12.1.17)

Inger, R. F. (1969). Organization of communities of frogs along rain forest streams in Sarawak. *J. Anim. Ecol.* 38: 123-148. (12.6.16)

Inger, R. F. (1980). Densities of floor dwelling frogs and lizards in lowland forests of Southeast Asia and Central America. *Am. Nat.* 115: 761-770. (12.6.17)

Inger, R.F. and Greenberg, B. (1966). Ecological and competitive relations among three species of frogs (Genus *Rana*). *Ecology* 45: 746-759. (12.6.15)

IPB/Team Biologi Perairan (1975). Laporan Survei Biologi perairan Delta Upang-Banyuasin. PUSDI-PSL, IPB, Bogor.

Iskandar, Dj. T. (1978). Kunci identifikasi kura-kura. *Terubuk* 4: 1-5. (12.6.26)

Jacobs, M. (1958). Contribution to the botany of Mt. Kerintji and adjacent area in West Central Sumatra - I. Ann. *Bogorienses* 3: 45-79. (12.10.38)

Jacobs, M. (1980). Significance of the tropical rain forest on 12 points. *BioIndonesia* 7: 75-94. (12.10.73).

Jacobs, M. (1981). Review of Rheophytes of the World by C.G.G.J. van Steenis. *Flora Malesiana Bull.* 34: 3617-3620. (12.1.113)

Jacobs, M. (1982). The study of minor forest products. *Flora Malesiana Bull.* 35: 3768-3782. (9.2.12)

Jacobs, M. and de Boo, T.J.J. (1983). *Conservation Literature on Indonesia: Selected Annotated Bibliography.* Rijksherbarium, Leiden.

Jacobson, E. (1918). De kambing oetan. *Trop. Natuur* 7: 129-134.

Jacobson, E. (1921). Notes on some mammals from Sumatra. *J. Fed. Malay. States Mus.* 10: 235-240. (12.8.14)

Jacobson, E. and Kloss, C.B. (1919). Notes on the Sumatran Hare. *J. Fed. Malay. States Mus.* 7: 293-297. (12.8.15)

James, F. (in prep.). *The late Quaternary vegetational history of the upper Kerinci valley, Sumatra, Indonesia.* MSc thesis, Univerity of Hull, Hull.

Jancey, T.E. (1973). Holocene radiocarbon dates on the 3 meter wave cut notch in northwestern peninsular Malaysia. *Geol. Soc. Malaysia Newsl.* 45: 8-11.

Janos, D.P. (1980a). Mycorrhizae influence tropical succession. *Biotropica* 12 (*Tropical Succession Supplement*): 56-64. (12.1.62)

Janos, D.P. (1980b). Vesicular-arbuscular mycorrhizae affect lowland tropical rain forest plant growth. *Ecology* 61: 151-162. (12.1.57)

Janos, D.P. (1983). Tropical mycorrhizas, nutrient cycles and plant growth. In *Tropical Rain Forest: Ecology and Management* (ed. S.L. Sutton, T.C. Whitmore and A.C. Chadwick), pp. 327-345. Blackwell, Oxford.

Janzen, D.H. (1970). Herbivores and the number of tree species in tropical forests. *Amer. Natur.* 104: 501-528. (12.10.119)

Janzen, D.H. (1973a). Dissolution of mutualism between *Cecropia* and its *Azteca* ants. *Biotropica* 5: 15-28.

Janzen, D.H. (1973b). Rate of regeneration after a tropical high-elevation fire. *Biotropica* 5: 117-122. (12.10.42)

Janzen, D.H. (1974a). Tropical blackwater rivers, animals and mast fruiting by the Dipterocarpaceae. *Biotropica* 6: 69-103. (12.10.59)

Janzen, D.H. (1974b). Epiphytic myrmecophytes in Sarawak: Mutualism through the feeding of plants by ants. *Biotropica* 6: 237-259. (12.1.18)

Janzen, D.H. (1975). *Ecology of Plants in the Tropics.* Edward Arnold, London. (12.1.93)

Janzen, D.H. (1976a). Why tropical trees have rotten cores. *Biotropica* 8: 110. (12.1.125)

Janzen, D.H. (1976b). Why bamboos take so long to flower. *Ann. Rev. Ecol. Syst.* 7: 347-391. (12.1.134)

Janzen, D.H. (1977a). Promosing directions of study in tropical animal-plant interactions. *Ann. Missouri Bot. Gard.* 64: 706-736. (12.13.16)

Janzen, D.H. (1977b). Why fruit rots, seeds mold and meat spoils. *Am. Nat.* 111: 691-713.

Janzen, D.H. (1978a). Complications in interpreting the chemical defenses of trees against tropical arboreal plant-eating vertebrates. In *The Ecology of Arboreal Folivores* (ed. G.G. Montgomery), pp. 73-84. Smithsonian Institution, Washington, D.C. (12.1.63)

Janzen, D.H. (1978b). A bat-generated fig seed shadow in rain forest. *Biotropica* 10: 121. (12.1.60)

Janzen, D.H. (1979a). Why food rots. *Nat. Hist.* 88: 60-64. (12.13.13)

Janzen, D.H. (1979b). How many babies do figs pay for babies? *Biotropica* 11: 48-50. (12.1.59)

Janzen, D.H. (1979c). How many parents do the wasps from a fig have? *Biotropica* 11: 127-129. (12.4.35)

Janzen, D.H. (1979d). How to be a fig. *Ann. Rev. Ecol. Syst.* 10: 13-51. (12.1.58)

Janzen, D.H. (1980). Specificity of seed-attacking beetles in a Costa Rican deciduous forest. *J. Ecol.* 68: 929-952. (12.4.34)

Janzen, D.H. (1981a). The defenses of legumes against herbivores. In *Advances in Legume Systematics* (ed. R.M. Polhill and P.H. Raven), pp. 951-977. Royal Botanic Gardens, Kew. (12.1.68)

Janzen, D.H. (1981b). Digestive seed predation by a Costa Rican Baird's tapir. *Biotropica (Reprod. Bot. Suppl.)*, pp. 59-63. (12.8.45)

Janzen, D.H. (1981c). Lectins and plant-herbivore interactions. *Recent Advances in Phytochemistry* 15: 241-258.

Janzen, D.H. (1982a). Variation in average seed size and fruit seediness in a fruit crop of a guanacaste tree (Leguminosae: *Enterolobium cyclocarpum*). *Amer. J. Bot.* 69: 1169-1178. (12.1.61)

Janzen, D.H. (1982b) Dispersal of small seeds by big herbivores: foliage is the fruit. *Amer. Nat.* 123: 338-353.

Janzen, D.H. (1983a). Seed and pollen dispersal by animals: convergence in the ecology of contamination and sloppy harvest. *Biol. J. Linn. Soc.* 20: 103-113.

Janzen, D.H. (1983b). Physiological ecology of fruits and their seeds. *In Physiological Plant Ecology III* (ed. O.L. Lange, P.S. Nobel, C.B. Osmond and H. Ziegler), pp 625-655. Springer-Verlag, Berlin. (12.1.9)

Janzen, D.H. (1983c). Food webs: who eats what, why, how and with what effects in a tropical forest? In *Tropical Rain Forest Ecosystems* (ed. F.B. Golley), pp. 167-182. Elsevier, Amsterdam. (12.13.14)

Janzen, D.H. (1983d). Tropical agroecosystems. *Science* 182: 1212-1219. (11.4.23)

Janzen, D.H., Miller, G.A., Hackforth-Jones, J., Pond, C.M., Hooper, K. and Janos, D.P. (1976). Two Costa Rican bat generated seed shadows of *Andira inermis* (Leguminosae). *Ecology* 57: 1068-1075. (12.1.56)

Jaspan, M.A. (1973). A note on Enggano. *Sumatra Res. Bull.* 3: 54-63.

Jenkins, P.D. (1982). A discussion of Malayan and Indonesian shrews of the genus *Crocidura* (Insectivora: Soricidae). *Zool. Med.* 56: 267-269. (12.8.15)

Jennings. J.N. (1972). The character of tropical humid karst Z. *Geomorph.* 16: 336-341. (23.3.3)

Jochems, S.C.J. (1926). De Westenenkpaadjes van Brastagi. *Trop. Natuur* 15: 65-71. (10.4.21)

Jochems, S.C.J. (1927). De boeging der vloedbosschen van Noord Sumatra. *Trop. Natuur* 15: 145-150.

Jochems, S.C.J. (1928). De begroeiing der tabakslanden in Deli en hare betehenis voor de tabakscultur. Med. *Deli-Proefstation* No. 59. (11.4.22)

Johansen, H.W. (1981). *Coralline Algae: A First Synthesis.* CRC, Florida.

Johns, A.D. (1981). The effects of selective logging on the social structure of resident primates. *Malays. appl. Biol.* 10: 221-226. (12.8.76)

Johns, A.D. (1982). Observations on nesting behaviour in the Rhinoceros hornbill, *Buceros rhinoceros. Malay. Nat. J.* 35: 173-177. (12.7.46)

Johns, A.D. (1983). *The Effects of Selective Logging on a Malaysian Rain-forest Community.* Ph. D. thesis, University of Cambridge, Cambridge.

Johnson, A. (1960). Variations in *Sphagnum junghuhnianum* subsp. *junghuhnianum* Dz. & Molk. *Trans Br. bryol. Soc.* 3: 725-728. (12.1.100)

Johnson, A. (1970). Blue-green algae in Malayan rice-fields. *J. Sing. natn. Acad. Sci.* 1:30-36.

Johnson, D.S. (1956). Systematic and ecological notes on the Cladocera of Lake Toba and the surrounding country, North Sumatra. *J. Linn. Soc. Zool.* 43: 79-91. (12.2.13)

Johnson, D.S. (1957). A survey of Malayan freshwater life. *Malay. Nat. J.* 12: 57-65. (3.2.16)

Johnson, D.S. (1960). Some aspects of the distribution of freshwater organisms in the Indo-Pacific area, and their relevance to the validity of the concept of an Oriental region in zoogeography. In *Proc. Centenary & Bicentenary Cong. Biol. S'pore* 1958. (ed.

R.D. Purcheon). Oxford University Press, Kuala Lumpur. (3.6.3)

Johnson, D.S. (1961a). Freshwater life in Malaya - its conservation. *Malay. Nat. J. CN*: 232-236. (12.13.24)

Johnson, D.S. (1961b). An instance of sudden large-scale mortality of fish in a natural habitat in S. Malaya. *Malay. Nat. J.* 15: 160-162. (3.2.34)

Johnson, D.S. (1961c). The food and feeding of the mudlobster *Thalassina anomala* (Herbst). *Crustaceana* 2: 325-326. (12.3.35)

Johnson, D.S. (1962). Water fleas. *Malay. Nat. J.* 16: 126-144. (12.3.33)

Johnson, D.S. (1967a). Distributional patterns in Malayan freshwater fish. *Ecology* 48: 722-730. (12.5.26)

Johnson, D.S. (1967b). On the chemistry of freshwaters in southern Malaya and Singapore. *Arch. Hydrobiol.* 63: 477-496. (3.3.5)

Johnson, D.S. (1968a). Malayan blackwaters. In *Proc. Symp. on Recent Adv. Trop. Eco.* (ed. R. Misra and B. Gopal). Int. Soc.Trop. Ecol., Varanasi. (3.2.22)

Johnson, D.S. (1968b). Water pollution in Malaysia and Singapore: some comments. *Malay. Nat. J.* 21: 221-222. (21.1.15)

Johnson, D.S. (1973). Equatorial forest and the inland aquatic fauna of Sundania. In *Nature Conservation in the Pacific* (ed. A.B. Costin and R.H. Groves), pp. 111-116. IUCN, Gland. (3.2.26)

Johnson, D.S., Soong, M.H.H. and Wee, B.T. (1969). Freshwater streams and swamps in the tree country of southern and eastern Malaya with special reference to aquarium fish. In *Natural Resources of Malaysia and Singapore* (ed. B. Stone), University of Malaya, Kuala Lumpur. (3.2.33)

Johnstone, I.M. (1981). Consumption of leaves by herbivores in mixed mangrove stands. *Biotropica* 13: 252-259. (12.9.36)

Joly, C.A. and Crawford, R.M.M (1982). Variation in tolerance and metabolic responses to flooding in some tropical trees. *J. Exp. Bot.* 33: 799-809. (12.10.9)

Jones, D.A. (1979). The ecology of sandy beaches in Penang, Malaysia, with special reference to *Excirolana orientalis* (Dana). *Est. Coast. Mar. Sci.* 9: 677-682. (2.4.37)

Jonker, H.A.J. (1933). De vloedbosschen van den Riaw Lingga Archipel. *Tectona* 26: 717-741 (12.9.51)

Jordan, C.F. and Farnworth, E.F. (1982). Natural vs. Plantation forests: a case study of land reclamation strategies for the humid tropics. *Env. Manage.* 6: 485-492. (10.3.13)

Justessen, P.Th. (1922). Morphological and biological notes on *Rafflesia* flowers, observed in the highlands of mid-Sumatra (Padangsche Bovenlanden). *Ann. Jard. bot. Buitenzorg* 32: 64-87. (12.1.135)

Kartawinata, K. (1974). *Report on the State of Knowledge on Tropical Forest Ecosystems in Indonesia.* Herbarium Bogoriense LBN-LIPI, Bogor. (12.10.87)

Kartawinata, K. (1978a). Biological changes after logging in lowland dipterocarp forest. *BIOTROP Spec. Publ.* 3: 27-34. (9.1.32)

Kartawinata, K. (1978b). The 'kerangas' heath forest in Indonesia. In *Glimpses of Ecology* (ed. J.S. Singh and B. Gopal), pp. 145-153. (12.10.17)

Kartawinata, K. (1980a). Classification and untilization of Indonesian forest. *BioIndonesia* 7: 95-106. (10.4.20)

Kartawinata, K. (1980b). The environmental consequences of tree removal from the forests in Indonesia. In *Where Have All the Flowers Gone? Deforestation in the Third World*, pp. 191-214. College of William and Mary, Williamsburg. (10.4.17)

Kartawinata, K., Adisoemarto, S., Soemodihardjo, S., and Tantra, I.G.M. (1979) Status pengetahuan hutan bakau di Indonesia. In *Prosiding Seminar 'Ekosistem Hutan Mangrove'* (ed. S. Soemodihardjo, A. Nontji, A. Djamali), pp. 21-39. Lembaga Oseanology Nasional, Jakarta. (12.9.2)

Kartawinata, K., Rochadi, A. and Tukirin, P. (1981). Composition and structure of a lowland dipterocarp forest at Wanariset, East Kalimantan. *Malay. Forest.* 44: 397-406.

Karunakaran, L. and Johnson, A (1978) A contribution to the rotifer fauna of Singapore and Malaysia. *Malay. Nat. J.* 32: 173-208. (12.2.11)

Kato, R., Tadaki, Y. and Ogawa, H. (1978). Plant biomass and growth increment studies in Pasoh forest. *Malay. Nat. J.* 30: 211-224. (12.10.76)

Keeley, J.E. (1978). Malic acid accumulation in roots in response to flooding: evidence contrary to its role as an alternative to ethanol. *J. Exp. Bot.* 29: 1345-1349. (12.1.129)

Keeley, J.E. (1979). Population differentiation along a flood frequency gradient: physiological adaptation to flooding in Nyssa sylvatica. *Ecol. Monogr.* 49: 89-108. (12.1.131)

Keeley, J.E. and Franz, E.H. (1979). Alcoholic fermentation in swamp and upland populations of *Nyssa sylvatica*: temporal changes in adaptive strategy. *Am. Nat.* 113: 587-592. (12.1.126)

Kettlewell, H.B.H. (1955). Selection experiments on industrial melanism in the Lepidoptera. *Heredity* 10: 287-301. (12.4.24)

Kettlewell, H.B.H. (1959). Darwin's missing evidence. *Sci. Amer.* 200: 48-53.

Khobkhet, O. (1980). Research on bats in Thailand: A bird's eye view. *BIOTROP Spec. Publ.* 12: 109-120. BIOTROP, Bogor.

Khoo, S.G. (1974). Scale insects and mealy bugs: their biology and control. *Malay. Nat. J.* 27: 124-130 (11.1.13)

Kiew, B.H. (1970). The ecology of *Myrmeleon celebensis* McLachlan (Neuroptera: Myrmeleontidae). *Malay. Nat. J.* 24: 11-15. (12.4.47)

Kikkawa, J. and Williams, W.T. (1971). Altitudinal distribution of land birds in New Guinea. *Search* 2: 64-65. (12.7.12)

King, B., Woocock, M. and Dickinson, E.C (1975). *A Field Guide to the Birds of South-East Asia.* Collins, London. (12.7.45)

Kira, T. (1978). Primary productivity of Pasoh Forest. *Malay Nat. J.* 30: 291-297. (12.10.75)

Kitchener, H.j. (1951). Diatoms. *Malay. Nat. J.* 8: 152-156. (12.2.11)

Koesoebiono and Chairul, M. (1980). *Prospects for Fisheries Development in the Musi Banyuasin Coastal Zone of Southeast Sumatra.* PSPSL Research Report 006, Institut Pertanian Bogor. (2.2.3)

Koesoebiono, Collier, W.L. and Burbridge, P.R. (1982). Indonesia: Resources' use and management in the coastal zone. In *Man, Land and Sea* (ed. C. Soysa, L.S. China and W.L. Collier) pp. 115-133. Agricultural Development Council, Bangkok. (2.4.35)

Koopman, M.J.F. and Verhoef, L. (1938). Eusideroxylon zwageri T. and B. Het ijzerhout van Borneo en Sumatra. *Tectona* 31: 381-399. (9.1.22)

Kostermans, A.J.G.H. (1958). Secondary growth on areas of former peat swamp forest. In *Proc. Symp. Humid Tropics Veg., Tjiawi,* 1958, pp. 155-169. UNESCO. Jakarta. (12.10.90)

Kostermans, A.J.G.H. (1979). Potocylon, a new Bornean genus of Lauraceae. *Malay. Nat. J.* 32: 143-148. (12.1.41)

Koyama, H. (1978). Photosynthesis studies in Pasoh Forest. *Malay. Nat. J.* 30: 253-258. (12.10.74)

Kramer, D.L. and McClure, M. (1982). Aquatic surface respiration, a widespread adaptation to hypoxia in tropical freshwater fisher. *Env. Biol. Fish.* 7: 47-55.

Kramer, D.L. and Mehegan, J.P. (1981). Aquatic surface respiration, an adaptive response to hypoxia in the guppy, *Poecilia reticulata* (Pisces, Poeciliidae). *Env. Biol. Fish.* 6: 299-313. (12.5.34)

Kramer, F. (1933). De natuurlijke verjonging in het Goenoeng-Gegdeh-complex. *Tectona* 26: 155-185. (12.10.40)

Krutilla, J.V. and Fisher, A.C. (1975). *The economics of natural environments: Studies in the Valuation of Commodity and Amenity Resources.* John Hopkins University Press, Baltimore. (23.8.1).

Kushwaha, S.P.S., Ramakrishnan, P.S. and Tripathi, R.S. (1981). Population dynamics of *Eupatorium odoratum* in successional environments following slash and burn agriculture. *J. appl. Ecol.* 18: 529-535. (12.1.91)

Lahmann, E.J. and Zuniga, C.M. (1981). Use of spider threads as resting places by tropical insects. *J. Arachnol.* 9: 339-341. (12.4.42)

Lam, T.M. (1983). Reproduction in the rice field rat *Rattus argentiventer. Malay. Nat. J.* 36: 249-282.

Langham, N. (1980). Breeding biology of the edible-nest swiftlet *Aerodramus fuciphagus. Ibis* 122: 447-461. (12.7.5)

Langham, N.P.E. (1982). The ecology of the Common Tree Shrew *Tupaia glis* in peninsular Malaysia. *J. Zool.* 197: 323-344.

LAPI-ITB (1980). *Environmental impact analysis of the Koba tin mining operation on Bangka Island, Indonesia.* Lembaga Afiliasi Penelitian and Industri, Institut Teknologi Bandung, Bandung. (20.2.3)

Laumonier, Y., Gadrinab, A. and Purnajaya (1983). *Southern Sumatra: Map of Vegetation and Environmental Conditions 1:1,000,000.* Institut de la Carte International du Tapis Végétal / SEAMEO-BIOTROP, Bogor.

Law, A.T. and Mohsin, Moh. A.K. (1980). Environmental studies of Kelang River. I. Chemical, physical and microbiological parameters. *Malay. Nat. J.* 33: 175-187. (3.2.24)

LBN (1980). *Laporan Teknik.* Bogor.

LBN (1983). *Laporan Teknis 1980-1982.* Lembaga Biologi National, Bogor.

Lee, D.W. and Lowry, J.B. (1980). Solar ultraviolet on tropical mountains: Can it effect plant speciation? *Am. Nat.* 115: 880-883. (12.1.23)

Lee, K.B. and Wee, Y.C. (1982). Algae growing on walls in Singapore. *Malay. Nat. J.* 35: 125-132. (12.1.77)

Leigh, E.G. (1975). Structure and climate in tropical rain forest. *Ann. Rev. Ecol. Syst.* 6: 67-86. (12.10.78)

Leighton, M. (1982). *Fruit Resources and Patterns of Feeding, Spacing and Grouping among Sympatric Bornean Hornbills.* Ph.D. thesis, University of California, Davis. (12.7.38)

Leighton, M. and Leighton, D.R. (1983). Vertebrate responses to fruit seasonality within a Bornean rain forest. In *Tropical Rain Forest Ecology and Management* (ed. S.L. Sutton, T.C. Whitmore and A.C. Chadwick), pp. 181-196. Blackwell, Oxford.

Lekagul, B. and McNeely, J.A. (1977). *The Mammals of Thailand.* Association for the Conservation of Wildlife, Bangkok (12.8.21)

Lenton, G. (1980). Biological control of rats by owls in oil palm and other plantations. *BIOTROP Spec. Publ.* 12: 87-94. (11.1.6)

Lenton, G. (1983). Wise owls flourish among the oil palms. *New Scient.* 97: 436-437. (11.3.5)

Lever, R.J.U. (1955). Notes on a mangrove caterpillar from Port Dickson. *Malay. Nat. J.* 10: 13-14. (12.4.2)

Lieftinck, M.A. (1950). Further studies of S.E. Asiatic species of Macromia Rambur. *Treubia* 20: 657-716. (12.4.3)

Lieftinck, M.A. (1984). The Odonata of Enggano. *Treubia* 19: 279-300. (12.4.53)

Liew, T.C. (1973). Occurrence of seeds in virgin forest top soil with particular reference to secondary species in Sabah. *Malay. Forester* 36: 185-193. (12.10.42).

Likens, G.E., Bormann, F.H., Pierce, R.S., Eaton, J.S., and Johnson, N.M. (1977). *Biogeochemistry of a Forested Ecosystem.* Springer-Verlag, New York. (12.10.93)

Lim, B.H. and Sasekumar, A. (1979). A preliminary study on the feeding biology of mangrove forest primates. Kuala Selangor. *Malay. Nat. J.* 33: 105-112. (12.8.73)

Lim, B.L. (1974). Small mammals associated with rice fields. *MARDI Res. Bull.* 3: 25-33 (12.8.83)

Lim, B.L. (1976). Snakes as natural predators of rats in an oil palm estate. *Malay, Nat J.* 27: 114-117. (11.1.9)

Lim, C.F. (1963). A preliminary illustrated account of mangrove molluscs from Singapore and S.W. Malaya. *Malay. Nat. J.* 17: 235-240. (12.3.17)

Lim, M.T. (1978). Litterfall and mineral nutrient content of litter in Pasoh Forest Reserve. *Malay. Nat. J.* 30: 375-380. (12.10.79)

Lim. B.L., Hadi, T.R., Sustriayu and Gandahusada (1980). A preliminary survey of rodents in two transmigration schemes in South Sumatra. *BIOTROP Spec. Publ.* 12: 67-78. (11.1.6)

Lind, E.M. and Morrison, M.E.S. (1974). *East African Vegetation.* Longman, London. (Referred to in Rabinowitz, 1978).

Lloyd, M., Inger, R.F. and King, F.W. (1968). On the diversity of reptile and amphibian species in a Bornean rain forest. *Am. Nat.* 102: 497-515. (12.6.8)

Lock, M.A. and Reynoldson, T.B. (1976). The role of interspecific competition in the distribution of two stream-dwelling triclads. *J. Anim. Ecol.* 45: 581-592. (12.3.31)

Löeb, E.M. (1972). *Sumatra: Its History and People.* Oxford Univ. Press, Oxford. (23.7.13)

Loveless, A.R. (1961). A nutritional interpretation of schlerophylly based on differences in the chemical composition of schlerophyllous and mesophytic leaves. *Ann. Bot.* 25: 168-184. (12.1.20)

Loveless, A.R. (1962). Further evidence to support a nutritional interpretation of schlerophylly. *Ann. Bot.* 26: 551-561. (12.1.21)

Lucas, J.R. and Brockmann, H.J. (1981). Predatory interactions between ants and ant lions (Hymenoptera: Formicidae and Neuroptera: Myrmeleontidae). *J. Kans. Entomol. Soc.* 54: 228-232.

Lugo, A.E. and Snedaker, S.C. (1974). The Ecology of Mangroves. *Ann. Rev. Syst. Ecol.* 5: 39-64. (12.9.9)

Lugo, A.E. Cintron, G. and Geonaga, C. (1978). Mangrove ecosystems under stress. In *Stress Effects on Natural Ecosystems* (ed. G.W. Barrett and R. Rosenberg), pp. 1-32. (12.9.19)

Lugo. A.E., Evink, G., Brinson, M.M., Broce, A., and Snedaker, S.C. (1975). Diurnal rates of photosynthesis, respiration and transpiration in mangrove forests of south Florida. In *Tropical Ecological Systems* (ed. F.B. Golley and E. Medina), pp. 335-350. Springer-Verlag, New York. (12.9.37)

Luythes, A. (1923). De vloedbosschen in Atjeh. *Tectona.* 16: 575-591. (12.9.38)

Macan, T.T. and Kitching, A. (1972). Some experiments with artificial substrata. Verb. Int. Verein theor. angew. *Limnol.* 18: 213-220.

MacIntosh, D.J. (1979). Predation of fiddler crabs (*Uca* spp.) in estuarine mangroves. In *Mangrove and Estuarine Vegetation in Southeast Asia* (ed. P.B.L. Srivastrava, A.M. Ahmad, G. Dhanarajan and I. Hamjah). BIOTROP Special Publication No. 10, pp. 101-110. BIOTROP, Bogor. (12.9.26)

MacKenzie, M.J.S. and Kear, J. (1976). The White-winged Wood Duck. *Wildfowl* 27: 5-17. (12.7.5)

MacKinnon, J.R. (1974). The behaviour and ecology of wild orang utans (*Pongo pygmaeus*). *Anim. Behav.* 22: 3-74. (12.8.53).

MacKinnon, J.R. (1977). A comparative ecology of Asian apes. *Primates* 18: 747-772. (12.8.61).

MacKinnon, J.R. and MacKinnon, K.S. (1980). Niche differentiation in a primate community. In *Malayan Forest Primates: Ten Years' Study in Tropical Rain Forest* (ed. D.J. Chivers), pp. 167-190. Plenum, New York. (12.8.25).

MacNae, W. (1968). A general account of the fauna and flora of mangrove swamps and forests in the Indo-Pacific region. *Adv. mar. Biol.* 6: 73-270. 612.9.32)

Maiorana, V.C. (1978). What kinds of plants do herbivores really prefer? *Amer. Nat.* 112: 631-635. (12.1.86).

Maiorana, V.C. (1979). Non-toxic toxins: the energetics of coevolution. *Biol. J. Linn. Soc.* 11: 387-296. (12.13.23)

Maiorana, V.C. (1981). Herbivory in sun and shade. *Biol. J. Linn. Soc.* 15: 151-156. (12.1.69).

Malley, D.F. (1977). Adaptations of decapod crustaceans to life in mangrove swamps. *Mar. Res. Indonesia.* 18: 63-72. (12.3.10)

Maloney, B.K. (1980). Pollen analytical evidence for early forest clearance in North Sumatra, Indonesia. *Nature*, Lond. 287: 324-326. (23.6.6)

Maloney, B.K. (1981). A pollen diagram from Tao Sipinggan, a lake site in the Batak Highlands of North Sumatra. Modern Quatern. *Res. S.E. Asia.* 1981: 57-76.

Maloney, B.K. (1983a). The plant ecology of forest remnants on the Toba Plateau of North Sumatra. *Indonesia Circle* 31: 23-37.

Maloney, B.K. (1983b). The terminal Pleistocene in Sumatra, Indonesia. *Quatern. Newsl.* 39:1-9

Maloney, B.K. (1983c). Man's impact on the rain forests of West Malesia: the palynological record. In *Proc. 7th Aberdeen-Hul Syamp.* Malesian Ecol.

Mani, M.S. (1972). General Entomology. Oxford IBH Publishing Co., New Dehli.

Mani, M.S. (1980a). The vegetation of highlands. In *Ecology of Highlands.* (ed. M.S. Mani, and L.E. Giddings), pp. 127-139. Junk, The Hague. (12.10.30)

Mani, M.S. (1980b). The animal life of highlands. In *Ecology of highlands.* (ed. M.S. Mani, and L.E. Giddings), pp. 141-159. Junk, The Hague. (12.13.5)

Mann, K.H. (1975). Patterns of energy flow. In *River Ecology* (ed.B.A. Whitton), pp. 248-263. Blackwell, Oxford. (3.2.29)

Manning, A., McKinnon, E.E. and Treloar, F.E. (1980). Analysis of gold artifacts from the Kota Cina site near Medan, Sumatra. *J. Malay. Br. R. Asiatic Soc.* 53: 102-113. (23.7.4)

Manokaran, N. (1978). Nutrient concentration in precipitation, throughfall and stemflow in a lowland tropical rain forest in Peninsular Malaysia. *Malay. Nat. J.* 30: 423-432. (12.10.107)

Manokaran, N. (1979). Stemflow, throughfall, and rainfall interception in a lowland tropical rainforest in Peninsular Malaysia. Malay. *Forester* 42: 174-201. (12.10.44)

Manokaran, N. (1980). The nutrient contents of precipitation, throughfall and stemflow in a lowland tropical rainforest in Peninsular Malaysia. *Malay Forester* 43: 266-289. (12.10.46).

Manullang, B.O. (1980). Kantong semar, tumbuhan pemakan serangga. *Suara Alam* 9: 45-47. (12.1.114)

Marcellini, D.L. (1977). The function of a vocal display of the lizard *Hemidactylus frenatus* (Sauria: Gekkonidae). *Anim. Behav.* 25: 414-417. (12.6.31)

Marsden, W. (1811). *The History of Sumatra.* Republished by Oxford Univ. Press, Kuala Lumpur, 1966.

Marsh, C.W. (1981). Primates in Malaysian peat swamp forest. In *Conservation Inputs from Life Sciences.* (ed. M. Nordin, A. Latiff, M.C. Mahani and J.C. Tan), pp. 101-108. University Kebangsaan Malaysia, Bangi. (12.8.10)

Marsh, C.W. and Wilson, W.L. (1981). *A Survey of Primates in Peninsular Malaysian Forests.* Universiti Kebangsaan Malaysia, Bangi. (12.8.9)

Marsh, C.W., Johns, A.D. and Ayres, J.M. (1984). Effects of disturbance on rain forest primates. In *Primates Conservation in Tropical Rain Forest* (ed. S.J. Gartlan, C.W. Marsh and R.A. Mittermeier). (12.8.77)

Marshall, A.G. (1971). The ecology of *Basilla hespida* (Diptera: Nycteribiidae) in Malaysia. *J. Anim. Ecol.* 40: 141-154. (12.4.22)

Marshall, A.G. (1980). The comparative ecology of insects ectoparasitic upon bats in West Malaysia. In *Proc. 5th Int. Bat Res. Conf.* (ed. D.E. Wilson and A.C. Gardner), pp. 135-142. Texas Tech. Press, Lubbock, Texas. (12.4.23)

Marshall, A.G. (1982) The ecology of insects ectoparasitic on bats. In *Ecology of Bats* (ed. T.H. Kunz), pp. 369-397. Plenum, New York.

Marshall, A.G. (1983). Bats, flowers and fruit: evolutionary relationships in the Old World. *Biol. J. Linn. Soc.* 20: 115-135.

Marshall, J.T. (1977a). Family Muridae. In *Mammals of Thailand* (ed. B. Lekagul and J.A. McNeely), pp. 397-494. Association for the Conservation of Wildlife, Bangkok. (12.8.21)

Marshall, J.T. (1977b). A synopsis of Asian species of *Mus* (Rodentia, Murridae). *Bull. Amer. Mus. nat. Hist.* 158: 173-220. (12.8.82)

Martosubroto, P. (1979). Sumbangan hutan mangrove terhadap perikanan. In *Prosiding Seminar Ekosistem Hutan* (ed. S. Soemodihardjo, A. Nontji, and A. Djamali), pp. 109-113. Lembaga Oseanologi National, Jakarta. (12.9.2)

Martosubroto, P. and Naamin, N. (1977). Relationship between tidal forests (mangroves) and commercial shrimp production in Indonesia. *Mar. Res. Indonesia* 18: 81-86. (2.2.7)

Mathias, J.A. (1977). The effect of oil on seedlings of the pioneer Avicennia intermedia in Malaysia. *Mar. Res. Indonesia* 18: 17. (25.5.33)

Matsumoto, T. (1976). The role of termites in an equatorial rain forest ecosystem of West Malaysia I. Population density, biomass, carbon, nitrogen, and calorific content and respiration rate. *Oecologia* 22: 153-178.

Matsumoto, T. (1978a). The role of termites in the decomposition of leaf litter on the forest floor of Pasoh Study area. *Malay. Nat. J.* 30: 405-413. (12.4.45).

Matsumoto, T. (1978b). Population density, biomass, nitrogen and carbon content, energy value and respiration rate of four species of termites in Pasoh forest Reserve. *Malay. Nat. J.* 30: 335-351. (12.4.44).

McClure, H.E. (1961). Garden birds in Kuala Lumpur, Malaya. *Malay. Nat. J.* 15: 111-135. (12.7.36)

McClure, H.E. (1965). Microcosms of Batu Caves. *Malay. Nat. J.* 19: 65-74. (12.4.7)

McClure, H.E. (1966). Flowering, fruiting and animals in the canopy of a tropical rain forest. Malay. *Forester* 29: 182-203. (12.10.116)

McClure, H.E. (1967). The composition of mixed-species flocks in lowland and submontane forests of Malaya. *Wilson Bull.* 79: 131-154. (12.7.47).

McClure, H.E. (1978). Some arthropods of the dipterocarp forest in Malaya. *Malay, Nat. J.* 32: 31-51. (12.4.43).

McClure, H.E., Lim B.L., and Winn, S.E. (1967). Fauna of the Dark Cave, Batu Caves, Kuala Lumpur, Malaysia. Pac. *Insects* 9: 399-428. (12.4.10)

McComb, A.J., Cambridge, M.L., Kirkman, H. and Kuo, J. (1981). The biology of Australian seagrasses. In *The Biology of Australian Plants* (ed. J.S. Pate and A.J. McComb), pp. 258-293. University of Western Australia Press, Perth.

McConnell, W.J. (1968). Limnological effects of organic extracts of litter in a southwestern impoundment. *Limnol. Oceanogr.* 13: 343-349. (3.1.15)

McCoy, E.D. (1982). The application of island-biogeographic theory to forest tracts: Problems in the determination of turnover rates. *Biol, Conserv.* 22: 217-227. (12.13.27)

McIntyre, A.D. (1968). The microfauna and macrofauna of some tropical beaches. *J. Zool., Lond.* 156: 377-392. (2.4.34)

McKey, D. (1978). Soils, vegetation, and seed-eating by black colobus monkeys. In *The Ecology of Arboreal Folivores* (ed. G. Montgomery), pp. 423-437. Smithsonian Institution Press, Washington, D.C. (12.8.27)

McKey, D. Waterman, P.G., Mbi, C.N., Gartlan, J.S., and Struhsaker, T.T. (1978). Phenolic content of vegetation in two African rain forests: ecological implications. *Science* 202: 61-64. (12.1.34)

McKinnon, E.E. (1974). A brief note on the current state of certain of the kitchen middens of East Sumatra. *Sumatra Res. Bull.* 4: 45-50.

McKinnon, E.E. (1982). A brief note on Muara Kumpeh Hilir: An early port site on the Batang Hari? *SPAFA Digest* 3: 37-40. (36.6.18)

McLay, C.L. (1970). A theory concerning the distance travelled by animals entering the drift of a stream. *J. Fish. Res. Board Can.* 27: 359-370. (3.2.20)

McNeely, J.A. (1978). Dynamics of extinction in Southeast Asia. *BIOTROP Spec. Publ.* 8: 137-158. (12.11.2)

McVean, D.N. (1974). Mountain climates of the south-west Pacific. In *Altitudinal Zonation of Forests in Malesia.* (ed. J.R. Flenley), pp. 47-57. University of Hull, Hull. (1.1.2)

Medway, Lord (1957). Birds' nest collecting. *Sarawak Mus. J.* 8: 252-260. (9.2.14)

Medway, Lord (1962). The swiftlets (*Collocalia*) of Niah Cave, Sarawak. *Ibid* 104: 228-245. (12.7.9)

Medway, Lord (1966a). Animal remains from Lobang Angus. *Sarawak Mus. J.* 14: 184-216.

Medway, Lord (1966b). Field characters as a guide to the specific relations of swiftlets. *Proc. Linn. Soc. Lond.* 177: 151-172. (12.7.4)

Medway, Lord (1966c). The monkeys of Sundaland: ecology and systematics of the cercopithecids of a humid equatorial enviroment. In *Old World Monkeys: Evolotion, Systematics and Behaviour* (ed. J.R., and P.H. Napier), pp. 513-553. Academic Press, London.

Medway, Lord (1967). A bat-eating bat, *Mag-*

aderma lyra Geoffrey. *Malay. Nat. J.* 20: 107-110. (12.8.26)

Medway, Lord (1968). Field characters as a guide to the specific relations of swiftlets. *Proc. Linn. Soc. Lond.* 177: 151-172 (12.7.4)

Medway, Lord (1969a). Studies on the biology of the edible nest swiftlets of South-east Asia. *Malay. Nat. J.* 22: 57-63. (12.7.10)

Medway, Lord (1969b). The diurnal activity cycle among forest birds at Ulu Gombak. *Malay. Nat. J.* 22: 184-186. (12.7.29)

Medway, Lord (1969c). *The Wild Mammals of Malaya.* Oxford University Press, Oxford. (12.8.98)

Medway, Lord (1971). The importance of Taman Negara in the conservation of mammals. *Malay. Nat. J.* 24: 212-214.

Medway, Lord (1972a). Phenology of a tropical rain forest in Malaya. *Biol. J. Linn. Soc.* 4: 117-146. (12.10.81).

Medway, Lord (1972b). The Gunong Benom Expedition 1967. 6. The distribution and altitudinal zonation of birds and mammals on Gunong Benom. *Bull. Brit. Mus. Nat. Hist. (Zool.)* 23: 105-154. (12.13.6)

Medway, Lord (1972c). The Quaternary mammals of Malesia. In *The Quaternary Era in Malasia* (ed. P. Ashton and M. Ashton), pp. 63-98. Univ. Hull, Hull. (23.6.15)

Medway, Lord (1973a). The antiquity of domesticated pigs in Sarawak. *J. Malay. Br. R. Asiatic Soc.* 46: 169-178.

Medway, Lord (1973b). A ringing study of migratory barn swallows in West Malaysia. *Ibis* 115: 60-86. (12.7.43)

Medway, Lord (1974a). Food of a tapir, *Tapirus indicus.* Malay. *Nat. J.* 28: 90-93. (12.8.52).

Medway, Lord (1974b). Migratory birds. In *Birds of the Malay Peninsula* (by Lord Medway and D.R. Wells), pp. 35-55. Witherby, London. (12.7.42)

Medway, Lord (1977a). The ancient domestic dogs of Malaysia. *J. Malay. Br. Royal Asiatic Soc.* 50: 14-27. (12.13.37)

Medway, Lord (1977b). The Mammals of Borneo. *Monogr. Malay. Br. Royal As. Soc.* No. 7: (12.8.23)

Medway, Lord (1979). The Niah excavations and an assessment of the impact of early man on mammals in Borneo. *Asian Perspectives* 20: 51-69. (23.6.5)

Medway, Lord (1980). Tropical forest as a source of animal genetic resouces. *BioIndonesia* 7: 55-63. (12.10.80).

Medway, Lord and Wells, D.R. (1971). Diversity and density of birds and mammals at Kuala Lompat, Pahang, Malay. *Nat. J.* 24: 238-247. (12.13.19)

Meijer, W. (1958). A contribution to the taxonomy of *Rafflesia arnoldi* in West Sumatra. *Ann. Bogorienses* 3: 33-44 (12.1.108)

Meijer, W. (1970). Regeneration of tropical lowland forest in Sabah, Malaysia, forty years after logging. *Malay. Forester* 33: 204-229. (9.1.30)

Meijer, W. (1974). *Field Guide for the Trees of West Malesia.* Univ. Kentucky Press.

Meijer, W. (1981). Sumatra as seen by a botanist. *Indonesia Circle* 25: 17-27. (10.12.89)

Meijer, W. (1982a). Plant refuges in the Indo-Malesia Region. In *Biological Diversification in the Tropics* (ed. G. Prance), pp. 576-584. Columbia University Press, New York. (23.6.10).

Meijer, W. (1982b). *Provisional checklist of the flora of Mt. Sago near Payakumbuh - West Sumatra.* University of Kentucky, Kentucky. (12.10.79).

Meijer, W. (1982c). *Checklist of the flora of Mt. Sago, West Sumatra.* Unpubl. Report, University of Kentucky. (12.10.33)

Meijer, W. (1983). *Tentative key to the species of* Rafflesia. Unpubl., University of Kentucky, Kentucky. (12.1.111).

Meijer, W. and Vangala, S. (1983). *A bibliography of literature on Rafflesiaceae.* University of Kentucky, Kentucky. (27.1.17).

Meijer, W. and Withington, W.A. (1981). A map of vegetation and land use in Sumatra. *Indonesia Circle* 24: 29-37. (10.4.12)

Mendoza, V.B. (1978). Some ecological considerations in the rehabilitation of Philippine grassland areas. *Philippine Lumberman* 24: 8-11. (10.3.12)

Merrill, E.D. (1940). Botanical results of the GeorgeVanderbilt Sumatra Expedition, 1939. Plants from Mt. Leuser. *Notula Naturae* 47: 9 (12.10.42)

Merton, F. (1962). A visit to the Tasek Bera. *Malay Nat. J.* 16: 103-110. (12.10.17)

Miksic, J. (1979). *Archaeology, Trade and Society in Northeast Sumatra.* PhD thesis, Cornell University. (23.7.9)

Miller, C.S. Jr. (1942). Zoological results of the George Vanderbilt Sumatran Expedition, 1936-1939. Part V - Mammals colected by Frederick A. Ulmer Jr. on Sumatran and Nias. *Proc. Acad. Nat. Sci. Phil.* 94: 107-165. (12.8.64)

Milner, A.C., MacKinnon, E.E. and Luckman, S. (1978). Aru and Kota Cina. *Indonesia* 26: 1-42. (23.6.23)

Mitchell, A.H. (1981). *Report on an survey of Pulau Simeulue, Aceh, with a proposal for a Suaka Margasatwa.* World Wildlife Fund, Bogor. (10.1.12)

Mitchell, B.A. (1963). Forestry and tanah beris. *Malay. Forester* 26: 160-170. (9.1.19)

Mizuno, T. and Mori, S. (1970). Preliminary hydrobiological survey of some southeast Asian inland waters. *Biol. J. Linn. Soc.* 2: 77-117. (3.1.13)

Mochsin, Moh. A.K. (1980). Ecology and morphology of the freshwater fishes of Selangor. Part 1. Cyprinoid fishes of the subfamilies Abraminae, Rusborinae and Garrinae and Family Homalopteridae and Cobitidae. *Malay. Nat. J.* 34: 73-100. (12.5.23)

Mohr, E.C.J. and van Baren, F.A. (1954). *Tropical Soils, Mouton,* The Hague. (23.4.6)

Mohsin Moh. A.K. and Law, A.T. (1980). Evironmental studies of Kelang River. II. Effects on fish. *Malay. Nat. J.* 33: 189-199. (3.2.25)

Mohsin, Moh. A.K. and Ambak, Moh. Azmi. (1982). Cyprinoid fishes of the subfamily Cyprinidae in Selangor. *Malay. Nat. J.* 35: 29-35. (12.5.25)

Mohsin, Moh. A.K. and Law, A.T. (1980). Environmental studies of Kelang River. II. Effects on fish. *Malay. Nat. J.* 33: 189-199. (3.2.25)

Morley R.J. (1980). Changes in dry-land vegetation in the Kerinci area of Sumatra during the late Quaternary period. *Proc. IV Intern. Palynol. Conf., Lucknow* (1976-77) 3: 2-10.

Morley, R.J. (1981). Development and vegetation dynamics of a lowland ombrogenous peat swamp in Kalimantan Tengah, Indonesia. *J. Biogeog.* 8: 383-404.

Morley, R.J. (1982). A palaeoecological interpretation of a 10,000 year pollen record from Danau Padang, central Sumatra, Indonesia. *J. Biogeog.* 9: 151-190.

Morley, R.J., Flenley, J.R. and Kardin, M.K. (1973). Preliminary notes on the stratigraphy and vegetation of the swamps and small lakes of the central Sumatran highlands. *Sumatra Res. Bull.* 2: 50-60.

Morley, R.J., Flenley, J.R., and Kardin, M.K. (1973). Preliminary notes on the swamps and small lakes of the central Sumatran highlands. *Sumatra Res. Bull.* 2: 50-60.

Morton, E.S. (1978). Avian arboreal folivores: Why not? In *The Ecology of Arboreal Folivores.* (ed. G.G. Montgomery), pp. 123-130. Smithsonian Institution, Washington, D.C. (12.7.31).

Moss, B. (1980). *Ecology of freshwaters.* Blackwell, Oxford.

Moyle, P.B. and Senanayake, F.R. (1984). Resource partitioning among the fishes of rainforest streams in Sri Lanka. *J. Zool.* 202: 195-223.

Mukhtar, A. and Sonoda, S. (1979). Ecologi elasmobranchii air tawar 1. Tentang ikan pari S. Indragiri, Riau. *Terubuk* 5: 2-11. (12.5.22)

Muller, J. (1972). Palynological evidence for change in geomorphology, climate and vegetation in the Mio-Pliocene of Malesia. In *The Quaternary Era in Malesia* (ed. P. and A. Ashton), pp. 6-34. University of Hull, Hull. (23.6.15)

Musper, K.A.F.R. (1934). Een bezoek aan de grot Soeroeman Besar in het Goemaigebergte (Palembang, Zuid Sumatra). *Tijds. Kon. Ned. Aardr. Gen.* 51: 521-531.

Musser, G.G. (1979). Results of the Archbold Expeditions. Part 102. The species of *Chiropodomys* arboreal mice of Indochina and the Malay Archipelago. *Bull. Am. Mus. Nat. Hist.* 162: 377-446.

Musser, G.G. and Newcomb, C. (1983). Malaysian murids and the giant rat of Sumatra. *Bull. Amer. Mus. nat. Hist.* 174: 327-598. (12.8.107)

Musser, G.G., Marshall, J.T., and Boedi (1979). Definiton and contents of the Sundaic genus *Maxomys* (Rodentia, Muridae). *J. Mammal.* 60: 592-606. (12.8.28)

Muul, I., and Lim, B.L. (1978). Comparative morphology, food habits and ecology of some Malaysian arboreal rodents. In *The Ecology of Arboreal Folivores.* (ed. G.G.Montgomery), pp. 361-368. Smithsonian Institution, Washington, D.C. (12.8.59).

Myers, M.L. and Shelton, R.L. (1980). *Survey Methods for Ecosystems Management.* Wiley, New York. (19.5.1)

Myers, N.M. (1979). *The Sinking Ark: A New Look at the Problem of Disappearing Species.* Pergamon, New York. (12.12.8)

Myers, N.M. (1980). Deforestation in the tropics: Who gains who loses? In *Where Have All the Flowers Gone? Deforestation in the Third World,* pp. 25-44. College of William and Mary, Williamsburg. (10.4.17)

Myllymaki, A. (1979). Importance of small mammals as pest in agriculture and stored products. In *Ecology of Small Mammals* (ed. E.M. Stoddart), pp. 239-279. Chapman and Hall, London. (11.1.14)

Nadkarni, N.M. (1981). Canopy roots: convergent evolution in rain forest nutrient cycle. *Science* 214: 1023-1024. (12.10.82).

Ng, F.S.P. (1980). Germination ecology of Malaysian woody plants. *Malay. Forester* 43: 406-437.

Ng, F.S.P. (1983). Ecological principles of tropical lowland rain forest conservation.

In *The Tropical Forest* (ed. S.L. Sutton, T.C. Whitmore and A.C. Chadwick), pp. 359-375, Blackwell, Oxford.

Niemitz, C. (1979). Outline of the behaviour of *Tarsius bancanus*. In *The Study of Prosimian Behavior* (ed. G.A. Doyle and R.D. Martin), pp. 631-660, Academic, New York. (12.8.62).

Ninkovich, D., Shackleton, N.J., Abdel-Monem, A.A., Obradevich, J.D. and Izett, G. (1978). K-Ar age of the later Pleistocene eruption of Toba, North Sumatra. *Nature, Lond.* 276: 574-577.

Nisbet, I.C.T. (1968). The utilization of mangrove by Malayan birds. *Ibis.* 110: 345-352. (12.7.3)

Nisbet, I.C.T. (1974). The eastern palearctic migration system in operation. In *Birds of the Malay Peninsula* (by Lord Medway and D.R. Wells), pp. 57-69. Witherby, London. (12.7.44)

Nixon, S.W., Furnas, B.N., Lee, V.,Marshall, N., Ong, J.E., Wong, C.H., Gong, W.K. and Sasekumar, A. (1980). The role of mangroves in the carbon and nutrient dynamics of Malaysian estuaries. Paper presented at the UNESCO Symposium on *'Mangrove Environment: Research and Management'.* Universiti Malaya, Kuala Lumpur. (2.4.19)

Nontji, A. and Setiapermana, D. (1980). Phytoplankton productivity and related aspects of the Pari Islands Lagoons. Paper given at 'Symposium on Recent Research Activities on Coral Reefs in South-east Asia', BIOTROP, 6-9 May, 1980. (12.2.15)

Nordin, C.F. and Meade, R.H. (1982). Deforestation and increased flooding of the upper Amazon. (Followed by reply by A.H. Gentry and J. Lopez-Parodi). *Science* 215: 426-427. (10.4.19)

Nur, N. (1976). Studies on pollination in Musaceae. *Ann. Bot.* 40: 167-177. (12.1.25)

Nursall, J.R. (1981). Behaviour and habitat affecting the distribution of five species of sympatric mudskipper in Queensland. *Bull. Mar. Sci.* 31: 730-735. (12.5.16)

Oates, J.F., Swain, T. and Zantovska, J. (1977). Secondary compounds and food selection by colobus monkeys. *Biochem. Syst. Ecol.* 5: 317-321. (12.8.104).

Oates, J.F., Waterman, P.G. and Choo, G.M. (1980). Food selection by the South Indian leaf-monkey, *Prebytis johnii,* in relation to leaf chemistry. *Oecologia* 45: 45-56. (12.8.89).

Obdeyn, V. (1941). Zuid Sumatra volgens de oudste berichten. I. de geomorfologische gesteldheid van Zuid Sumatra in verband met de opvatting der ouden. *Tijd. K. ned. Aardrijksk.* 58: 190-216. (2.4.40)

Obdeyn, V. (1942). De oude zeehandelsweg door de straat van Malaka in verband mat de geomorfologie der selateilanden. *Tijd. K. ned. Aardrijksk.* 59: 742-770 (2.4.41).

Odum, W.E. (1976). *Ecological Guidelines for Tropical Coastal Development.* IUCN Publ. N.S. No. 42, IUCN, Gland. (2.4.29)

Ogawa, H. (1978). Litter production and carbon cycling in Pasoh Forest. *Malay. Nat. J.* 30: 367-373. (12.10.106).

Ogawa, H., Yoda. K., Kira. T., Ogino, K., Shidei, T., Ratanawongse, D. and Apasutaya, C. (1965). Comparative ecological studies on three main types of forest vegetation in Thailand. II. Plant biomass. *Nature Life in S.E. Asia* 4: 49-80. (Referred to in Whitmore, 1975).

Ogilvie, G.H. (1929). Bison eating bark. *J. Bombay nat. Hist. Soc.* 33: 706-707 (12.8.94).

Ohsawa, M. (1979). Tentative report of forest ecological survey in Mt. Kerinci, Central Sumatra. Unpubl. Report, BIOTROP, Bogor. (12.10.102)

Oka, I.N. (1979). Cultural control of the brown planthopper. In *Brown Planthopper: Threat to rice production in Asia.* pp. 357-369. International Rice Research Institute, Los Banos.

Oka, I.N. (1980). Feeding population of people versus population of insects: The example of Indonesia and the rice brown planthopper. *BioIndonesia* 7: 121-134. (11.1.10)

Okuma, D. and Kisimoto, R. (1981). Airborne spiders collected over the East China Sea. *Jap. J. appl. Ent. Zool.* 25: 296-298. (12.4.49)

Oldeman, L.R., Las, I. and Darwis, S.N. (1979). An Agroclimatic Map of Sumatra. *Contr. Centr. Res. Inst. Agric. Bogor* 52: 1-35. (1.1.1)

Oldfield, M.L. (1981). Tropical deforestation and genetic resources conservation. In *Blowing in the Wind: Deforestation and Long-range Implications,* pp. 277-345. College of William and Mary, Williamsburg. (10.4.16)

Olivier, R.C.D. (1978a). *On the Ecology of the Asian Elephant* Elephas maximus *with particular reference to Malaya and Sri Langka.* Ph.D. thesis, University of Cambridge, Cambridge. (12.8.55).

Olivier, R.C.D. (1978b). Distribution and status of the Asian elephant. *Oryx* 14: 379-424.

Olson, S.L. and James, H.F. (1982). Fossil birds from the Hawaiian Islands: Evidence for wholesale extinction by Man before Western contact. *Science* 217: 633-635. (23.6.1)

Ondara (1969). *Laporan survey perikanan di Danau Toba.* Laporan Penelitian No. 3, Lembaga Penelitian Perikanan Darat, Palembang. (3.1.2)

Ong, J.E., Gong. W.K., and Wong, C.H. (1980a). Ecological Survey of the Sungai Merbok Estuarine Mangrove Ecosystem. Universiti Sains Malaysia, Penang. (12.9.21)

Ong. J.E., Gong, W.K., Wong, C.H. and Dhanarajan, G. (1980b). Contribution of aquatic productivity in a managed mangrove ecosystem in Malaysia. Paper presented at the UNESCO Symposium on *'Mangrove Environment: Research and Management'.* University Malaya, Kuala Lumpur. (12.9.20)

Palm, B.T. (1934). A *Mitrastemon* Rafflesiaceae from Sumatra. *Acta Horti. Gotoburgensis.* 9: 147-152.

Palmieri, M.D., Palmieri, J.R. and Sullivan, J.T. (1980). A chemical analysis of the habitat of nine commonly occurring Malaysian freshwater snails. *Malay. Nat. J.* 34: 39-45. (12.3.28)

Parkhurst, D.F. and Loucks, O.L. (1972). Optimal leaf size in relation to environment. *J. Ecol.* 60: 505-537. (12.1.116)

Parrish, J.D. (1980). Effects of exploitation upon reef and lagoon communities. In Marine and Coastal Processes in the Pacific: Ecological Aspects of Coastal Zone Management. UNESCO, Jakarta. (2.4.33)

Partomihardjo, T. (1982). Sepuluh hari di Anak Krakatau. *Suara Alam* No. 15: 32-33. (23.3.11)

Pattee, E., Lascombe, C. and Delolme, R. (1973). Effects of temperature on the distribution of turbellarian triclads. In *Effects of Temperature on Ectothermic Organisms* (ed. W. Wieser), pp. 201-207. Springer-Verlag, Berlin. (12.2.14)

Payne, J.B. (1979). *Synecology of Malayan tree squirrels with special reference to the genus* Ratufa. PhD thesis, University of Cambridge, Cambridge. (12.8.56)

Payne, J.B. (1980). *Competitors. In Malayan Forest Primates: Ten Years' Study in Tropical Rain Forest* (ed. D.J. Chivers), pp. 261-277. Plenum, New York. (12.8.25)

Pearce, W.J.H. and MacDonald, F.D. (1981). An investigation of the leaf anatomy, foliar mineral levels, and water relations of trees of a Sarawak forest. *Biotropica* 13: 100-109. (12.1.33)

Pearson, D.L. (1982). Historical factors and bird species richness. In *Biological Diversification in the Tropics.* (ed. G. Prance), pp. 441-452. Columbia University Press, New York. (12.7.39)

Peck, S.B. (1981). A new cave-inhabiting *Ptomaphaginus* beetle from Sarawak (Leiodidae: Cholevinae). *Syst. Entomol.* 6: 221-224. (12.4.12)

Peckarsky, B.L. (1979). Biological Interactions as determinations of distribution of benthic invertebrates within the substrate of stony streams. *Limnol. Oceanogr.* 24: 59-68. (3.2.24)

Pendelbury, H.M. (1936). An expedition to Kerinchi Peak, Sumatra, Carried out in 1914 by Messrs. H.C. Robinson and C. Boden Kloss. *J. Fed. Malay States Mus.* 8: 1-31.

Penny, D.H. and Ginting, M. (1980). House gardens - the last resort? Further economic arithmatic from Sriharjo. In *Indonesia: Dualism, Growth and Poverty* (ed. R.G. Garnaut and P.T. McCawley), pp. 487-499. Australian National University, Canberra. (15.2.5)

Petersen, R.C. and Cummins, K.W. (1974). Leaf processing in a woodland stream. *Freshwat. Biol.* 4: 343-368.

Phillipson, J. (1966). *Ecological Energetics.* Studies in Biology No. 1. Edward Arnold, London. (23.2.8)

Pianka, E.R, Huaey, R.B. and Lawlor, L.R. (1979). Niche segregation in desert lizards. In *Analysis of Ecological Systems* (ed. D.J. Horn, G.R. Stairs and R.D. Mitchell), pp. 67-115. Ohio State University Press, Columbus. (12.6.5)

Pianka, E.R. and Huey, R.B. (1978). Comparative ecology, resource utilization and niche segregation among gekkonid lizards in the southern Kalahari. *Copeia* 1978: 691-701.

Pichon, M. and Morrissey, J. (1981). Benthic zonation and community structure of South Island Reef, Lizard Island (Great Barrier Reef). *Bull. Mar. Sci.* 31: 581-593. (12.3.20)

Piggott, A. (1979). *Common Epiphytic Ferns of Malaysia and Singapore.* Heinemann, Singapore. (12.1.105)

Pitelka, F.A., Tomich, P.Q. and Treichel, G.W. (1955). Ecological relations of jaegers and owls as lemming predators near Barrow, Alaska. *Ecol. Monogr.* 25: 85-117.

Piyakarnchana, T. (1981). Severe defoliation of *Avicennia alba* Bl. by larvae of *Cleora injectaria* Walker. *J. Sci. Soc. Thailand* 7: 33-36.

(12.9.43)
Pocock, R.I. (1934). The monkeys of the genera *Pithecus* (or *Presbytis*) and *Pygathrix* found to the east of the Bay of Bengal. *Proc. Zool. Soc. Lond.* 1934: 895-961.
Pócs, T. (1982). Tropical forest bryophytes. In *Bryophyte Ecology* (ed. A.J.F. Smith), pp. 59-104. Chapman and Hall. London, (12.10.83)
Polak, E. (1933). Uber Torf und Moor in Niederlandisch Indian. *Verh. Akad. Wet.* 30: 1-85. (23.4.11)
Polunin, N.V.C. (1980). Guidelines for a BIOTROP project on reef fishes and fisheries. Paper given at the *'Symposium on Recent Research Activity on Coral Reefs in South-east Asia'*, BIOTROP, 6-9 May, 1980.
Poniran, K. (1974) Elephants in Aceh. *Oryx* 12: 576-580.
Poore, M.E.D. (1968). Studies in Malaysian rain forest. I. The forest on triassic sediments in Jengka Forest Reserve. *J. Ecol.* 56: 143-196. (12.10.114)
Porter, J.W., Porter, K.G. and Batac-Catalan, Z. (1977). Quantitative sampling of Indo-Pacific demersal reef plankton. *Proc. Third Intnl. Coral Reef Symp.* 1: 105-112. Referred to in Parrish 1980.
Praseno, D.P. and Sukarno (1977). *Mar. Res. Indonesia* 17: 59-68.
Prendergast, H.D.V. (1982). Pollination of *Hibiscus rosa-sinensis. Biotropica* 14: 287. (12.1.76)
Primack, R.B. and Tomlinson, P.B. (1978). Sugar secretions from the buds of *Rhizophora. Biotropica* 10: 74-75. (12.9.52)
Proctor, J., Anderson, J.M., Chai, P., and Vallack, H.W. (1983b). Ecological studies in four contrasting rain forests in Gunung Mulu National Park, Sarawak. I. Forest environment, structure and floristics. *J. Ecol.* 71: 237-260. (12.10.22).
Proctor, J., Anderson, J.M., Fogden, S.C.L. and Vallack, H.W. (1983a). Ecological studies in four contrasting lowland rain forests in Gunung Mulu National Park, Sarawak. II. Litterfall, litter standing crop and preliminary observations on herbivory. *J. Ecol.* 71: 261-283. (12.10.21)
Prowse, G.A. (1957). An introduction to the desmids of Malaya. *Malay. Nat. J.* 11: 42-58. (12.2.9)
Prowse, G.A. (1958a). The Euglenidae of Malaya. *Gdns' Bull. Sing.* 16: 136-204.
Prowse, G.A. (1958b). Fish and food chains. *Malay. Nat. J.* 12: 66-71. (3.4.5)
Prowse, G.A. (1962a). Further Malayan freshwater Flagellata. *Gdns' Bull. Sing.* 19: 105-145. (12.2.20)
Prowse, G.A. (1962b). Diatoms of Malayan freshwaters. *Gdns' Bull. Sing.* 19: 1-104. (12.2.21)
Prowse, G.A. (1964). Some limnological problems in tropical fish ponds, *Verb. Internat. Verein. Limnol.* 15: 480-484. (3.4.6)
Prowse, G.A. (1968). Pollution in Malayan waters. *Malay. Nat. J.* 21: 149-158. (25.1.16)
PSL-IPB (1977). *Laporan Akhir.* Penelitian daerah Upang-Banyuasin.
Pugh, D.T. and Rayner, R.F. (1981). The tidal regimes of three Indian ocean atolls and some ecological implications. *Estuar. Coast Shelf Sci.* 13: 389-407. (12.3.34)
Purchon, R.D. and Enoch, I. (1954). Zonation of the marine fauna and flora on a rocky shore near Singapore. *Bull. Raffles Mus. S'pore* No. 25. (2.4.38)
Putz, F.E. (1980). Lianas vs. trees. *Biotropica* 12: 224-225. (12.10.84)
Rabinowitz, D. (1978). Early growth of mangrove seedlings in Panama and an hypothesis concerning the relationship of dispersal and zonation. *J. Biogeogr.* 5: 113-133. (12.9.46)
Raemaekers, J.J. (1978). The sharing of food sources between two gibbon species in the wild. *Malay. Nat. J.* 31: 181-188. (12.8.53)
Raemaekers, J.J. (1979). Ecology of sympatric gibbons. *Folia primatol.* 31: 227-245. (12.8.54)
Raemaekers, J.J., Aldrich-Blake, F.P.G., and Payne, J.B. (1980). The forest. In *Malayan Forest Primates: Ten Years' Study in Tropical Rain Forest* (ed. D.J. Chivers), pp. 29-61. Plenum, New York. (12.8.25)
Rakoen, M.P. (1955). Beberapa pendapat mengenai daerah hutan Bengkalis/Kampar. *Rimba Indonesia.* 4: 223-248. (9.1.18)
Rand, A.S (1978). Reptilian arboreal folivores. In *The Ecology of Arboreal Folivores* (ed. G.G. Montgomery), pp. 115-122. Smithsonian Institution, Washington, D.C. (12.6.17)
Rankin, J.M. (1978). *The influence of seed predation and plant competition on tree species abundances in two adjacent tropical rain forest communities in Trinidad, West Indies.* PhD thesis, University of Michigan, Michigan.
Rao, P.S. Jacob, C.M. and Ramasastri, B.V. (1969). Nutritive value of bamboo seeds. *J. Nutr. Diet.* 6: 192. (12.1.119)
Rappart, F.W. (1936). Een schildpaddenstrand in het wildreservaat Zuid Sumatra. *Trop. Natuur* 25: 124-126.
Ratnasabapathy, M. (1974). *Acacia auriculae-*

formis and *Casuarina equisetifolia* - the urban invaders. *Malay. Nat. J.* 28: 18-21. (12.1.78)

Relevante, N. and Gilmartin, M. (1982). Dynamics of phytoplankton in Great Barrier Reef Lagoon. *J. Plank. Res.* 4: 47-75. (12.2.16)

Ribi, G. (1981). Does the wood boring isopod *Sphaeroma terbrans* benefit red mangroves (*Rhizophora mangle*)? *Bull. Mar.Sci.* 31: 925-928. (12.9.4)

Ribi, G. (1982). Differential colonization of roots of *Rhizophora mangle* by the wood-boring isopod *Sphaeroma terebrans* as a mechanism to increase root density. *Mar. Ecol.* 3: 13-19. (12.9.42)

Richards, D.G. and Wiley, R.H. (1980). Reverberations and amplitude fluctuations in the propagation of sound in a forest: implications for animal communication. *Am. Nat.* 115: 381-399. (12.13.33)

Richards, K. (1982). Introduction. In *The Krakatoa Centenary Expedition* (ed. J.R. Flenley and K. Richards), pp. 1-8. University of Hull, Hull. (12.1.92)

Richards, P.W. (1941). Tropical forests podzols. *Nature*, Lond. 148: 129-131

Richards, P.W. (1952). *The Tropical Rain Forest: An Ecological Study*. Cambridge Univ. Press, Cambridge.

Richards, P.W. (1983). The three-dimensional structure of tropical rain forest. In *Tropical Rain Forest: Ecology and Management* (ed. S.L. Sutton, T.C. Whitmore and A.C. Chadwick), pp. 287-309 Blackwell. Oxford.

Ricklefs, R.E. (1977). Environmental heterogeneity and plant species diversity: a hypothesis. *Amer. Nat.* 111: 376-381. (12.10.85)

Ricklefs, R.E. (1979). *Ecology*. Chiron, London. (23.2.1)

Rijksen, H.D. (1978). *A Field Study on Sumatran Orang Utans (*Pongo pygmaeus abellii *Lesson 1827). Ecology, Behaviour and Conservation.* Veenman and Zonen, Wageningen. (12.8.88)

Riswan, S. (1981). Natural regeneration in lowland tropical forest in East Kalimantan, Indonesia (with refence to kerangas forest.) *BIOTROP Spec. Publ.* 13: 145-152. (10.4.18)

Riswan, S. (1982). *The Ecology of heath forest (kerangas) in East Kalimantan, Indonesia.* PhD thesis, University of Aberdeen, Aberdeen.

Robbins, R.G. and Wyatt-Smith, J. (1964). Dry land forest formations and forest types in the Malay Peninsular. *Malay. Forester* 27: 188-217. (12.10.48)

Robertson, J.M.Y. (1982). Behavioural ecology of *Macaca n. nemestrina*. Final progress report to LIPI. (12.8.90)

Robertson, J.M.Y. and Soetrisno, B.R. (1981). Penebangan hutan, lerengan, erosi, bendungan dan kematian di Lembah Alas. Unpubl. (10.4.14)

Robertson, J.M.Y. and Soetrisno, B.R. (1982). Logging on slopes kills. *Oryx* 16: 229-230. (8.1.2)

Robinson, H.C. and Kloss, C.B. (1918). Results of an expedition to Kerinchi Peak, Sumatra, l. Mammals. *J. Fed. Malay. St. Mus.* 8: 1-80. (12.8.20).

Rollet, B. (1981). Bibliography on Mangrove Research 1600-1975. UNESCO, Paris. (27.1.14)

Rosen, B.R. (1983). The recovery of the corals of the Krakatoa region. Paper given at the symposium *'Krakatoa: 100 Years On'*, British Museum (Natural History), 21 September 1983.

Roth, L.M. (1980). Cave dwelling cockroaches from Sarawak, with one new species. *Syst. Entomol.* 5: 97-104. (12.4.13)

Ruttner, F. (1931). Hydrographische und hydrochemische Beobachtungen auf Java, Sumatra und Bali. *Arch. Hydrobiol. Suppl.* 5: 197-454. (3.1.11)

Ruttner, F. (1943). Beobachtungan iiber die tagliche Verbikal wanderung des Planktons in tropischen Seen. *Arch. Hydrobiol.* 40: 474-492. (12.2.14)

Saanin, H. (1968). *Taksonomi dan Kuntji Identifikasi Ikan.* I and II vols. Binatjipta, Jakarta. (12.5.1 and 12.5.2)

Sabar, F., Djajasasmita, M., and Budiman, A. (1979). Susunan dan penyebaran moluska dan krustasea pada beberapa hutan rawa payau: suatu studi pendahuluan. In Prosiding Seminar *'Ekosistem Hutan Mangrove'* (ed. S. Soemodiharjo, A. Nontji, A. Djamali), pp. 120-125. Lembaga Oseanologi Nasional, Jakarta. (12.9.2)

Saenger, P., Hegerl, E.J. and Davie, J.D.S. (1981). *First Report the Global Status of Mangrove Ecosystems.* Commission on Ecology, IUCN, Gland. (12.9.23)

Sakai, K.I., Budhiyono, B.E. and Iyama, S. (1979). Studies on pioneer trees and the pioneer index in tropical forest. Unpubl. report, BIOTROP, Bogor. (12.10.79)

Sale, P.F. and Douglas, W.A. (1981). Precision and accuracy of visual census technique for fish assemblages on coral patch reefs. *Environ. Biol. Fish.* 6: 333-339. (12.3.25)

Salick, J. (1983). Natural history of crop-related wildspecies: Uses in pest habitat management. *Env. Manage.* 7: 85-90.

(11.1.15)

Samingan, M.T. (1980). Notes on the vegetation of the tidal areas of South Sumatra, Indonesia, with special reference to Karang Agung. In *Tropical Ecology and Development* (ed. J. Furtado), pp. 1107-1112. University Malaya, Kuala Lumpur.

Sani, S. (1976). Notes on dustfall in and around Batu Caves, Selangor. *Malay. Nat. J.* 29: 168-175. (25.7.10)

Saravanamuthu, J. and Lim, R.P. (1982). A preliminary limnological survey of a eutrophic pond. Taman Jaya Pond, Petaling Jaya. *Malay. Nat. J.* 35: 83-97. (3.1.14)

Sardar, Z. (1980). Quarrying blasts Malaysia's underground past. *New Scient.* 88: 352. (5.1.3)

Sasekumar, A. (1974). Distribution of macrofauna on a Malayan mangrove shore. *J. Anim. Ecol.* 43: 51-69. (12.9.16)

Sastrapardja, S., Soenartono, A., Kartawinata, K. and Terumingkeng, R.C. (1980). The conservation of forest animal and plant genetic resouces. *BioIndonesia* 7: 1-42. (12.13.10).

Sastrapradja, S. (1979). *Tanaman Pekarangan.* Lembaga Biologi Nasional, Bogor. (11.4.7)

Sastrapradja, S. (ed.). (1977). *Sumber Daya Hayati Indonesia.* Lembaga Biologi Nasional, Bogor. (9.2.3)

Sastrapradja, S. (ed.). (1978a). *Tanaman Obat.* Lembaga Biologi Nasional, Bogor. (9.2.5)

Sastrapradja, S. (ed.). (1978b). *Tanaman Industri.* Lembaga Biologi National, Bogor. (9.2.4)

Sastrapradja, S. and Kartawinata, K. (1975). Leafy vegetables in the Sundanese diet. In *South-east Asean Plant Genetic Resource* (ed. J.T. Williams), pp. 166-170. Lembaga Biologi Nasional, Bogor. (9.2.15).

Sastrapradja, S. and Rifai, M.A. (1975). Exploration and conservation of the undeveloped genetic resources in Indonesian forests. *BioIndonesia* 1: 13-83.

Sastrapradja, S., Soenartono, A., Kartawinata, K. and Tarumingkeng, R.C. (1980). The conservation of forest animal and plant genetic resources. *BioIndonesia* 7: 1-42 (12.13.10)

Sato, H. and Kondo, T. (1981). Biomass production of water hyacinth and its ability to remove inorganic minerals from water. I. Effect of the concentration of culture solution on the rates of plant growth and nutrient uptake. *Jap. J. Ecol.* 31: 257-267. (25.1.14)

Schaeffer, F.A. (1970). *Pollution and the Death of Man: The Christian View of Ecology.* Tyndale House, Wheaton. (23.2.5)

Schefold, R. (1980). *Spielzeug für die Seelen.* Museum Rietberg, Zurich. (15.1.10)

Schemske, D.W. and Brokaw, N. (1981). Treefalls and the distribution of understory birds in a tropical forest. *Ecology* 62: 938-945. (12.7.22)

Schindler, D. (1974). Eutrophication and recovery in experimental lakes: implications for lake management. *Science* 184: 897-899. (3.1.12)

Schnitger, F.M. (1938). The archaeology of Hindoo Sumatra. *Int. Archiv. v. Ethnogr.* 35 (Suppl.) 1-44. (23.7.5)

Schnitger, F.M. (1939). *Forgotten Kingdoms in Sumatra.* Brill, Leiden. (23.7.7)

Scholz, U. (1983). *The Natural Regions of Sumatra and Their Agricultural Production Pattern: A Regional Analysis. Two vols.* Central Research Institute for Food Crops, Bogor.

Schurmann, H.M.E. (1931). Kjökkenmöddinger und paläolithicum in Nord Sumatra. *Tijd. Kon. Ned. Aard. Gen.* 48: 905-923.

Schuster, W.H. (1950). Comments on the importation and transplantation of different species of fish into Indonesia. *Contr. gen. agr. Res. Stn. Bogor* 111: 1-31. (12.5.15)

Schuster, W.H. (1952). *Local Common Names of Indonesian Fishes.* W. van Hoeve, Bandung. (12.5.7)

Searle, A.G. (1956). An illustrated key to Malayan hard corals. *Malay. Nat. J.* 11: 1-28.

Seow, R.C.W. (1976). *The Effects of a Mixed Organic Effluent on the Distribution of Pelagic Macrofauna at Sungai Puloh with Species Reference to Water Quality.* BSc thesis, University of Malaya, Kuala Lumpur (2.5.14)

Setten, G.G.K (1953). The incidence of buttressing among Malayan tree species when of commercial size. *Malay. Forester* 16: 219-221. (12.1.120)

Sewandono, M. (1938). Het veengebied van Bengkalis. *Tectona* 31: 99-135. (12.10.67)

Sharma, R.E. and Fernando, C.H. (1961). Leeches and their ways. *Malay. Nat. J.* 15: 152-159. (12.3.29)

Sharma, Y.M.L. (1980). Bamboos in the Asia-Pacific region. In *Bamboo Research in Asia* (ed. G. Lessard and A. Chouinard), pp. 99-120. International Development Research Centre, Ottawa. (12.1.94)

Sheppard, C.R.C.(1981). Coral populations on reef slopes and their major controls. *Mar. Biol.* 7: 83-115. (12.3.23)

Siagian, Y.T. and Harahap, R.M.S. (1981). *Preliminary Result of Species Trial on Tin Mine Tailing Land in Dabo Singkep. Balai Penelitian Hutan,* Report No. 384, Bogor. (10.1.4)

Sinnappa, S. and Chang, E.T. (1972). Identification of rotenone in fish poisoned by derris root resin. *Malay. Agric. J.* 48: 20-24. (6.1.4)

Smith, A.C. and Briden, J.C. (1977). *Mesozoic and Cenozoic Palaeocontinental Maps.* Cambride University Press, Cambridge.

Smith, A.P (1972). Buttressing of tropical trees: a descriptive model and new hypotheses. *Am. Nat.* 106: 32-46. (12.1.122)

Smith, A.P (1979a). Buttressing of tropical trees in relation to bark thickness in Dominica, B.W.I. *Biotropica* 11: 159-160. (12.1.47)

Smith, A.P. (1979b). Function of dead leaves of *Espeletia schultzii* (Compositae), an Andean caulescent rosette species. *Biotropica* 11: 43-47. (12.1.38)

Smith, J.M.B. (1970). Herbaceous plant communities in the summit zone of mount Kinabalu. *Malay. Nat. J.* 24: 16-29. (12.1.30)

Smith, J.M.B. (1975). Notes on the distribution of herbaceous angiosperm species in the mountains of New Guinea. *J. Biogeogr.* 2: 87-101. (12.1.29)

Snedaker, S. (1982a). A perspective on Asia mangroves. In *Man, Land and Sea* (ed. C. Soysa, L.S. Chic, W.L. Collier), pp. 65-74. Agricultural Development Council, Bangkok. (12.9.45)

Snedaker, S. (1982b). Mangrove species zonation: why? In *Tasks for Vegetation Science, vol. 2* (ed. D.N. Sen and K.S. Rajpurohit). Junk, The Hague. (12.9.44)

Sobrado, A. and Medina, E. (1980). General morphology, anatomical structure, and nutrient content of schlerophyllous leaves of the 'Bana' vegetation of Amazonas. *Oecologia* 45: 341-345. (12.1.28)

Soderstrom, T.R. and Calderon, C.E. (1979). A commentary on the bamboos (Poaceae: Bambusoideae). *Biotropica* 11: 161-172. (12.1.99)

Sody, H.J.V. (1940). On the mammals of Enggano. *Treubia* 17: 391-401. (12.8.30)

Soedibja, R.S. (1952). Penyelidikan tentang tumbuh dan ekologi kaju besi (*Eusideroxylon zwageri* T. and B.) di lingkungan hutan Semandai (Palembang). *Rimba Indonesia* (1952): 215-223. (9.1.21)

Soegiarto. A. and Polunin, N.V.C. (1980). *The Marine Environment of Indonesia.* IUCN/WWF, Gland. (2.5.3)

Soekarna, D., Partoatmodjo, S., Wirjosuhardjo, S. and Boeadi. (1980). Problems and management of small mammals in Indonesia with special reference to rats. *BIOTROP Spec. Publ.* 12: 35-54. (11.1.6)

Soepadmo, E. (1972). Fagaceae. *Flora Malesiana I* 7: 265-403.

Soepadmo, E. and Eow, B.K. (1977). The reproductive biology of *Durio zibethinus. Gdns' Bull. S'pore* 29: 25-34. (12.1.68)

Soepraptohardjo, M. and Barus, A. (1974). *PRA-Survey untuk penelitian tanah kolong (tintailing soil).* Lembaga Penelitian Tanah, Bogor. (10.4.11)

Soeriaatmadja, R.E. (1979). Plant inventory and vegetational analysis at proposed tidal reclamation project areas in Siak and Rokon, Riau, Sumatra, Indonesia. In *Mangrove and Estuarine Vegetation in Southeast Asia* (ed. P.B.L. Srivestrava, A.M. Ahmad, G. Dhanarajan and I. Hamzah). BIOTROP Species Publication No. 10., pp. 155-179. BIOTROP, Bogor. (12.9.26)

Soerianegara, I. (1970). Soil investigation in Mount Hondje Forest Reserve, West Jaya. *Rimba Indonesia* 15: 1-16. (23.4.5)

Soerjani, M. (1980). Status and future needs of small mammal research and training in Southeast Asia. *BIOTROP Spec. Publ.* 12: 237-240. (11.1.6)

Soong, M.K. (1948). Fishes on the Malayan padi fields. I. Sepat Siam (*Trichogaster pectoralis*). *Malay. Nat. J.* 3: 87-89.

Soong, M.K. (1949). Fishes of the Malayan padi fields. II. Aruan: Serpent-headed fishes. *Malay. Nat. J.* 4: 29-31.

Soong, M.K. (1950). Fishes of the Malayan padi fields. II. Keli. Cat-fishes. *Malay. Nat. J.* 5: 88-91.

Sopher, D.E. (1977). *The Sea Nomads.* National Museum, Singapore. (2.5.15)

Southern, H.N. (1970). The natural control of a population of tawny owls (*Strix aluco*). *J. Zool.* 162: 197-285.

Southern, H.N. and Lowe, V.P.W. (1982). Predation by tawany owls (*Strix aluco*) on Bank voles (*Clethrionomys glareolus*) and Wood mice. (*Apodemus sylvaticus*) *J. Zool.* 198: 83:-102. (12.8.84)

Specht, R.L. (1979). Heathlands and related shrublands of the world. In *Heathlands & Related Shrublands* (ed. R.L. Specht), pp. 1-18. Elsevier, Amsterdam. (12.10.37)

Specht, R.L. and Womersley, J.S. (1979). Heathlands and related shrublands of Malesia (with particular reference to Borneo and New Guinea). In *Heathlands and Related Shrublands* (ed. R.L. Specht), pp. 321-378. Elsevier, Amsterdam. (12.10.20)

St. John, T.V. and Anderson, A.B. (1982). A reexamination of plant phenolics as a source of tropical black water rivers. *Trop.*

Ecol. 23: 151-154.

Start, A.N. (1974). More cave earwigs transported by a fruit bat. *Malay. Nat. J.* 27: 170. (12.4.61)

Start, A.N. and Marshall, A.G. (1975). Nectarivorous bats as pollinators of trees in West Malaysia. In *Tropical Trees : Variation, Breeding and Conservation* (ed. J. Burley and B.T. Styles), pp. 141-150. Academic, London. (12.8.37)

Stebbins, R.C. and Kalk, M. (1961). Observations on the natural history of the mudskipper, *Periophthalmus sobrinus. Copeia* 1961: 18-27. (12.5.31)

Stephenson, A.G. (1981). Flower and fruit abortion: proximate causes and ultimate functions. *Ann. Rev. Ecol. Syst.* 12: 253-279. (12.1.48)

Stern, V.M., Smith, R.F., van den Bosch, R. , and Hagen, K.S. (1959). The integrated control concept. *Hilgardia* 29: 81-101.

Steup, F.K.M. (1941). Botanische aanteekeningen wit Riouw. *Trop. Natuur* 30: 88-94. (9.1.35)

Stocker, G.C. (1981). Regeneration of a North Queensland rain forest following felling and burning. *Biotropica* 13: 86-92. (12.10.100)

Stone, B.C. (1983). The genus *Pandanus* in Sumatera. *Fed. Mus. J.* 28: 101-122. (12.1.101)

Strahler, A.N. (1957). Quantitative analysis of watershed geomorphology. *Trans. Amer. Geophys. Union* 38: 913-920. (23.3.17)

Sukardjo, S. (1979). Hutan payau di Kuala Sekampung, Lampung Selatan, Sumatra. In *Prosiding Seminar 'Ekosistem Hutan Mangrove'* (ed. S. Soemodihardjo, A. Nontji, A. Djamali), pp. 59-68. Lembaga Oseanologi National, Jakarta. (12.9.2).

Sukardjo, S. and Kartawinata, K. (1979). Mangrove forest of Banyuasin, Musi River Estuary, South Sumatra. In *Mangrove and Estuarine Vegetation in Southeast Asia* (ed. P.B.L.Srivestava, A.M. Ahmad, G. Dhanarajan, and I. Hamzah). BIOTROP Special Publication No. 10., pp. 61-79. BIOTROP, Bogor. (12.9.8)

Sumadhiharga, O.K. (1977). A preliminary study on the ecology of the coral reef of Pombo Island. *Mar. Res. Indonesia* 17: 29-49. (2.5.18)

Suselo, T.B. (1981). Preliminary report on ecological studies of *Eusideroxylon zwageri* T. and B. in Jambi, Sumatra. BIOTROP, Bogor. (12.1.40)

Sustriayu, N. (1980). Zoonotic diseases transmissible from Indonesian rodents, *BIOTROP Spec. Publ.* 12: 101-108. (11.1.6)

Swadling, P. (1976). Changes induced by human exploitation in prehistoric shellfish population. *Mankind* 10: 156-162. (23.6.11)

Symington, C.F. (1933). The study of secondary growth on rain forest sites in Malaya. *Malay. Forester* 2: 107-117. (12.10.118)

Symington, C.F. (1943). *Foresters' Manual of Dipterocarps.* Malayan Forest Records No. 16. Republished by Penerbit Universiti Malaya, Kuala Lumpur, 1974. (12.1.110)

Tanimoto, T. (1980). Vegetation of the Alang-alang grassland and its succession in the Benakat District of South Sumatra, Indonesia. *Bull. For. For. Prod. Res. Inst.* 314: 11-19. (10.3.14)

Taniuchi, T. (1979). Freshwater elasmobranches from Lake Naujan Philippines, Perak River Malaysia and Indragiri River Sumatra, Indonesia, South East Asia. Japan. *J. Ichthyol.* 25: 273-277. (12.5.24)

Tanner, E.V.J. (1983). Leaf demography and growth of the tree fern *Cyathea pubsecens* in Jamaica. *Bot. J. Linn. Soc.* 87: 213-227. (12.1.32)

Tanner, E.V.J. and Kapos, V. (1982). Leaf structure of Jamaican upper montane rainforest trees. *Biotropica* 14: 16-24 (12.1.74)

Tay, S.W. and Khoo, H.W. (1980). The distribution of coral reef fishes at Pulau Salu, Singapore. Paper given at the *'Symposium on Recent Research Activities on Coral Reefs in Southeast Asia',* BIOTROP, 6-9 May, 1980. (12.5.32).

Taylor, K.D. (1978). Range of movement and activity of common rats (*Rattus norvegicus*) on agricultural land. *J. appl. Ecol.* 15: 663-677. (12.8.85)

Tee, A.C.G. (1982a). Some aspects of the ecology of the mangrove forest at Sungai Buloh, Selangor I. Analysis of environmental factors and the floral distribution and their corelationship. *Malay. Nat. J.* 35: 13-18. (12.9.40)

Tee, A.C.G. (1982b). Some aspects of the ecology of the mangrove forest at Sungai Buloh, Selangor II. Distribution pattern and population dynamics of tree-dwelling fauna. *Malay Nat. J.* 35: 267-278. (12.9.47)

Thienemann, A. (1932). Die Tierwelt der Nepenthes-Kannen. *Arch. Hydrobiol. Suppl.* 11:1-54. (12.4.30)

Thienemann, A. (1934). Die Tierwelt der tropischen Pflanzengewasser. *Arch Hydrobiol., Suppl.* 13: 1-91.(12.1.45)

Thienemann, A. (1957). Die Fische der Deutschen Liminologischen Sunda Expedition. *Arch. Hydrobiol.* 23: 471-477. (12.5.18)

Thompson, K. (1977). Why organisms feed on the cores of tropical trees. *Biotropica* 9: 144. (12.1.124)

Tilson, R.L. (1977). Social behaviour of simakobu monkeys and its relationship to human predation. *J. Mammal.* 58: 202-212. (12.8.101)

Tilson, R.L. and Tenaza, R.R. (1976). Monogamy and duetting in an Old World monkey. *Nature* 263: 320-321. (12.8.100)

Tinal, U.K. and Palinewan, J.C. (1978). Mechanical logging damage after selective cutting in the lowland dipterocarp forest at Beloro, East Kalimantan. *BIOTROP Spec. Publ.* 3: 91-96. (9.1.32)

Tjia, H.D. (1980). The Sunda Shelf, Southeast Asia. *Z. Geomorph.* 24: 405-427. (23.3.6)

Tobing D.H. (1968). *A Preliminary Bibliography on Fauna, Flora, Vegetation and Conservation of Nature and Natural Resources of Sumatra.* Nature Conservation Dept., Agricultural University, Wageningen. (27.1.16)

Tomlinson, P.B., Primack, R.B. and Bunt, J.S. (1979). Preliminary observations on floral biology in mangrove Rhizophoraceae. *Biotropica* 11: 156-277. (12.9.34)

Towsend, C.R. (1980). *The Ecology of Streams and Rivers,* Edward Arnold, London. (3.2.28)

Towsend, C.R. and Hildrew, A.G. 91976). Field experiments of the driffting colonisation and continous redistribution of stream benthos. *J. Anim. Ecol.* 45: 759-772. (3.2.38)

Troll, D. and Dragendorf, O. (1931). Uber die Luft wurzeln von Sonneratia Linn.f. und lhre biologische Bedeutung. *Planta* 13: 311-473.

Turner, R.E. (1975). Referred to in Soegiarto and Polunin (1980).

Turner, R.E. (1977). Intertidal vegetation and commercial yields of penaeid shrimp. *Trans. Am. Fish. Soc.* 106: 411-416. (2.2.26)

Tuttle, M.D. and Ryan, M.J. (1981). Bat predation and the evolution of frog vacalizations in the Neotropics. *Science* 214: 677-678. (12.8.29)

Tuyt, P. (1939). Oorspronkelijke Bijdragen. Schaduwrijen cultuur van ijzerhout in de Resedensie Palembang. *Tectona* 32: 808-828. (9.1.20)

Tweedie, M. and Harrison, J.L. (1970). *Malayan Animal Life.* Longman, Kuala Lumpur. (12.13.2)

Tweedie, M.W.F. (1961). On certain Mollusca of the Malayan limestone hills. *Bull. Raffles Mus.* 26: 49-65. (12.3.5)

Tweedie, M.W.F. and Harrison, J.L. (1970). *Malayan Animal Life.* Longman, Kuala Lumpur. (12.13.2)

Uhl, C., Clark, H., Clark, K. and Maquirno, P. (1982). Successional patterns associated with slash-and-burn agriculture in the upper Rio Negro region of the Amazon basin. *Biotropica* 14: 249-254. (12.10.99)

Uhl, C., Clark, K., Clark, H. and Murphy, P. (1981). Early plant succession after cutting and burning in the upper Rio Negro region of the Amazon basin. *J. Ecol.* 69: 631-649. (12.10.98)

Umbgrove, J.H.F. (1947). Coral reefs of the East Indies. *Bull. Geol. Soc. Am.* 58: 729-778. (23.3.13)

Ungemach, H. (1969). Chemical rain studies in the Amazon region. In *Simposio y Foro de Biologia Tropical Amazonica,* pp. 354-358. Association pro Biologia Tropical.

Vaas, K.F., Sachlan, M. and Wiraatmadja, G. (1953). On the ecology and fisheries of south-east Sumatra. *Contr. Inl. Fish. Res.* Sta. 3: 1-23. (3.2.23)

Valeton, TH. (1980). Lindeniopsis. Een nieun subgenus der Rubiaceae. *Verh. Kon. Akad. Wet. Amst.* 17: 120-126. (12.10.57)

van Alphen de Veer, E.J. (1953). Plantations of *Pinus merkusii* as a means of reafforestation in Indonesia. *Tectona* 43: 119-130. (10.1.3)

van Baren, F.A. (1975). The soil as an ecological factor in the development of tropical forest areas. In *The Use of Ecological Guidelines for Development in Tropical Forest Areas of South East Asia,* pp. 88-98. IUCN, Gland. (20.1.14)

van Beek, C.G.G. (1982). *A Geomorphological and Pedological study of the Gunung Leuser National Park, North Sumatera, Indonesia.* Wageningen Agricultural University, Wageningen. (23.3.5).

van Beers, W.F.J. (1962). *Acid Sulphate Soils.* Veenan and Zonen, Amsterdam. (10.3.3)

van Bemmelen, R.W. (1929). The origin of Lake Toba. In *Proc. 4th Pac. Sci. Cong. Vol. IIA (Physical Papers),* pp. 115-124. (3.1.1)

van Benthem Jutting, W.S.S. (1959). Catalogue of the nonmarine mollusca of Sumatra and of its satellite islands. *Beaufortia* 7: 41-191. (12.3.30)

van Bodegom, A.H. (1929). De Vloedbosschen in het gewest Riouw en onderheerigheden. *Tectona* 22: 1302-1332. (12.9.31)

van der Bosch, A.C. (1938). Het verdwalen van dieren op bergtoppen. *Trop. Natuur* 27: 169-172. (12.8.17)

van der Laan, E (1925). De boschen van de

Zuider en Oosterafdeeling van Borneo. *Tectona* 18: 925-952. (12.10.62)
van der Meer Mohr, J.C. (1928). Poelaoe Berhala. *Trop. Natuur* 17: 85-97.
van der Meer Mohr, J.C. (1936). Faunistisch onderzoek van eenige grotten op Sumatra's Oostkust en Tapanoeli. *Trop. Natuur Jub. Vitg.* : 60-67.
van der Meer Mohr, J.C. (1941a). Biologisene waarnamingen bij Pantai Tjermin (III). *Trop. Natuur* 30: 49-55. (2.4.31)
van der Meer Mohr, J.C. (1941b). On two species of Malaysian King Crabs. *Treubia* 18: 201-205. (12.3.39)
van der Pijl, L. (1937a). Biological and physiological observation on the inflorescence of *Amorphophallus. Rec. trav. bot. neerl.* 34: 157-167. (12.1.106)
van der Pijl, L. (1937b). Disharmony between Asiatic flower-birds and America bird-flowers. *Ann. Jard. Bot. Buitenzorg.* 48: 2-26. (12.7.2)
van der Voort, M.(1939). De merkwaardige Padang Bolak. *Trop. Natuur* 28: 201-209.
van Emden, H.F. (1974). *Pest Control and Its Ecology*. Edward Arnold. London. (11.18)
van Heekeren, H.R. (1958). *The bronze-iron age of Indonesia.* Verh. Kon. Inst. Taal-Land-Volkenkunde 22. (23.7.8)
van Heurn, F.C. (1937). Eerste registratie van de kudden olifanten die voorkomen op Sumatra's Oostkust, Westkust en Tapanoeli en van gegevens, die daarop betrekking hebben. *Ned. Comm. Intern. Natuurbesch. Med.* 11: 66-78.
van Noordwijk, M.A. and van Schaik, C.P. (1983). A study concerning the relation between ecological and social factors and the behaviour of wild long-tailed macaques (*Macaca fascicularis*). Final Progress Report to LIPI, Jakarta. (12.8.105).
van Steenis, C.G.G.J. (1938). Exploraties in de Gajo-Landen. Algemeene resultaten der Löser-expeditie 1937. *Tijdschr. Kon. Ned. Aardr. Genootschap* 55: 728-801. (12.10.39)
van Steenis, C.G.G.J. (1932). Botanical results of a trip to the Anambas and Natoena Islands. *Bull. Jard. Bot. Buitenzorg, ser. III* 12: 151-211. (12.10.53)
van Steenis, C.G.G.J. (1934a). On the origin of the Malaysian mountain flora. Part 2. Altitudinal Zones, general considerations and renewed statement of the problem. *Treubia* 13: 289-417. (12.10.34)
van Steenis, C.G.G.J. (1934b). On the origin of the Malaysian mountain flora. Part I. *Treubia* 13: 135-262. (12.10.35)
van Steenis, C.G.G.J. (1935). Maleische vegetatiescheten. *Tijd Kon. Ned. Aard. Gen.* 52: 25-67, 171-203, 363-398.
van Steenis, C.G.G.J. (1938). Exploratie in de Gajo-Landen. Algemeene resultaten der Loser-expeditie 1937. *Tijdschr. Kon. Ned. Aardr. Genootschap* 55: 728-801. (12.10.39)
van Steenis, C.G.G.J. (1950). The delimitation of Malaysia and its main plant geographical divisions. *Flora Malesiana 1*: 70-75. (12.1.97)
van Steenis, C.G.G.J. (1952). Rheophytes. *Proc. R. Soc. Queensland* 62: 61-68.
van Steenis, C.G.G.J. (1957). Outline of vegetation types in Indonesia and some adjacent regions. *Proc. Pac. Sci. Cong.* 8: 61-97. (12.10.51)
van Steenis, C.G.G.J. (1959). Discrimination of tropical shore formations. In *Proc. Symp. Humid Tropics Vegetation, Tjiawi, 1958*, pp. 215-217. UNESCO, Paris. (2.4.30)
van Steenis, C.G.G.J. (1962). The mountain flora of the Malaysian tropics. *Endeavour* 21: 183-193. (12.10.31)
van Steenis, C.G.G.J. (1971). Plant conservation in Malaysia. *Bull. Jard. Bot. Nat. Belg.* 41: 189-202. (12.12.9)
van Steenis, C.G.G.J. (1972). *The Mountain Flora of Java.* Brill, Leiden. (12.10.43)
van Steenis, C.G.G.J. (1978). Rheophytes in South Africa. *Bothalian* 12: 543-546.
van Steenis, C.G.G.J. (1981). *Flora untuk Sekolah.* Pradnya Paramita, Jakarta, (12.1.75)
van Steenis, C.G.G.J. and Ruttner, F. (1933). Die Pteridophyten und Phanerogramen den Deutschen Limnologischen Sunda Expedition. *Arch. Hydrobiol. Suppl.* 11:231-287.
van Strien, N.J. (1974). *The Sumatran or Two-horned Asiatic Rhinoceros - A Study of Literature.* Veenman and Zonen, Wageningen. (12.8.95)
van Strien, N.J. (1977). *Sumatran Birds: Supplement to the Field Guide to the Birds of South-east Asia.* Agriculture Dept., Wageningen. (12.7.45)
Verbeek, R.D.M. (1897). Giologie van Bangka. *Jaarb. Mijnw. Ned. O. Indie* 26: 60-61. (23.5.1)
Verstappen, H.TH. (1960). Some observations on karst developoment in the Malay Archipelago. *J. Trop. Geogr.* 14: 1-10. (23.3.2)
Verstappen, H.TH. (1973). *A Geomorphological Reconnaisance of Sumatra and Adjacent Island (Indonesia).* Wolters-Noordhoft, Groningen. (23.3.1)
Versteegh, F. (1951). Proeve van een bedri-

jfsregeling voor de vloedbossen van Bengkalis. *Tectona* 41: 200-258. (9.1.27)

Vitousek, P.M. (1984). Litterfall, nutrient cycling and nutrient limitation in tropical forests. *Ecology* 65: 285-298.

Vitt, L.J., Congdon, J.D. and Dickson, N.A. (1977). Adaptive strategies and energetics of tail autonomy in lizards. *Ecology* 58: 326-337. (12.6.29)

Vogt, R.C. and Bull, J.J. (1982). Temperature controlled sex determination in turtles: ecological and behavioural aspects. *Herpetologica* 38: 154-164. (12.6.18)

Volz, W. (1909). *Nord-Sumatra. Band I - Die Bataklander.* Dietrich Reimer, Berlin.

Volz, W. (1912). *Nord-Sumatra. Band II - Die Gajolander.* Dietrich Reimer, Berlin.

von Hügel, A. (1896). The land of the Bataks. *Geog. J.* 7: 75-82, 175-183. (23.7.2).

Voris, H.K. (1977). Comparison of herpetofaunal diversity in tree buttresses of evergreen tropical forests. *Herpetologica* 33: 375-380. (12.6.14)

Voris, H.K. and Jayne, B.C. (1979). Growth, reproduction and population structure of a marine snake, Enhydrine schistosa (Hydrophiidae). *Copeia* 1979: 307-318. (12.6.10)

Voris, H.K. and Moffett, M.W. (1981). Size and proportion relationship between the beaked sea snake and its prey. *Biotropica* 13: 15-19. (12.6.9)

Voris, H.K., Voris H.H., and List, L.B. (1978). The food and feeding behavior of a marine snake, *Enhydrina schistosa* (Hydrophiidae). *Copeia* 1978: 134-146. (12.6.11)

Walker, D. (1982). Speculations on the Origin and Evolution of Sunda Sahul Rain Forests. In *Biological Diversification in the Tropics* (ed. G. Prance), pp. 554-575. Columbia Univ. Press, New York. (23.6.14)

Walker, J.R.L. (1975). *The Biology of Plant Phenolics.* Edward Arnold, London. (12.1.73)

Wallace, A.R. (1869). *The Malay Archipelago: The Land of the Orang-utan and the Bird of Paradise. A Narrative of Travel with studies of Man & Nature.* Reprint edition 1962, Dover, New York. (23.7.12)

Wallraf, H.G. (1978). Proposed principles of magnetic field perception in birds. *Oikos* 30: 188-194. (12.7.19)

Wallraf, H.G. and Gelderloos, O.G. (1978). Experiments on migratory orientation of birds with simulated stellar sky and geomagnetic field: method and preliminary results. *Oikos* 30: 207-215. (12.7.20)

Wallwork, F. (1982). The autecology of *Barringtonia asiatica* on Rakata. In *The Krakatoa Centenary Expedition* (ed. J.R.Flenley and K. Richards), pp. 127-141. University of Hull, Hull. (12.1.92)

Walsh, G.E. (1974). Mangroves: a review. In *Ecology of Halophytes* (ed. R.J. Reimold and W.H. Queen). pp. 51-174. Academic, New York. (12.1.98).

Walsh, G.E. (1977). Exploitation of mangal. In *Wet Coastal Ecosystems* (ed. V.J. Chapman), pp. 347-362. Elsevier, Amsterdam.

Walton, O.E. (1978). Substrate attachment by drifting aquatic insect larvae. *Ecology* 59: 1023-1030. (12.4.25)

Ward, P. (1968). Origin of the avifauna of urban and suburban Singapore. *Ibis* 110: 239-255. (12.7.34)

Ward, P. and Poh, G.E. (1968). Seasonal breeding in an equatorial population of the tree sparrow Passer montanus. *Ibis.* 110: 359-363. (12.7.35)

Wastermann, J.H. (1942). Snakes from Bangka and Biliton. *Treubia* 18: 611-619. (12.6.7)

Watanabe, K. (1981). Variation in group composition and population density of the two sympatric Mentawaian leaf-monkeys. *Primates* 22: 145-160. (12.8.70)

Waterman, P.G. (1983). Distribution of secondary metabolites in rain forest plants: toward an understanding of cause and effect. In *Tropical Rain Forest: Ecology and Management* (ed. S.L. Sutton, T.C. Whitmore and A.C. Chadwick), pp. 167-179. Blackwell, Oxford.

Waterman, P.G. and Choo, G.M. (1981). The effects of digestibility-reducing compounds in leaves on food selection by some Colobinae. *Malays. appl. Biol.* 10: 147-162.

Watson, J.G. (1928) Mangrove forests of the Malay Peninsula. Malay. *For. Rec. No. 6.* (12.9.1)

Webb, L.J. (1959). A physiognomic classification of Australian rain forests. *J. Ecol.* 47: 551-570.

Webb, L.J., Tracey, J.G. and Haydock, K.P. (1967). A factor toxic to seedlings of the same species associated with living roots of the non-gregarious sub-tropical rain forest tree *Grevillea robusta*. *J. appl. Ecol.* 4: 13-25. (12.1.127)

Wells, D,R. (1975). The moss-nest swiftlet *Collocalia vanikorensis* Quoy and Gaimard in Sumatra. *Ardea* 63: 148-151. (12.7.32)

Wells, D.R. (1971). Survival of the Malaysian bird fauna. *Malay. Nat. J.* 24: 248-256. (12.7.39).

Wells, D.R. (1974). *Resident Birds. In Birds of the Malay Peninsula.* (by Lord Medway and

D.R. Wells). Witherby, London. (12.7.23)

Wells, D.R. (1978). Numbers and biomass of insectivorous birds in the understory of rain forest at Pasoh Forest. *Malay. Nat. J.* 30: 353-362. (12.7.24)

Wells, D.R., Halls, C.J. and Hails, A.J. (1979). Birds of the forests of the lowlands and lower hill slopes on Gunung Mulu. Paper presented at Gunung Mulu Symposium, 1979, Royal Geographical Society, London (12.7.14)

West, K.L. (1979). *The ecology and behaviour of the siamang* (Hylobates syndactylus) *in Sumatra.* MSc thesis, University of California, Davis. (12.8.69)

Wheelwright, N.T. and Orians, G.H. (1982). Seed dispersal by animals: contrasts with pollen dispersal, problems of terminology and constraints on coevolution. *Am. Nat.* 119: 402-423. (12.1.49)

Whitmore, T.C. (1967). Studies in *Macaranga,* an easy genus of Malayan wayside trees. *Malay. Nat. J.* 20: 89-99. (12.1.118).

Whitmore, T.C. (1973). Frequency and habitat of tree species in the rain forest of Ulu Kelantan. *Gdns' Bull. S'pore* 26: 195-210. (12.10.71,

Whitmore, T.C. (1977). *Palms of Malaya.* Oxford Univ. Press, Oxford. (12.1.3)

Whitmore, T.C. (1980). Potentially economic species of Southeast Asia forest. *BioIndonesia* 7: 65-74. (9.1.26)

Whitmore, T.C. (1982a). On pattern and process in forests. In *The Plant Community as a Working Mechanism* (ed. E.I. Newman), pp. 45-59. Blackwell, Oxford. (12.10.70)

Whitmore, T.C. (1982b). Palaeoclimate and vegetation history. In *Wallace's Line and Plate Tectonics* (ed. T.C. Whitmore), pp. 36-42. Oxford University Press. Oxford. (23.6.12)

Whitmore, T.C. (1984). *Tropical Rain Forests of the Far East.* 2nd edition. Claredon, Oxford. (12.10.8)

Whitmore, T.C. (in press). Vegetation map of South-east Asia. *J. Biogeog.*

Whitmore, T.C. and Burnham, C.P. (1969). The altitudinal sequence of forests and soils on granite near Kuala Lumpur. *Malay. Nat. J.* 22: 99-118. (12.10.36)

Whittacker, R. and Flenley, J.R. (1982). The flora of Krakatau. In *The Krakatoa Centenary Expedition* (ed. J.R. Flenley and K. Richards), pp. 9-53. University of Hull, Hull. (12.1.92).

Whittaker, R.H. and Likens, G.E. (1973). Primary productivity: The Biosphere and Man. *Human Ecol.* 1: 357-367. (9.2.9)

Whitten, A.J. (1980a). *Arenga* fruit as a food for gibbons. *Principes* 26: 143-146. (12.8.3)

Whitten, A.J. (1980b). *The Kloss Gibbon in Siberut Rain Forest.* PhD. thesis, University of Cambridge, Cambridge. (12.8.60)

Whitten, A.J. (1981). Notes on the ecology of *Myrmecodia tuberosa* Jack on Siberut Island. *Ann. Bot.* 47: 525-526. (12.1.6)

Whitten, A.J. (1982a). Possible niche expansion in the spangled drongo on Siberut Island, Indonesia. *Ibis* 126: 122-193. (12.7.7)

Whitten, A.J. (1982b). *The Gibbons of Siberut.* J.M. Dent, London. (12.13.3)

Whitten, A.J. (1982c). A numerical analysis of tropical rain forest using floristic and structural data and its application to an analysis of gibbon ranging behaviour. *J. Ecol.* 70: 249-271. (12.10.19)

Whitten, A.J. (1982d). Home range use by Kloss gibbons (*Hylobates klossii*) on Siberut Island, Indonesia. *Anim. Behav.* 30: 182-198. (12.8.7)

Whitten, A.J. (1982e). Diet and feeding behaviour of Kloss gibbons on Siberut Island, Indonesia. *Folia primatol.* 37: 177-208. (12.8.6)

Whitten, A.J. (1982f). The ecology of singing in Kloss gibbons on Siberut Island, Indonesia. *Int. J. primatol.* 3: 33-51. (12.8.91)

Whitten, A.J. (1982g). The role of ants in selection of night trees. *Biotropica* 14: 237-238.

Whitten, A.J. (1983). Compendium of abstracts from papers concerning Sumatran invertebrates. (27.2.8)

Whitten, A.J. and Damanik, S.J. (1986). Mass defoliation of mangrove in Sumatra, Indonesia. *Biotropica* 18: 176.

Whitten, A.J. and Whitten, J.E.J. (1982). Preliminary observations of the Mentawai macaque on Siberut Island, Indonesia. *Int. J. Primatol.* 3: 445-459. (12.8.93)

Whitten, J.E.J. (1979). *Ecological Isolation of an Indonesian Squirrel* (Sundasciurus lowii siberu). M. Phil. thesis, University of Cambridge, Cambridge. (12.8.61).

Whitten, J.E.J. (1980). Ecological separation of three diurnal squirrels in tropical rainforest on Siberut Island, Indonesia. *J. Zool., Lond.* 193: 405-420. (12.8.5)

Widagda, L.C. (1981). An ecosystem analysis of West Javanese home gardens. Working paper, East-West Environment and Policy Institute, Honolulu. (11.4.21)

Widjaja, E.A. (1980). Indonesia. In *Bamboo Research in Asia* (ed. G. Lessard and A. Chouinard), pp. 63-68. International Development Research Centre, Ottawa. (12.1.94).

Wigley, T.M.L. and Jones, P.D. (1981).

Detecting CO_2-induced climatic change. *Nature* 292: 205-208. (1.2.3).

Wigley, T.M.L., Jones, P.D. and Kelly, P.M. (1980). Scenario for a warm, high CO_2 world. *Nature* 283: 17-21. (1.2.2).

Wilford, G.E. (1960). Radiocarbon age determinations of Quaternary sediments in Brunei and north east Sarawak. *British North Borneo Geological Survey Annual Report, 1959.* (Referred to in Whitmore, 1975).

Williams, D.D. (1981). Migrations and distributions of stream benthos In *Perspectives* in *Running Water Ecology* (ed. M.A. Lock and D.D. Williams), pp. 155-207. Plenum, New York.

Williams, J.T. (ed.) (1975). *Southeast Asian Plant Genetic Resources.* Lembaga Biologi Nasional, Bogor. (9.2.15)

Williams, K.D. (1979). Radio-tracking tapirs in the primary rain forest of West Malaysia. *Malay. Nat. J.* 32: 253-258. (12.8.39)

Williams, K.D. and Petrides, G.A. (1980). Browse utilization, feeding behaviour and management of the Malayan tapir in West Malaysia. *J. Wildl. Managt.* 44: 489-494.

Wilson, C.C. and Wilson, W.C. (1973). *Census of Sumatran primates. Final report to LIPI,* Jakarta. (12.8.40)

Wilson, C.C. and Wilson, W.L. (1975). The influence of selective logging on primates and some other animals in East Kalimantan. *Folia primatol.* 23: 245-274. (12.8.72)

Wilson, D.S. (1974). Prey capture and competition in the Ant Lion. *Biotropica* 6: 187-193. 912.4.48)

Wilson, J.M. (1977). Cave ecology in the Himalaya. *Studies Speleol.* 3: 66-69, 4: 84.

Wilson, J.M. (1981). Plants underground. In *Himalaya Underground* (ed. G.A. Durrant, C.M. Smart, J.E.K. Turner, J.M. Wilson). pp. 33-34. Southampton University, Southampton.

Wilson, W.L. and Johns, A.D. (1982). Diversity and abundance of selected animal species in undisturbed forest, selectively logged forest and plantations in East Kalimantan, Indonesia. *Biol. Conserv.* 24: 205-218. (12.8.79)

Wilson, W.L. and Wilson C.C. (1975). Species-specific vocalizations and the determination of phylogenetic affinities of the *Presbytis aygula-melalophos* group in Sumatra. *Contemp. Primatol,* pp. 459-463. Karger. (12.8.51)

Wiltschko, R. and Wiltschko, W. (1978). Relative importance of stars and the magnetic field for the accuracy of orientation in night migrating birds. *Oikos* 30: 195-206. (12.7.21)

Wint, G.R.W. (1983). Leaf damage in tropical rain forest canopies. In *Tropical Rain Forest: Ecology and Management* (ed. S.L. Sutton, T.C. Whitmore and A.C. Chadwick), pp. 229-239. Blackwell, Oxford.

Wisaksono, W. (N.D.). *Pencemaran minyak oleh kapal super tanker 'Showa Maru'.* Lemigas, Jakarta. (25.5.7).

Witkamp, H. (1925). De ijzerhout als geologische indicator. *Trop. Natuur* 14: 97-103. (23.4.7)

Wolters, O.W. (1970). *The Fall of Srivijaya in Malay History.* Oxford Univ. Press, Kuala Lumpur. (23.7.11)

Wong, P.P. (1978). The herbaceous formation and its geomorphic role, East Coast, Peninsular Malaysia. *Malay. Nat. J.* 32: 129-141.

Wood, B.J. (1969). Population studies on the Malaysian wood rat (*Rattus tiomanicus*) in oil palm, demonstrating an effective new control method and assessing some older ones. *Planter* 45: 510-526.

Wood, B.J. (1971). Investigations of rats in rice fields demonstrating an effective control method giving substantial yield increase. *PANS* 17: 180-193.

Wood, G.H.S. (1956). The dipterocarp flowering season in North Borneo, 1955. *Malay. Forester* 19: 193-201. (12.1.121)

Wood, T.G. (1978). The termite (Isoptera) fauna of Malesian and other tropical rainforests. In *The Abundance of Animals in Malesian Rain Forest* (ed. A.G. Marshall), pp. 113-132. University of Hull, Hull. (12.8.22)

Wratten, S.D., Goddard, P. and Edwards, P.J. (1981). British trees and insects: the role of palatability. *Amer. Nat.* 118: 916-919. (12.1.83)

Wratten. S.D. and Watt, A.D. (1984). *Ecological Basis of Pest Control.* George Allen and Unwin, London.

Wright, J.S. (1981). Intra-archipelago vertebrate distributions: the slope of the species-area relation. *Am. Nat.* 118: 726-748. (12.13.25).

Wright, J.S. and Biehl, C.C. (1982). Island biogeographic distributions: testing for random, regular, and aggregated patterns of species occurrence. *Am. Nat.* 119: 345-357. (12.13.26).

WWF (Malaysia)/Payne, J.P. and Davies, A.G. (1981). *Faunal Survey of Sabah.* World Wildlife Fund (Malaysia), Kuala Lumpur. (12.13.29).

WWF/ van Strien, N.J. (1978). Proposed Gunung Leuser National Park Management Plan 1978/9 - 1982/3. IUCN/WWF,

Bogor. (10.2.10)

Wyatt-Smith, J. (1951). Distribution of the Pied Imperial pigeon. *Malay. Nat. J.* 5: 208.

Wyatt-Smith, J. (1954). A note on the fresh-water swamp, low land and hill forest types of Malaya. *Malay. Forester* 24: 110-121. (12.10.115)

Wyatt-Smith, J. (1955). Changes in composition in early natural plant succession. *Malay. Forester* 18: 44-49. (11.4.24).

Wyatt-Smith, J. (1963). Manual of Malayan silviculture for inland forests (2 vols). *Malay. For. Rec.* No. 23. (9.1.17)

Wyatt-Smith, J. (1979). Pocket Check List of Timber Trees. Revised by K.M. Kochumen. *Malay. Forest Rec.* No. 17. (12.1.70)

Wyatt-Smith, J. and Kochummen, K.M. (1979). Pocket check list of timber trees. *Malay. Forest Rec.* No. 17. (12.1.70)

Wyrtki, K. (1961). *Physical Oceanography of the Southeast Asian Waters.* Scripps Institute for Oceanography, La Jolla. (2.5.7)

Yalden. B.W. and Morris, P.A. (1975). *The Lives of Bats.* David and Charles, London. (12.8.16)

Yamada, I. and Soekardjo, S. (1983). Ecological study of mangrove forests in South Sumatra. In *South Sumatra: Man and Agriculture* (ed. Y. Tsubouchi, Nasruddin, I.Y. Takaya, and A.R. Hanafiah), pp. 1-33. Center for S.E. Asian Studies, Kyoto. (12.10.16)

Yap, S.P. (1976). The feeding biology of some padi field anurans. B.Sc. thesis, University Malaya, Kuala Lumpur. (13.6.22)

Yoda, K. (1978a). Respiration studies in Pasoh forest plants. *Malay. Nat. J.* 30: 259-279. (12.1.50)

Yoda, K. (1978b). Organic carbon, nitrogen and mineral nutrients stock in the soil of Pasoh Forest. *Malay. Nat. J.* 30: 229-251. (23.4.13)

Yoda, K. and Kira, T. (1982). Accumulation of organic matter, carbon, nitrogen and other nutrients in the soil of a lowland rain forest at Pasoh, West Malaysia. *Jap. J. Ecol.* 32: 275-292.

Yoda, K. and Sato, H. (1975). Daily fluctuation of trunk diameter in tropical rain forest trees. *Jap. J. Ecol.* 23: 47-48.

Yoneda, T., Yoda, K. and Kira, T. (1978). Accumulation and decomposition of wood litter in Pasoh Forest. *Malay. Nat. J.* 30: 381-389. (12.10.72)

Yong, H.S. (1978). Mammals of virgin and logged forests in Peninsular Malaysia. *BIOTROP Spec. Publ.* 3: 153-158. (9.1.32).

Young, A. (1976). *Tropical Soils and Soil Survey.* Cambridge Univ. Press, Cambridge. (23.4.16)

Young, A.M. (1981). Temporal selection for communicatory optimalization: the dawn-dusk chorus as an adaptation in tropical cicadas. *Am. Nat.* 117: 826-829. (12.4.32)

Zimmerman, P.R., Greenberg, J.P., Wandiga, S.O. and Crutzen, P.J., (1982). Termites: a potentially large source of atmospheric methane, carbon dioxide and molecular hydrogen. *Science* 218: 563-565. (1.2.1).

Index